AUGUSTANA UNIVERSITY COLLEGE
LIBRARY

A COUNTRY SO INTERESTING

*. . . what a field to face the imagination, what a
number of ideas rushes in at once, all for the means
to investigate a Country so interesting.*
 Edward Smith to the Committee, March 1825

Rupert's Land Record Society Series
Jennifer S.H. Brown, Editor

1 The English River Book
 A Northwest Company Journal
 and Account Book of 1786
 Harry W. Duckworth

2 A Country So Interesting
 The Hudson's Bay Company and Two
 Centuries of Mapping, 1670–1870
 Richard I. Ruggles

A Country So Interesting

THE HUDSON'S BAY COMPANY
AND TWO CENTURIES OF MAPPING
1670 – 1870

Richard I. Ruggles

McGILL-QUEEN'S UNIVERSITY PRESS

Montreal & Kingston • *London* • *Buffalo*

© McGill-Queen's University Press 1991
ISBN 0-7735-0678-0
ISBN 0-921206-06-2 (limited edition, Rupert's Land Record Society)

Legal deposit first quarter 1991

Bibliothèque nationale du Québec

Printed in Canada on acid-free paper

This book has been published with the help of a grant from the Social Science Federation of Canada, using funds provided by the Social Sciences and Humanities Research Council of Canada, as well as with subventions from the Department of Geography, the Office of the Principal, the School of Graduate Studies and Research, and the Faculty of Arts and Sciences at Queen's University.

Composed by Typo Litho composition inc. in Palatino

Designed by Peter Moulding

Canadian Cataloguing in Publication Data

Ruggles, Richard I., 1923–

A country so interesting: the Hudson's Bay Company and two centuries of mapping, 1670–1870

Includes bibliographical references.
ISBN 0-7735-0678-0

1. Hudson's Bay Company – Maps. 2. Canada – Historical geography – Maps. 3. Hudson's Bay Company – Map collections. 4. Canada – Historical geography – Maps – Catalogs. I. Title.

GA473.7.R84 1991   912.71   C90-090274-4

TO MILDRED,
WITH ME FOR MANY MILES
AND IN MANY ARCHIVES

Contents

Location Maps and Figures ix
Abbreviations ix
Acknowledgments xi
Foreword xiii

PART ONE: INTRODUCTION 1

1 Mapping Policy and Records 3
2 The Explorers and Map Makers 10
3 Field and Office Methods and Equipment 14
4 Company Procedures in the Operation of Policy 18

PART TWO: INVESTIGATING A COUNTRY SO INTERESTING 23

5 London Chart Makers, 1669 to the Late 1720s 25
6 The Initiation of Company Mapping: The Late 1720s to 1754 32
7 Mapping Rivers and Barren Grounds Inland, 1754–1778 37
8 Mapping Inland from the Bay and Over to Athabasca, 1778–1794 49
9 Mapping to the Columbia and Behind the Eastmain, 1795–1821 60
10 Mapping Rupert's Land, the Mackenzie Basin, and the Arctic Shore, 1821–1849 74
11 Mapping West of the Mountains, 1821–1849 87
12 Pemberton and the Colony of Vancouver's Island, 1849–1859 96

13 Exploration and Mapping in the Northwest and the Establishment of Company Land Claims, 1849-1859 106
14 Mapping Company Land Claims and Exploring Inland Routes, 1859–1870 110
Afterword 118

PART THREE: PLATES 123

PART FOUR: CATALOGUES 191

A Manuscript Maps (1670–1870) in the Hudson's Bay Company Archives 193
B Hudson's Bay Company Manuscript Maps (1670–1870) in Archives Other than the Hudson's Bay Company Archives 237
C Hudson's Bay Company Manuscript Maps (1670–1870) Not Located 241

PART FIVE: APPENDICES, GLOSSARY, NOTES, BIBLIOGRAPHY, AND INDEX 257

1 The Hudson's Bay Company 259
2 General Characteristics of the Cartographic Records 261
3 General Inventory of the Hudson's Bay Company Map Archives 261
4 The Catalogues of the Hudson's Bay Company Cartographic Collection 261

5 Procedure for Obtaining Latitude on Land by Measuring the Double Meridian Altitude of the Sun 262

6 Procedure for Obtaining Longitude by Using the Method of Lunar Distances 263

7 Persons Who Prepared Hudson's Bay Company Maps or Charts, 1670–1870 264

8 Hudson's Bay Company Employees or Other Non-Native Persons Who Provided Sketches or Descriptions for Maps Prepared by Peter Fidler 266

9 Native Persons Who Provided Sketches or Descriptions for Maps Prepared by Peter Fidler and Philip Turnor 266

10 Native Persons Who Drafted Maps 266

Glossary 267

Notes 273

Bibliography 287

Index 290

Location Maps and Figures

LOCATION MAPS

1 West of Hudson Bay and the Great Lakes; the Arctic Shores *xv*

2 South and East of Hudson and James Bays *xvi*

3 The Manitoba Lakes and the Nelson and Churchill River Systems *xvii*

4 Between Lake Superior and James Bay *xviii*

5 Vancouver Island and the Southern Mainland of British Columbia *xix*

FIGURES

1 Manuscript maps and sketches drafted by Hudson's Bay Company employees, purchased by the company, or received at company headquarters in the course of business, by five-year periods between 1670 and 1870 *8*

2 Retreat of the unknown: European knowledge of Canadian territory in 1670 *27*

3 Retreat of the unknown: European knowledge of Canadian territory in 1763 *39*

4 Retreat of the unknown: European knowledge of Canadian territory in 1795 *73*

5 Retreat of the unknown: European knowledge of Canadian territory in 1870 *119*

Abbreviations

DCB Dictionary of Canadian Biography

DNB Dictionary of National Biography

HBC Hudson's Bay Company

HBCA Hudson's Bay Company Archives

HBRS Hudson's Bay Record Society

PAM Provincial Archives of Manitoba

PSAC Puget's Sound Agricultural Company

NWC North West Company

Acknowledgments

This study of Hudson's Bay Company maps began in the 1950s as an aspect of my doctoral research on the historical geography and historical cartography of the Canadian West to 1795. Initial study owed much to the aid of the late Alice M. Johnson, Hudson's Bay Company Archivist, in the London facilities of the company archives. Later, in London and in Winnipeg in the 1970s and 1980s, the fullest cooperation and guidance and the experience of Shirlee Anne Smith, Keeper of the Archives, and of her staff were proffered, were accepted, and are gratefully acknowledged. Most of my research has taken place at the archives in Winnipeg since 1981 when my study was accepted for future publication as a volume in the research series published by the Hudson's Bay Record Society. Access was given freely to all maps and records, and I was made welcome and encouraged to feel in a sense a member of the archives group, whenever I appeared to continue my research. Several initial manuscripts were edited in the highest professional manner by Dr Hartwell Bowsfield, then editor of the Hudson's Bay Record Society. He carefully pruned extensive versions into more acceptable lengths and into a more cohesive story. He has been of great aid and guidance, and his contribution is sincerely acknowledged at this time. Publication, under the auspices of the society, was cancelled due to the cessation of operation of the society when economic difficulties forced the Hudson's Bay Company to withdraw its financial support.

Mention must be made also of the assistance given by archivists and librarians of the National Archives of Canada, especially those in the National Map Collection, and by archivists and librarians of the Provincial Archives of Manitoba, the British Columbia Archives, the Royal Society of London, and the Guildhall Library, London. The staff of Surveys and Land Records, British Columbia, graciously allowed me to search through their extensive map collection. I also wish to acknowledge the aid given me in this long process by my wife, Mildred E. Ruggles, not only for her forbearance and support, as the dedication states, but more specifically for her research help in the company records for British Columbia – most enthusiastically in those of her home area, southern Vancouver Island.

Crucial financial support was forthcoming from a number of sources, particularly from the Hudson's Bay Record Society. I was honoured by the award of a senior Killam Research Fellowship which provided full funding for a year's research. I received a sabbatical leave fellowship and research grant from the Social Science and Humanities Research Council. Finally, Queen's University has been most supportive through an Arts and Science Faculty special research grant for a British sojourn, and a grant from the School of Graduate Studies and Research to aid in preparing the manuscript. Queen's also supported fully my requests for sabbatical and research leaves, without which this work could not have been completed.

I wish to express my deepest appreciation to the Hudson's Bay Company Archives, Provincial Archives of Manitoba, for permission to reproduce selected maps from the collection. Mrs. Smith was most gracious in allowing George Innes, photographic technician, Department of Geography, Queen's University, to examine all the photo negatives, and was most willing to allow and to arrange for some maps to be rephotographed in order to obtain greater clarity. It is not easy to reproduce map documents such as these, of considerable age, and of variable original size and drafting quality. In some cases even though it was not possible to obtain the degree of legibility desired, a map or chart has been included because of its intrinsic interest or historical importance. Also acknowledgment for permission to reproduce plates is given to the following: the British Museum (plate 1); the University of Minnesota, James Ford Bell Collection (plate 10); and Surveys and Land Records, Victoria (plate 52).

The location maps and figures were drafted by Ross Hough of the Cartographic Laboratory, Department of Geography, Queen's University. His aid in rendering the final design of the maps and the graph is much appreciated. A great debt is owed to Mrs. Helen Phelan, word

Acknowledgments

processor, who calmly and with a welcome sense of humour, transformed copy and manipulated changes and alterations again and again into the final computer disks and print-outs. Some time after the publication of the study by the Hudson's Bay Record Society was cancelled, I was encouraged to submit my manuscript to McGill-Queen's University Press. My colleague and friend, Dr. Peter Goheen, Queen's University, an editor of the press, has been a valued mentor in this parturient process. I also wish to express my sincere thanks to the two appraisers who, when offering their professional assessments, gave me much advice, and in one case, very full and perceptive editorial aid. Since then, Mary Norton, copy editor, with whom it has been a joy to work, and at the press, Joan McGilvray, Co-ordinating Editor, Susanne McAdam, Production Manager, and Peter Moulding, have refashioned the manuscript and employed their talents in preparing this publication.

Foreword

The "Governor and the Company of Adventurers of England tradeing into Hudsons Bay" were major contributors to the mapping of Canada and the Northwest Pacific region of the United States. In the two centuries, 1670 to 1870, they geographically defined and measured, and cartographically depicted the larger share of the territory of our nation. In effect, the Hudson's Bay Company was our first national mapping agency. Exploration, surveying, and mapping became vital elements of the company's trading enterprise. These basic "tools of the trade" were part of the training and daily life of many of its servants, significant cost entries in its financial ledgers, and fundamental requirements for the successful capture and management of its market area.

Much has been written on the history of the Hudson's Bay Company as part of the development of Canada, but only the tip of the iceberg of its exploratory and mapping program has so far been revealed, and little has been said about the scope and character of its map documents. My inspection of the Hudson's Bay Company Archives map collection for this volume represents the first full-scale investigation.

I have had several objectives during my research. The first has been to prepare the history of the contribution of the company to the cartography of our nation. Thus the roles of exploration, surveying, and mapping in company operation, in the formulation of trading policies and methods, and in the geographical pattern of trade are analysed. A further purpose has been to bring to light the significant contribution of the company and its employees, such as Philip Turnor, Peter Fidler, David Thompson, Henry Kelsey, and Anthony Henday among many others of lesser renown, to the unfolding of geographical knowledge, mensuration, and the cartographic depiction of a vast region. I have also attempted to reconstruct the geographical conceptions of the west and north held by company traders as reflected in maps and written documents and to relate these to business decisions made by the successive governors and committees. Finally, but most importantly, my purpose has been to reveal to the scholarly community the unique and rich treasure which is held in the Hudson's Bay Company Archives, to make manifest the authenticity and the cartobibliographic quality of these documents, and to make them more useful as a research resource.

As this study is concentrated on the role of the Hudson's Bay Company in map making, the contributions of well-known fur traders who did not work for the company is not of concern, even though their activities were of considerable significance. For example, the early, less substantial part of David Thompson's career, which took place while he was employed by the company, is described, but not his later, more significant exploration and map making for the North West Company nor his work in the international boundary survey.

Nor does this inquiry analyse land survey, land tenure, or territorial-administrative systems, although map making touched on these. Spatial organization *was* depicted on many maps, particularly of the colony of Vancouver Island, where, for some years, the company was the responsible agent in surveying and mapping, a role it did not have in the Red River colony. There, although the company was not administratively in charge, company employees, Peter Fidler and George Taylor Jr, helped lay out survey lines for the colony's directors.

The following Hudson's Bay Company document series have been of fundamental value to this study. (There are far too many individual items within each series to list them all separately here.)

Series A: Headquarter's Records (London)

A1 Minute Books (Committee Minutes)
A2 Minute Books (General Court)
A4 Agenda Books
A5 London Correspondence Books Outwards – General Series
A6 London Correspondence Books Outwards – Official
A10 London Inward Correspondence – General
A11 London Inward Correspondence from Hudson's Bay Company Posts

Foreword

A12 London Inward Correspondence from Governors of Hudson's Bay Company Territories
A14 Grand Ledgers – (Accounts)
A15 Grand Journals – (Outfits)
A30 Company Servants Lists
A64 Miscellaneous Note Books

Series B: Post Records
Example:

B3/a Albany Post Journals
B3/b Albany Correspondence Books (Letters between Posts)
B3/c Albany Correspondence Inward to London
B3/d Albany Account Books
B3/e Albany Reports on District
B3/z Albany Miscellaneous Items

Series C: Ships' Logs, Seamen's Wages, Portledge Books

Series G: Catalogue of Maps, Plans, Charts, etc. (Manuscript and Published)

Class 1 Small manuscript maps, etc. stored flat
Class 2 Large manuscript maps, etc. rolled
Class 5 Atlases
Class 7 Bound volumes of manuscript maps

The documentation approach used in this study is intended to assist readers in finding sources while avoiding excessive endnotes. Thus most archival references for manuscript documents are placed in parentheses in the body of the text. (B135/a/10,fo.11d) is an example of one such HBCA reference in the text; it indicates a record as follows:

B – records of company posts
135 – Moose post records
a – the annual journals of the post
10 – the journal for 1740–41
fo – a page in the journal
11d – the page number, "d" being the dorse or back of the page

Three catalogues of Hudson's Bay Company manuscript maps have been prepared for this volume. Catalogue A lists manuscript maps found in the Hudson's Bay Company Archives. Catalogue B lists maps found in archives other than that of the Hudson's Bay Company or available only in published sources. Catalogue C lists maps that, although there is evidence in archival documents for their existence at one time, have not been located. When maps in the catalogues are mentioned in the text, the catalogue number is given so that readers may turn to the catalogue entry for further information regarding the map under discussion. Superscripts [A], [B], [C], are used with map numbers to distinguish respective catalogues. At the end of the catalogue entry for each map, the text page (or pages) on which this map is mentioned is given.

The location maps at several scales focus on the eastern and western areas of company operation. On these the reader will find, within the constraints of scale, those place names mentioned in the text. Names of locations on these maps may be given in either the contemporary or the modern form. In several cases where both names are shown, the contemporary toponym is the one enclosed in parentheses. This is in contrast with usage in the text where the modern forms are the ones placed in parentheses.

The glossary includes definitions of terms which are associated particularly with the Hudson's Bay Company, as well as some technical terms and certain words with regional application.

Ten appendices provide additional specific information. Appendices one through four supply background on the history and structure of the company and discuss the Hudson's Bay Company Archives map records and catalogues. For readers who wish to understand the basic steps involved, appendices five and six describe the procedures normally used by company personnel when observing for latitude and longitude. Appendices seven through ten list persons who prepared maps or who provided sketches or information to map makers.

Location Maps

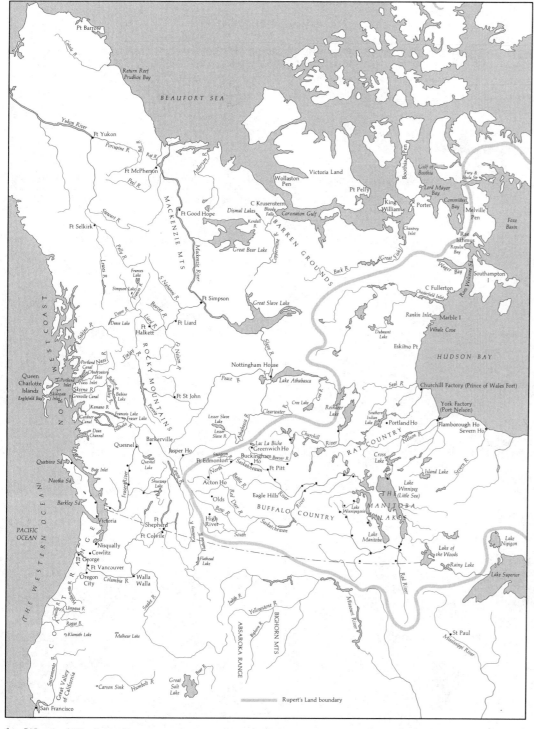

1 West of Hudson Bay and the Great Lakes; the Arctic Shores

Location Maps

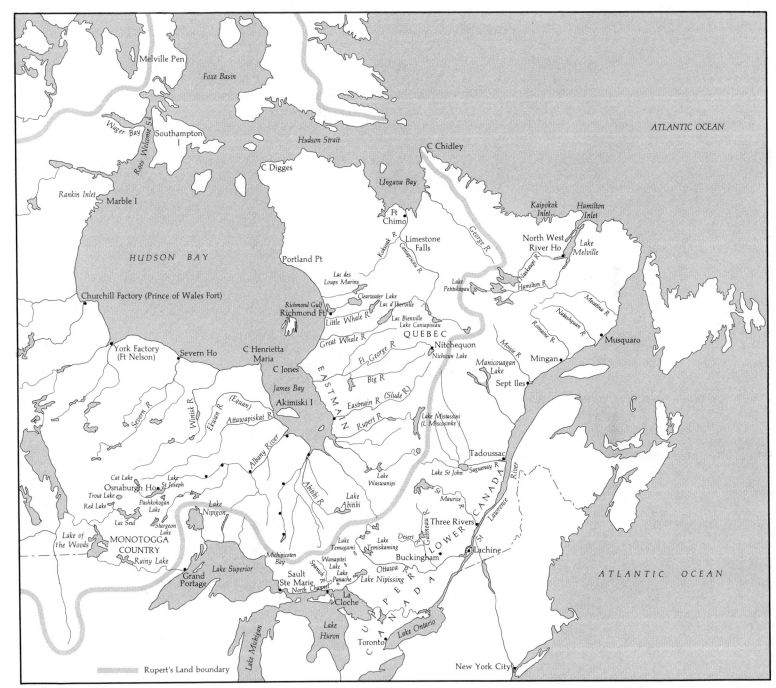

2 South and East of Hudson and James Bays

xvii

Location Maps

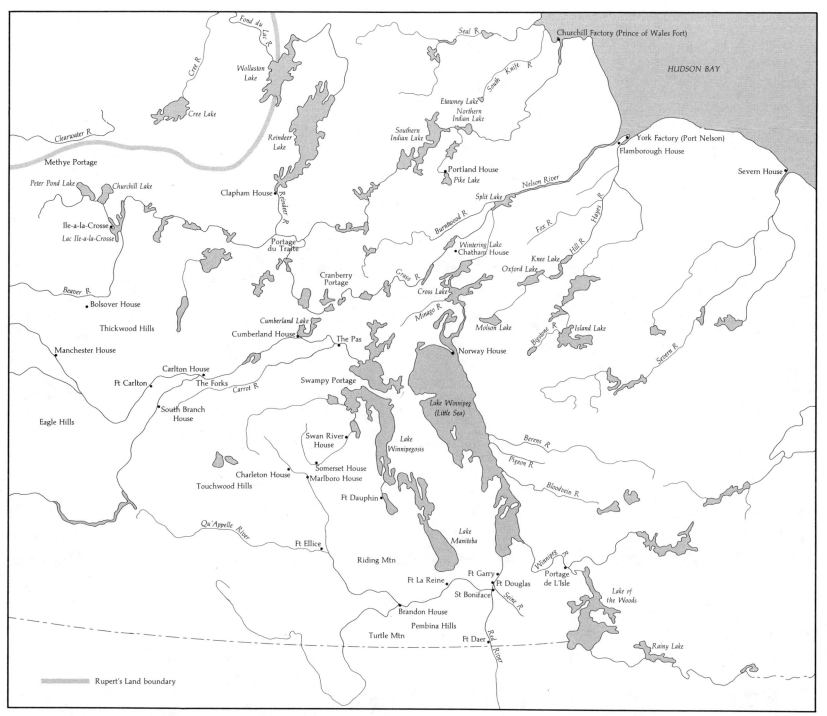

3 The Manitoba Lakes and the Nelson and Churchill River Systems

xviii

Location Maps

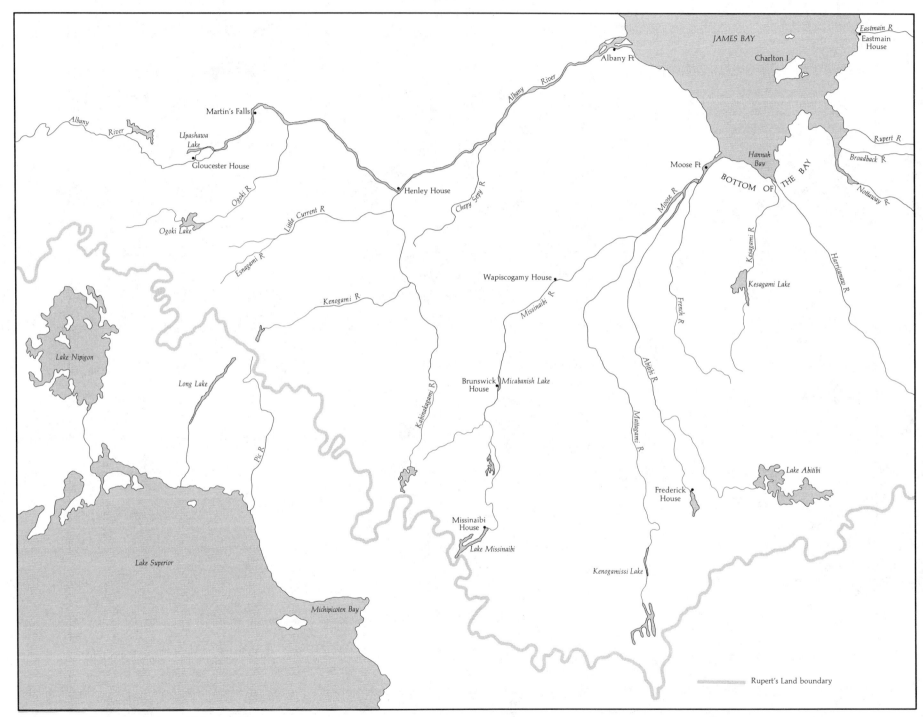

4 Between Lake Superior and James Bay

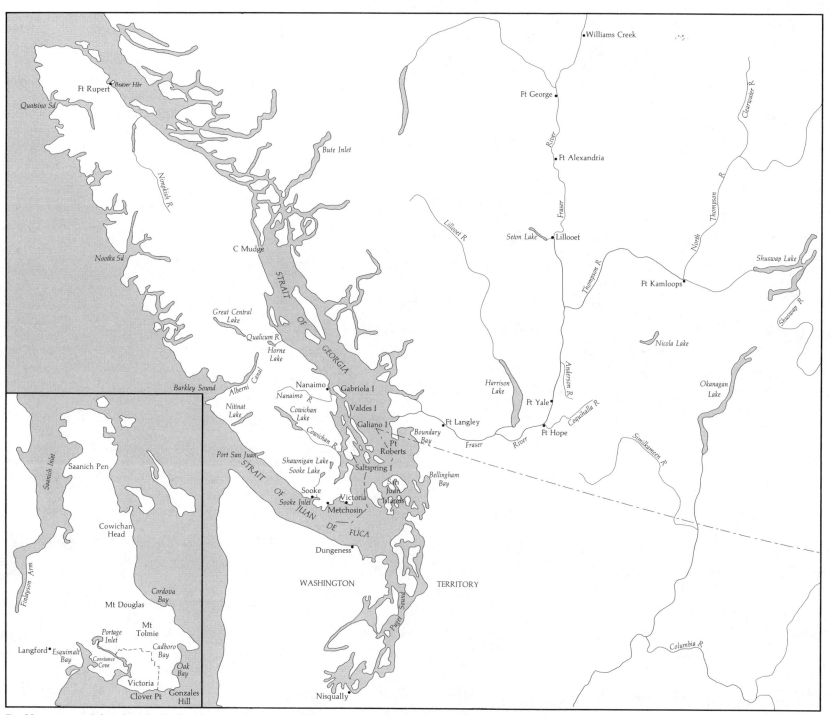

5 Vancouver Island and the Southern Mainland of British Columbia

PART ONE

Introduction

1 Mapping Policy and Records

"The English made maps because they had to, and conversely they seldom made maps for which they did not feel an immediate need. Nothing was done for pleasure, nor was there much in the way of pure geographical scholarship. It was a serious business."[1] Although the writer of these lines was discussing the English Atlantic-coast colonies, her description applies equally well to that commercial component of the British colonial empire called the Hudson's Bay Company.[2] The company's chartered territory of Rupert's Land, with an extended area where it had trading privilege, encompassed the larger share of mainland Canada and of the American Northwest.[3] Maps of this immense – and largely uncharted – area were crucial to the pursuit of profit there and became valuable business documents, as did the nautical charts that provided an intimate view of the paths ships took across hazardous seas into the bays and rivers where the company established its posts.[4]

The company did not begin its commercial operations with any systematic plan for exploration or mapping in mind, but it immediately began to buy or have made the maps and charts it required. Commercial firms were applied to, and company servants were encouraged to observe, measure, and make sketches in the course of their trading tasks. Before too long, some employees were being paid to explore, to survey, and to map as part of their career with the company. The company regularly provided them with surveying instruments, pens, brushes, inks, drafting paper, mathematical tables, text books, and map chests, and paid over and over again for the repair of their instruments. Although a century elapsed before an official cast was given to the program – with the hiring of a trained surveyor-cartographer – a mapping policy, albeit a piecemeal one, did evolve.[5] And as this policy was carried out, the great inland empire of the Hudson's Bay Company was forged, and northern and western Canada were unveiled.

THE ROLE OF MAPS AND CHARTS IN THE COMPANY

Exploration and mapping became an indispensable concomitant of the fur trade. However, they were not "to be the only Object of ... attention, but to be made Subservient to promote the Company's Interest and increase Their Trade."[6] Thus it is not surprising that the company did not have at Hudson's Bay House in London a chief surveyor, a cartographer, or geographer responsible for overall exploration programs, for surveying in the field, for drafting maps, for maintaining map files, or for managing public relations with cartographers or scientific societies at home. There was no central drafting room or map room; some incoming maps were maintained in a semblance of order in several map cases and portfolios at different locations in Hudson's Bay House in London. Others, sketched on the pages of journals, or folded and attached to them, were kept in the office files. Still others, drawn on separate sheets, were attached to wooden rails and used as wall maps in the company's offices. Those drafted on multiple sheets and glued together on a linen or canvas backing might also be used as wall maps before they were stored away.

Maps were examined when a journal, letter, or report was being read at committee meetings. On 13 March 1750, for example, two maps of the Eastmain by Captain William Coats and a letter relating to his 1749 expedition were "taken into Consideration and after some time spent therein [the committee] Resolved that for extraordinary Trouble and Care ... a Gratuity be allowed ... of £80" (A1/38,fo.331).[7] Maps were also consulted by successive governors and other officials from time to time as the areas they circumscribed came under discussion. They became an essential tool in making decisions about trading strategy, transport routing, and the movement of goods since they provided the locational and spatial information necessary to the development of an inland trading system. When the committee decided to send employees in search of more efficient water routes between fur trading posts or between posts and native groups, or when it wanted to change trading strategies, it needed to see the configuration of the river and lake networks, their crucial crossing-points, their relationships with major terrain features, the difficulties posed by waterfalls and rapids, the numbers, locations, and the potential risk in portages. Larger scale maps were used to compile smaller-scale regional maps that made the intricacies and interconnections of these waterway trading patterns more apparent. Given the territory, it is understandable that the larger number of maps required for such consultation was planimetric or topographic and dominated by water features.

The company also wanted to be informed about the characteristics of its establishments in North America and requested detailed architectural plans of its factories and posts. It was concerned with the mapping of the locales of the fur posts, the siting of their buildings, and with other land use on company property. In later years, as colonial administrations in Canada and British Columbia required clear indication of property boundaries, many of the Hudson's Bay Company's land claims across the continent were surveyed. Company personnel, or hired surveyors and draftsmen, also helped measure and subsequently map the property lines in newly established settlements, especially in British Columbia. Survey plats were drawn for the registry of lots. At larger scales were a variety of maps that the officers of a related operation, the Puget's Sound Agricultural Company, required in the Esquimalt District of Vancouver Island, and at Nisqually and Cowlitz in Washington Territory.[8] These depicted field patterns, the location and nature of crops grown, the farmsteads and farm boundaries. All of these maps were the subject of extensive examination and interchange of correspondence between London and local officers.

Some maps served special purposes: they were drawn to illustrate specific company projects. The beaver conservation preserves on islands in James Bay are one example. Beaver houses and ponds were mapped and inventoried on several occasions during the decade the preserves were maintained. A variety of maps resulted from the exploration and examination of coal deposits on Vancouver Island, from the coal mining operations at Nanaimo, and from the sale of the operation to another coal mining company.[9] Other groups of maps reflected the company's business dealings and its participation in various facets of political, administrative, and economic life. Among these were the maps required for colonial land surveying and subdivision, colonial administration, native land treaties, and for the telegraph line from Red River to the Pacific. Nautical charts were essential for company ships on the high seas as well as on smaller vessels engaging in coastal transport along the shores of Hudson and James bays and the north Pacific. River mouth charts were vital for ocean-going sailors; the drafts of these charts were continually updated with more accurate soundings and position markers.

PUBLIC RELATIONS OF THE COMPANY AND THE DISSEMINATION OF GEOGRAPHICAL KNOWLEDGE

The Hudson's Bay Company was both a contributor to and a guardian of the geographical knowledge that flowed annually into London through the media of correspondence, post journals, journals of expeditions, reports, maps, charts, plans, and through the discussions of its officers, captains, and other employees returning from abroad, with members of the committee. Geographical information thus obtained was transmuted into business data, crucial to the enterprise. Not unexpectedly, given the competition in the seventeenth and eighteenth centuries, such information was kept within the company and not laid open to the public gaze: restricted access was an important defence of the company's monopoly trade.[10]

Within prescribed limits, the company had been willing to share certain types of knowledge and to provide examples for exhibits of fauna, flora, and local cultures, especially for the Royal Society, with which it had maintained a long, and at times, close association. This association began in 1668–9 when several members of the society helped finance Captain Zachariah Gillam's voyage to Hudson Bay, the voyage that led to the incorporation of the Hudson's Bay Company in 1670.[11] Both in 1669 and again in 1673, on his return from another voyage to the bay, Captain Gillam was interviewed by the society and gave valuable intelligence on geographical subjects in the form of answers to written questions. Since the Royal Society served also as a geographical and an anthropological association before the founding of the Royal Geographical Society in 1830, and the Anthropological Institute in 1843, data on weather and climate, on compass variations, tidal conditions, the natural environment, and on the physical features, nature, and mores of native peoples were presented at society meetings. Several company officers who became noted authorities on environmental topics were elected to fellowships and read papers or sent reports to the society on subjects ranging from observations on the magnetic needle to the dressing of animals skins by native people.[12] Yet, information about the interior of the country, its hydrographic patterns, and all else considered vital to trade was sacrosanct and lay locked in the records at Hudson's Bay House. Very little of this inner core of knowledge reached the cartographic and geographical world as a glance at maps of the north and west of the continent, prepared by the leading cartographers of Britain and the rest of Europe before the 1780s, will demonstrate; not even the geographer of His Majesty the King was given access.

In the latter half of the eighteenth century, the attitude of the company toward its hoard of information began to change as its "habit of secrecy" came to be seen as detrimental to company well-being.[13] The effects of this change, of co-operation and activity within the geographical, cartographic, and natural science worlds of the nineteenth century were far reaching. The company gained considerable public repute and esteem, as did a number of its officers. Secondly, there was, after 1795, a great flowering of published maps, for which the eighteenth-century fur traders, explorers, and navigators had provided the geographical framework. The earliest impulse for the company's reversal of policy was the unsettling controversy after 1741 with an English group led by Arthur Dobbs, whose underlying aim – to break the company's monopoly of trade in the northern reaches of the continent – was partly hidden by its more publicized avowal of interest in the Northwest Passage.[14] This controversy led to the parliamentary investigation in 1749 that confirmed the company's monopoly in its chartered area. It also stimulated the company in 1754 to encourage actively selected employees to winter inland in the home areas of native groups where they would be in a position to induce the natives to come again in the spring to the bayside posts.[15] In addition, the company sponsored further summer voyages along the coast of Hudson Bay north of the Churchill River in 1754, 1758, 1760, 1761, and 1762 in order to prove to the world at large that the entrance to the Northwest Passage would not be found on this coast but farther northward. (And just in case it was wrong about this, the company wished to make the historic discovery itself.)

Although not the only initiator of decisive changes, the main architect of this alteration in attitude was Samuel Wegg, whose family had been involved with the company since 1697. His personal association, begun in 1748 when as a fledgling lawyer he gained his first stock in the company, only ended with his death in 1802. After 1760 when he was elected to the committee and through his term as deputy governor, but most critically during his years as governor, significant decisions were taken in regard to obtaining and disseminating geographical information. A number of apprentices with some education in mathematics, surveying, navigating, and drafting were put under contract. Training was provided in the field for several eligible young men, and certain of these were provided with surveying instruments, drawing tools, and materials. Increasingly, the committee demanded that charts, plans, and geographical information be forwarded to them. And most significantly, the company decided to hire several inland surveyors, although only one was finally engaged.

Wegg encouraged officers of the company to become active in the Royal Society and the Society of Antiquaries, and he set an example. He was elected a fellow of the Royal Society in 1753, and for thirty-four years was its treasurer. The company and the society had numerous contacts over these years with Wegg as intermediary. In the first cooperative venture in which Wegg was involved, the observation of the transit of Venus in 1769, William Wales and Joseph Dymond of the society manned an observation station at Hudson Bay after a mutual agreement on financial and logistical details.[16] The company also increased its gifts of fauna and flora to the society's museum, a form of munificence continued in the nineteenth century, with societies and museums in both Britain and the United States as recipients.

Wegg widened his own horizon and those of his business colleagues through his participation for over forty years in the Royal Society's weekly dining club. In this gustatorial atmosphere, where conversation was undoubtedly at par with, and as varied as, the menu, the leading scientific figures of the day gathered. Even more significant, from the company's point of view, were the guests on some of these occasions: Captain James Cook and other leading explorers including Louis Antoine de Bougainville, John Albert Bentinck, James A. King, William Bligh, and William Robert Broughton were numbered among the distinguished company. As were Alexander Dalrymple, geographer, hydrographer, and cartographer (and a close friend of Wegg),[17] members of the Cassini family who were astronomers, surveyors, and cartographers, and Thomas Pennant, the leading authority on British and Arctic biology. Somewhat more novel, though eminently practical from the company's point of view, were the visits of Chief Thayendenega of Upper Canada and two Inuit of Rupert's Land.[18]

Thus it was during the period when Wegg was a senior official in the Hudson's Bay Company and wielding great influence in the Royal Society that the secretive atmosphere of the company gradually dissipated. Company employees were encouraged to publish books and allowed to search in company records for data. Alexander Dalrymple used company maps, charts, and other research material in the preparation of his notable map, book, and pamphlet publications.[19] Of more consequence, the company developed a business relationship with Aaron Arrowsmith, a young cartographer of London, that made it possible for him to use map, sketch, and journal resources as he compiled successive editions of maps of British North America, of America, and of segments of the company's trading region. He was succeeded by his son Samuel and his nephew John, making this eminent cartographic family unofficial cartographers of the Hudson's Bay Company until the late nineteenth century.[20]

THE CARTOGRAPHIC RECORDS OF THE COMPANY

The cartographic history of the Hudson's Bay Company from the granting of its charter in 1670 to the Deed of Surrender of Rupert's Land in 1870 – the period of this study – may be divided into two broad time spans, one before the union of the Hudson's Bay Company with the North West Company in 1821 and one after. Although maps depicting the exploration and increasing geographical knowledge of the north and west of the continent dominated throughout the entire span of time, the later stage from the 1820's to 1870 is characterized by an increasing diversity in mapping operations and map types. In the earlier period, maps depicting routes followed in exploration prevailed; in the later, surveys of settlements and special purpose maps increased in volume and variety. Within the two periods, 1670 to 1821 and 1821 to 1870, I have distinguished eight subordinate periods of cartographic activity. These eight form the framework for my discussion in chapters five through fourteen of the history of the cartography of the Hudson's Bay Company.

The full extent of the company's cartographic archives is not exactly known.[21] A few cartographic documents of the period may still be uncatalogued and therefore uncounted; the number of manuscript maps of the post-1870 period has not yet been absolutely determined.[22] Nevertheless, the figures presented here represent an almost complete enumeration of the company's collection.[23] Any estimate of the total production of cartographic materials, or of the number of maps and charts purchased by the company for various purposes is unwarranted. Nor is it possible to estimate how many maps, charts, and plans, known from documentary evidence to have existed, are in other collections.

The most significant cartographic heritage of the Hudson's Bay Company are those manuscript maps, charts, and plans drawn by company servants in the course of their duties, those prepared for the use of the company by private cartographers, and those received by the company and entered in its records in the course of business negotiations. These manuscript items are the unique heart of its cartographic resources, and distinguish this collection from others.

From the period I am discussing, there are 581 maps and charts and 557 segmental sketches in the company archives. These are almost exclusively manuscripts. There is another group of maps, amounting to at least 256, which are not now in the collection, but which did exist and are recorded in the company's written records.[24] A few may be viewed in other locations, but most cannot be seen because they have not survived. But their 'ghostly' presence must be included in this analysis for they were drawn by company employees, they were the basis for discussion and action, and they are a significant part of the Hudson's Bay Company mapping history.

It is sad to note the extent of attrition in the collection during the two centuries under analysis. It amounts to no less than thirty percent of the total. This is a heavy enough loss overall, but when one compares the situation before 1778, when the first official surveyor-cartographer was appointed by the company, with the period after his appointment, it is apparent that the survival rate of the earlier material was only about thirty-five percent. Attrition, fortunately, decreased over time. For example, none of the seven maps drawn before 1700 is found in the archives.[25] Fifty percent of the manuscript maps and charts drawn before 1800 have been lost to the collection, but about seventy-five percent of the materials prepared between 1800 and 1870 are still there. These losses may be attributed to a number of causes. Materials deteriorated; marine charts were subjected to the wear and tear of use on the high seas; maps were lost or spoiled in inclement weather or when canoes overturned; others were destroyed in fire or rain; obsolete charts were discarded; storage was haphazard both on the bay and in London. Of course, some of the maps mentioned in journals or letters may never have left the bay or ever reached London if they did. A post master, who had reported in a journal or a letter that a map was being sent home, may have neglected to enclose it, and the document may have silently vanished. If an annual post journal in which a map had been inserted disappeared, so also did the map slipped inside it. Since the governor and committee did not acknowledge receipt of every map supposedly forwarded, it is difficult to trace map movement. It is possible that committee members who borrowed items from the map chests for perusal at home never returned them. There are also grounds for believing that the Arrowsmith map firm, which became a constant borrower of items from the company after 1790, may not have returned all the maps it was loaned.[26] Finally, there is an unfortunate tendency among cartographers (decried by historians of cartography) to destroy compilation materials once the fair copy of a new map has been completed.

It was during the nineteenth century that the Hudson's Bay Company accumulated about eighty percent of its pre-1870 manuscript maps and charts. This is not unexpected since the company had only begun its inland expansionary post-building schemes in the 1770s, and as already noted, did not hire its first official surveyor-cartographer until 1778. About fifteen percent of the cartographic records belongs to the eighteenth century, with some ninety-three percent of this group being produced between 1740 and 1800. The distribution of map production over the two centuries is elaborated in figure 1.

Ninety percent of the manuscript maps were drafted by employees of the company as part of their regular duties. About four percent were prepared by private cartographers at the request of the company,[27] and approximately five percent came into the company's files from a variety of other sources.[28] In all, about 160 men have been identified as being involved in the preparation of these manuscript maps and charts during the period 1670 to 1870.[29] The records refer to over fifty Indians and Inuit who drew original maps or provided sketches and information to the Europeans who used these as the basis for other maps. The most prolific individual cartographer was Peter Fidler, who was responsible for eighty separate maps, as well as the larger part of the segmental sketches previously noted. Joseph Despard Pemberton and his colleagues in the combined company and Vancouver Island colonial mapping office in Fort Victoria contributed the largest single block of maps.[30]

There was a considerable diversity of map types, although, as could be expected in an expanding commercial operation, the largest group (thirty-seven percent) include the detailed sketches and intricate tracery, drawn at many different scales, of the rivers and lakes which formed the trading networks through the country. The second largest class of documents (thirty-five percent) were those that showed, in larger scales, the immediate environs of fur posts and factories, or the surveyed details of new settlements, the roads, building locations, lot subdivisions, and property patterns of villages and towns, and rural field patterns and farmstead details. The coasts of Hudson and James bays and Hudson Strait, with their gulfs, bays, inlets, and river estuaries, were charted in some detail. About fourteen percent of the total cartographic effort was designed for use at sea, or for coasting during trading expeditions, or for the investigation of harbours and the establishment of port sites at river mouths. Six percent were composite maps, drawn at medium and small scales, of regions or larger parts of the continent. Four percent were specialized economic maps, such as those showing the coal mining operations on Vancouver Island.

It is natural to expect that the larger share of the mapping would be undertaken west of Hudson and James bays where the bulk of the chartered territory lay, and beyond which, to the west and north, trade later expanded. Less than eight percent of the maps and charts relate to the Eastmain, to Quebec, and Labrador. About thirty-three percent of the cartographic materials depict aspects of the western interior, from the Missouri River basin in the south, through the Yukon and Northwest Territories mainland to the Arctic shores and islands. The largest block, just under fifty percent, represents the territories west of the Rocky Mountains, although much of this total (especially maps after 1821) are survey plats of land on Vancouver Island, particularly of the Victoria and Esquimalt districts through the 1850s and 1860s.

Seen from a chronological perspective, the map documents reflect changes in company priorities through the two centuries. Surveying and mapping proceeded first in Hudson and James bays and Hudson Strait, as the newly chartered company examined the coasts and established the shipping lanes in this great embayment during the formative years of the trade. Because the most accurate charting possible was essential to ships and boats, coastal charts, sea charts, and river mouth surveys continued to be important throughout the two centuries. Sketches derived from coastal exploration by trading expeditions, especially after 1740 and into the mid-nineteenth century, were also vital. Inland exploration and mapping of the basins of the rivers of Rupert's Land flowing into the bay, of the Nelson-Saskatchewan, and of the Albany and Moose systems predominated during the decades of the 1770s through the 1820s; the Churchill system became important in the 1790s and later. The Mackenzie basin and the Yukon, although explored and charted in the 1770s by other than company people, was dominated from 1790 to 1860 by Hudson's Bay Company men, particularly during and after the 1820s. East of the bay, inland movement was initiated much later. Mapping of Quebec's interior rivers and lakes, from Eastmain to Ungava, of Labrador, and south across to the St. Lawrence north shore, was undertaken through the five decades after 1810. The mapping of the region west of the mountains proceeded at first in the 1820s and 1830s in the Columbia and Snake country, the Thompson River district, and along the Northwest coast. In the 1840s the Puget Sound and southern Vancouver Island regions became the locus of detailed land surveys and map production that continued through the last two decades under review. With the establishment of colonial government on the British Columbia mainland, the Hudson's Bay Company played some part in the delineation of the interior and the plotting of early town sites there. Finally, in 1827, the company's last frontier, the Arctic shores, were breached. From that time, for some thirty years, various Hudson's Bay Company officers tested their skill, fortitude, and daring as they investigated, surveyed, and mapped that remote and forbidding region.

MAP DESIGN AND CHARACTERISTICS

The East India Company, another chartered trading company of the same period, like the Hudson's Bay Company, used maps as an important aid to business. That firm, however, was a much larger and richer concern. It established a map office, appointed chief cartographers, and had its surveyor-cartographers trained to the standards of the British military mapping agency of the period. As a result, the maps

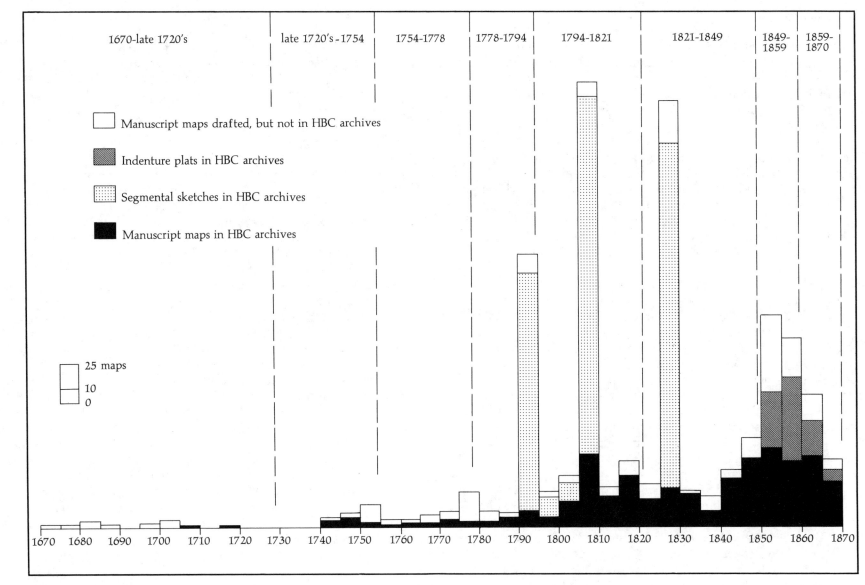

Figure 1 Manuscript maps and sketches drafted by Hudson's Bay Company employees, purchased by the company, or received at company headquarters in the course of business, by five-year periods between 1670 and 1870.

of the East India Company were of a high quality in design, draftsmanship, and precision. Except for the few charts and maps that reflected contemporary standards and styles prepared for the Hudson's Bay Company in the early years of its operations by the commercial map makers in Britain, its map documents were not remarkable either for significant innovations in cartographic design or drafting techniques. Nor was the company responsible for inaugurating and sustaining a distinctive school of fur trade cartography. A large number of the documents may be classed as sketch maps. Most were black ink line maps, dominated by sea, lake, and river outlines. The draftsmen did not attempt to show much physical or social detail with symbols. More meaningful for the trade were river obstructions such as the rapids and waterfalls, which were often labelled "portages." Only now and then was there an attempt at a simple depiction of terrain, a reference to vegetation, or the plotting of native tribal locations. The maps were essentially reference maps, similar in form to those sketched by other fur traders working in the continental interior, and by other travellers passing across previously uncharted land in other places. Few of the company's map makers used colour on their drafts or added decorative elements such as cartouches. Many of their maps did not provide scale information or give the date of preparation, the area depicted, or the map maker's name.

The main exception to this general characterization were those large-scale topographic maps drafted in the mid-nineteenth century in the drafting room in Fort Victoria under the direction of Joseph Despard Pemberton, the company-colonial surveyor-cartographer on Vancouver Island. But these maps were drafted for a very different purpose: they were an aid in the subdivision of settlements and the sale of land, rather than a contribution to the operation of the fur trade. As far as originality is concerned, the most important innovator in the history of company cartography was Peter Fidler, who developed a mapping method for use during expeditions. He drew large-scale sketches of successive sections of river and lake networks in his journals as he travelled, noting distances and compass directions, as well as particulars of the shorelines, rapids, falls and portages. Later, he used these sketches and the information to make composite maps drawn to a smaller scale.

The special cartographic contribution of the Hudson's Bay Company, therefore, was not the evolution of unique cartographic designs or of methodologies applied to the special purpose of the fur trade. Rather it was the continuing support the company gave to map making as an essential element of its business operations. In reaching out for new trading partners, company men were often the first Europeans to see, to explore, and to survey the northern and western lands that the company claimed as its territory. At the cost of wearisome, difficult, and often dangerous travail, whether on the open sea or the great plains, in the intricate water tangles of the northern forest or up the high passes of the Rocky Mountains, these men, mostly amateur map makers, did, all unknowingly, provide Canada with the largest private collection of original, primary map documents depicting the geographical form of her emerging national territory.[31] In truth, the Hudson's Bay Company was Canada's first national mapping agency.[32]

Introduction

2 The Explorers and Map Makers

The men who explored, observed, and made the maps and charts used by the Hudson's Bay Company varied widely in background, education, and profession. A few were commercial chart makers or cartographers; several were trained in surveying; a number had received instruction in navigation and the use of instruments; and several had some background in drafting maps. However, most had little or no experience in measuring distances, in taking bearings, or in making astronomical observations. Only a few were proficient in assessing the geography of the land they traversed or in keeping proper field notes. A handful were hardly literate although most had some years of elementary education. A few had professional education, and a very few had graduated from university. Several had been apprenticed to a trade; a number were engaged in or had fulfilled their apprenticeships in the company as writers or bookkeepers; others had advanced through the ranks in the sea service of the company. Some were indigenous people.

The abilities and training of the professional chart makers and cartographers may be omitted from this discussion because they practiced their professions in their workshops in London. Their products were purchased by the company as a result of specific needs, or they were asked to undertake particular cartographic assignments. There were also some surveyors and draftsmen in private practice or government offices who supplied the company with maps and survey plats during the transaction of business affairs. Their competence was important to the company in these restricted fields, but they were not involved in exploration.

Except for the few men hired specifically for these tasks, company servants did not come to North America to explore or make maps. They were occupied, first and foremost, as company officers and traders, as doctors and bookkeepers, carpenters and clerks, each primarily concerned with his own particular assignment. They explored and made maps only incidentally in the course of their regular duties. When, for example, the chief at York Fort, James Isham, at the committee's request, took on the task of measuring distances from York Fort to the various landmarks on the estuarine shores and along the banks of the lower Nelson and Hayes rivers, he was thanked for his effort by the committee and advised that he should continue when opportunity presented itself, but he was also admonished that he should do so only – "without breaking into the necessary Business of the Factory" (A6/8,fo.96d).

With this caveat in mind, one may consider the personal qualities, the abilities, and the training that company officials emphasized were advantageous for, or apparent in, those employees undertaking the difficult tasks of exploration and mapping. When the company described one employee as "a very Sober, honest, and Diligent young man (A11/114,fo.79), it was aware of the particular importance of the last quality for those who would be encouraged to seek out new trade routes or to measure and record the details of unknown coasts. In other instances, officials noted that "good sense" was the quality needed during times of both physical and social stress. "Very active" men were desirable, as were those with "youthful vigour" (A11/4,fo.52d). Some capacity for organization and command was needed by any man who would rise higher in the ranks of the trade, although such qualities were also called for in those who assumed leadership of an expedition or who combined exploration with their regular trading duties. The exigencies of inland travel required men who could learn to shoot, who could become expert in the management of canoes and boats and in "other exercises of the country" (A11/4,fo.46). The company from the beginning emphasized the necessity of hiring individuals who would be willing to move away from the bayside posts, who would have the ability to observe, and who would be able to keep records of their surroundings as they travelled. For example, in 1753, the committee suggested to the York factor that "if a proper Person were sent a great way up into the Country with presents to the Indians it may be a means of drawing down many of the Natives to Trade." The factor should assure this person that "we will sufficiently reward him for any Service he may do ... the Company by such a Journey" (A6/8,fo.118d).

The ability to use instruments, to measure distance and direction, and to compile and sketch maps was especially prized. The committee, after concluding that Anthony Henday, who had been chosen for such a journey, was "not very expert in making Drafts with Accuracy or keeping a just Reckoning of distances other than by Guess which may prove Erroneous" (A6/9,fo.33d), directed the York factor to choose "any person at the Factory skilled in the above particular who is willing to accompany him in his future Travels ... by which means we may be furnished with such further lights as we may with certainty depend on" (A6/9,fo.34).[1]

Young recruits with linguistic ability were always being sought; high praise was implied when an officer or servant was described as being proficient in Indian languages.

James Sutherland, hired as a tailor in 1770, was an epitome of desired qualities. His chief described him as "intelligent" (B3/b/21,fo.17d); a colleague characterized him as "a man of great prudence" (B3/b/21,fo.26). Sutherland, soon accepting and carrying out much wider responsibilities than expected, was respected for his combined abilities as canoeist, net maker, hunter, carpenter, trader, and linguist.[2] Of more importance, however, he became a proven traveller, observer, and maker of maps of the inland waterways he investigated. On these journeys he kept records of daily travel, describing the rivers and lakes he passed, and noting the compass bearings of the courses he followed. Such a servant would have been declared a "Proper person" by the committee (A6/8,fo.118d).

The committee also signified the need for employees who could read, write, and understand at least a little arithmetic. Being proficient in "the Mathematicks" was a benefit to the company, even more so if this proficiency included the use of instruments like the compass and quadrant or sextant to determine location, distance, and direction. The apprentices hired from the charity hospitals were supposed to have had such training, and, as important, were presumed to have had some map drafting experience.[3] Even so, most of them needed further training. It was hoped that a literate, intelligent man with some knowledge of mathematics could be given lessons in the field in the use of instruments and in working out latitude and longitude. Ideally, he could also be taught how to record his observations. When an Albany Fort employee was sent north up the James Bay coast as far as the Equan (Ekwan) River in 1777, it was "in order to learn him how to make remarks in case ... he should send Inland upon Discovereyes" (B3/b/14,fol.15d), as well as to train him in "the manner of taking his Course by the Compass, computing his Distance, & observing the Sun's altitude; that he may be the more expert at it when he goes inland" (A11/4,fo.52d).

ENCOURAGEMENT OF INLAND TRAVEL

Some company employees needed little or no urging to engage in land service or to travel away from the established posts. The lure of adventure, a delight in physical exercise, the chance to observe untamed wilderness, and the relief of being away from the more mundane life at the bayside posts were attractive to some even though the reality was much less enticing. Returning travellers' tales of long hours of paddling and tracking canoes, of trudging loads over portages and hauling boats up steep grades, of dragging loaded sleds in all extremes of weather, and of fighting off insects in summer were effective in discouraging the fainter-hearted from undertaking inland service. Many canoes overturned in rough water or rough weather, and lives and supplies were lost. Long periods of poor hunting could reduce travellers to near-starvation. Reports of possible danger from French or Canadian traders and from native groups, added to the fear of the unknown that made other employees reluctant to undertake exploration. Even some among those who were hired because of their seeming potential for exploration turned out to be unwilling to leave the established posts. Indeed, the company had to maintain pressure on its factors to promote inland service. Recalcitrant servants who refused to accept such assignments might be fined or have their work contracts broken or not renewed. Inducements – rewards, gratuities, premiums, and promises of advancement – were used to make inland work attractive. These were also offered to men who were zealous in exploration or in preparing maps or charts, both as rewards and as incentives to further effort.[4] However, albeit reluctantly, the company had to recognize that some employees were "best adapted for" serving as traders rather than travellers (A11/117,fo.46).

TYPES OF MAP MAKERS

The first group of map and chart makers includes over half of the long list of people responsible for one or more of the sketch maps or the more sophisticated maps or charts in the collection. The maps they produced were of varying grades of accuracy and refinement. The individuals in the group could in no way be characterized as having had any education or training for such work before their employment with the Hudson's Bay Company. Their educational attainments covered a range exemplified at one end by a little-educated man of mixed blood and at the other by a well-educated businessman and member of the company committee, but none had any specific surveying or cartographic training before or after becoming company employees.[5] Nor did most of these ever get any training from their more experienced colleagues during their careers. Nevertheless, some became adept at such tasks as estimating distances on the move. Some developed into superb inland travellers, although they may never have become skilled at portraying the land they crossed.

The second group, like the first, had had no surveying or cartographic education previous to employment with the company, but unlike the first, they were given the opportunity during their careers to learn the rudiments from and to get experience by working with more experienced men. Several doctors were in this group. Being well-educated, they became skilled in the use of surveying instruments and in making astronomical observations without much difficulty. Usefully, they could also instruct others with whom they came in contact. Dr John Rae, who was appointed to lead the company's Arctic expedition in 1846–7, went to the Red River settlement in 1845 to learn surveying techniques from George Taylor, Jr, an experienced surveyor there (A12/2,fos.517d,518).[6] The most successful practitioner in this group was Peter Fidler, a novice when he was chosen, at the age of twenty, to become a member of the first Athabasca expedition. After his initiation into the art of surveying and mapping at Cumberland House, Fidler went on to become a master of field observation, notation, and map sketching.

The third group of employees who explored, surveyed, and drafted maps and charts for the company were hired as apprentices from the Grey Coat and Blue Coat hospitals when they were young, usually fourteen or fifteen years of age. Most were chosen because they had been members of the mathematical classes of these institutions, which provided an education suitable for the company's needs.[7] All of the apprentices who passed through the mathematical classes of either hospitals could do basic arithmetic, were taught to write properly, to sketch, to make basic measurements, and to use a compass. All had some experience with several other instruments and in reading, compiling, and drawing simple maps and charts. Practical navigation methods were included as major elements of the curriculum. The boys had a reasonable acquaintance with the construction and operation of sailing ships. Most were, in fact, slated to be entered into sea service, in the Royal Navy or merchant marine, which was the main reason for the provision of such special subjects. Some became members of the crews of company ships, and several became veteran captains.

Company association with the Blue Coat School began in 1680; from then until 1717 some nine apprentices were taken. Then after a hiatus of about fifty years, the company, under the influence of Wegg and after the controversy over the Northwest Passage had focused its attention inland again, reinstituted its apprentice program. Between 1717 and 1800, twelve Grey Coat boys were hired. Of these, seven went into land service. Although all of these became engaged to some degree in inland travel and exploration, several were more interested in becoming traders. Only three contributed in reasonable measure to the geographical and cartographic interests of the company.

The fourth group includes seamen other than hospital apprentices. It was to be expected that a sailor of some years' service, especially when he had risen in rank to be second or first mate, to be master of a sloop or other company coasting boat, or to be captain of a larger vessel, could handle nautical instruments, could fix the ship's position, would know something of simple hydrographic surveying, and could prepare navigational charts. At least twenty-eight mariners, including several who were Royal Navy captains, prepared charts or maps which eventually entered the company's records.

The fifth group were the most specialized; they were trained originally as surveyors, engineers, architects, or draftsmen. At least a dozen of these either worked for the company, were hired to undertake certain projects, or sent survey plats or maps to London in the course of business activities. Philip Turnor was the first employee hired strictly as a surveyor and cartographer. The most highly educated and proficient person appointed during these two centuries to carry out such tasks was Joseph Despard Pemberton. He was trained as an engineer in Ireland, was in the company's service for eight years from 1851, on Vancouver Island, and was simultaneously the first colonial surveyor for the colony of Vancouver's Island.[8]

There was another group of persons who with differing abilities and in diverse roles proved cartographically useful to the company. The managers of the Hudson's Bay Company's coal operations on Vancouver Island, for example, prepared geological cross-sections, plans of coal mining works, and related maps and diagrams to accompany reports to company officials. Another example of this group is the young scientist, A.W. Schwieger, who was hired in 1864 to be assistant to Dr John Rae in making a detailed examination of the country from the Red River to the Yellowhead Pass, in order to advise on and locate the route of an overland telegraph line. Schwieger prepared a working plan of the proposed line across the prairies.

THE ROLE OF NATIVES IN EXPLORATION AND MAPPING

Attention must also be given to the role of native peoples in exploration and surveying. Native peoples were not simply the source of supply for the fur trade. They were necessary, if unequal and sometimes reluctant, frustrating and frustrated, participants in this complex operation. Without the cooperation of the natives, without their assistance as guides and interpreters in exploration, without the adoption of many tools, methods, and other items from their material culture, the fur trade could not have developed as it did. Nor would European geographical exploration and discovery have taken the direction that the history of our western and northern cartography reveals. Exploration of much of the coastal and interior country by the fur traders was in reality guided examination of the established pathways of original peoples; discovery was only a Eurocentric claim to priority in viewing homelands long familiar to their indigenous inhabitants. In this joint adventure, mutual adjustments were made, at times cooperatively, and at times hostilely. The tendency in traders' reports was, perhaps naturally, to lay greater stress upon those actions which interfered with normal business or which resulted in delays or dissatisfaction.[9] Positive support was frequently taken for granted and mentioned less often. Moreover, journals and reports were, of course, written from the social and economic viewpoint of the European traders and expressed their cultural biases and outlook. While native reactions were sometimes frustrating and upsetting to company personnel, from the point of view of the natives and in light of their habits and mores, theirs was the greater frustration.

The contribution of native peoples was evident first of all in the transmission of their geographical knowledge to company personnel. Throughout the years of the trade, many officers asked natives about their home areas, about what lay beyond the regions they knew at first hand, what routes to take, about trading possibilities with various tribes, about the environment, and about the cultures of different groups of people. Natives were queried about the number and difficulty of portages and the navigability of rivers; they were invited to sketch maps of the pattern of waterways, or to comment on maps prepared by the traders.[10] On the whole, company men respected natives for their detailed knowledge. However, when the traders expected too much and urged their informants to describe areas about which they knew little or nothing, they would be surprised and disappointed when the information turned out to be wrong. One traveller, who had several years of contact with the Indians during his exploration of the northern mainland and Arctic coast, was "fully convinced that their peculiar faculty of finding their way over pathless wilds has its origins in memory, in the habitual observation and retention of local objects, even the most trifling, which a white man, less interested in storing up such knowledge, would pass without notice." On the other hand, when an expedition was obliged to pursue a devious course in an unknown area, and the European members lost "all idea of true directions ... Our Indian companions were quite as wide of the mark as the rest."[11]

While serving as guides brought natives into the closest relationship with officers and men of the company out in the field, it was also the activity that generated the most tension and disagreement and was the source of continual comments and criticisms in letters, journals, and reports. It must be borne in mind that the company's plans for trade in general, and exploration and mapping activity in particular, were developed to fulfil its own purposes, needs and organization; that is, to fit its timetables, ship arrival and departure times, personnel arrangements, and so on. Native American habits of mind, attitudes and way of life were not well understood, or necessarily considered at all. Thus when requests or demands for guides were made, refusal or evasion on the part of the indigenous people could annoy, frustrate, and infuriate company men. However, although hiring of guides was often fraught with problems, and officers thought that sometimes they had spent far too much time and energy on "intreaties, Persuasions, Presents and promises" – occasionally to no avail (B86/b/2,fo.13), agreements worked out with natives were generally carried out successfully with reasonable rapport between the two parties.

3 Field and Office Methods and Equipment

The nature of the Hudson's Bay Company's business and of its market area in North America ensured its growing involvement with the prerequisites and the idiosyncrasies of the sciences and practical arts of astronomy, navigation, surveying, cartography, and geography. Many company employees were concerned, willy-nilly, and in varying degrees, with the methods and instruments employed in measurement, observation, and delineation. As has been noted, a few came to these tasks well-trained, but most were ill-prepared and had to learn through experience. Certainly, it was imperative that the company officers gain an understanding of the geography of the territory they controlled. For this purpose they constructed over the years, an increasingly precise projection grid or graticule from latitude and longitude observations. Onto this graticule, the growing volume of geographical detail, based on ever more accurate astronomical observations, was interpolated.[1]

THE MEASUREMENTS REQUIRED

The company especially needed three kinds of measurements: location, distance, and direction. Measurement of distance was a continuous activity on sea and land. Dead reckoning, the art of calculating position by recording one's continuously changing speed and course, was a normal procedure on ships, but a form of it was also used by travellers in canoes and boats, passing along the interior waterways. Company ship captains, while investigating the sea coasts, combined dead reckoning and astronomical observation to provide locational and distance data for coastal charting. Direction, determined by observing the compass bearings of coastal features, was a necessary part of the records.

Not as important were measurements of area and elevation. However, depth measuring techniques for sounding the seabed, that is, for obtaining the enantiomorph of elevations, were understood by mariners, and were applied by a number of company officers along coasts, particularly in bays and river mouths. These soundings were frequently depicted on coastal and harbour charts. Land elevation measurements were not essential for most of the trade area and through much of the period involved.[2] The great plains and the Precambrian shield have such low relative relief that exceptional heights were rarely a problem. Of course, in the Cordillera and along the Pacific coast this was not true. There traders were searching for routes with the least grades, along river valleys, lake basins, and over the interior uplands. The heights of mountain ranges nearby, although of general interest, were not of professional concern.

Area measurement was associated in the main with the surveying of lots in the Red River colony, in the Columbia Department, and in British Columbia in the nineteenth century, although cadastral measurements had been carried out earlier at the posts on Hudson and James bays. Some employees might be classed as cadastral surveyors and were involved at times in laying out property lines, but they had been hired because they were acquainted with the instruments and methods of astronomical surveying used to determine latitude and longitude rather than for their ability to plot boundaries. This was true even for those surveyors, in the professional sense of the term, who worked for the company. It was only at Fort Victoria from 1851 to 1859 when the company established a surveying department under Joseph D. Pemberton, with a small staff of surveyors and draftsmen, that company employees had as one of their explicit tasks the mapping of property lines.

EQUIPMENT AND FACILITIES

At the beginning of their venture the gentlemen of the committee had to purchase instruments and appropriate sea charts of the North Atlantic, Hudson Strait, and Hudson and James bays. They first turned to John Seller, the most prominent maker of instruments and sea charts practicing in London. Seller was an obvious choice, for he had just received the Royal Privilege and was Hydrographer in Ordinary to King Charles II, who had issued the company's charter. Seller was especially noted for the hand-crafting and repairing of compasses, but he also made and sold other nautical, surveying, and drawing instruments. The committee authorized the payment of over £145 to Seller during its first decade of operation for instruments and "other things."[3] These other items probably included charts. The instruments purchased in the latter half of the seventeenth century were largely those of a navigational type, and these were quite adequate for mapping the land around the bay. Various compasses and spare compass cards for taking bearings would have been part of the order. The company undoubtedly purchased the

Davis quadrant which was dominant at this period for position fixing. A form of portable sun dial, part of Seller's stock-in-trade, which was capable of indicating local time in any latitude, might also have been ordered. In addition, it was necessary to provide ships with telescopes for long-distance viewing, with a set of compasses, dividers and parallel rulers for plotting courses, and appropriate drawing materials and tools. On land, sets of instruments were provided to the officers at the factories for local use and for the use of travellers it expected to send inland. The words "box" or "set" of surveying instruments were employed in company records; unfortunately, they do not specify what types came in the box or set. However, in 1715, for example, a quadrant (probably a Davis) and stand were sent to York Fort for the use of inland travellers. A surveying measuring wheel or perambulator was available at York and perhaps elsewhere, in the 1740s and early 1750s. Some form of compass would also have been contained in the sets.

The number, diversity, and quality of instruments used in measurement increased in the latter half of the century as witnessed clearly in the records of the equipment the company purchased. In the first place, the quadrant was gradually replaced by a much better tool, the sextant, which, although it went through various modifications, became the universal instrument through the rest of the period under discussion. The basic changes made were twofold: brass rather than wood construction made it stronger, and reduction in size made it more easily portable. Both improvements were, of course, very welcome to company travellers, for whom sturdiness, compactness, and lighter weight were of high priority. Second, of special value to land observers were two new pieces of easily transportable equipment that could be used with either quadrant or sextant in observing the sun, moon, or stars. These were the artificial horizon, packed in a rugged box, and parallel glasses. The first usually required quicksilver (mercury), in heavy bottles, as the reflecting medium, and the latter was set over the horizon to protect the quicksilver surface from wind, rain, or debris during observation.

The third useful innovation came in the 1750s when John Dollond of London invented the achromatic lens for use in the telescope.[4] Requests for telescopes so equipped soon appeared in the records of the company. Fourth, beginning in the late 1770s, inland travellers demanded a boat compass especially adaptable to the conditions under which they took bearings. It was small and designed to lessen oscillations caused by the pitching, rolling, and yawing of canoes and boats. Fifth, and of supreme importance to the determination of longitude particularly, was the development of exceptionally precise chronometers in the late eighteenth century. The company, however, did not use these until the nineteenth century when it began to equip its ships with them. In the 1850s and 1860s, smaller portable or watch chronometers were mentioned as being available, notably on the west coast. Pocket watches became the almost universal aid for the determination of longitude on land. There was a constant demand for better watches as the records provide evidence of an increasing number sent back and forth across the sea for repair. The more experienced explorer-surveyors usually carried two or three watches with them in case of loss or damage but also so they could check for accuracy. The men became used to the rate variations in their watches and learned to compensate for them.

Thermometers were also critical since extremes of temperature, especially low readings, affected the operation of instruments. Correct temperature readings were important because temperature was also an aspect of the equation when the observations were being calculated. Specific requests for thermometers appeared after 1778, and especially in the 1790s, when the request was usually for instruments that held a liquid which would withstand the extremes of low temperatures.

Finally, there appeared in 1766 the first annual edition of a famous series of tables for calculating longitude, which were to be used along with astronomical observations and a chronometer or watch. This *Nautical Almanac and Astronomical Ephemeris* was accompanied by a handbook of *Requisite Tables* to show the method of using the almanac.[5] After 1778, surveyors requested new almanacs yearly.[6] Without an undated version, they were unable to obtain accurate longitudes, since they were not versed in the very complicated (albeit seldom accurate) methods employed before the advent of such aids. By the time the company set about hiring a professional surveyor and bringing in apprentices from the Grey Coat School, surveying and navigational training were providing the basic methods and an understanding of the various surveying instruments. Instruments were improved a great deal more during the nineteenth century, and the records of the Hudson's Bay Company are replete with notations of requests for such equipment. The largest requisitions for such items came from Pemberton after 1851 when he was establishing the company-colonial surveying department in Fort Victoria, the only company establishment with a drafting room. Outside Fort Victoria, the best drafting aid that was available was a drawing board constructed by a post carpenter. But even in Victoria there were complaints that boards were not large enough to make an efficient drafting table. The most common drawing tools for drafting maps included a set of drafting instruments, a ruler of some type, several pens, pencils, several camel hair brushes, sticks of India ink, water colours, and paper. The drawing paper ranged from regular correspondence stationery in sheets or bound in the regular post journal books, to the more common Whatman's paper. Cartridge paper was provided for rough work; tracing paper was also available.[7]

To aid in their surveying and mapping endeavours some men had the company purchase and dispatch to them books on these subjects. John Robertson's *The Elements of Navigation* was the most sought-after, various editions being sent out over the years.[8] Peter Fidler was the most persistent importer of reading material, ordering books and articles on eclipses, astronomy, trigonometry, navigation, and geography, along with publications dealing with the use of solar and lunar tables. Parenthetically, practical volumes, on surgery and the nature and cure of wounds, on fauna and flora, arrived, to provide information useful in emergencies or in promoting familiarity with the local environment.

INSTRUCTIONS FOR THE EXPLORERS AND MAP MAKERS

Most explorers and map makers followed instructions sent out directly by the committee or set down for them by chief factors or other immediate superiors. These instructions might be of a general nature or drawn up in more detail for the special circumstances at hand. General directions might be along the pattern of those sent in 1777 to Albany: the committee urged the factor to dispatch an expedition of active and diligent men, among whom at least one should be able to use instruments to obtain the latitude and longitude of places, so that it "may be furnished with intelligible Plans of the Country" (A5/2,fo.23). Or instructions might be only a terse order such as "take a Draft" of a river "in Your going up if Possible to be Complied with" (A11/44,fo.44). Or the instruction might be amplified: "be particular in remarking the Courses, Connections, Scource, & Depths of the Rivers, Creeks, & Lakes, and take as exact a plan of them as the nature of the Service will admit" (B86/b/1,fo.13). Or really amplified: in his short excursion inland from York in 1754, the young tyro explorer, Anthony Henday was expected to keep a journal, to convince Indians to serve as bearers, guides, and sources of geographical information. He was instructed to be out on the trail by 6:00 AM each morning, urged to speak the Indian language, to record notes each day, to observe the winds and the weather every two hours, to take the depth and width of rivers, to note islands, woods, lakes and creeks, to take samples of the soil, and to record the type of game available. While the guides were preparing camp, he was to make notes on the surrounding flora (A11/114,fos.166–167d).

Complicated sets of directions were also issued to two experienced sea captains, coasting south along the Eastmain shore in 1749. They were told to obtain "a more perfect knowledge of some parts of Hudsons Streights and Bay and places adjacent." Additionally, they were to make a new "settlement" on the Eastmain, to search out and examine all "harbours, rivers and creeks," to explore Richmond Gulf, to draft a chart of the Eastmain coast, to make a map of the site of a proposed new post, to collect geographical and other data, to note the trees, shrubs, and grasses, and to bring back to London samples of the plants, seeds, and uncommon animal species (A6/8,fos.19–19d).

OBSERVATIONAL METHODS

A large share of the observations was a combination of compass bearings for direction, dead reckoning estimations for distances, and geographical observations of the environment, especially of rapids and waterfalls, of water depths, of portages, of the terrain, and of weather conditions. Observations of vegetation and of animal populations provided valuable information about indigenous food supplies as well as trading potential. Some travellers made rough sketches as they proceeded. Those men who could handle the more complex instruments and had the opportunity to do so, noted latitude readings and calculations and, to a lesser extent, those for longitude.

The observations made in the field to calculate latitude and longitude co-ordinates helped to add reasonably precise locations of specific points to those obtained by other observers elsewhere. Eventually, a considerable network of points was added to the projection graticule that was being constructed for the northern part of the continent. The points located over the years were key positions – company posts, river mouths, important river junctions, main waterfalls or rapids, sea and lake shore features, mountain passes and mountain peaks. As more precise observations were provided, the main geographical lineaments of the landscape were added to the maps, and the lesser detail could be plotted in from the more rudimentary surveys.

The co-ordinates of latitude and longitude of land positions were obtained by two processes. The first, observation, required the use of several pieces of equipment, normally a sextant (earlier, a quadrant), an artificial horizon, parallel glasses, a watch, a thermometer, a magnifying glass, and likely a telescope. The second, computation, was based on the figures recorded and the data from a current *Nautical Almanac* and the *Requisite Tables*. In observing for both latitude and longitude, four sets of errors had to be considered: personal errors made during observation; errors in recording or computing the figures produced from the observations; instrumental errors; and atmospheric refraction. If the observer was well-trained, experienced and careful, if he used good quality, well-maintained instruments, and if he made all the corrections required, then the co-ordinates obtained would have been close to the more modern, more precisely observed and accepted figures. Otherwise – and not infrequently under the circumstances – readings varied to a greater or lesser extent.

Since the process of obtaining latitude was much less complicated than that of longitude, observers obtained the latitudes of many points on their travels, but worked out longitudes less often, usually at major locations where they stayed a longer time and could take a greater number of readings. Although latitude may be found by various means, the best and easiest method used was based on the double meridian altitude of the sun; that is, the sun was observed when it had passed over the local meridian. At times, double altitude observations of the sun were made in the morning and the evening for local time, to determine noon time exactly, and thus to achieve greater accuracy in the sun observations.[9]

Several methods of obtaining longitude had been developed over the years, but the most common method used in the eighteenth and nineteenth centuries, and that used by Hudson's Bay Company observers, was the method based on lunar distances.[10] All methods required mathematical ability, skill, and patience, since all were complicated and open to personal as well as instrumental errors. The observer needed the additional aid of a companion for simultaneous readings; and the computations were time-consuming. Fortunately, if all measurements had been taken and recorded fully at a particular time, the calculations could be made in more comfortable surroundings later.

OBSERVATIONAL DIFFICULTIES

Whether the instructions provided were terse or detailed, neither officers of the company nor the committee in London could appreciate fully the difficulties encountered by men taking observations in the field. Under the conditions prevailing in Hudson's Bay Company territory, exploration and mapping could not be portrayed as a light-hearted occupation, in spite of Chief Factor John Newton's avowal in 1749 that he would have liked to have made some measurements above York Fort, as it "would have been an agreeable amusement at some times" (A11/14,fo.135). Certainly, a few exceptional employees showed by their lifelong devotion to geographical observation that this was a metier from which they received satisfaction. But accounts of the occupational difficulties and stresses of travel and field survey occur repeatedly throughout the records. Travel, exploration of new regions, any service away from the more established posts always involved fatigue, sometimes overwhelming in its effects, especially with hunger an ever present spectre. Miserable weather not only made travel disagreeable and sometimes dangerous, it presaged cold, wet camp sites or temporary lodgings. One traveller remonstrated about the variability of mid-June weather: the "Tent in the night was so hot that ... [they] could not sleep – yet ... [the next] day was so cold ... [they] had to heap clothes on ... [themselves] to keep ... from shivering in the Canoe" (B198/a/58b,fo.11d).[11]

In the forests and out on the grasslands, there was the risk of being overtaken by summer fires. One post master at the Liard River entered in his journal that "the fire raging at a furious rate" put his post "in imminent danger." He was "kept busy all day endeavouring to keep it off, and had not a kind Providence changed the wind the Fort must have been inevitable reduced to ashes" (B85/a/9,fo.8d).[12]

In the still forests, away from clearing breezes, mosquitoes and black flies were the bane of existence. At Richmond Gulf in 1744, an exasperated employee reported that "ye musketoes Like to Pick out our Eyes" (B59/a/9,fo.10). A trader complained that he and his companions "were like to be eaten with Moschettos last night and this morning before we embarked" (B198/a58b,fol.11).[13]

In addition to the hardships shared by all employees travelling inland, certain problems were peculiar to carrying out field observation. Frostbitten fingers, for example, were an occupational hazard. In extreme temperatures, holding metal instruments and putting them to the eye was painful and potentially damaging. Often enough, the intense cold caused the liquid in bulbs and tubes on thermometers and other instruments such as quadrants or sextants to expand and burst. Low temperatures made the metal components of instruments, especially watches and chronometers, expand and contract differentially. Other adverse weather conditions had a bearing on the taking of observations; heavy rains, freezing rain, snow, grey overcast skies, and dense clouds which obliterated the moon, sun, or stars could make observation impossible. Deep snow made it difficult to travel; soft snow slowed down surveyors who were pulling the thin-wheeled perambulators. Sun glaring on snow could cause snow blindness.

Local magnetic aberrations, likely due to iron concentrations, affected compass needles and made accurate reading difficult. For those explorers in the northern Barren Grounds and along the Arctic shore, the nearer they approached the magnetic pole, the less they could rely on their compasses; at some point, instruments could not be used at all.

A further observational difficulty obtained in the mapping of lake shores, particularly when these were island-strewn and had many hidden bays. For the most part, travellers, hurrying on their way, passed along the main axis of a lake and were not able to search carefully along the main shores. The usual method was to take compass bearings to obvious points, to estimate distances, and to make some reasoned guesses as to the character of many of the unseen details. This was also the procedure when sea coasts were being charted from coasting ships without sending in boats to examine the shore more closely. When explorers were examining major rivers, they tended to follow the shores rather than steering a middle course. Then because of the convolutions of the shoreline, there was a tendency to overestimate the length of river channels.

On more occasions than not, observations were not taken, or were taken with difficulty, or were inexact. Vital instruments were frequently lost, broken, of poor construction, or unavailable for some reason. They were often difficult to transport, and being delicate, were always at risk when loaded in or out of canoes and boats, or packed on horses or sleds, or carried across rocky portages. During storms or in rough water, canoes were frequently upset or driven against rocks and over dangerous rapids, putting men, records, and instruments at risk. A recurring complaint was the unavailability of the proper instrument, either because it had not been purchased, had not arrived from London, or had not reached the potential user over the long distances in North America. Or perhaps, an instrument, sent to London for repair, had not yet been returned; sometimes after a repaired piece of equipment had reached its destination, it was found to be useless because it had been improperly calibrated. At times, nautical almanacs for the current year were not at hand because the edition had been sold out in England before the company put in its order. The transport lag over much of this period was an irksome matter; between the loss or breakage of an instrument, or the ordering of an instrument and its receipt in the field, one year and more often two years, could intervene. When loss or breakage occurred or an order was decided upon in the summer, the request could be put in the packet at ship time in late August or early September, and, ideally, the item in question sent back from London the following spring for arrival in late summer, a year later. Of course, it had still to be transported to the observer, wherever he was located. If, however, the order or the loss or breakage occurred after ship's sailing in the autumn, two years could elapse between the time of the request and its fulfillment.[14]

4 Company Procedures in the Operation of Policy

Maps and charts arrived at the Hudson's Bay Company's office in London over the years, at the request of the governor and the committee, from unsolicited sources, and from business contacts. Those prepared by company employees were normally dispatched by the chief officers at posts in the annual packet for Hudson's Bay House. Occasionally, a map maker would himself forward his maps. No specific document in London outlined procedures to be followed in the cycle of exploration, surveying, and mapping. However, a system of sorts did evolve and was more or less followed during the two centuries.

EXECUTIVE DECISIONS ABOUT MAPPING

The governor, deputy governor, and other members of the committee met every Wednesday to conduct the affairs of the company with the secretary in attendance at all meetings. Of course, various members of the committee would have met for other business engagements, informally at their clubs, at coffee houses, or during social events. No doubt, often enough on these informal occasions, the discussion came around to the company's trade and exploration policy, to the need for maps and charts. However, most of the ideas, proposals, and requests for particular discussion about exploration, about trade and exploration policy, about specific maps, or about changes in a surveying and mapping arrangement would probably have originated in the more formal atmosphere of committee meetings. After discussion and decision, the latter timed to catch the annual voyage to North America, the committee would send off a directive to its representatives at Hudson Bay.

The decision to hire several surveyors for inland service and then to employ Philip Turnor in 1778 provides a good example of the procedures the committee followed in making and executing decisions about map making. It would appear that there had been private conversations on this matter before it first surfaced on the agenda of a committee meeting in March 1778. It also seems that an informal approach had been made to William Wales, Mathematics Master at Christ's Hospital, for his aid in engaging surveyors, for reference to such assistance occurs in the minutes, even before a formal letter of request had been sent to Wales. He was the obvious person to ask for assistance in the matter, since he was well known to the company as leader of the Royal Society expedition to Hudson Bay to observe the transit of Venus. As already mentioned, he was an acquaintance of Samuel Wegg, the company's deputy governor in 1778, and like Wegg, a fellow of the Royal Society. Wegg had negotiated the agreement between the company and the society for the expedition, and it was probably Wegg who first broached the matter of surveyors to Wales.

The agenda of the committee meeting of 25 March 1778 included the item, "Astronomers to be procured," and the minutes recorded that the committee passed a "Resolution respecting 3 Proper persons to go In the Co's Service to HB with the Title of Land Surveyor" (A64/1,fo.18). The secretary was asked to write a formal letter requesting Mr Wales's aid, which he did the following day (A5/2,fo.32). The minutes made reference again to the mathematics master the next week, 1 April, with the cryptic entry, "Mr Wales" (A4/s,fo.34d). By 8 April, two men had been recommended by Wales although, after that date, they are not mentioned again nor is there any indication of the reasons why they were not hired (A4/3,fo.34d).[1] A third applicant was named the following week. On 15 April, a letter entered in the minutes as "Mr Wales Letter Philip Turnor" (A4/3,fo.34d) was summarized by the secretary: "Mr Philip Turnor of Laleham Middx. 27 yrs age not marry'd brot up in a farming business propos'd by Mr Wales as a Land Surveyor (A64/1,fo.18d).[2]

Terms of agreement were under discussion for several weeks, as the agenda noted on 22 April, "Philip Turnor – Observer" (A4/3,fo.34d). By 29 April, Turnor and the company had come to an agreement because mention is made of the need of "A Sextant for Mr. Turnor" (A64/1,fo.19d). The company agreed to hire Turnor for a three year period at £50 per annum. His salary was to commence with his arrival at York, and he was to eat at the captain's table while enroute and with the chief factor at York (A1/140,p.3)[3] Arrangements for his transport, his accommodation and reception at York Fort, and for necessary equipment were merged into the general organization of the ships' departure for the bay later in the spring.

MAPPING DECISIONS SENT TO COMPANY OFFICERS IN NORTH AMERICA

The governor and committee included the decisions on mapping in the annual instructions to the chief officers of the company at the factories in Hudson and James bays. The company ships bound for North America normally cleared the Thames River estuary in late May or early June. However, arrangements for this vital event began months earlier; for the most part in February. A roster of activities for 1778 when Turnor was sent to Rupert's Land will indicate the complexity of the task. The committee had to settle the number of ships to be fitted out, decide on their masters and mates, have the ships placed in dry dock for inspection and repairs, arrange for a naval convoy for the ships, work out the signals to be used by the captains while at sea and entering ports, approve the requisitions (or indents as they were called) for ships' provisions and stores, and purchase new instruments for use on board. During these months, the committee also had to work out the outfits for the several factories, prepare the drafts and final wording of instructions to the captains, and ready the general letters for the factors and private letters for officers and servants.

By late May 1778, financial arrangements had been completed, and on 29 May, Turnor set off. Committee minutes recorded later that summer indicate than an allowance of £10.10.0 was paid to the captain for Turnor's food on board ship. The secretary also noted that Turnor was provided with a sextant and watch. The committee meeting of 24 June ordered that a bill for £3.15.0 be paid for "Mathematical Books" that had been sent out with Turnor (A1/140,p.19), although these were described as "Astronomical Books" in the Agenda Book (A4/3,fo.36). This money was due William Wales as he had provided the books. Also, the committee ordered the secretary to present Wales with £5.5.0 for his trouble in helping procure a surveyor (A1/140,p.19). An entry for 15 July 1778 shows that the committee paid £1.7.6 for bedding for Turnor to use on ship board and while at the bay. On the same date, the secretary recorded that Turnor had been advanced £5.5.0 on account for his wages during his first year (A1/140,p.27).

COMPANY OFFICERS REPORT BACK TO THE COMMITTEE

Philip Turnor arrived in Rupert's Land at the roadstead off York Factory on board the *King George* on 24 August 1778, and, after the proper signal was given by the ship and answered, came ashore to be "entertained" by Chief Factor Humphrey Marten.[4] The reason for his arrival, the nature of his task, and the instructions to the chief at York as to his requirements had not been known to Marten, but all this was explained to him in the official letter from the governor and committee. As inland surveyor for a three-year period, Turnor was expected to report to London on his progress by the next "shipping," that is, by late summer 1779. Turnor was immediately put to work by Marten; in the presence of the council of York Factory, he called on Turnor to make an accurate survey of the post and of the adjacent ground, and, along with two carpenters, to make an examination of the physical fabric of the buildings in order to send a report on their condition directly back to the committee by the *King George*. Turnor set to work at once and drafted a plan as the result of his survey.[5]

Before the ships returned to England, Marten decided to retain on short contract two York employees who had been scheduled to leave for home; they were good inland men, a quantity always in short supply, who could travel with Turnor. The ships arrived back in England in the autumn with the packet from York announcing that Turnor had reached the fort and promising that the chief and council would give him all the assistance he required. The packet also included Turnor's plan of the fort and a report in which he recommended structural repairs to the factory.

On 13 January 1779, Turnor's plan and report on the factory were considered by the committee (A4/3,fo.38). The secretary then wrote to the officials at York indicating committee approval of these documents, sending their letter on the summer ship of 1779, along with letters both for the council and for Turnor, reminding the former that the committee expected a report on progress made during the 1778–79 season and that the members wished for a map from Turnor laying down the locations and "the several distances of the places ... [he] may have been to from York Fort" (A5/2,fos.43,43d). The letters reached York – indirectly via Churchill since massive ice floes had prevented the ships from getting nearer – in late autumn 1779.

PREPARATION OF MAPS FROM FIELD INFORMATION

The maps of Turnor's first journey, inland to the Saskatchewan River, were completed by the early autumn of 1779 (G1/21;G1/22,1779). Turnor would have had well over a month at York between his return from the interior and the expected ship arrival to prepare his materials, using whatever facilities were available at the factory. These maps were put into the factory packet by Marten, along with a covering letter from Turnor, awaiting ship arrival (A11/116,fo.57).

TRANSPORT OF MAPS TO LONDON AND DELIVERY TO THE COMMITTEE

The 1779 packet from York did not reach London that autumn, the ship being unable to reach York because of ice. It was not until a year later that both the 1779 and the 1780 packets were put on board and carried to London. The ship captain reported to the committee, delivered the packets, which were then considered by a subcommittee on 10 and 14 November 1780. Philip Turnor's maps would have been perused, although there was no specific mention of them having been received, as there was of the plan of York Factory.[6] He informed the committee in his accompanying letter of 15 September that "I have had my health exceeding well and went through the inland journey with good spirits though with some difficulty as expected." Moreover, he admitted, "I cannot say I have a great aversion to the country." He asked for a boat's compass, a small telescope, and a nautical almanac for 1781. He was also worried that the members of the committee might think his astronomical observations to be too few, "but by the loss of my quicksilver ... [one bottle had been lost when his canoe overturned and one had been left on a portage] I was rendered incapable of getting more in the winter and the weather proving so exceeding cloudy this summer prevented my getting more at York Fort." He urged that a new supply of quicksilver be sent over on the next ship (A11/116,fo.57). Chief Factor Marten's letter informed the committee that Turnor, having done all that he could at York, was to go on board the Severn sloop to begin surveying down the bay.

In the letters sent in 1781, the committee refers to the 1779 and 1780 packets, which had included Turnor's maps, journals, letters, and his requests (A5/2,fos.61–61d). Turnor, in the autumn of 1780, had complained about the great inconvenience he had to endure in using his instruments for want of the 1781 almanac, and noted that he was still handicapped by the lack of quicksilver although he could report that his instruments were still in good condition. He also reminded the committee that his three-year contract would expire in 1781 (A11/44,fos.106–7). In reply to these accumulated comments and to the cartographic and journal information sent back by Turnor, the committee indicated in the 1781 letters that it was well satisfied with his work and was sending him the quicksilver and almanacs for both 1781 and 1782. The book, *Requisite Tables*, provided by William Wales, was enclosed, but the compass and telescope he had requested were not mentioned. His contract was renewed for a further three years at £50 per annum, he was placed on the council at Moose Fort, and he was definitely promised an early promotion. The committee asked him to proceed further to survey the Moose area, the Missinaibi tributary of the Moose River, and other locales. He was also encouraged to make himself a "Master of the Indian Language" in order to qualify for "taking Charge of any Settlement" (A5/2,fo.61).

Turnor's response to this was contained in the autumn packet, 1781, from Moose Fort, where he was then located. "I have now sent ... my Journal and Observations taken at the different parts of the Country." He thanked the committee for the quicksilver, almanacs, and tables, and promised "next year, as likewise, a Draught of the Rivers and lakes through which I passed in my way to and from Lake Superior and to contain the bottom of the Bay from Albany to Eastmain." Unfortunately, he had to report that his watch was now broken and the index glass on his sextant had been damaged. He further desired the almanacs for 1783 and 1784 and a light tent for his travels (A11/44,fo.125).

After Turnor's maps were examined by members of the committee, they would have been stored away in the manner mentioned in chapter 1, that is, in chests, or at a later date, in portfolios. The letters and journals he sent along would have been filed on shelves and drawers. Both maps and written material would have been available to committee members for further consultation as needed.[7]

For the better part of the two centuries after 1670, most of the maps and charts delineating the geographical characteristics of the country or concerned with the main lineaments of the company's economic operation were handled in a manner illustrated by the Turnor episode recounted here. However, as the Hudson's Bay Company operations were extended west of the Rocky Mountains to the Pacific shore, and as new transport arrangements came into force, correspondence, packets, trade goods, and trade returns also began passing through the west coast ports. These were Fort George, and then Fort Vancouver on the Columbia River, Fort Langley on the lower Fraser, and especially Fort Victoria, the entrepot on Vancouver Island. In the east, correspondence and materials were, in later days, shipped from Lachine, the company's North American headquarters after 1826, or through New York City.

PART TWO

*Investigating
a Country
So Interesting*

5 London Chart Makers, 1669 to the Late 1720s

> To Jno Thornton for 2 Mapps of hudsons Bay ...
> £3:0:0
>
> Company records, July 1701

By the time British seamen began to reappear in the Hudson and James bays around 1670, there had been a lapse of some thirty years in maritime activity in this great embayment. The noteworthy voyages of captains Luke Foxe and Thomas James in 1631 and 1632 concluded an epoch of discovery that had commenced with George Weymouth's venture in 1602. Henry Hudson's voyage in 1610 and the journeys of Thomas Button, Robert Bylot, William Hawkridge, and Jens Munck in the intervening years further helped to trace almost the entire outline of Hudson Strait and the two bays. A generally realistic shape had been sketched by 1635, as witness the maps of Foxe and James, the critical ones for this region.[1] These two explorers had depicted the west coast of James Bay and rounded out the western bulge of Hudson Bay, north of Cape Henrietta Maria. At the time the Company of Adventurers undertook its first expedition to the Bottom of the Bay in 1668–9, there was still great hope that further investigation of openings on the western side of the bay and on the north, such as that into Foxe Basin, might reveal some channel to the sought-after Western Ocean, either through the land mass, or around the northern shores of the continent. This geographical wishful thinking became epitomized as "the search for the Northwest Passage."

Thus, in 1670, the year the Adventurers were granted their Royal Charter by King Charles II, direct knowledge of the near western regions was essentially confined to the shorelines (see figure 2). Except for the purported journey of Pierre-Esprit Radisson and Medard Chouart, Sieur des Groseilliers from the Great Lakes to James Bay,[2] explorers had only sailed along the coast, with actual landings confined to a few spots on the shore, notably at river mouths. The width of the land mass was unknown; the bulk of the north and west still lay shrouded in obscurity. Even speculative notions about the west, displayed principally on French maps, were exceedingly vague and inconsistent.

MAPS APPEAR IN HUDSON'S BAY COMPANY RECORDS

The new company proprietors were undoubtedly influenced somewhat in their thinking by the information on French maps; they certainly had examined Captain James's map. An entry in the general ledger on 3 May 1669, the first reference in company records to mapping or the use of maps, shows that a Mr Norwood was paid the sum of £2 "for making a Map of Hudsons bay according to Capt. James's Description" (1ᶜ) (A14/1,fo.108). Although there is no further mention of this map, it was probably used by the Adventurers in planning their commercial enterprise. The first indication of company interest in having its servants engage in map making occurs in a 29 May 1680 letter of instruction the committee sent to Governor John Nixon[3] in the bay. In the letter the governor was introduced to a young man, Brian Norbury, who, the committee believed, could be useful to the business if he proved a diligent employee. One of his special attributes was that he had "been entred in the Mathematicks, and ... had a peculiar Genius for making of Landskips" (A6/1,fo.7). There is no evidence that Norbury used this ability to make any maps, although he served successfully for seven years with the new company. However, he was the first in a long line of servants whose predilection for making sketches would be encouraged by the company.[4]

There is archival evidence that at least sixteen maps and charts were drafted for the company during the first half-century of trading activity. As might be expected, most of them were sea charts or maps of Hudson Strait, Hudson and James bays, and of their chief harbours. Several were of the north east of the continent. Only three of these maps are found in the Hudson's Bay Company Archives today; two others are in the British Museum.

The company ordered fourteen of these sixteen maps from leading chart makers in London. Only two were made by company servants. One, assembled by Governor James Knight shortly before 1719, will be discussed later in this chapter. The other, by Thomas Moore, a company seaman and trader, is the earliest map attributable to a company servant and the earliest detailed chart of the west coast of James Bay (1ᴮ) (plate 1). It is also the first English map to conjecture the course of any of the great bay rivers, in this instance, the Albany, and its presumed connection through a large, island-strewn lake, probably Lake Nipigon, to Lake Superior. Moore's map includes inland elements such as rivers and lakes, although, not unexpectedly, the style is typical of sea charts at that time.

Although there were several company servants by the name of Thomas Moore, the map maker is likely the seaman who was in the service from 1671 to 1678.[5] He appears to have been a member of a 1674 party that, after exploring along the west coast of James Bay past Cape Henrietta Maria to the Severn River, stayed for some time at Albany River. Moore may have known the Cree language and obtained information on this river network from the Indians. He may also have gathered details from Radisson and Groseilliers with whom he was in contact at the Bottom of the Bay for some years. However, he apparently drafted the map after his return to London in 1678 and after he had left the company's employ. It is possible that he intended to use it as evidence of his skill and familiarity with the territory when he applied – unsuccessfully – for re-entry to the service. The records do not show why he left the service, why he was refused another job, or how his map made its way into Sir Hans Sloane's collection instead of the company's. The following year he drew the remainder of his wages.[6]

JOHN AND SAMUEL THORNTON

Except for Norwood's map, all other known charts and maps prepared by commercial cartographers for the company during this period were the work of John Thornton and his son, Samuel.[7] The elder Thornton succeeded John Seller as the most notable of the chart makers of London. Apprenticed in 1656 to an experienced chart maker of the Drapers' Company,[8] he was well enough known by 1680 to become the main supplier of maps and charts to the new company, and later to be appointed hydrographer to the king and to the East India Company. Between 1680 and 1702, he drafted at least ten, possibly eleven, maps and charts, for the company, although unfortunately they have not survived in the archives. Samuel Thornton provided two almost similar maps in 1709.

The first contracts for Thornton's services were likely for three "Mapps of the bottome of the Bay" (2^C); one to be given to each of two ship captains, and the third to be held in reserve (A1/3,fo.30). This note in the minutes of the committee meeting of 5 May 1680, was followed by one in the record of the meeting of 7 May, ordering that "John Thornton mappmaker be paid for his worke," and noting that he was to "fill up ye names of ye Rivers etc. into ye bargaine" (A1/3,fo.30d). Doubtless, he was being paid for the three copies of the chart of southern James Bay.

Thornton's second map, completed in 1685, is the best-known of the maps he produced for the company as it is the only one extant (2^B) (A1/8,fo.20).[9] It is a map of Hudson Bay. The company would have provided him with as much data as possible on the characteristics of the coasts and various river mouths, including some soundings. The several harbours where factories had been built are depicted on larger-scale insets. Representative of the maps of its time, the chart is drawn in India ink and coloured inks, with colour washes on parchment. A further map, that of 1686, of the Nelson River mouth, may well have been drafted by Thornton also (3^C). No further details are available, except that the company secretary was ordered to show Captain Portin this 1686 map, as well as any other in his custody that would aid the captain in a later voyage to the bay (A1/84,fo.27). Thornton may have simply drawn a larger-scale version of the "Port Nelson" inset in his 1685 map, for it would have been the most up-to-date and detailed version available.

Thornton was paid £3 for the delivery of two "mapps of Hudsons Bay" to the secretary (4^C), the minutes of 4 October 1700 record (A1/22,fo.19). As the only known Thornton map of this area at this time is one for 1699, these two copies were probably prepared by him in that year.[10] Evidence for this is found in the minutes of a series of meetings of the general court of the company from 10 May 1700 to 29 January 1701, at which France's territorial claims in North America were discussed (A2/1,fos.19–21). The French were disputing the company's proprietorship of all the shores and immediate territory of Hudson and James bays. The French insisted on a boundary extending across James Bay at 52°30′ north latitude, with their ownership confirmed to the south of this line. This would have disbarred the company from the mouths of the Albany, Moose, Rupert, and Eastmain rivers. In rebuttal, the company claimed its charter rights in presenting its claims to the officials of the government and to William III (to the latter in a petition). At least on two occasions, when these matters were discussed with the Lords Commissioners of Trade and Plantations, the company representatives took along a "Mapp of Hudsons Bay."[11] The two copies of this map may have been taken by the committee to a meeting with the Lords Commissioners on 12 June 1700 and on one other occasion to be used as evidence during the discussions. In any event, the copies were probably of the 1699 map; on it, a line running along what was represented to be the Hudson-James Bay watershed through Lake Mistassini, northeast to the Labrador coast just south of 60° North latitude, illustrated the company's claim. There is a possibility that one of the maps may have been included with the company's petition to King William, and the other may have accompanied a paper concerning the national limits in the bay, "according to Mr Secr Vernons desier" (A1/22,fo.17).[12]

During 1701 and 1702 at least four maps of Hudson Bay were delivered to the company by John Thornton (5^C, 6^C, 7^C).[13] Some of them may have

London Chart Makers

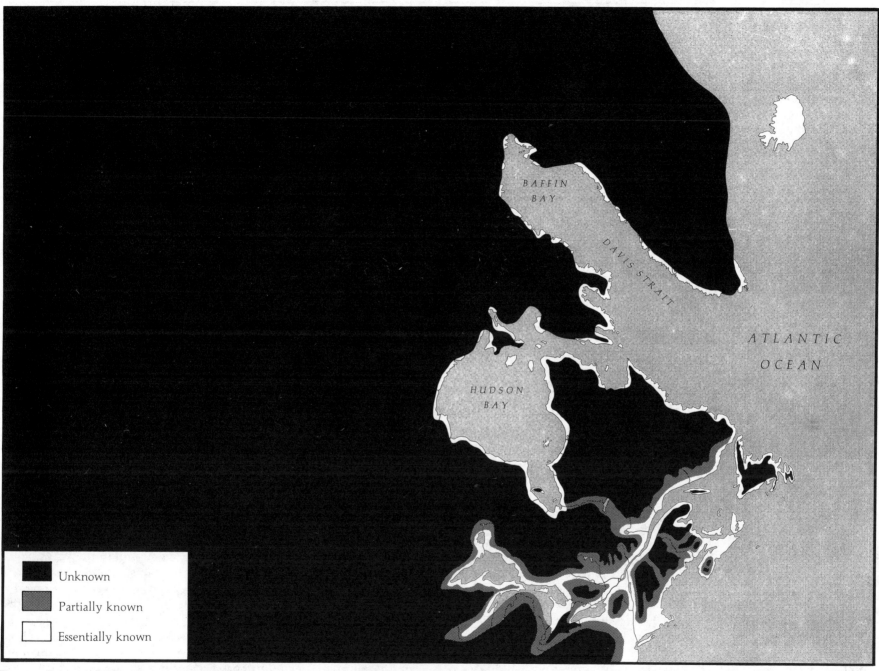

Figure 2 Retreat of the unknown: European knowledge of Canadian territory in 1670

been charts of Hudson Bay similar to that of 1685, although updated; others may have been versions of the 1699 map. The last of these, the map done in 1702, was to be paid for after a committee meeting on 18 February 1702 (A1/24,fo.9). It was likely prepared to aid discussions about general military security in Hudson Bay held during the period 14 January and 25 February among the governor and committee, the Lords Commissioners of Trade and Plantations, and Lord Marlborough, who was responsible for British military affairs.[14] The company had been asked to relay its thoughts on security matters in the form of a deposition to the Lords Commissioners. A copy of the deposition was also presented to Marlborough by the governor and committee at a meeting on the matter. The company's deposition was sent to the House of Commons by the Lords Commissioners for debate. The map, most useful in this affair and suitable for defence talks, would have been a more up-to-date version of the 1685 map, with details of the strait, sea, harbours, and posts (A1/24,fos.6,7,7d,10).

Samuel Thornton carried on his father's mapping business from the same shop, at the Signe of the Platt in the Minories,[15] adjacent to the Tower of London. He was overshadowed by his father in this craft; he drafted only two maps for the Hudson's Bay Company, both in 1709, and both extant (1^) (plate 2). He used his father's 1699 map as the prototype in drawing the two quite similar maps, but they are inferior in quality of production, even though drawn in the typical, colourful style of the Thames-side cartographers.[16] The differences between the two copies are largely those of the spelling of several place-names, some details left unnamed on one of the maps, and certain variations in colouration. The maps were used to illustrate those proprietary rights claimed by the governor and committee in northeast America in various memoranda submitted to the Lords Commissioners during the negotiation of the Treaty of Utrecht. Other such copies may have been drawn up by Samuel Thornton during this period since it appears that at least on four occasions maps were sent to accompany various memorials and depositions. The first one mentioned was sent by the governor with a letter and other evidence to Marlborough in Holland on 2 April 1709 (A9/4,fo.54). In May, two committee members were sent to Holland to represent the company personally. Was one of these maps sent with them? The symbol of primary interest on the map was the "Line drawn Cross the Grand Lake Miscosinke [Mistassini] twixt Hudsons Bay & Canada, which may serve as A Boundary between the two Nations, viz, the French not to goe to the westward of that line by wood Runners or otherwise, or make any settlement from the same towards Hudsons Bay, nor no English in Like maner to the Eastward of the said Line towards Quebec" (A9/4,fo.54). Actually, the line drawn does approximate the Hudson Bay-St. Lawrence watershed generally, although in the Labrador region it is incorrect, extending to the coast rather than north to Cape Chidley.

"TO PENETRATE INTO THE COUNTREY": KELSEY, STEWART, AND NORTON

For the first twenty years of the company's existence, the focus of geographical exploration was on coastal areas, as the new firm was occupied with maneuvering into advantageous positions along the shores of the bay. A number of elements combined to delay the company's taking fuller advantage of the apparently unlimited inland opportunities granted by its charter: its almost complete ignorance of the geography of the north and west, the novelty of commerce in such a difficult environment, the lack of experience of the early employees in trading with the native peoples of this new land, and the absence of a cadre of experienced overland travellers. However, as early as 1682, the committee sent a letter of instructions to John Bridgar, governor of the new fort he was in the process of establishing at the mouth of the Nelson River, directing him "to use yor dilligence to Penetrate into the Countrey to make what discoveries you can, and to gett an Acquaintance and Comerce with the Indians therabts, which wee hope in time may turne to Acct" (A6/1,fo.14d).[17] This was followed the next spring by a letter to Henry Sergeant, governor at Albany, instructing him "to choose out from amongst our servants such as are best quallified wth. strength of body and the Country Language to travaile and to penetrate into the Country" (A6/1,fo.29).[18] Bridgar and Sergeant attempted to follow these directives, but it is apparent from the correspondence that there were no employees willing to "travaile," even though they were promised suitable rewards. The few men who did venture more than several score miles up the rivers, did so mainly to hunt or to search for firewood.[19]

After 1682, more pressing than the need for inland exploration was the company's struggle with the French for control of the very shores of the bay. During this time, the enterprise could not have supported a major program of inland travel.[20] Instead, it sent explorers along the west shore of Hudson Bay to investigate the suitability of the mouths of the Severn, Nelson, and Churchill rivers as sites for trading factories. Coastal exploration was also extended several hundred miles north of Churchill when, in 1689, the committee commissioned Captain John Ford to sail from England to York Factory in a small vessel suitable for continuing north of the Churchill River to search for the "great river," the Deer or Dering River. There he was to establish a post to winter in; the following summer he was to try to send a party inland to explore the land and encourage Indians to trade. This was part of a plan for expansion that was to include building a post at Churchill, a plan that had to be deferred when the ship was captured by a French privateer before it sailed beyond the shores of England. The same year a trading expedition was sent by land in order to find natives in the north who could be encouraged to come south to a company factory to trade. A young servant, Henry Kelsey, travelled with an Indian companion on a route that ran a few miles back from the coast. They journeyed north of the Churchill River for over two hundred miles, approximately to Eskimo Point. Kelsey failed in his mission for the party met no Inuit whom they might have persuaded to come south to trade.[21]

But it was useful training for a pioneering inland journey Kelsey took between 1690 and 1692, when he accompanied a band of Indians as they returned from York Factory to their wintering grounds. They crossed from the northern forests in the Nelson River basin out onto the open grasslands, south of the forks of the North and South Saskatchewan Rivers. Although this trip projected one ray of geographical light far southwest into the interior plains, Kelsey did not make a map. Nor did his journal, written in verse, provide more than superficial information.[22] However, this journey foreshadowed what came to be an important trading method, that of inland wintering.

Kelsey's journey was also in accord with the policy, noted earlier, that the company had developed from the commencement of its trading activities. It would build factories at the bay shore and encourage the interior natives to come from their winter trapping-grounds to these posts in the summer. Employees were urged to "Draw downe the Indians by fayre & gentle meanes to trade" with the company (A6/1,fo.29). Traders were also instructed to denominate certain Indians as "captains" who could be given special consideration and favours with the expectation that they would lead their native followers regularly to the bay posts. Thus when inland exploration was undertaken, it was not with the idea of establishing posts inland. Such trips were primarily diplomatic missions to maintain peace among the tribes and to persuade them to trap and come annually to the coast to trade.

When the Treaty of Utrecht ended the war in 1713 Albany was the only coastal factory still under company control, but by the terms of the treaty, the company regained command of the entire bay area. For some time thereafter, it concentrated its activities at its two major establishments, Albany and the reopened York Factory where, in 1714, there were greater opportunities for the expansion of trade. Not only was York adjacent to the untouched regions of a seemingly vast, if vague, interior, it also had the advantage at this time and for some years following of being free from French competition in its catchment area. The company also began at this time to turn its attention toward the Barren Grounds and the northwest coast of Hudson Bay.

Although its "Home" Indians, the Crees of the Nelson and Severn region, had for many years been habitués of this post and middlemen in the exchange of furs from the interior, and although the Crees and Assiniboine of the Manitoba lakes area also travelled to this fort via the Nelson-Hayes route, it had had no trade, or even contact, with the various groups that resided in the forests and tundra generally north of the Nelson and Saskatchewan River basins. In fact, contact and direct trade with these distant tribes had been made increasingly difficult as a result of the strong position and strategic locations of the intervening groups. Armed with the weapons the company had traded for their furs, the Cree had been able to push out from the Nelson and Saskatchewan country, north and northwest into the Churchill basin, driving the Northern or Athapaskan Chipewyans from this region into more distant and less desirable territory. And they were loathe to give up any of the advantages which they had previously gained.

Immediately upon taking over command of York after the war, the two experienced officers, James Knight and Henry Kelsey, began to hear more of the Northern Indians. Information came not only from some of the Cree, who brought in furs that had been plundered from these people in forays into their territory, but also from several Chipewyan women who had been captured on the Cree expeditions. Besides building up the confidence of the inland Indians in the trade, Knight and Kelsey had to attempt to restore peace among the various interior tribes and to break down the barrier between the Northern Indians and the coast by persuading the Crees to cease hostilities against them and to allow them to pass unharmed through the intervening lands. They set out first to renew the wintering system begun by Kelsey himself. In 1715–16, William Stewart, who had been in company service since 1691 and, could, like Kelsey and Knight, speak Cree, travelled northwest across to the Barren Grounds, specifically to try to create peaceful conditions between the Cree and Northern Indians and to encourage an annual trading visit from the distant tribes to the bay. Stewart's track described somewhat of an arc from York Fort some six to seven hundred miles to the area east of the Slave River, south of Great Slave Lake, and north of Lake Athabasca.[23]

A second step toward the goals of peace and trade was achieved when a more strategically located entrepot at the mouth of the Churchill River was opened in 1717. Being at the southern edge of the tundra, this post would act as a destination for the Northern Indians, thereby allowing them to bypass the main Cree territory. The trader, Richard Norton, a member of the party from York that founded the Churchill post, was sent inland by Knight in July 1717.[24] How far he and his Indian companions travelled is not known; it may have been to the Athabasca-Slave River country. He returned to the new post about Christmas 1717, bringing with him a group of Chipewyan Indians. Although neither of the main goals were completely realized in this period – the expansion-bent and aggressive Crees were reluctant to be bypassed – bringing young Chipewyans to the bay shore to learn the Cree language and to perform interpreting duties were useful steps toward their eventual accomplishment.

THE COMPANY'S FIRST MAP OF THE NORTHWEST: KNIGHT'S GEOGRAPHICAL IMAGE OF THE INTERIOR

Kelsey, Stewart, and Norton did not prepare any maps based upon their extensive journeys. Nor were they skilled in geographical observation, although, of course, they did provide their superiors with oral reports on what they had seen. One of these, James Knight, by then governor of the Company's territories, was, more than any other local officer, engaged in the collection of such geographical data; apparently he took every opportunity to question Stewart on his return and the Indians during their trade visits, as to the physical geography and human occupants of their home areas and of the broader regions of the interior. Knight converted this information into a geographical conception of the north and west of the continent, a conception which was eventually portrayed on a most unusual – if crude – company map.[2A].[25] In the spring of 1716, Knight had asked some Indians to make several sketch maps for him. On another occasion, a Cree chief drew a map of the region where friends of his, on a war party in 1715, had killed some fourteen tents of Copper Indians to the "WNW" of their own home area (B239/a/2,fo.32). Then when a group of young Northern Indians arrived at York with Stewart, Knight "had abundance of Discourse wth. them about there Country and did gett them to lay down there Rivers along Shore to the Norward they chalkd 17 Rivers some of them very Large" (B239/a/2,fo.29d).

The substance of Knight's conception of the northern part of the continent, only partly depicted on this map, may be summarized as follows. He believed that the north of the continent stretched hundreds of miles across from Hudson Bay to the yet-unknown northern shores of the Western Sea at the Strait of Anian. The midlands and western part of the continent were very high and mountainous, beyond which, just a little way, lay the sea. Between these mountains and the bay were several rivers and a great many lakes. Many Indians lived on the west coast and in the mountainous areas. They garnished themselves with a white metal, which they hung on their noses and ears and about their necks. They also had a yellow metal; and a great deal of another metal, washed down by the heavy rains from the cliffs, was to be found along the sides of the rivers. Lumps of the metal were so big that the local people could hammer out knives and lances and even fashion dishes from it.

Further, Knight believed there were some seventeen rivers reaching the continental coast beyond Churchill River. The last five flowed through wooded country, with the woods growing larger from the thirteenth to the seventeenth rivers. These last rivers must flow to the warmer western ocean; the thirteenth river was still in the barren deserts, which extended southwestwards to the third river beyond the Churchill. The Northern Indians inhabited the region of forests beyond the thirteenth river. The Copper Indians inhabited the region of the thirteenth river, which is the Copper River (the Coppermine?) which must flow into the sea where the coast bent south into the western sea and southeast towards Hudson Bay. The Inuit lived to the seaward of the Northern and Copper Indians, dwelling all along the shores at the mouths of the rivers. To the south of the Northern Indian country were great lakes (Athabasca and Great Slave?) in which there were large stocks of all sorts of fish, especially large salmon. In the Northern Indian country, there was a great river, out of the banks of which gum or pitch flowed down into the river (Athabasca and the tar sands?). The Copper River ran by the side of a great mountain, one side of which contained the other metal – natural copper which did not need melting. It could be hammered between two stones into flat pieces and then pounded into whatever shape was needed.

The Northern Indians, with whom Stewart travelled between the bay forts and their northwestern home region, would certainly not have touched upon the Arctic coast nor upon the bay shores until they were reasonably near the forts. Their course, within reason, would have been an oblique transect northwest to southeast (or the reverse), passing out from the northern forest in the Athabasca-Great Slave Lake area, across the tundra, and back into the taiga behind York Fort and Churchill. When they "chalked out" their routes, they most likely would have described an essentially straight line, crossing a number of rivers between the termini of their travels. In developing his overall draft, using the Indian sketches and his combined impressions, Knight extended these rivers to the sea-coast and added others, to a total of seventeen, because, as he reported "they [the Northern Indians] still persist in it there is 17 Rivers from Churchill River to ye Norwards" (B239/a/2,fo.31d). He oriented the entire map essentially north-south, since the Indians characterized this as the cardinal direction. The major change in sea-coast direction at Melville Peninsula, between the Bay shore and the Arctic coast, was unknown. Nor were the great protuberances of Melville and Boothia peninsulas yet discovered; these major elements do not appear on the map. Knight indicated the Indians' track to Churchill in a great arc from the middle course of the eighth river to the fort. From the sixth to the eighth river was an extensive wooded area with many deer, which the Indians came to hunt. The Usquemay (Inuit) territory was shown as beginning at the seventh river beyond Churchill and extending north along the mainland shore of Roe's Welcome Sound. In the far north, the most northerly of the copper mine locations was depicted.

If this map was not penned in by Knight over a pencil base, it was someone else's attempt later to lay out Knight's ideas. It was probably finished after 1719, after Knight, resigning from the Hudson's Bay Company, embarked upon a search for the ore lodes of the northwest. In fact, two other pieces of information are on the map, which could be dated from the late 1730s and 1740s. The first is the symbol and name, Prince of Wales Fort, the stone fortress at the mouth of the Churchill River, the construction of which was not begun until 1731. The second is the symbol of a ship, situated in Roe's Welcome Sound, between Southampton Island and the mainland, with a notation that this was the farthest point reached by Captain Christopher Middleton on his voyage of discovery in 1742. Of course, these details could have been added after this date to a map drafted earlier, or the entire map could have been drawn after that date. Since the map illustrates information which Knight had been receiving in the period 1713 to 1719, a post-1719 date is appropriate.

More than anyone else at this time, Governor Knight was absorbed in the possibilities supposedly unfolded by the reports and rumours that he had taken pains to collect. He was persuaded that the copper and yellow and white metals could be reached from the sea on rivers which flowed to the sea coast. Following the shore along the northwest coast of the bay would mean forging a way to the Northwest Passage. His decision to press for a full-scale northern expedition and a mining enterprise, as a private citizen, but with the financial backing of the company, was based on the belief that he had untangled the intricate geography of the region and that he would be able to solve the mystery of the Northwest Passage. As a newly elected member of the company committee and as a result of the esteem in which he was held, Knight was able to convince the company to support his plan. The company's immediate attraction was the discovery of metal deposits: the attempt to find a passage through the northern part of the continent must be assessed with this more specific goal in mind.

Upon his resignation from the company in 1718, Knight plunged into the organization of his expedition. In March 1719, he put his proposals forward; by 1 June the sailing orders had been delivered to the captains of two vessels assigned to his command: and very soon afterward they were on their way from London. Not one of the group was seen alive after this by any European. There was no particular anxiety felt during the rest of the year, or even during the following, for it might well have been that the party had passed successfully through the passage and the Strait of Anian and was currently somewhere on the Western Ocean, returning home.

Meanwhile in 1719–20, unaware of the Knight expedition, Kelsey at York was directing more expeditions northward with instructions to improve the trade with the Northern Indians, to open trade with the Inuit, to promote peace between the two groups, and to obtain information on mines and minerals. Kelsey himself went north beyond Churchill to approximately 62°40' in the summer of 1719 and traded with the Inuit. It was not until 1722 when John Scroggs, master of the sloop *Whalebone*, returned to Churchill from an expedition northward that the fate of Knight and his party became known. Scroggs reported that he had been to Marble Island where the two ships of the Knight expedition had been wrecked. "Every man," he reported, "was killed by the Eskimoes" (B42/a/2,fo.51).[26]

After Scroggs' voyage, until the mid-1730s, no further attempts were made by the Hudson's Bay Company or by other interested parties to find the mines or the long-sought-after strait to the Western Sea. Up to 1725, correspondence was exchanged across the Atlantic about the possibility of reaching the copper mines; after that date the company assumed that these lay too far away to be reached overland in a reasonable time and the idea receded into the background for some years. No further inland exploration was undertaken until that by Anthony Henday in 1754. His expedition inaugurated an extensive period of inland wintering upon the interior plains.

6 The Initiation of Company Mapping: The Late 1720s to 1754

> Wee are very Confident in that large Bay, there are several Creeks and Harbors, undiscovered wch. may prove exceeding advantageous toward the advancemt. of our Trade & Interest.
> Committee to Nixon, May 1682

The exploration and charting of coastal areas, especially of river estuaries and their lower courses, predominated during these three decades, the period in which company employees can be said to have begun mapping. At least six servants sketched or drafted maps and charts while engaged in their regular activities. Most of these maps and charts were based on rapid field work and the use of simple instruments; they were not expected to be absolutely accurate nor of superior quality. They were, however, useful locally: coastal and harbour navigation became safer; the depiction of the intricacies of certain lower river courses would be helpful later on for inland travel. Certainly, they also provided reference data for members of the committee. Early in this period, when the greatest emphasis was placed on the pursuit of regular trade and the building or rebuilding of coastal trading posts, some plans for forts were also prepared. Although the committee had repeatedly admonished its officials in the bay to organize inland trading journeys or at least encourage individual employees to travel beyond the coastal strip, there was a frustrating lack of interest at the bay in the interior. After 1731, the company had to be concerned with the dual attack, on the one hand from the French, and on the other from the English group, led by Arthur Dobbs, on its monopoly of the Hudson and James bays' drainage basin.

FRENCH ENCROACHMENT ON RUPERT'S LAND

A continuous flow of reports, warnings, and surmises concerning increasing French trade and competition in the company's chartered territory reached the committee from employees in the field. Men serving in the southern reaches of the bay were kept on edge by the rumours; even innocuous incidents seemed to bode ill for their factories. Yet the company did not send out any investigators to assess the situation, did not establish any inland trading posts to counter the competition, nor did it change its overall trading strategy. In a minor initiative during the summer of 1743, Joseph Isbister established Henley House near the confluence of the Ogoki with the Albany River; it was thought of as a subordinate and defensive outpost, not a trading station.[1] Meanwhile, the French had extended their trade west and northwest from the Lake of the Woods region into the Manitoba lakes and the Saskatchewan River basin. For example, La Verendrye, head of western posts for the French trade, erected Fort Maurepas in this region in the summer of 1734. Possibly the first direct reference to the new and enlarged threat to English monopoly in this part of the west was given by some Indians to Post Master Thomas White at York in the winter of 1735.[2] He reported in his post journal, "we under stand that ye french woodrunners, has made a new Setlement, to ye: Northward near Port Nelson, on this side, wht. they call ye Little sea [Lake Winnipeg], wch: if true I fear will be some check to ye: Small furr trade, it being Just in ye: Middle of those Indians, in whom Lyes the greatest Dependance" (B239/a/17,fo.15)[3]

THE DOBBS CONTROVERSY

The contacts with the Inuit north of Churchill, which had been developed under Kelsey's leadership particularly, were carried on after 1722 by sporadic summer visits of the Churchill sloop which put in at such gathering places as the Seal River. These journeys were concerned with trading and had little or no bearing upon the expansion of geographical knowledge, a practice that would apparently have continued longer than it did, had not the company in 1733 been approached by Arthur Dobbs and induced to support another search farther north for the Passage. He had become convinced after examining all the journals related to the exploration of Hudson Bay that there was a strait somewhere north of Cape Eskimo, leading to the Western Ocean. His choice for the site of the supposed passage was Wager Bay. In the summer of 1737, a jointly organized expedition reached as far north as Rankin Inlet, established a rendezvous at Cape Eskimo, and sailed back to Churchill. But that was the end of it for some unexplained reason. Disappointed by the company's apparent lack of interest in his vision, Dobbs broke away to seek support elsewhere, and, by 1741, he had been able to gain the backing of the Admiralty, which put two ships at his command for another trial. The two vessels under captains Christopher Middleton and William Moor wintered at Fort Churchill in 1741–2 and set off in June 1742 for the northern coast.[4] They entered the bight behind Marble Island, naming it Rankin Inlet, sailed by Chesterfield Inlet, passed Cape Fullerton, and entered the hoped-for passage, which they later named Wager Bay. Their hopes were dashed when Middleton found that this was only an extended arm of the sea. They sailed on to the entrance of the bay they aptly named Repulse, when ice blocked them from pushing farther north. Then the ships swung south around Southampton Island and sailed for home. That Middleton's voyage eliminated both Rankin and Wager inlets as sea passages was not comprehensible to Dobbs who, casting aside his role as a reasonable and scientific advocate of an objective approach to the search, assumed increasingly a position of hostility to the Hudson's Bay Company and its North American trade.[5]

Dobbs was next able to secure an Admiralty inquiry into a charge that Middleton had not been truthful in his report. Then a barrage of books and pamphlets pressured Parliament into posting a £20,000 award for the first British subject to discover the Northwest Passage. Dobbs also managed to form a company to make a final trial expedition to the bay in 1746. Although this expedition, too, was unsuccessful, Dobbs and others continued to clamour against the company's chartered position. In 1748, his group petitioned the Crown for incorporation as a company to be granted exclusive right to lands that they had discovered or should discover adjoining Hudson Bay and Strait in America. With the decision of the Crown against this petition, they campaigned successfully for a public investigation in Parliament into the Hudson's Bay Company's charter. In 1749, the evidence by both sides was presented to a select committee of the House of Commons. In its testimony to the parliamentary committee the company asserted that it had engaged in numerous searches of the coast, had lost considerable money earlier in backing Knight, and was of the opinion that if the passage did exist it must be sought much farther to the north. The company insisted that Dobbs's two main expeditions had substantiated its opinion, even though Dobbs would not agree that this was so. After considering the committee's report, the Commons upheld the charter and the rights of the company, and the challenge finally came to an end.

Apparently, the company decided to have a map printed to accompany its evidence to the parliamentary committee (8A) (plate 3).[6] Copies were to be distributed to the committee membership and to certain members of Parliament. A curious affair, this 1748 map – the outline of the west coast of America so highly speculative, the configuration of the north Pacific region so patently false, the outline of the Great Lakes so unusually crude – the governor and the committee must have been appalled. They could not have approved of its distribution, for it would not have supported the company's claim that it was a reputable, conservative business concern. There are nine copies of this map in the company's archives, and another was given to the British Museum in 1957. These would appear to be all that remain of the initial run, delivered by the engraver, R.W. Seale, to the company office.[7] Probably, the other copies were quietly disposed of.

ON THE NORTHWEST COAST OF THE BAY

The extension of geographical knowledge resulting from the Dobbs controversy was considerably less imposing than the effort expended and the turmoil resulting might lead one to expect. Two of the entrances to possible straits leading to a Northwest Passage had been explored and eliminated. The third, although only partially examined, seemed not to merit further attention. The explorers had gone but a few miles farther up the coast than previous travellers. On the other hand, the coastline had become better known and various natural harbours located. After the unsettling experience of a full-scale enquiry, the company began to be less cautious as it took careful note of the opinions expressed in evidence, particularly those of company employees, whether these were retired, discharged, or still engaged. This can be seen in a renewed interest in inland exploration; the company also continued to sponsor summer voyages along the northwest coast of the bay.

The Churchill sloop, under its master, James Walker, was ordered in 1751 to make trading stops to the north as far as the ship could safely go in the season. Along the way, Walker was to explore as many bays and coves as possible and prepare charts and make notes on tides and safe harbours. Although he sailed north as directed in the years 1751–4, he did not send any maps or charts to London, nor did his journals contain any illustrations.[8] Undeterred, the company continued to sponsor voyages along this coast.

CHARTING ALONG THE EASTMAIN

By 1740, Indian reports of the existence of three large lakes, inland from the Eastmain coast at about sixty degrees north, on whose shores both Inuit and trading Indians were said to reside, had been transmitted to London. The possibility of an easy connection between the lakes and the coast was suggested. That same year the Slude River post became a permanent establishment, a move based on the optimistic expectation of increased trade from the Eastmain. The committee, hoping to increase returns from the peoples living northeast of the post, urged that some attention be paid to these reports (B59/a/4,fos.27d–28,37),[9] and reinforced its repeated requests with the provision of a new sloop, the *Eastmain*. In 1744 the *Eastmain* and the sloop, *Phoenix*, from Moose Fort, with Thomas Mitchell and John Longland as their masters, were instructed to search the Eastmain coast as far north as 60°. Undoubtedly the insistence of the company on this northern terminus was also related to its desire to investigate the large, inviting opening just south of this latitude, repeatedly depicted on some contemporary maps.[10] Captain Middleton's "Chart of Hudson's Bay and Straits," published in 1743,[11] would have been the most respectable map evidence available to the governor and committee as he was an experienced and reliable bay navigator. If such a passage existed, it could possibly lead to one of the hypothetical openings on the Labrador coast or in Ungava Bay and provide an easier passage than that through Hudson Strait. However, the two ships sailed along the Eastmain to only about 57°. No concrete reason was given for not proceeding farther north, but Mitchell's account mentioned several factors which may have been germane to this decision. By then, the group had discovered "Several Places that a Sloop may winter in safety if requird," and also they had travelled "further north then any of our Northern Indians Do winter" for the reason that "there is little or no wood to be got" (B59/a/10,fo.13d). Moreover, in the summer the Indians "Come as fur as this Gulph or lake [Richmond] to kill deer & as ye winter Comes on they Return" south (B59/a/10,fo.14).

During the journey, Mitchell sketched three maps in his journal. He promised a fourth map that would show several of the islands along the coast which they had passed on the third day of sailing from Albany Fort, but it was not included. On 13 July they reached the entrance of "ye great river," the Fort George River (B59/a/9,fo.3d). Two days later he reported that he had drawn "a Plain of this River … Whose Lattd; is By My observations 53° = 51′ North" (4A) (plate 4) (B59/a/9,fos.4,4d). The second sketch is of the Great White Whale River (Great Whale River), "which Layeth in Latt.d 55° = 35′ North & about 02° 00′ Ey Longitude from Cape Jones" (5A) (B59/a/9,fo.8d). Both river mouth sketches were drawn in pencil and then inked in; both show the channels several miles upstream from the mouths, and include an extensive array of soundings; both also indicate shoals. Neither are finished maps, however, only hastily-drawn, rough sketches.

After arriving off the entrance to Richmond Gulf, the boats had to wait until the awesome channel, which Mitchell called the "Gulph of Hazard," into the mountain-girt bay had been investigated and the strong currents and tidal race observed (B59/a/9,fo.11d). On 1 August, they passed through and began to explore the large body of water to its full extent. Mitchell promised to "give ye Honble Company a more Planer Acct: on a Plain of all ye Is Lands & Bays" so that he could "Go into with Safety in this Gulph" (B59/a/9,fo.10). This 'Plain' is the sketch of the "Gulph called by ye Natives Wenipegg" (6A) (B59/a/9,fo.12).[12] It is oriented with the east shore to the top of the page. Richmond Gulf had no water passage leading east toward the interior, although as discovered later, river connections existed with an assemblage of quite large lakes, Clearwater, Lower Seal, Lac d'Iberville, and others. Along with the gulf, these are very likely the lakes described by the Indians and sought by the company. Although pleased to receive these first renditions of the gulf and major river estuaries on the Eastmain, the committee regretted that the party had not gone north to at least sixty degrees latitude as it had been ordered. The desirable alternative passage to the Atlantic Ocean had not been reached; neither Mitchell nor Longland had mapped the entirety of the Eastmain coast north to the gulf entrance. Another expedition would be required.

And orders for another were dispatched in 1748. The governor and committee directed that the Moose and Eastmain sloops were to sail as far as Cape Diggs (at about 63° North), exploring the coast north from the point reached by Mitchell in 1744 on their way (A6/7, fos.126,127). These orders, not carried out in 1748, were repeated for the 1749 season, but this time a two-ship expedition was sent out directly from Britain with detailed instructions. Captain William Coats, an experienced company navigator, sailed in the *Mary*, and Captain Thomas Mitchell, who had gone back to Britain, brought out the newly-constructed sloop, *Success*. Individually, the two ships were to sail along the coast from Cape Diggs, making detailed hydrographic and geographical observations, stopping here and there to make collections of natural materials. They were also to try to decide on a suitable new post site, if not on the coast, then in Richmond Gulf. Each was instructed to draft a coastal chart and a plan of the chosen post site (A6/8,fos.19–24d). In 1749, the two ships reached the entrance to Richmond Gulf after traversing the shore as ordered. They used the standard method for coastal surveying, that of the running traverse, with dead reckoning providing the coastwise distances and compass bearings being read to the various coastal and island headlands and river mouths. Soundings were also taken along the coast, especially from just south of Cape Diggs to Portland Point. Each of the several harbours scrutinized along the route were found wanting, and so they decided to search out a location within the gulf. Their choice was a site on the south shore of Metesene (Cairn) Island, which lies off the hilly south coast of the Gulf.[13] Captain Coats sailed from the gulf for Britain in late August; Mitchell, in the *Success*, went south to Albany.

There are six extant charts related to this expedition, all drawn by Coats. Mitchell did not fulfil this part of his instructions, which specifically ordered him to make a "Draught of ye Coast from Cape Diggs to the Intended Settlement" (A6/8,fo.24d). Five of Coats's charts are in the Hudson's Bay Company Archives; one has found its way into the Hydrographic Department of the Admiralty. Another chart may have been prepared and discarded. After his return to London in the autumn of 1749, Captain Coats successfully petitioned the governor and committee for a reward above his salary for the services he had rendered. About this time he handed in at least two, and possibly three, charts. One was the "chart of Artiwinipeck," dedicated to the governor and committee, and having on its face an extensive "Description of Artiwinipeck" (9[A]) (plate 5). The second map would have been a coastal chart from Cape Diggs south to the Whale River (10[C]). Apparently, the committee was not satisfied with the particulars of this chart, for after his return from Richmond Fort as captain of the supply ship there in the summer of 1750,[14] he was directed, on 5 December 1750, "to make a new Draft of the Coast" (A1/38,p.285).[15] The third map is one which extends from the southern shore of the gulf to the Little Whale River, and is endorsed on the reverse side as "Capt Coats's Drt of Gulph Hazard & Artiwinipeke" (10[A]). Captain Coats delivered his new or revised coastal chart to the company office (13[A]), and a copy of his original Richmond Gulf map (14[A]). The third of the maps which concentrate on the shores of Richmond Gulf can be called a sketch map (11[A]) in comparison with those already mentioned. It may have been a first draft of that accepted by the committee in 1749.[16] Or it may have been a later rough part-copy by Coats, used for other purposes. The final map in this suite, now located in the British Admiralty, extends from the gulf south to the Great Whale River (3[B]).[17]

There is some continuity of style in these five maps in the company archives, the most obvious being the use of ink and grey wash lines to depict the steep hillsides fronting on the gulf, sea, rivers, and lakes. Where extensive vegetation is shown, the tree symbol is the same on all the maps. (It is interesting to note that a deciduous tree symbol is used, not one representing the dominant coniferous forest.) Also colour wash in yellow green is employed; "ye Colouring shews the Verdure and Woods."[18] Captain Coats used a liberal hand in naming the local topographical features after his own friends and acquaintances and well-known London locales. Very few of these toponyms have survived. The marine portion of the charts use the normal convention of compass roses with a circle of apices of intersecting rhumb lines, set in a framework of a latitude graticule.

John Yarrow, named master of the *Success* after it arrived at Albany, went north with Mitchell in the summer of 1750 to the gulf with supplies and the prefabricated parts of a frame house built by the Albany carpenter the preceding winter (A6/7.fo.154d). They spent from 10 to 26 July establishing the building at Richmond Gulf before Yarrow sailed for Albany, leaving Mitchell to await the arrival of Coats in the supply vessel. Yarrow sailed up the Albany estuary to the fort on 8 August 1750, in time to complete his journal and to draft his chart, presumably of the Eastmain coast and Richmond Gulf area as he had been instructed to do (11[C]). The committee notified him in the next packet that the journal and chart had been received.

MEASURING AND MAPPING LOWER RIVER COURSES

The committee, still concerned that so little was known of the country even a few miles behind the coast, continued to prod its officers and servants to try to learn more about their surroundings. It especially urged them to measure and map the estuaries and lower courses of the main rivers flowing into the bay. Although in 1727 the Albany sloopmaster had paddled around the Moose River estuary in a canoe searching for the best site to rebuild the fort, it was not until the 1740 season that more extensive investigation was undertaken. At that time George Howy, the sloopmaster at the new Moose Factory, completed the examination of sixteen miles of the estuary below the fort and approximately twenty-four miles of the river channel above this establishment (B135/a/10,fo.11d). He drafted two maps. The first was a chart of the entrance to Moose River, likely drafted in 1740, that details the islands, sand bars, depths and harbourage in the estuary, extending past the upper ship hole (the farthest upstream ship anchorage area) to Moose outer roadstead, at a scale of one inch to one mile (3[A]). Included are annotations concerning the sailing directions within the estuary and the latitude, longitude, and compass variation at five locations in this stretch of water. Howy spent a good deal of time up and down the river and estuary from the fort in the open-water periods during 1740. His chief, Richard Staunton, affirmed in 1741, upon Howy's return from surveying above the fort, that the sloopmaster would have drafts "both upwards & downwards for 40 odd miles" (B135/a/10,fo.11d).

The second map drafted by Howy was prepared from the data he obtained between 26 and 28 November 1740, when he and two others, after the river had frozen, took "a Surva of ye River for 24 Miles upwards"

(8ᶜ) (B135/a/10,fo.11d). The committee approved of these two maps and acknowledged his "care and diligence in performing the same" (A6/7,fo.4d). Earlier, in 1740, the company had asked Staunton to have Howy survey the lower Albany River after he completed his Moose assignment. Although the committee repeated the request to Howy himself in the 1742 letter of congratulations, there is no evidence Howy followed these instructions (A6/7,fo.4d).[19]

Joseph Robson, a stonemason and surveyor, took time out in the middle of building York Fort, from 1744 to 1746, for a similar measurement and mapping expedition on the lower Nelson and Hayes rivers.[20] On 5 August 1745, the York factor wrote to London that he had given Captain Fowler of the *Sea Horse* "a draught of ye 2 rivers taken by Mr. Robson" (9ᶜ) (A11/114,fo.121). Three copies of this original chart were made in London in 1750 by the company for reference purposes, two being sent back to York and one being provided to Captain Spurrell (12ᶜ). Later, in the summer of 1745, and during the early winter of 1746, Robson surveyed the Nelson River upstream as far as the Pennycutaway tributary by boat and then by foot, dragging the measuring wheel, and sighting with compass. The completed survey map is very likely that drawn on the back of two plans of York Fort (7ᴬ).[21]

In the committee's official letter of 1750 detailed instructions were sent to John Newton at York Factory to have a series of ten specific distances measured in the vicinity of the factory, on both the Nelson and Hayes rivers. To aid in this enterprise the committee also sent over a "Compleat Set of surveying Instruments and Measuring Wheel ... [supposing] what were formerly sent are either Broke lost or mislaid" (A6/8,fos.45d–46d). Moreover, a copy of Robson's first chart of the Nelson-Hayes estuary was to be used by Newton and Spurrell, in the shiptime of 1750, to be amended from their knowledge, to be remarked upon, and to be returned to London by the captain. The committee also enjoined Newton to have the list of measurements carried out and reported upon the following year. Newton died before carrying out the assignment, and James Isham, who succeeded him, reported that none of them would have time until the late autumn and winter, 1750–1, when he himself would go. He also sent several small parties up the rivers that season. In 1751 he was able to send back in the autumn ship "a draft, a Plan, Book marked P:Q:U" (A11/114,fos.148d,149), in which a series of measurements were recorded (13ᶜ). In 1751–2, he made further measurements up the Hayes River, as far as the Steel River, some fifty-two miles from the factory. The results of this journey were sent to Britain "in a book M" (B239/a/35,fo23). Possibly a map based on the measurements for this season was included. It appears from the post journal that this program was not elaborated upon during 1752–3.

In 1752, Isham suggested to the committee that a new post be established at the mouth of the Severn River, noting that Indians had reported there was a French trading house up the river. He urged the members to examine an unidentified map which showed the Severn's connection with other interior water bodies.[22] Although the committee differed with him in the interpretation of the map data, in 1753, it directed Isham to send a boat down coast to the Severn. The boat was to proceed up river as far as it could safely go, past, it was hoped, the supposed French post. The captain should search for a suitable building site above this post, if there was one, and to "make an Exact draught of the said River and its Outlets" (A6/8,fo.116). In July 1754, the York whaling sloop under Christopher Atkinson, with Richard Ford, the factory carpenter, sailed to the Severn River. After a month spent examining the twelve navigable miles of the river, they located a desirable post site, gave a favourable report on the river's lower course, and noted that there was no French post that low on the river. As directed, they delivered a chart of the river, drafted by Ford (15ᶜ). The committee was inclined to think that the Indians had used a report of a French post as a pretext to get a company house placed in their territory, but agreed that a post should be built at the site on the lower river.

"GOING UP INTO THE COUNTRY"

Since the pre-1720 Stewart and Norton expeditions, the company neither supported further inland journeys nor developed an extensive inland post system. As noted earlier, it was depending upon exhortations, gifts, and its Indian trading captains for the maintenance of trade with inland peoples. After the 1749 parliamentary investigations, however, the committee came to the decision to bolster its trading system with a renewal of inland wintering. They agreed with Isham's proposals for continuing the old policy: they would maintain good relations with the native people by kindly treatment and encourage them to come to the bayside posts by sending company representatives on missions of "peaceful penetration up rivers." Only after careful inland exploration should settlement away from the bay be contemplated. Thus the committee resumed its earlier pleas to the chief factors to search out "proper persons" to "go up into the Country." Isham accepted the challenge, and, in 1754, he found a protégé, Anthony Henday, who had been at York Factory as labourer and netmaker since 1750, and spent considerable energy training him for his first major journey inland.

Henday's apprenticeship included a trial expedition up the Hayes River in the depth of winter, February 1754.[23] He and three Indian companions, dragging a measuring wheel, took compass bearings some

100 miles up the Hayes River, crossed to the Nelson, and walked about 140 miles back down the river on the ice to the factory. Since the journal Henday kept during this exercise is short on detail, that part of the training was not too successful. No map was specifically asked for, and perhaps none made by Henday, for the committee's spring letter of 1755 to Isham referred only to a journal of distances and of observations. However, Isham must have included a sketch of his own, for in the same letter the committee referred to "your said draught" having details on it about the Nelson River, which the members thought he must have obtained from Indian information (14^c) (A6/9,fo.10). The committee wondered how Isham could have supported the idea that the Nelson could become a regular transport route, when the river had not been travelled over by company servants in either canoe or boat with full soundings made. Isham should, therefore, send a party up river to take depth measurements and prepare a map of the river system, at least to the point where Henday had reached the Nelson. But by the time these instructions arrived, Isham had inaugurated the annual wintering program from York Fort, with Henday setting off for the Saskatchewan country on 26 June 1754. Further exploration of the Nelson would have to wait.

7 Mapping Rivers and Barren Grounds Inland, 1754–1778

Observe the Course you take daily ... taking a Draught of the rivers you pass, Remarking the Names of bays, Bluff etc, both up and down.
Isham to Smith and Waggoner, August 1756

During the quarter century after 1754, the Hudson's Bay Company intensified its mapping activity. Geographical information accumulated at the bayside factories and in London as company traders who wintered inland returned annually from the interior, and as the company, in a major alteration of its trading strategy after 1774, opened its first inland trading posts in the Saskatchewan, Albany, and Moose basins. If the committee had seemed undecided earlier about the merits of erecting posts in the interior compared to the advantages of inland wintering, this was a reflection of the general division prevailing among the influential senior servants in Rupert's Land, although at mid-century, the weight of opinion was still against inland settlements. James Isham's advice typified that offered by those opposed to building inland. He was convinced that as long as the French had the geographical advantage, which holding Canada gave them, the company should support extensive inland travelling solely for the purpose of drawing the Indians to the bay; only when the French were finally driven from Canada should the company change its strategy.

The forty-seven charts and maps produced during this period doubled the number from the previous one.[1] These were drawn by fifteen company employees and by several Indians at the request of employees. Three regional maps were compiled by commercial cartographers in London using data provided by Hudson's Bay House. The largest category of maps was of the river and lake networks west and south of the bay. These were based on information gleaned from inland winterers or, after 1774, on the reports of expeditions sent inland especially to bring back descriptions of the routes suitable for trade. They included a range of regional compilations; nine depicted elements of the complex web of waterways behind Churchill, York, Severn, Albany, and Moose forts. Most were simple line-work sketches at various scales. A smaller group included small-scale charts of Hudson and James bays and large-scale delineations of major inlets in the northern littoral of Hudson Bay or of river estuaries. Unfortunately, few were accurate enough for the committee's requirements for, by now, London officials were fully aware that understanding the geography of Rupert's Land was essential to conducting business successfully. Thus they required more precisely detailed maps: maps that showed the location of their posts in relation to major lake and river junction, and maps that gave more explicit configurations of the water networks and represented distances accurately. They decided, mid-way through this period, to employ several Grey Coat Hospital apprentices, hoping that men trained in a modicum of mathematics, surveying, and cartography would be able to provide what they needed.

MAPS RELATED TO INLAND WINTERING FROM YORK AND SEVERN

It is difficult to assign any boundaries to the journeys of the traders who wintered inland. From the few extant journals and scanty references found in company records, it is apparent that the country in the vicinity of the North Saskatchewan and the main Saskatchewan rivers to Lake Winnipeg had been entered, and also that the regions of Lake Winnipegosis and Swan Lake and the upper Assiniboine River area had become the wandering grounds of English as well as French traders. York Factory was the company's first center of inland wintering operations, and Anthony Henday made the inaugural sojourn from there in 1754–5. His journey expanded the knowledge of the plains well to the west of Kelsey's activity, since he kept a reasonably careful journal of events witnessed and areas passed through. His record lends credence to the view that he was the first European to sight the Rocky Mountains front from Canadian territory.[2] He was also the only inland winterer from York who produced a map.[3] Unfortunately, Henday's original map showing his route to the west of the Red Deer River, into the vicinity of the present-day settlement of Olds, Alberta, is not in the collection (16C). Perhaps, James Isham discarded it at York in 1755 after he had finished making a fair copy "from Hendy's Original," to send to the committee (17C) (A11/114,fo.197). When the copy was received in London, the secretary reported that the members were surprised to see a version of the courses of the Hayes and Seal rivers different from anything they had seen before. They conceded that the discrepancy probably resulted from Henday's inexperience in making maps and measuring distances and awarded him a gratuity of £20. Isham was also responsible for a 1757 map depicting the trip of Joseph Smith and Joseph Waggoner who formed the second party to go inland (18C). Their daily journal, replete with descriptions and compass directions, was detailed enough to depict their route into the Cedar, Winnipegosis Lake, and Swan River region, and west across the Assiniboine River into the grassland plains (B39/a/43).

Andrew Graham occupies an honourable position among the officers of the Hudson's Bay Company for his landmark compendium of observations on the natural history, native peoples, and the fur trade of Rupert's Land.[4] Furthermore, he was the first of the map makers of the bay to depict the complicated network of interior waterways that company exploration had revealed. As a youth, Graham served on the Churchill sloop for some years where he would have become acquainted with charts and nautical instruments and their use. From 1754 to 1761, he was writer and book-keeper at York Fort. Then, except for short periods relieving the York factor, Graham worked at Severn Fort from 1761 until 1774. During the first years of inland wintering, he copied many of the journals and letters arriving at York and so would have been well aware of the geographical and anthropological information in them. He would have been familiar with all the wintering journeys, certainly with those of William Tomison in 1767–8 and 1769–70, of Matthew Cocking in 1772–3, and of William Falconer in 1767. Thus it is probably Graham who produced two maps, similar in coverage and characteristics, whose major subjects are the routes inland of these men, especially those of Tomison and Cocking (21A, 23A). Although neither map was dated or signed, some clues in the records, as well as his experience and proximity, make Graham the likely author.

The gradual increase in Pedlar trading in the later 1760s in the Manitoba lakes country, reported by Indians to the Severn factor and commented on in letters between Severn and other posts, precipitated several wintering journeys from this post. The trader William Tomison kept detailed accounts of both his inland trips. Unfortunately, he lost the journal of the first when his canoe overturned on Lake Winnipeg; happily, he described his journey to Graham, who wrote a general account of it in a letter, dated 8 July 1768, to York Fort after Tomison's return (B198/a/9). Graham does not specify the exact route, but it would appear that Tomison reached Lake Winnipeg's east shore via the Severn and either the Berens or Pigeon River and probably returned to the fort via the same route in the spring. In the 1769–70 season, Tomison returned to Lake Winnipeg, but on this occasion he crossed over into the upper end of Lake Manitoba and struck west over the Riding Mountain into the prairies of the Assiniboine River valley, where he wintered. His

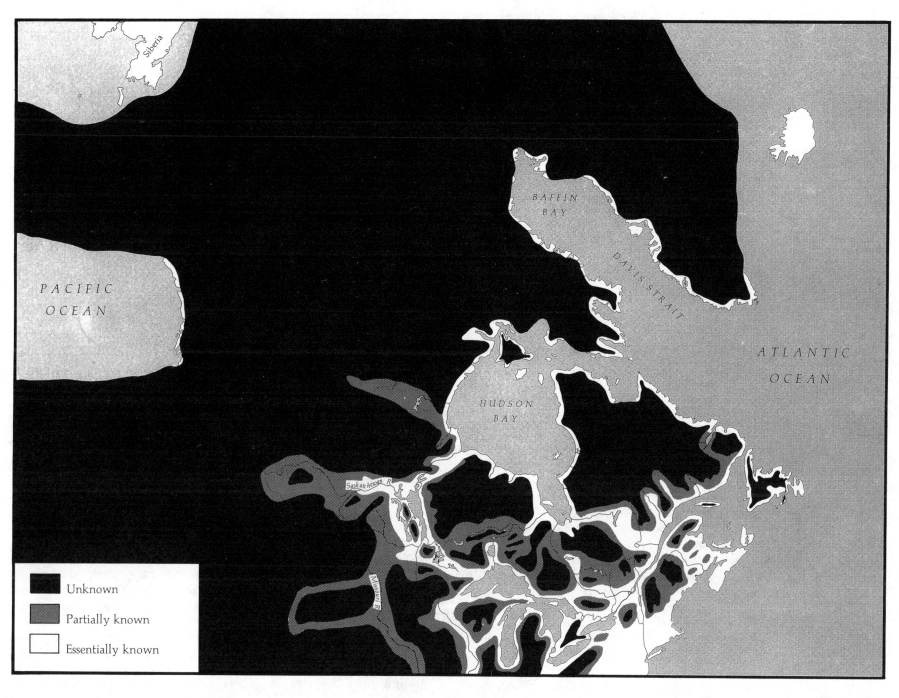

Figure 3 Retreat of the unknown: European knowledge of Canadian territory in 1763

return course, after reaching the east shore of Lake Winnipeg, was then along the southern trade track through Oxford and Knee Lakes to the Hayes River and then to York Fort. Tomison's two journeys are laid out in generalized fashion on the first map.

The geographical detail on the map attributable to William Falconer, sloop master at Severn, resulted from a journey he had made in 1767 along two adjacent tributaries of the Severn and Hayes rivers. In break-up that year, flood waters and ice had severely damaged the sloop. After Falconer canoed with two Indians to York for supplies to repair it, he depicted the route they had taken on a map, dated the same year (23C).

Graham was very much aware of the paucity of written and mapped information from the inland winterers, especially of the Saskatchewan country. He remarked in a letter to the committee on 26 August 1772: "I have often reflected that the Accounts given us by Men sent Inland were incoherent and unintelligible." He had therefore chosen a "sensible Person" who "might answer the Purpose much better, and make many observations which may be of utility" (A11/115,fo.144). This person, Matthew Cocking, a very competent writer at York since 1765, had no inland experience.[5] However, he was instructed in the use of instruments so he could make more accurate observations and sent off on 27 June 1772 for a trek that took him southeast of the forks of the Saskatchewan, using the established Hayes River-Cross Lake water route to that river. He returned to York Fort the following summer. In July 1774, he was again sent inland into the Red Deer River area west of Lake Winnipegosis, returning to York on 27 June 1775.

The opportune time for Graham to have worked on his first map would have come during the 1771–2 period when he was acting chief at York. It was during this year that he prepared a revised version of his *Observations on Hudson's Bay*, and spent much time taking records of the weather and renewing his knowledge and practice of astronomy. He also would have had with him the information provided by Tomison from his two trips and by Falconer's from his. Graham definitely sent a map of the "Tracts between Basquiau (The Pas) and York Fort" to London on the 1772 supply ship (21A), for it was copied there and this copy sent back in the 1773 ship to Samuel Hearne (26C), who had just arrived at York Fort from Churchill, for his use in planning his expedition to open Cumberland House, the first post in the interior (A1/44,fo.79).

The map does have a number of cartographic "signatures" which can be used to identify Graham's hand, some of which appear on two additional charts, identified with Graham – of Hudson Bay and Strait (24A) and of the Hayes River mouth (22A). One of these characteristics is the generalized coastal and interior hydrographic detail, particularly that of the shoreline which is drawn in a style typical of the sea charts Graham would have seen early in his career. The identifying colouration on river and lake systems, the design and content of the title cartouches, the complexity of the compass rose, and the style of lettering all point to Graham as author. The most intriguing feature of the map is the extreme disorientation of lakes Winnipeg and Manitoba. Their axes are canted far off to the southwest, approximately double their real angles, (i.e. 60° rather than 30° from North). Another problem is Graham's attribution of the discovery of the Hayes River-Cross Lake water route to the Saskatchewan River to Matthew Cocking. In a note on the map Graham had indicated that the rivers and lakes outlined in yellow were discovered by Tomison, those in red by Cocking, and those in black by Falconer. The Tomison and Falconer credits were basically correct, but the red outline of the Hayes River and Cross Lake route is problematic. In the first place this route was well-known to the wintering servants long before Cocking used it. Secondly, he did not follow it until the summer of 1772 on his way inland, and he did not retrace it to York until the summer of 1773, when Graham was at Severn and the map in England. It is possible that Graham did not complete the map until after 17 June 1772 when Cocking left York Fort, taking, as Graham would have known, that particular water route. Thus the colouring may mean that Graham was simply recording that Cocking went via the standard route. The map does not include Cocking's pioneering travels southeast of the Forks of the Saskatchewan during that trip.

The second regional map (23A) (plate 6), attributable to Graham, is more or less an enlarged-scale version of that of 1772. It is also the first map to show tribal regions on the Canadian prairies and southern boreal forest. In a rather geometric fashion, the boundaries of the main Indian groups of this area were inserted. Such ethnographic details were also present in Graham's *Observations*. Most of the cartographic techniques used in the 1772 map are also present. There is no indication, however, that he had seen the journal Cocking wrote during his second journey for there are no changes in the northern end of the Manitoba lakes nor is the Red Deer River introduced. Data from Cocking's first journey was included in the form of the lower section of the North Saskatchewan River west of the forks, the lower portion of the South Saskatchewan, and his overland route (shown in a dotted line) between the two branches around the Eagle Hills. Also Lake Dauphin and River, Mantouapau Hills (Riding Mountain), and the prairie grassland are depicted, the first known mention of the latter on English maps. A notation records that William Tomison had been past the wooded hills and out into the "Barren ground. Buffalo plenty in winter."

This second map could have been drafted between June 1773 and late summer 1775. Cocking and Graham were not in contact, however, until 16 February 1774 when Graham arrived back at York from Severn on his way through to take charge at Churchill. He remained at York for a week and would have been able to see Cocking's first journal and to converse with him. He may in fact have started his map at Severn in January of that year for he had sent to York for any available drafting paper. Graham could have shipped the map to London in the autumn of 1774, or he may have kept it at Churchill until he set out for London and retirement in 1775. Possibly, he presented it to the committee when he met with them on 11 October 1775, for on that occasion he gave them a manuscript volume of his *Observations on Hudson's Bay*. Although the minutes do not make mention of a map appearing then, it is a reasonable deduction, since the tribal regions identified on the map would have served as an illustrative adjunct to the manuscript.

CHARTING COASTAL INLETS OF NORTHERN HUDSON BAY

In the mid-1750s and early 1760s, a number of coastal voyages by company ships along the northwest coast of the bay were referred to as "voyages of discovery and trade northward" (A11/13,fo.142; A6/10,fo.97). Although the ship captains were not instructed to look for a sea passage to the west, the committee did expect them to be "diligent" in looking out for a large river, the Kisk-stack-ewen (A6/8,fo.120d),[6] which was said to run from a bay on the north coast a long distance into the country. In the previous decade Indian reports about this river had been relayed to London. Many natives, whose trade would be invaluable, reportedly lived along it. In 1760, Moses Norton, the factor at Churchill, had drafted a map based on these reports (16[A]) (plate 8), a map he carried along to London when he went on leave in 1761.[7] On his map, the Kisk-stack-ewen River is depicted draining two very large lakes into the sea or ocean far up the coast from Churchill. The committee told Norton that he would be paid a gratuity of £40 when he found and surveyed the river.[8] With this incentive, in 1761, he sent William Christopher, the Churchill sloop master, with instructions to look for the mouth of Kisk-stack-ewen. If he found the river, he was to go as far up it as he could, making a population count and mapping as he went.

Christopher was successful in reaching Chesterfield Inlet and sailed up it about one hundred miles, where he still measured twenty-five fathoms of water in the channel. Following directions, Christopher handed in a map of the "Straight or River call'd by the Indians Kiskcatch-ewen" (19[C]) (A11/13,fo.165). The committee applauded this accomplishment and were impatient for more. The following summer, Norton and Christopher were to "do everything in their power ... to Discover whether the same be a Streight or passage, or not" (A6/10,fo.31d). In 1762 the two men sailed north together, along with two Northern Indians to act as translators if they encountered natives. They passed slowly up Chesterfield Inlet (also called Bowden's Inlet), naming various islands, capes, and bays in honour of company personnel and national personages as they went. Finally, they reached a large, freshwater body, which they christened Baker Lake, and after tacking back and forth from side to side in the boat's cutter, they reached its northwest end. Having ascertained that this last remaining unexplored bight in the coast was not the desired strait, they turned back eastward to the sea and thence homeward. A pencil and ink chart (17[A]), sent to London in the autumn packet, was catalogued by the company under Norton's name, and the company refers to his "draft" (A64/52,Chest 1,#10,p.177). However, Christopher also brought the committee's attention to his "Journal and Draught for Particulars" (A11/13,fo.172). I have assigned the chart to both men, although it is more likely to have been the work of Christopher, for, as captain of the sloop, he had used both a Hadley and Elton quadrant and a theodolite to make observations during the expedition; in addition, he took soundings. On the chart, the course of the vessel was pecked out, and soundings and anchorages noted. Cartographically, the chart is somewhat disappointing; the shore details were drawn in with a soft pencil and are fairly difficult to decipher. In addition, the map, originally folded in quarters, has been creased and re-creased over the years.

Although moderately let down at finding that Chesterfield Inlet did not "afford the least Shadow of being Beneficial to the Nation and the Companys Interest" (A6/10,fo.59d) the governor and committee commended Norton and Christopher. At the same time they instructed Norton to continue the investigation of openings shown on the chart, especially one near Whale Cove and one northwest of Marble Island. In the summer of 1763 Christopher found the Whale Cove inlet to be a shallow bay "full of ridges" (A11/13,fo.189). A year later, Magnus Johnston, by then the Churchill sloop master (the mate on earlier voyages), set out, on Norton's orders, to explore the second opening. He found "nothing but a Dead Water Bay with many Shoals etc" (A11/14,fo.3), but, as directed, he produced a chart of this bay, Rankin Inlet, which was sent to London (20ᶜ). Neither it nor Johnston's second chart (21ᶜ), prepared in 1765 on his arrival back at Churchill from further trading and exploration along the northern coast can be traced in the collection.

The geographical and hydrographic information from this series of voyages was assembled on a chart, probably drawn after 1764, called a "Map of Bowden's & Ranken's Inlett's" (18ᴬ) (plate 7). Although the map was not signed, the Hudson's Bay Company Archives have suggested it was the work of Andrew Graham. Surely it is not. The cartographic style is not similar to the earlier maps that were certainly by Graham. His manner is much more crude and generalized, and his colouration emphasizes the land, rather than the water. Nor are the print and script similar. Finally, the compass rose and cartouche are much more complex on this map than are those on the Graham maps. Moreover, Graham was not connected with Churchill during these years. It is more likely that this map was prepared by Samuel Hearne who was intimately associated with Churchill and the northwest coast, as will be seen in the following section. Two other maps by Hearne in the archives are stylistically similar: they have the same delicacy in the pencil and ink work, the ornate cartouches and the lettering and script, especially many of the numbers, are similar. Another element in this story is that Hearne included in his published volume of 1795, three plans of harbours in the Nelson-Hayes, Moose, and Slude rivers, which were not prepared for the company but were engraved and printed for his own use. Since there is no indication in company records of his having drawn the manuscript originals, or having sent them to London, it may be that he also produced the chart of Rankin and Bowden inlets as part of this series. If he prepared the original charts in 1774, he could have done them at Churchill, where he was posted, after his trip to the Coppermine, from late June 1771 to late August 1773. Or they could have been drafted at York Fort, where he was from 5 September 1773 to 23 June 1774, after which he went inland to open Cumberland House. At York, the chief fort on Hudson Bay, he would have had access to a range of materials and information. Also, he would have had access at Churchill to data on the northern bay coast. And he was familiar with some portion of this coast as seaman and mate of the Churchill sloop, which coasted from this harbour. This chart then is probably Hearne's, drafted sometime after 1765 and perhaps in 1774.

MAPS RELATED TO INLAND WINTERING FROM CHURCHILL

Three motives lay behind the inland wintering program from the Prince of Wales Fort: the hope of drawing an increasing number of Indians from the northwest across to Churchill,[9] the possibility of finding ore, and the general desire to be acquainted with the hinterland of the factory. Moses Norton had marked out Samuel Hearne as an intelligent and adventurous man soon after he entered company service as a young sailor attached to the Churchill sloop. He had served previously in the Royal Navy as a captain's servant, and from this experience he would have gained some insight into mapping and the use of basic instruments. In the autumn of 1769, Norton assigned Hearne to attempt an inland journey with an Indian group. By this time, the purposes for such trips had been clarified and Norton's instructions were specific (A11/14/fos.130–2). Hearne was to try to increase trade by seeking out various Indian bands, to go to the Athabasca country and on to the Coppermine River to see whether it was navigable and communicated with Hudson Bay, to assess this area as a site for settlement, and to take possession of the region on behalf of the company. While there, he was to look for evidence of ore deposits, determine the course of the river and the character of its mouth, and note the location of woods in relation to the river and the quality of the soils. Along the way, he was to make a reconnaissance of other possible rivers that might be suitable for settlement. If he was unable to find the Coppermine River, he was to search instead for the upper reaches of Wager Bay and Chesterfield Inlet in order to assess their suitability for the location of trading posts. Finally, he was to ascertain whether there was a strait leading west from Hudson Bay. All this was a considerable responsibility to place upon the shoulders of a young man of about twenty-four.

Why did Norton give Hearne an alternative to the search for the Coppermine River? Norton would have based his directions on two maps, the first, his 1760 draft of the region, already mentioned above (i.e., 16ᴬ, plate 8). On this map, the Arctic Ocean-Hudson Bay angle was not shown, but the Nelson-Saskatchewan and the Churchill River systems were easily recognizable. So too were the large lakes in the north: the Northern Indian Lake (likely Great Slave), connected with the lake in the "Athapeeska" Indians' country (Athabasca), with the Peace, Athabasca, and Beaver rivers, all displaying very reasonable courses. The second map (19ᴬ), the most amateurish of the maps during this period, was drafted by Norton from a sketch made by two Northern Indian leaders.[10] In 1762, he had encouraged a Northern Indian captain, Idotlyazee, to "Trace to ye mouth of ye Largest River to ye Northward" (A11/14,fo.18d), and to produce a map of the large territory which he passed. On 9 August 1767, Idotlyazee and another captain, Matonabbee, arrived at Churchill with a manuscript map on deer-skin (22ᶜ). Norton reported that he had "taken it off on paper" (B42/a/64,fo.63). He may have started to do so, but in his September 1767 letter he informed London that the draft was not quite completed, and that he would keep it until the 1768 ship. The map can be dated 1767–8 as he probably finished it before he took ship for London on leave in 1768, since the committee had requested he bring it with him. No doubt he referred to it when he described the geography of the northern bay area to assembled committee members on 18 January 1769 (A1/43,fo.76d). The following week he wrote to the committee to outline his ideas on the development of inland trade northward of Churchill, ideas "agreeable to the Draught presented to the last Committee" (A1/43,fo.78). On this map the coast of Hudson Bay extends due north to the Coppermine River from the mouth of the Churchill River. There are still thirteen rivers shown beyond the fort. The great bend of the continental coastline at Melville Peninsula is not depicted, for it had not been observed by any of the parties involved. It would likely exist between Bowden's Inlet and "ye Grand Fish River," the name which was earlier applied to the Back River. Instead of flowing into the bay somewhere near Wager Bay as shown on the map, this river, Norton claimed, turned north into the Arctic Ocean. Thus, it is easy to see why he believed that the great river, which the natives located "far to ye Northward where ye Sun dont Set," flowed into Baffin Bay (A11/14,fo.78d). Inland was a great lake said to be in Athabasca country and from which the river "Kiscachewan" flowed into the Western Ocean.[11] This lake would be Great Slave Lake, not only because of its shape, but because of its relationship to the Coppermine River. The "Kiscachewan" is likely the Mackenzie, draining from the northwest end of the great lake, as it does in fact. It is obvious Norton thought that if Hearne did not find his way completely through to the Coppermine along the general route taken by the Indians in a great arc to Great Slave Lake, then, by veering closer to the bay shore on his way back, he would cut across the route of any potential strait and strike the western ends of Wager and Chesterfield Inlets. And thus Norton's alternative set of instructions.

In all, Hearne started off three times for the Coppermine River, in the late autumn of 1769, in the winter of 1770, and in the autumn of 1770. Only on the last attempt was he successful. The first trial lasted only a little over a month, commencing after the waterways had been frozen over and the spongy moss and watery muskeg hardened to a more passable state. They reached the Seal River valley and beyond to the northwest. But the group was larger than their ability to supply themselves with food in the Barrens; they were abandoned by their Indian guides and forced to return to Churchill. Since Hearne was still willing to carry on the expedition – and Norton greatly desirous that he should – he was refurbished, supplied with five Indian companions, and sent off again on 23 February 1770. This time their path lay west into the middle Seal River valley, west to Dubawnt Lake, and beyond some miles on the Barrens. In late July, Hearne's guide decided it was too late to strike out for the Coppermine; his plan was to winter in the general area and start out again in the late spring after break-up. In August, however, Hearne's quadrant was smashed in an accident, and because he was determined to take astronomical observations throughout his journey, he turned back, reaching Churchill on 25 November 1770. Before heading off again, he completed a map of his previous journey, which was forwarded by Norton to London in September 1771 (25ᶜ).

The third journey, during which he reached the Coppermine River, the Arctic Ocean, as well as Great Slave Lake, was a success not only because of Hearne's personal capacities, but also because he had as a guide the experienced Northern Indian leader, Matonabbee, who having promised to "conduct him, to the Copper Mines and River, the shortest and best way," remained loyal and of invaluable aid throughout the expedition (B42/a/80,fos.20d,21). Leaving the fort on 20 December 1770, they took a course generally south of the previous ones, eventually reaching the long-sought river several score miles from its mouth. After this, the Indian party, with a reluctant Hearne unable to talk them out of it, made a murderous attack on a small camp of Inuit at the lower rapids in the river and slaughtered all they could catch. Immediately after, they moved downstream to the Arctic Ocean shore that Hearne had viewed from the rocky ridge over which the river plunged. Hearne named the cascade, Bloody Falls, to commemorate the atrocity. The return trip began with a visit to a part of the Coppermine Hills where the mines of copper were said to be located. However, Hearne found only a small lump of native copper in the area and no indication of anything worth more attention. In their southward course, they retraced their earlier route until they branched south from the Barrens into the taiga, to reach the northern shore of Great Slave Lake. They passed across the lake on the ice, moving from island to island, then wandered southwest to the east bank of the Slave River, following it upstream for perhaps thirty miles. There they struck east to Dubawnt River and back to Churchill Factory in late June 1772, touching most of the same lakes and rivers as on their outward journey. In his last two journeys, Hearne had essentially accomplished his objectives. He had reached the Coppermine River and Arctic Ocean, had observed part of the river's course, and having cut across the continent from the bay to the Arctic Ocean east of this river, he could report that this "large river" had no connection with Hudson Bay. The copper mines were apparently of little value, there was no soil of any quality to speak of, and timber of building size lay far away to the south. Because he had successfully completed the first objective, he did not attempt to reach the upper ends of Wager and Chesterfield bays. The hope of meeting other Indian groups had been partly achieved. From a geographical standpoint, the most significant result was the end of all hope that there was a strait leading across the continent from the bay to the Western Ocean.

In the autumn letter of 1772 Norton indicated that Hearne's journal and map were on their way to London. In fact, two maps had been sent, a detail verified by the committee's spring 1773 letter. The first map, however, is the only Hearne map from the three journeys in the company archives.(20^A) (plate 9). It does not show the course of the second trip in 1770. Since he had already sent the 1770 map to the committee, there was no reason for him to add that data to his map; his final journey lay mainly south and west of the earlier one. It also may be that he did not have a copy of the 1770 detailed notes with him during the 1772 summer, for he regretted that his old journal had been given inadvertently to a companion as scrap paper in December 1770, when he had been in such a hurry to leave on his final journey. (He never recorded that it was returned to him.) Later, Hearne explained in his book the cartographic procedures he had used in his last trip.[12] Because he knew how difficult it would be to prepare base maps while travelling, he had provided himself with a projection base on which the west coast of Hudson Bay had been located. The base was drawn on parchment, with the grid extending twelve degrees of latitude north of Churchill and thirty degrees west of the post. The interior was blank. Then he cut out, on a larger scale, a map for each degree of latitude and longitude. On these separate pieces, while en route, he pricked off his daily courses and distances, putting in as much detail of the rivers and lake shores as he could see or get information on from his Indian companions. He did not know how to observe and calculate longitude, but whenever possible he corrected his position by instrument observation of latitude. His method was essentially that of dead reckoning, using compass readings to determine direction and his watch to estimate mileage (converting the time taken to cover ground into mileage). There are a number of gross errors in astronomical position in the western section of the map, particularly of the location of the Coppermine River mouth. His quadrant, an Elton, was not accurate, and he had broken it when he was north of Great Slave Lake. He was not aware of the complexities of the magnetic declination of his compass, which was affected also by local magnetism, and anyway, the compass was broken by the time he reached Great Slave Lake. When he began the compilation of this map for publication, he attempted to adjust the northwestern latitudes and some longitudinal positions. Hearne's map is the most professional-looking of the maps extant from this period, nothwithstanding the positional inaccuracies. Having established a latitude-longitude graticule from the data available to him, he plotted the geographical information in considerable and careful detail and provided many place names.

The second map from this last expedition, handed in by Hearne at Churchill, was a large scale plan of the lower forty or so miles of the Coppermine river, from the point where he joined it, to its mouth (4^B) (plate 10). Since he had been asked to survey the river for its transport value, he had travelled all the way along it to the mouth.

MAPPING THE ROUTE TO THE FIRST INLAND POST

The momentous alteration in the historic trading procedures of the Hudson's Bay Company, the move into the interior, began in 1774, although it had been in the process of administrative gestation for several years. Warnings had come to the committee as far back as 1768 of the decreasing flow of furs; suggestions that posts be opened inland included recommendations for sites. Two in particular received the most attention, one at the Grand Falls near the mouth of the Saskatchewan River in Lake Winnipeg and the second at Pasquia (Basquiau, The Pas). Cocking had proposed, while recommending the latter in his 1772 report, that a post farther up the river might be even better. Graham forwarded Cocking's full proposals to the committee and advised the members to offer the leadership of the post-building party to Samuel Hearne (A11/15,fo.143d), with Hearne adding a plea for his own selection (A11/14,fos.174–5). The committee approved the proposal in 1773 (A6/11,fo.173d), and the expedition was able to get away from York Fort on 23 June 1774, after Hearne had spent almost a year there in preparation. Since Hearne and Cocking had become convinced that a greater concourse of tribes would be met farther upstream at Cumberland Lake (Pine Island Lake), where a post would tap not only the Upper Saskatchewan River but also the southern prairies and the Churchill country, the idea of a post at Pasquia was abandoned. Hearne was back at York a year later, having essentially completed the building of the first inland trading post, Cumberland House. In the process of moving back and forth to Cumberland House, Hearne surveyed and mapped the routes the followed, and prepared a map of "the principal Lakes River's etc leading from YF to Basquiaw" (25^A). It depicted the Nelson River to Split Lake and the Hayes-Foxe rivers, with the Bigstone River connection to the Nelson. Beyond this he portrayed the Grass River route through Wintering Lake and Cranberry Portage, to the new post, and from there, the Saskatchewan River downstream to The Pas. Since Hearne had more difficulty with longitude than latitude determination, his map is extended two degrees too much east and west, while being out only one degree of latitude between York and Cumberland. There is a delicately drawn title cartouche, typical of Hearne's earlier work. Although exactly when Hearne drew this map is uncertain, it seems reasonable to assign it a 1775 date: it had been copied by 19 June 1776 in London by Henry Hanwell (Sr) (see below) where it could have arrived only on the return of the autumn supply ship in 1775. Hearne had been back in York Fort for two weeks in June-July 1775 and could have drafted the fair copy then from his notes and perhaps from a version worked on during the previous winter. On the other hand, he might have had his materials and pen with him inland and worked on it during the long winter evenings so as to have it ready to be handed in on his return to York. In either case, it was finished by the late summer of 1775.

The utility of this map was immediately recognized by the company since it handed it over to be copied by a young employee, Henry Hanwell, (Sr) educated in the Mathematical Class of the Grey Coat Hospital and the first young apprentice hired from this school in 1766. Hanwell, a gunner on the *Prince Rupert*, was in Britain between ship sailings in the winter and spring of 1776 and thus available to prepare two copies of the map during that time. On 19 June 1776 the committee approved for "Henry Hanwell ... a Gratuity for Drawing Charts £3-3-." These two maps were listed as #39 (33^C) in the 1796 catalogue and as having been drawn under the inspection of Captain Spurrell, an experienced officer. Another map, #38, of James Bay (34^C), was catalogued as having been drawn by Henry Hanwell (Sr) and was likely paid for also by the emolument noted.

INLAND MAPS FROM THE BOTTOM OF THE BAY

Exploration in the region inland from the southern end of the bay was mostly non-existent up to the year 1774. Thus the bayside posts there had not developed a cadre of resourceful travellers. Yet, in the years following the introduction of the new company policy, the officers and servants at the Bottom of the Bay rapidly and, perhaps, unexpectedly, developed an efficient program of exploration. What is also interesting is that employees in the Albany and Moose basins, in the course of their normal work, produced more maps than did those in the northern bases. There is no evidence that the company or the chiefs at Albany had a particular plan for the occupation of the region, but it is obvious that they had a broad geographical pattern in mind to which individual movements were linked. In 1777, this pattern was concretely outlined: "The Rainy, Woody, Assinibouels and Winnipeg Lakes [were] ... to be the principal Objects" (B3/b/14,fo.9).[13] The Albany River network was to be used as a major means of approaching the interior plains. Preliminary plans called for forestalling the expansion of Canadian traders toward Albany and intercepting their trade route to the interior by establishing posts at Rainy Lake, Lake of the Woods, and the Winnipeg River. More specifically, there were the two immediate tasks of reaching two lakes, St. Anns and Mepiskawaucau, whose locations were unknown at that time.

The first few years after the 1777 decision to move inland in the south were essentially ones of preparation. The company had to determine what they knew and what they needed to know. Surprisingly enough, two of the unknowns were the distances from the mouth of the Albany River to Henley House, and the location of the various islands, points, tributaries, falls, rapids, and portages along the river's course. In March 1774, John Martin proceeded up the river, keeping an exact journal of measurements and record of geographical features along the way. He recorded some sixty-six place-names in the distance of 134.5 miles (B3/a/67). Although he did not draft a map, the company expressed its pleasure with his detailed description.

Edward Jarvis, a young surgeon at Albany, declared himself in 1774 to be ready to explore inland (A11/3,fo.199). In the spring of 1775, he went up to Henley, planning to look for a route to Lake Mepiskawaucau. When his plan was foiled by the refusal of any Indians to guide him, Chief Factor Hutchins decided to initiate him "into the manner in which ... [he was] to prosecute the grand discovery inland" (B3/b/13,fo.12d) by having him examine the Chepy Sepy, a south bank tributary of the Albany. Through part of October and November 1775, Jarvis and the two Indian companions who were willing to go this far with him passed up this stream to the Missinaibi River, a major western tributary of the Moose River. From there they passed down the rivers to Moose Fort, and back to Albany on December 13, "half starved, & greatly fatigued" (B3/b/13,fo.20). Jarvis drafted a "plan" of his journey, which Hutchins told the company would "sufficiently recommend him" (29^C) (A11/4,fo.25). In January 1776, the factor suggested that a young Grey Coat apprentice, John Hodgson, might also make a "draft" of this journey, although there is no evidence that he did so.

Hodgson, at age fourteen or fifteen, had arrived at Albany, fresh from the Grey Coat Hospital, in the autumn of 1774. He was assigned to this district where he remained throughout most of his career. At the urging of the committee who were anxious that Hodgson become more expert, Hutchins sent him to survey and make a map of the Albany River up as far as Henley. As early as October 1774, Hutchins told the committee he planned to have a small boat built that could be used both for fishing and for surveying the river. Hodgson, teamed up with the master of the Eastmain sloop, Thomas Moore, began working from this boat at the mouth of the river. No doubt Hodgson had John Martin's detailed measurements to aid him. Hodgson was then put on shore to plot their data; by the end of the month his plan of the mouth of the Albany was sent to London.(30^C) Young Hodgson was next placed under the superintendence of John Favell, master at Henley, who took him up river to that post. Hodgson surveyed on the way, and then during October 1775, was hard at work "making a Draft." (31^C) This was sent back to Albany Fort in late October and drew praise from Hutchins, although he noted several errors in it.

Hodgson continued surveying and drafting during the following year also. This included assisting Martin in surveying for a few miles above Henley in October 1775 and preparing a map (32^C). They went farther up river the following summer and compiled another map apparently (37^C). In the mid-summer of 1776, after returning downstream to Albany, he teamed up with Moore again to go "to Nodaway River on discovery," his job being to survey and "to make Plans" (B3/a/70,fo.34d). Although no map is mentioned nor located in the archives, it seems likely that a map was prepared since that was the assignment (38^C). Moreover, Hutchins reported that on the four days immediately after Hodgson's arrival back at the fort he was busy "making drafts." It may be that his sketch was kept at Albany or Henley, to be used by Hodgson for the compilation of the extensive regional map which he drew later.

By 1776, although reaching Lake Superior had not been the stated goal, plans were being organized at the fort to explore the southern Albany tributaries and across the water divide into the Great Lakes basin; the intent was to counter the operations of Canadian traders south toward Lake Superior and to keep open the Indians' routes of communication with the bay post. The search for Lake St. Anns was part of the plan. There was confusion over the identity of this large lake which was said to lie between Lake Superior and the Albany River. It was at times called Lake St. Anns, but it was also confused with a body of water named Lake Meshippicoot. By mid-May 1776, Indian guides had been hired, and in late May, Jarvis set off for Henley and the south. On 19 June, the party had reached Lake Meshippicoot, which turned out to be Michipicoten Bay, part of Lake Superior. Their route had brought them up the Kenogami River, (the south branch of the Albany River).

They had then followed its branch, the Kabinakagami, to its source lake, and over the divide to the Esnagamie River which led to Lake Superior. Back at Albany Fort in early July, Jarvis set to work to complete his journal and to prepare his map (35^C). In his journal, Jarvis had inserted various marginal sketches to aid him in the final drafting.[14] Hutchins reported that the map he was sending to London by autumn packet was "executed" by Jarvis, and also that John Hodgson was employed in making a copy of it (36^C) (A11/4,fo.28). In the letter Jarvis wrote to the committee at that time, he called attention again to the two journeys he had made earlier (A11/4,fo.37). He apologized for the inaccuracies in the maps he had sent them, inaccuracies due to the rapidity and difficulty of travelling, but also because of his "own imperfections." He also declared that he had found the travelling so taxing that he declined making any further expeditions (A5/2,fo.25).

During 1777, the committee was pressing Hutchins for more exploration, mapping, and post-building. In particular, the committee wanted maps, and the Albany chief was asked to use employees with the ability to make astronomical observations wherever they went. Hutchins in reply regretted that there was no one else at present but Hodgson who could do this. He tried to remedy this situation by training one of his employees, George Sutherland, personally, sending him north up the James Bay shore as an exercise. The first major step toward Lake Winnipeg took place in the summer of 1777, when John Kipling led an expedition, with John Hodgson going along, up river to Upashewa Lake, where he proposed to winter at least for the first year. By 25 June, they were starting to build a new post, Gloucester House, on the lake shore. A map of the route from Henley House to the lake was drafted, and although its authorship cannot be definitely determined, it was most likely by Hodgson (45^C). In the first place, there is no evidence that Kipling was cartographically inclined. Secondly, Kipling was completely engrossed in his own duties until 28 June when Hodgson headed back toward Albany with the journal of the expedition in his hand. Hodgson reached Henley on 1 July, and the post journal records that he was busy writing for at least five days there before leaving for Albany on 19 July. Certainly, Hodgson could have been making a map during this time. Such a map existed because a letter of 19 August 1777, to the committee by Hutchins, stated that it "will judge, by the Plan, of the convenience of its [Gloucester's] *situation between the two large lakes* which have, as we suppose, great communications farther inland" (A11/4,fo.55). The only puzzle about the map is why the committee, in its return letter of May 1778, asked Hutchins for Gloucester's longitude and general distance from Henley. Why couldn't the committee obtain this data from the map? The answer undoubtedly was that the map was not clear on overall distances, because Hodgson had not taken any locational observations. This can be deduced from the Hutchins's complaint in September 1777 that the quadrant sent to London to have its burst bubble fixed had not been returned. Without the quadrant, Hodgson could not have measured longitude.[15] The extreme hardships the crew underwent may also explain the lack of observation. Kipling reported that Hodgson was half-dead when they reached Upashewa Lake, and what with the "want of Victuals" there, he had decided to send Hodgson back to Henley. Hodgson and his three companions arrived late at night at Henley "in a miserable Condition, half-starved" (B86/a/30,fo.43d).

For the next few years, the action moved some two hundred miles up the Albany from Henley House, centering at Gloucester House, from which expeditions struck out upstream, eventually reaching Lake Winnipeg. Up to the time Gloucester was established, only one Canadian house was reported to exist in the upper Albany basin – on the shores of Mepiskawaucau Lake. George Sutherland was sent up to Gloucester House in the late summer of 1777 to evaluate the reports firsthand. From Gloucester, he was instructed to strike west as far as he could and to tell as many Indians as he could find about the convenient position of this new company post. Sutherland pioneered the route from Upashewa Lake, past Lake St. Joseph to Lac Seul, then north to Red Lake and the Bloodvein or the Pigeon River leading to Lake Winnipeg. He returned on 27 June 1778 in wretched physical condition, having travelled over nine hundred miles from Albany by his reckoning, to report that there was considerable Canadian activity in the western Albany waterways. Sutherland's journal was characterized by Hutchins as "defective very much in Orthography and Grammar and in many places more whimsical than usefull, yet contains several Observations worthy of consideration" (A11/4,fo.73). From Hutchins's further description, Sutherland's map appears to have been merely "a rough Sketch," which included not only the route across to Lake Winnipeg, but, in a general way, also to have shown the position of York and Severn forts (46^C). Hutchins hoped that, although the map and journal would require the committee's greatest indulgence, it would be agreeable to them. The committee deemed them acceptable and expressed the hope that Sutherland would be kept on such service. Doubtless this approval was partly because Sutherland had been supplied with a quadrant, a compass, and a watch, and before his quadrant was broken, had calculated two latitude fixes, one being at the shore of Lake Winnipeg (52°31'N.). This was the type of information the company wanted. Furthermore, at Lake Winnipeg, local Indians had offered Sutherland a special set of distances from that point; they reported it took five days' hard paddling to reach Cumberland House, WNW; thirteen days to York Fort, NbE; twelve days to Severn Fort, ENE; and twenty-one days to Albany Factory, Esoutherly (B3/a/73,fo.24).

The southern James Bay watershed was also investigated more thoroughly during the 1770s. The Nottaway and Harricanaw rivers were followed inland from the bay shore in order to look for routes to Lakes Abitibi, Mattagami, and Waswanipi, upon which the French and later the Canadians were very active. John Thomas, for example, travelled east from the lower Moose River to Lake Abitibi and in 1775 went up the Nottaway River some miles. Moore and Hodgson, as noted earlier, were in the region in 1776. In 1777 and 1778 especially, Moose Fort personnel were actively exploring the inland reaches of Moose River. Starting in October 1776, an expedition led by George Atkinson, surgeon at Moose Fort, pressed up the river to establish a post on a lake called Wappiscogamy. He was accompanied by George Donald, a Grey Coat apprentice, as surveyor and draftsman. Donald had arrived at Moose Factory in 1774, at the same time as Hodgson had arrived at Albany. Donald's job among others was to chart the river from Moose to the lake. After several difficult winter journeys up and down the river, one journey at least complicated by his snow blindness, Donald completed a map in the summer of 1777 (41^C). That same summer, he was attached to John Thomas, the leader of a group dispatched up the Moose River to cross over the divide to Lake Superior at Michipicoten. Jarvis's journey to this locale earlier from Albany had proved that this area should be left to Moose, and there were no later trips made from Albany to Lake Superior. To aid the planning at Moose, Hutchins had sent Eusebius Kitchin, the chief factor there, a copy that Hodgson made of the Jarvis map (i.e., the map of Jarvis's journey). The Thomas party left Wappiscogamy House on 24 June, and after a tiring journey reached the lake situated at the head of the Missinaibi River. Donald wrote a letter to Kitchin, and with it included his map of the route from Wappiscogamy House to Lake Missinaibi (42^C). He expressed regret for its inaccuracies but explained that these could not be helped as the canoe containing his instruments had been lost in a stretch of turbulent waters. Since earlier the level in his quadrant had burst due to the extreme cold, he was poorly equipped having "not so much as a Pair of Compasses left to draw a circle or Make the Scale" (B135/b/5,fo.39). He was unable to take any astronomical observations at Missinaibi Lake, though he had earlier calculated that Wappiscogamy House was at 49°59' North. The map reached Kitchin who sent it to London in the fall packet. While at the lake, and upon asking Indians about the route to Lake Superior,

Donald reported that one of the Indian captains, Pequatisahaw, had drawn a map of the route, for the use of guides, along with instructions (43^C). No further references was made to this map.

Another house was opened in the Moose basin at Mesackame Lake by George Atkinson, who left Moose in October, in the late autumn of 1777. By early January 1778, Kitchin had dispatched a letter to Atkinson in which he reproved him for not sending back the "sun's altitude" and the course and distance of the route to this new post. He had wanted these details so that George Donald could "have made a fair draft for the Company's inspection" (B135/b/6,fo.10). Atkinson sent the information down river in a letter of 21 February 1778, but if Donald ever made a map based on it, there is no mention of it being sent to London or having been received there.

All during these years, as the governor and committee tried to get an accurate sense of their trading territory in order to plan the pattern of trade expansion, they were hampered by the lack of adequate latitude and longitude data. When the Grey Coat Hospital apprentices were not able to supply them with the information they had hoped for, the committee decided to have several maps of Rupert's Land compiled commercially in London in 1776. One of these was a map of the region west and south of the Albany and Moose locales, one copy of which was sent to Kitchin at Moose and another to Hutchins at Albany (39^C). In their discussions during the preparation of this map, the secretary recorded that Albany and Moose rivers, Lake Superior, Lake Winnipeg, the Grand Portage, Rainy Lake, and Lake of the Woods were specifically mentioned by the members (A5/2,fos.13,14,16; B3/b/14,fo.2). In addition, the committee had a new regional map prepared "from the best Evidence and information collected from the Journals" (47^C) (A6/12,fo100; A11/4,fo.87). Presumably it included the region south and west of Hudson and James bays and may also have included the southern Eastmain area. The committee sent a copy in May 1778 to Thomas Hutchins and asked that future maps prepared in North America covering the same account of area be drafted at the same scale. This would enable them to collate new data on their own general map. This second map may have been based partly on an earlier map by William Falconer, which was referred to on 4 May 1775 by the committee. The letter stated that it was pleased with the "Chart ... [Falconer] sent us of the Coast, the interior part of the Country, and the principal Rivers communicating with our Factories on the Westward and Southern Coasts" (27^C) (A5/1,fo.169). The map had probably been sent in the autumn packet 1774, since the committee, when it acknowledged a map, usually did so by the next supply ship. The content of the map can only be deduced from the reference, but it could have had some similarity to the Graham maps of 1772 and 1774, portraying the inland wintering trips from Severn and York, with perhaps some reference to the southern rivers. Information about the latter area would have been sparse at that date. Perhaps the idea surfaced sometime during the discussion of these maps that what they really needed was a professionally developed framework into which they could fit the data coming into London. Whenever the discussion began, by March 1788, they had decided a full-time surveyor and cartographer was what they required. Philip Turnor was hired and set out for the bay in the summer of 1778.[16]

8 Mapping Inland from the Bay and Over to Athabasca, 1778–1794

> Turnor ... is return'd ... from Hudson's Bay, where he has been employ'd in surveying different parts of that Country, & is now engaged in drawing Maps & making Observations thereon.
> Governor and Committee, December 1792

With the opening of Cumberland, Gloucester, and Wappiscogamy houses, the Hudson's Bay Company began the long occupation of the interior of the continent, carrying its goods to and from a network of posts and into the midst of the Indians themselves. When Philip Turnor arrived on the scene as inland surveyor at York Fort on the autumn ship in 1778, the need for trading posts farther west of Cumberland, up the Saskatchewan River, was being discussed. Those in favour argued that the Indians travelling to the bay were being enticed aside along the way by Canadian traders who talked them into trading their furs for alcohol. Another argument was that a new post up the Saskatchewan River could obtain provisions from the Buffalo Country, especially pemmican from the Assiniboine Indians. Basic to all this was the obvious fact that Cumberland was losing out because the spearhead of trade was moving farther up river. Opposition posts had increased in number and expanded their trade areas during the years since the retreat of the French. Not only had rival trading houses been built up the Saskatchewan, but the traders had infiltrated the Churchill system, developing a route almost to the Athabasca divide. The company decided to build a temporary hut first in the autumn of 1778, near the farthest upstream Canadian house; and in the spring of 1779, to erect, under the superintendency of William Tomison, a permanent post on the north bank of the North Saskatchewan River, upstream from the Forks of the Saskatchewan, and at the beginning of the trail leading north to the Thickwood Hills. From 1779 to 1786, this post, Hudson House, was the only company post beyond Cumberland and the only one in direct contact with the Buffalo Country.

As one might surmise, during this period of inland expansion the mapping was essentially of riverine and lacustrine features, although several coastal charts of the bay shores, by then well delineated, were also completed.[1] Seventeen of the nineteen men who prepared maps between 1778 and 1794 were company employees. Most were responsible for only one map, some for two. Seven of these cartographers took care to locate their data more exactly than the others who did not know how to use the necessary instruments. Even so, these less accurate sketches of smaller segments of the terrain could be fitted into the larger regional maps that were an important innovation during this period. Turnor prepared ten maps including his most famous, the large composite map of the north and west that he presented to the governor and committee when he retired in 1794 (57A). Peter Fidler, Turnor's protege, drew 144 segmental sketches between 1790 and 1794. His sketches, all of which have survived, were also used as the basis for maps. Unfortunately, however, half of the total of thirty-eight maps drafted during these years are no longer in the company archives.

TURNOR MAPS MAIN RIVERS AND THE BOTTOM OF THE BAY, 1778–1787

During his first four years with the company, Turnor fulfilled his designated role to the committee's satisfaction. He prepared three original maps and several copies. He observed and calculated the astronomical locations of all the factories and posts, of the significant junctions of the interior waterways, and of important geographical features along the sea coasts, all of which he had visited. On most of his journeys he followed the established routes or did not go far afield from them. Turnor's initial itinerary called for him to go inland from York Fort to survey and map the main tracks to Cumberland House and beyond, following which he was to proceed down the bay to map the main river routes inland, and finally to complete his perambulations at Churchill Factory. After he completed his first official task by surveying and preparing a plan of the ground on which York stood as well as that adjacent to it, Turnor headed inland before freeze-up in 1778 via the northern route along the Nelson and Grass rivers and across Cranberry Portage to Cumberland House. He wintered there with his travelling companion, George Hudson, where they were joined by Joseph Hansom. Both young men, apprentices from the Grey Coat School, received further instruction in practical astronomy from Turnor. The following spring, while the ice was still on the Saskatchewan River, Turnor and Hudson snow-shoed up river to the temporary house built in the previous fall. Turnor estimated this was one hundred miles by river course above the Forks of the Saskatchewan. After about a month's stay, during which he decided not to proceed upstream farther because of difficulties between the Indians and Canadians, he turned back to Cumberland House, arriving on 26 May 1779. During the short visit there he spent two days examining the west side of Cumberland (Pine Island) Lake. Within two weeks he was off to York Factory by way of a different route, through Cedar Lake and Lake Winnipeg, then by Cross, Oxford, and Knee lakes to the Hayes River. In a tragic accident on the way down, Hansom was drowned while attempting to shoot a steep rapids.

Having arrived back at York in mid-July, Turnor was able to work on the map of his journey, on which both of the routes of travel were delineated (27A).[2] On this map, based entirely on his own observations, Turnor established the drafting style he followed on most of his later maps, a style best exemplified by his 1794 map. In the first place, the data was placed on a detailed projection graticule indicating subdivisions of degrees and minutes. While there are some discrepancies from exact positions, ranging from one to four minutes, these reflect not lack of skill but difficulties en route. He had lost some of his instruments and books and half of his quicksilver when a canoe overturned on the way inland. Unfortunately, the other half of the quicksilver had been left behind at a portage. Thus the discrepancies may have resulted from the difficulty of taking sights without quicksilver to use in the artificial horizon. His sextant had also had rough usage on the trip. Turnor's maps have a simplified appearance since he generalized the complexity of features of the watercourses when he prepared his smaller scale maps. Only the main channels, larger islands, and embayments are depicted. He applied a grey wash, using thinned India ink, to outline the waterways, thus accentuating these topographic elements.

To complete the next phase of his assignment, Turnor sailed on the Severn sloop to the Severn post and proceeded to Moose Fort, arriving on 21 October 1779. He remained at Moose until late December, when with a small party he walked north along the coast to Albany before travelling inland. He had intended to travel to Gloucester House and farther by snow-shoe, but his party only reached Henley House, where the expedition had to be abandoned because Turnor was suffering from snow-blindness and extreme fatigue. Accompanied by John Hodgson, who had been to Gloucester twice before, Turnor went up river the following summer taking only rough measurements, leaving detailed observations for the return trip. Although he had hoped to go farther, Indian guides were unobtainable. The Albany River map (28A) (plate 11), on which he used the same cartographic approach as before, was probably sent out by the 1780 packet from Moose Factory, for in his letter to the governor and the committee, he expressed the hope that his "Observations, Remarks, Draughts and past conduct" would meet with their approbation (A11/44,fo.106). Although having traversed a known river course, and one upon which Hodgson had previously taken some measurements, Turnor followed instructions that he provide more trustworthy data, as he did with the surveys he made in late 1780 and 1781.

Between 19 December 1780 and 12 January 1781, Turnor walked along the coast from Moose to Albany and back in order to "take a stretch of the coast as it appears in winter" (B135/a/64,fo.1). Ten days later he set off along the coast from Moose Fort, headed east around the Bottom of the Bay for the Eastmain post, to undertake more coastal observations and also to determine the position of the post. With him and a small party was George Donald as assistant. Their sledge journey was a moderately expeditious one since they arrived at Eastmain by early March. On their return to Moose Fort, they made a more accurate survey of the route. Turnor did not make a separate map of the coast but used the data he had collected, along with that from his Moose-Albany trip, when he drafted his next composite map some time later.

Turnor continued his assignment by mapping several of the rivers south of Moose Fort and also the main route across the drainage divide to Michipicoten Bay. His first goal was to reach Kesagami Lake, southeast of Moose near the head of the Kesagami River. Leaving the fort on 2 April, 1781, he chose a route up the French River, a Moose tributary, whose upper waters lay across a short portage from this lake. The first attempt failed: the sled party was unable to proceed far up the river valley without breaking through the snow's hard crust and floundering in the soft, deep snow "up to the middle": indeed, they broke three sleds trying. Retracing their course to Moose, they waited until the ice broke up before starting off again on 11 May. This time, they went up the Moose River, toward Wappiscogamy House, leaving the earlier route for a later trial, using instead the route with which Donald was familiar: to Michipicoten via the Missinaibi River to Missinaibi Lake and over the watershed. They returned to Moose Fort on 13 July, completing an accurate survey during the return trip. By determining the astronomical location of Michipicoten Bay, Turnor obtained an overall scale and distance framework into which the estimated details could be fitted. Unfortunately, his sextant was damaged in another canoe accident just after he left Missinaibi Lake on his way back down river. But since Turnor had earlier on the journey upstream obtained two latitude fixes along this river, he was able to map this part of the course to his satisfaction. The map was not drafted until much later for he was determined to get to Abitibi and Kesagami Lakes before starting it.

Turnor started again up the French River, but, impeded by shoal water, he returned to Moose Factory and left the Abitibi trip for the following season. Although he sent his journal of observations for these journeys on the 1781 ship, he did not include a map, for, as he explained, he had intended to draw a map of the rivers and lakes through which he had passed on his way to and from Lake Superior, and "to contain the bottom of the Bay from Albany to Eastmain ... but the want of the Almanack to calculate my Observations" prevented him from accomplishing it that year (A11/44,fo.125). In the 1781 packet, he received the 1781–82 almanac and a new supply of quicksilver.[3] He was also sent a new contract for a further three-year period, along with the promise of an assignment to a trade position when an appropriate opening became available. Meanwhile, he was to carry on the surveying tasks (A6/13,fo.6), which he did after the intervening winter and the completion of several smaller assignments. One of these was to draw a new plan for a structure to replace the recently burned Henley House.

In the next journey from 23 May to 2 August 1782, Turnor reached Abitibi and Kesagami lakes in spite of the "irreparable" loss of his sextant, when a canoe was swamped about ninety miles up the Abitibi River, and the necessity of sending George Donald back to Moose to pick up replacement quadrants. Turnor also broke his watch. The party followed the Abitibi River to the lake, along the length of Abitibi Lake to its east end, and hence by a river and lake network northwest to their goal, Lake Kesagami. Then the French River route was taken on the return to Moose River and the factory. They must have met with further difficulties on the road, for Turnor wrote that even his guides did not know the Abitibi River, and, had he had previous knowledge of it, he would not have been so easily persuaded to go that way. Although he had time to work up his observations before ship-time, no map was sent back by the autumn 1782 packet, nor was mention made of the receipt of a map in the committee's next spring letter.

A third Turnor map, however, depicts the round trip to lakes Abitibi and Kesagami, as well as the Moose to Michipicoten and the Albany to Eastmain journeys of 1781 (30^A).[4] Included in this map is a reduced scale version of Turnor's Albany River map of 1780. The probable date for the completion of this map is 1782 or 1783; certainly he finished it before 1784. In the autumn of 1782, he was appointed to a trading position and transferred as master to Wappiscogamy House, which had been renamed Brunswick House. He remained there or at Moose Factory until April 1784. He would have had all his observations and calculations with him and must have had a considerable amount of time at Brunswick House for compiling and drafting his map through the winter of 1782–3, as he was suffering from rheumatism and unable to do much travelling. Moreover, this map does not depict any details of the country observed by Turnor in the years 1784 to 1787, that is, the country southeast to Lake Timiskaming, or the area of Frederick House, which he had established on a southern tributary of the Abitibi River in 1783–4. (These features did not appear on a Turnor map until 1794.) Turnor returned to Britain on the September 1787 ship where he stayed on furlough until 1789, when he had a new contract to lead the company's first expedition across Methy Portage into Athabasca country.

THE COMPANY EXPANDS IN THE SASKATCHEWAN AND CHURCHILL RIVER REGIONS

The advance of independent traders like Peter Pond and of Canadian traders later up to and across the Methy Portage into the Mackenzie system after 1778 greatly affected the company's planning for expansion both in the Saskatchewan valley and in the Churchill system. There was a proliferation of Canadian trading posts above Hudson House on the North Saskatchewan and a growing conviction in the company that not only did it not know the Churchill system adequately, but that this region's trade was being captured by rival traders. Accordingly, the company decided to extend its range of influence far up the Saskatchewan and southwest into the great plains. In 1785, orders were given for the erection of several new houses up this river, and in the same spring the traders James Gaddy and Isaac Batt were sent to winter with Indian bands to the southwest.[5] Gaddy apparently travelled for about one hundred fifty miles along the Rocky Mountain foothills.[6] The summer of 1786 saw the building of two new company posts well up the Saskatchewan: Manchester House, about sixty miles above the mouth of the Battle River, and South Branch House, sixty miles up the South Saskatchewan from the Forks.

In 1785, the committee also resolved that an experienced inland trader should be given the task of exploring "the Course & Connection of Churchill River & examining the adjacent country ... to establish Trading houses ... for the purpose of ... regaining the Trade" (A1/46,fo.41). Robert Longmoor, a superior canoeist but not a trained surveyor, was chosen. A new recruit, the Grey Coat apprentice, George Charles, supplied with "proper Instruments & ... perfectly well qualified to determine the Latitude & Longitude of ... places" (A5/2,fo.129d), was instructed to join him.[7] Longmoor was to make the daily estimates of distances and observations of direction, while Charles was to make astronomical determinations "at all Convenient or remarkable places ... entering them in his Journal, and reserving his Calculations to be done at his Leizure when sitting in his tent." Longmoor was given responsibility for the apprentice, to "take every Care, both of his person & Morals" (A5/2,fo.129d). Although Hearne had been asked by the committee to tutor Charles in practical astronomy, he had been ill and unable to do so. Fortunately, Longmoor was replaced by Malcolm Ross as leader of the expedition: the latter was reasonably experienced in the use of instruments.[8] Ross remarked in his journal in the summer of 1786 that, while Charles did not make any sextant observations because he had never used that instrument before, he was able to use a quadrant to observe for altitude (B49/a/18,fo.5). In spite of the committee's directions, Ross and Charles avoided the lower Churchill River. They bypassed north to Pauqua a-thacow a scow a (South Knife) River and up it to Etawney Lake, then back to the Southern Indian Lake portion of the Churchill River. They then cut southwest to Cumberland House. Among various reasons given for their decision to leave the Churchill River was the conviction of Samuel Hearne at Churchill and of York Factory officials that inland expeditions could not be carried out by going up the Churchill River and that the upper Churchill system would be best tackled from a base at Cumberland House.

The information gathered by Ross and Charles was the basis for a new map, probably drafted by Turnor, as it has the characteristics of his drafting style (34^A) (plate 12). On this map, the "Rivers and Lakes above Churchill Fort joining the Same taken from a Journal kept by Malcom Ross" were appended, at the same scale, to the configuration of the Nelson-Saskatchewan rivers that had first been delineated by Turnor on his 1779 map. The journal and survey observations would have reached York Fort too late for the 1786 autumn ship to Britain. Thus, they would not have been forwarded until the late summer of 1787. Turnor, reaching England at the same time as the packet, proceeded to work part of the time on maps at his home in Laleham, Middlesex. He was there for two years between contract periods. The committee authorized a payment of twenty guineas for his "Draught of several Inland Settlements belonging to the Company" (A1/46,fo.162). This sum, a very large one for the production of one map, may also have been a form of gratuity for his previous surveying and mapping endeavours.

TURNOR LEADS A MAPPING EXPEDITION INTO ATHABASCA COUNTRY, 1789–1792

Although their emphasis was on the Saskatchewan River area, the committee members, despite controversy over the matter, continued to refine proposals for entry into the Athabasca region. As early as 1781, they had decided to send an expedition there to let the Indians know that a post would be built (B239/b/42,fos.4,4d). Because of various problems, the expedition was postponed until 1789. The plan then called for sending Philip Turnor back to Rupert's Land that autumn ready to go inland from York to Cumberland House immediately. Supplies, trade goods, and a party of men were to be on hand there so the expedition could leave as soon as Turnor arrived. The supplies and party did not materialize as intended, however, and Turnor found it necessary to winter at Cumberland House.

Three young men, all trained in making astronomical observations, were assigned to Turnor's expedition. From among them Turnor was to choose a mapping assistant. In the autumn of 1789, Thomas Stayner had applied at Churchill for the assistant position, but by then he had fallen into the bad graces of the committee, having refused a year earlier to go with Robert Longmoor on a trial of the Churchill River as a route of entry to the Athabasca region. He claimed then that he was unwell and that he found such work too taxing. Admonished by the committee, Stayner agreed to go to Cumberland House. However, as the result of a change in plan, he worked in the autumn and winter of 1789–90 between Manchester and South Branch houses and, during this time he drew a map of the intervening areas, sending it to London in 1790 (54^C). Turnor had meanwhile assessed and rejected Stayner as a possible assistant, concluding that the young man was much better suited to the life of a trader than the "unsettled life of a Traveller" (A11/17,fo.46). Turnor also rejected George Hudson, with whom he had been associated in his first inland journey to the Saskatchewan valley in 1778. Hudson had not explored or surveyed since then, had been increasingly ill, and, in fact, died at Cumberland House in the spring of 1790.

The last of the three candidates for the Athabasca party, David Thompson, was still convalescing from an accident in which his leg had been crushed. Thompson had been trained in the mathematical class at Grey Coat Hospital, and, since his arrival at Churchill in 1784 at the age of fourteen, had been tutored in practical surveying by Hearne. The next year he trekked with Indians to York Factory. In 1786, he was attached to a party going inland to build South Branch House and winter there; in 1787, he was assigned to Manchester House, and that autumn he wintered with James Gaddy and others as far south as the Bow River among the Piegan Indians. From the spring of 1788 to the following Christmas, he was attached mainly to Manchester House – it was during this time he had had the accident. The one useful outcome was that Thompson was quartered during the winter of 1789–90 at Cumberland House, where he benefitted from Turnor's surveying instruction. In place of Thompson, Turnor chose an older assistant, Peter Fidler, to whom he gave instruction in preparation for the journey.

The reconstituted Athabasca party consisted of Turnor as surveyor, Fidler as assistant surveyor, Malcolm Ross as master or head of the trading expedition, and four other servants. Turnor and Fidler were to determine the true locations of the major lakes and rivers which they would reach beyond Methy Portage, and to observe the basic physical and human geography of the region, as these would relate to future trade possibilities. The expedition left Cumberland House on 13 September 1790 and reached York Factory on 17 July 1792. The path they followed had been established by Canadian traders; it lay northwest across to the Churchill River, then along this complicated system to Ile a la Crosse where the party stayed from October 1790 until 30 May 1791. Turnor's journal shows that he and Fidler were actively taking observations during the twenty-five day trek to Ile a la Crosse. On the average they made one estimation of the route distance each mile, or an average of eighteen per day; they took compass bearings of their course and the details along it at about the same rate. They also estimated the widths of rivers and lakes, the lengths and widths of islands, and the depths of bays, took compass readings of the orientations of rivers and lakes, and commented on river currents and on terrain and vegetation forms. In addition, Fidler made an almost continuous series of sketch maps of the waterways along which they moved. The two men collaborated on fifteen latitude determinations.[9] They determined the longitude of only one location, on Lac Geneau, making three sets of observations at that point. Tyrrell has observed that Turnor's latitude figure at that lake was reasonably correct, but that the three longitude observations were from one to seven or eight miles east of the correct location. However, these were reasonable results for that period.[10]

In the spring and summer of 1791, from 30 May to 26 July, they continued north and west across Methy Portage to Lake Athabasca and to the southern shore of Great Slave Lake. During this part of their journey, Turnor and Fidler paid much more attention to the determination of their exact course. They calculated latitudes for thirty-one locations and the longitudes of seven of these places within a grid of about six degrees east-west and the same north-south. Upon their return to Lake Athabasca from farther north, Turnor and Fidler made a circuit of this lake, working out fourteen further latitudes and three longitudes of sites along the north and south shores. The party began to retrace its steps to Cumberland House on 9 May 1792, with the two surveyors recording seven further latitudes and seven longitudes from Lake Athabasca to Lac Ile a la Crosse. They arrived at Cumberland on 27 June and almost immediately set off for York Factory.

Apparently, in the course of the expedition, Turnor did not draft maps, with the exception of a small sketch based on information a Chipewyan Indian, Shew ditha da, gave him at Great Slave Lake (42A). The sketch is a highly compressed delineation of a strip of territory from the west end of that lake to the entrance in Hudson Bay of what is likely Chesterfield Inlet. In contrast, Fidler developed the knack of sketching as much as possible while in passage. It may be that Turnor requested that, as his assistant, he do so. Whatever the stimulus, it became Fidler's lifelong habit.[11] There are twelve sketches by Fidler available from the journey to the shore of Great Slave Lake, starting at the Methy Portage (39A, 40A, 41A), and sixteen sketches illustrating most of the shoreline of Lake Athabasca (43A). During the winter of 1791–2, Fidler, wanting to become conversant in the Chipewyan language, was sent out by Turnor to travel with the Indians and to subsist, like them, off the land. They travelled around in the vicinity of the Slave River, and Fidler revisited Great Slave Lake. Six sketches of the shore of the lake eastward and westward of the river mouth resulted (44A). During the seven-month sojourn, Fidler determined the latitudinal location of some thirty-two places, based on fifty-six sets of observations. He had a sextant and artificial horizon with him, but because he lacked parallel glasses, he had to be very careful to observe in sheltered locations in order that the surface of the quicksilver not be disturbed. He had no watch with him, and therefore, although he tried some longitude calculations, they were not accurate, and he did not record any in his journal. He did have the necessary nautical almanac and *Requisite Tables*, and, in addition, a boat's compass, for taking directional readings. Since he did not bring enough European clothing, he learned to make skin clothing by watching the several women in his party. He had mastered the language well enough that he was certain he could carry on trade negotiations with the

Chipewyans (during the spring he reported having dreamed in the Chipewyan language for the first time). On the return trip, Fidler added some sixty-one more sketches to his journal, eight being from the Clearwater River across Methy Portage to la Crosse (45A), forty of them being of the lakes and rivers along the Churchill system to Portage du Traite,[12] and the remainder across to Cumberland House (46A) (plate 16).

Immediately after his return to York Fort, Turnor offered to lead a trading expedition back to the Athabasca, but his offer was declined. Instead, since it had proved difficult to organize a crew to carry supplies up the Nelson river to Chatham House (near the junction of the Nelson and Grass rivers), Turnor volunteered to lead a party that would include several of his own recently returned group. Turnor believed the Nelson to be an easier and safer river than the Albany; the journey would give him a chance to find out. His proposal was accepted, and he was asked to determine also whether it would be practicable to get "Boats of Burthen" any distance up the Nelson, something the committee had been wanting to know since 1754 when they had asked James Isham the same question. While Turnor was on the mission (from 31 July to 24 August 1792), he fell victim again to misfortune in rough water. When a canoe overturned, he lost his sextant and everything else of value, except his watch. He drafted no separate map as a consequence of the journey up the Nelson, but the details he made note of were useful for his later maps.

Turnor left Rupert's Land for the last time in September 1792, taking ship from York Factory for England. After settling at home, he began to work on the compilation of maps based upon his years of experience in North America. On 12 December 1792, the committee agreed that he should be paid retroactively from 24 October, the sum of one guinea per week until further orders (A1/47,fo.4). Apparently, it was during this time that he drafted two maps. The first was very likely the fourteen sheet map of the "Track from Cumberland House to Ile la Crosse with the Magnetic Bearings" (59C)[13]. It would be his rendering, at a larger scale, of the data that he and Fidler had obtained, from which then smaller scale versions could be made. The second map was in all probability the "Chart of Lakes and Rivers in North America" (47A). It, like the first, was probably drafted in 1792, since the waterways traversed in Turnor's recent journey to the northwest are the major element. This configuration would be a reduction of the fourteen-sheet compilation with further areas included. There is no information on the map related to events later than the summer of 1792; Buckingham House, for example, erected on the North Saskatchewan in the autumn of that year (news of which would not have reached London until the autumn of 1793), is not shown. Turnor indicated that his map was compiled from two types of sources, the shaded areas were based on his own "Actual Surveys," the unshaded outlines "from Canadian and Indian information." The map reflects his surveys from York Fort to Great Slave Lake; in addition there are reduced renditions of the Nelson-Hayes rivers, the Grass River, northern Lake Winnipeg, and the Saskatchewan River to Hudson House,[14] and the Albany River to Gloucester (28A). Finally, he has shown the Hudson and James Bay shores from Churchill to Albany River. The details from other sources include the lower Peace River, the Beaver River and lakes to the west, including Cold Lake and Lac la Biche, the Northern Saskatchewan from Hudson House past Manchester House, the Nelson River connection from Playgreen Lake to the mouth of the Grass River, and the Hudson Bay shore some distance north of Churchill.

On Turnor's last and most famous map, he incorporated most of the previous exploration carried out by the company, as well as cartographic production of the period from 1778 to the autumn of 1794. Turnor also took advantage of other map resources available at the time – information from explorers and cartographers outside the company's sphere. Thus the 1794 map is a composite of information, of his own maps and calculations as well as those of others, fitted onto a graticule that was the most precise provided to that date for the region, a graticule that he himself had done so much to perfect. The map extends from the east coast of Hudson and James bays to the Pacific coast, and from the Arctic Ocean at the mouth of the Mackenzie River to the upper Missouri River and the Lake Superior region. (For the area essentially known by 1795, see figure 4).

OTHER COMPANY SOURCES OF INFORMATION
FOR TURNOR'S 1794 MAP

Mapping in the Moose and Albany Regions

Considerable exploration and post-building occurred in the tributary regions of the Moose River, based on Frederick House in the east, and old and new Brunswick houses in the western tributaries. Only two maps were prepared, as far as is known, neither of which is now available in the company archives. George Donald had examined the Harricanaw River in 1788, along with the upper portions of the Mattagami River, including both lakes Kenogamissi and Mattagami. In spite of his surveying and drafting experience, the records do not indicate that he prepared any maps. Frederick House was also the starting-point from which a number of searches of the larger region were undertaken by John Mannall, who came to the Moose region to be a writer and was assigned eventually to Frederick House. In addition to local trips, his two main expeditions out of the Abitibi Lake area, south towards Lake Timiskaming and west into the Mattagami River valley, gave him the information for a map on which these journeys were outlined (63c). For several years, the Frederick House Lake area was visited by Canadian traders from the Soweawaminica settlement which lay south of Lake Abitibi on the way to Lake Timiskaming. These traders were interested in coming north to the district to compete for the trade. In 1793, Mannall decided to map the route between the two settlements, and was away from 13 to 23 June. After conferring with Chief Factor Thomas at Moose about the competitive situation in the region, Mannall set off from Frederick House in the spring, 28 April to 15 May 1794, across country to survey the rivers and lakes west to the Mattagami River and to Lake Kenogamissi, and to decide how to "most effectually counteract the Designs of the Canadians" (B135/b/23,fo.48d). On his return, Mannall was sent off to head a house-building crew, going back to Lake Kenogamissi to erect a post there. Both areas, south and west of Abitibi, were included in his map. Possibly other regional details were included.

To the west, competition with the Canadian traders was centering in the upper headwaters of the western tributaries of the Moose River and in the Lake Superior watershed. By 1788, it had become apparent that a new company post was needed in this region. Brunswick House was only about one-third of the way to the head of the Missinaibi River. The site chosen was at Micabanish Lake, about two hundred miles farther south, and the company started building a new post at this lake in the summer of 1788. Then Brunswick House was closed in 1791 and supplies were removed to Micabanish, which became known as New Brunswick House. But the company was convinced another post was essential somewhere to the south of this new house and nearer Lake Superior in order to compete with the Canadian traders, and Philip Good, a writer at Moose, was sent to investigate. Good was furnished with a sextant and the necessary books on his voyage out to Moose Factory in 1790. George Donald, then chief pro tem at Moose, was directed in the official letter to have Good practice taking observations for latitude and longitude and to have his calculations sent back to London so his progress could be checked. In 1791, Good was transferred to New Brunswick House to act as writer and to make astronomical observations. He was reported to have behaved well in his job but to be deficient in his understanding of astronomy. In May 1792, Thomas, chief at Moose Factory, instructed the master at New Brunswick House to send Good out to survey the country beyond the post and to "chart his journey" (B135/b/23,fo.18d). Because hiring guides proved difficult, Good did not leave until 6 June 1793. He returned to the post on 2 July after having surveyed "ye Picque River" (B145/a/6,fo.51). By 12 July he had finished drawing his map, which was sent off to Moose Fort later (64c). Good apologized for his draft, saying that it was not "laid down in so compleat a manner" as he "could wish," because he "had not the instruments for that purpose" (B135/b/23,fo.55).[15] The country covered on his map lay slightly southwest of New Brunswick House, across to Lake Kabinakagami, on to the shore of Lake Superior, and the lower course of the Pic River. It would also have shown such a complicated and difficult route, marked by eighty-two portages, that the company declared it useless. Missinaibi House, erected later on Missinaibi Lake, and the post built on Abitibi Lake, completed the chain of posts built inland from Moose Factory during this period.

While Turnor and these other men had been occupied in delineating some elements of the region to the south of Moose Factory, there were a number of attempts to extend company influence into the upper Albany River basin and adjacent areas, and particularly to find a route to Lake St. Anns (Nipigon). It was hoped at first that the trade from this lake could be attached to Gloucester House, but later Albany officials concluded that the better route might lie along the Albany south branch, the Kenogami, with the connection being through Henley House. Both possibilities were investigated between 1784 and 1791. In 1784, James Sutherland, after assessing Piskocoggan (Pashkokogan) Lake as a possible site for a new post, set out southwest from Gloucester House to search for Lake St. Anns with the hope that it would provide a more advantageous location (B78/a/11,fo.7d).[16] His party made the return journey from St. Anns' north shore, apparently following the outward path. Because Sutherland found the connections between the lake and Gloucester too forbidding, he could not advise that Lake St. Anns be attempted from Gloucester (B78/a/9,fo.26). Moreover, he had not met with a single Indian on the route he followed. Sutherland drafted a map in the late summer of 1784 (50c), which he transmitted to London that autumn with his journals.[17] In the spring, the committee expressed its pleasure at receiving the journals and the "draught" (A5/2,fo.123).

In the summer of 1786, James Sutherland was again chosen to search the region farther to the west than he had reached previously, and especially to go into the Monotogga country (Monotoggy, Monatai). This lay to the south towards Rainy Lake, through which he intended to swing on a route back toward the east. An Indian, who had promised to guide him, refused to do so when the group reached Lake Upishingunga (Lac Seul) in part because of "Panic fear" of the smallpox raging inland. He was therefore obliged to return via the same route. In his journal, he penned in on an attached sheet a simple outline of the complicated river, lake, and island-strewn waterway, from Lake St. Joseph to Lake Saul (Lac Seul) (31ᴬ).¹⁸ The project to examine the Albany south branch, the Kenogami, as a more suitable pathway to Lake St. Anns, did not fare well at first. Notwithstanding the fact that several company men were willing to take part in this venture, no Indian guides could be persuaded to lead a party. Finally, in 1788, a young writer at Albany, James Hudson, was chosen as leader. Two young Indians agreed to act as guides just a few days before he left. However, the guides refused to lead the party on the final leg of the trip, and Hudson had to stop short of the goal (B86/a/42).

Whether Hudson had the time in the four days he spent at Henley to draft his map (35ᴬ), or whether he did it at Albany upon his return, using the better facilities there, he had finished it by sailing time, for Edward Jarvis referred the committee to it in his official letter. He warned the members that distances on the map were exaggerated, but at the same time he commended Hudson, saying, "it is a pretty preface to a better performance … he will do the next better" (A11/5,fo.101d).¹⁹

In May 1788, the committee sent John Hodgson a letter, asking him to draft a "Chart of James Sutherlands Inland Journey" (A5/2,fo.183). The letter suggested he could easily do it by working out the details from Sutherland's journal, using the magnetic declination of Henley House if he did not have it for the areas to the west. The committee also stipulated that if any other journeys had been made in the meantime he was to include information from them. The members wanted the compilation finished in time for the ship's sailing in the autumn of 1789. Hodgson replied immediately by the 1788 ship that he would gladly do the map and would add James Hudson's "small excursion up Frenchman's (Ogoki) River" (A11/5,fo.89). Both in the Henley House journal on 19 August 1789, and in a letter of 4 September 1789, Hodgson confirmed that he had drawn the map "as well as circumstances would admit" (36ᴬ) (plate 13) (A11/5,fo.109). He was referring to the loss of all his instruments and books when Henley House was destroyed by fire, a loss which forced him to use a very old book of logarithms to work out the longitudes. The company thanked Hodgson the next year for the "Chart of Sutherland and Hudson Journeys" (A5/3,fo.42d), and in return the committee received a courtesy note from him on the return of the ship (A11/5,fo.131). This map is one of the best documented in the company's archives, except that Hodgson, as usual, did not place his name on the map. But the map he prepared is assuredly the "Chart of Rivers and Lakes Communicating with Albany River," for it contains the details of Sutherland's three journeys, from Henley House to Piskocoggan Lake in 1784, to Lake St. Anns in 1784, and from Piskocoggan to Lake Upishingunga in 1786. It includes James Hudson's route from Henley to Lake La Puew (Ogoki) and his return in 1788. Apparently, copies of Sutherland's and Hudson's maps were not at hand, for the configurations of the lakes are not similar to those on the extant maps. Hodgson, as could be expected, generalized their outlines on his smaller-scale map.

The urge to reach Lake St. Anns using the south branch of the Albany continued. Richard Perkin, a young gun maker of the Albany establishment and "an intelligent man" (B3/b/26,fo.24), according to Jarvis, volunteered to go inland with Hodgson who was organizing a new expedition. Hodgson was successful in hiring Indian guides to go with Perkin, and they set off on 22 May 1790. Perkin was back at Henley in one month exactly, claiming to have set foot on the shores of Lake St. Anns. Hodgson took Perkin's journal details and worked out a sketch map (53ᶜ). He believed the large lake described at the conclusion of Perkin's outward journey was Lake St. Anns and called it that on this map. Edward Jarvis voiced the general suspicion that Perkin was mistaken, that he must have been deceived by his guides (A11/5,fo.173d). He showed the lake to be part of the Hudson Bay drainage basin whereas Sutherland had claimed to have run *down* a river to get to the lake. In September 1791, Hodgson agreed that Perkin had been misled, and another party was sent out. Lake St. Anns was finally reached via the Kenogami River that same year when John McKay arrived at Gloucester House, via the Kenogami River from the lake. McKay, it can be surmised, had used one of the western tributaries of the Kenogami to reach St. Anns and perhaps had followed James Sutherland's route back to Gloucester House. Both McKay and his brother Donald had come into company service in the autumn of 1790 in Albany territory, coming north through the route from Sturgeon Lake to Osnaburgh and then down to Albany Factory. They had been Canadian traders previously and were familiar with Lake St. Anns, Lake Superior, Rainy Lake, Lake of the Woods, and the Winnipeg River region: There is only one piece of evidence that a map of John McKay's journey was drawn. This is found in John Hodgson's 17 September 1791 letter to the committee. He wrote about both McKay's expedition and that of John Knowles (who had tried to reach Lake Mepiskawaucau that same summer), noting that "their Remarks and draughts of their Journeys … were transmitted to your Honours" (55ᶜ, 56ᶜ). Whether the draftsmen of the McKay expedition map was McKay or Hodgson is impossible to determine absolutely, but it was quite likely the latter. He had had contact with McKay and seen the journal of his journey. Moreover, Hodgson was accustomed to drawing maps from verbal and written "Remarks." Chief Factor Jarvis had harsh words for Knowles's journal and map. The first he classed as a "heap of absurdities and nonsense," since Knowles had not only never reached his goal, Mepiskawaucau, but had written an essentially spurious journal. Jarvis exclaimed that Knowles had also handed him "a Map! – it is a serious thing but it is hardly possible to be so in perusing it" (A11/5,fo.174d). The map depicted the water running "from Lake Superior to Albany"! This sketch does not appear to have survived – it was likely not sent to London by Jarvis, or if it was, may have been deliberately expunged from the files. Jarvis benevolently gave Knowles a chance to redeem himself, since the man seems to have been a little ashamed of himself. The result was that in the early summer of 1792, Knowles reached Lake St. Anns from Albany. Unfortunately, Knowle's journal does not also make it easy to place the position of Lake Mepiskawaucau. It may well be the modern Long Lake at the source of the Kenogami River.

By 1786 or 1787 a number of Canadian traders were coming regularly into lakes lying in a semi-circle to the northwest, west, and south of Osnaburgh House, and company officers were undecided about the best geographic location for a post or posts to oppose them. The choice of sites had been narrowed down to Cat, Trout, or Seul lakes. Cat Lake to the northwest of Osnaburgh was chosen first, and in August 1788, John Best opened a house there. As an aid to understanding the hinterland to the north of Cat Lake House, Richard Perkin was sent in June 1789 to reconnoitre a stretch of country from the lake as far as Severn Fort. The details were made into a map by Peter Fidler at a much later date, 1815 (165A), but there is no indication that Perkin worked out a sketch of his own. By the end of 1790, cabins had been built also at Lac Seul and at Red Lake, which are about due west of Osnaburgh House. These acted somewhat as shields, but also as outlying feeders to Osnaburgh. The relationship of these cabins was depicted clearly on a map prepared by John Best, who was an active post master in this region for some years (51A). It indicates the rivers and lakes followed by the traders and the Indians between the main house and its outlying posts. However, there was no protective post to the south to stem the inroads of the Canadian trade there. Jacob Corrigal, an experienced trader, was sent to look for a route from Osnaburgh south to Sturgeon Lake. During his journey, from 16 May to 29 May 1794, he examined the lake for a potential site and on his return drew a sketch of the route between these two termini to accompany his journal description. (55A).

By 1790, the company was well on the way to reaching those "principal objects," described in 1777, for inland movement from Albany, namely, the areas of Rainy Lake, Lake of the Woods, and the Manitoba lakes. The years 1791 to 1795 witnessed the culmination of the movement west from the Albany corridor into the Manitoba lakes country and beyond into the Assiniboine River and Swan Lake regions. Accordingly, the company, wanting to realign its trade in this complex area, needed a better understanding of its geography. Edward Jarvis was a prime mover toward such a realignment and better defined procedures. Fortunately, he found a strong supporter and a willing agent in Donald McKay.

MAPPING THE ALBANY CORRIDOR INTO MANITOBA LAKES COUNTRY

McKay's trading plan was a very direct one: go straight to the centre of the Canadians' trade, cutting right across their main supply route, that is, to the main portage on the Winnipeg River route that led into the Red-Assiniboine valley. From the time of McKay's arrival at Albany on 11 October 1790, when he met Jarvis, until his departure for the interior the following January, the two discussed at length the proposal to be made to the committee, the plans for the next season, and the map or maps of the region that ought to be forthcoming (A11/5,fos.123–124d,125,167d). The main features of their plan were: in 1791, a small group would travel from Red Lake post to the Lake of the Woods, via Portage de l'Isle, in the vicinity of the junction of the Winnipeg and English rivers. Indians were to be informed that a house would be built and stocked at Portage. Then, once based both at Osnaburgh and Portage de l'Isle, the company could quickly extend itself to the Red and Assiniboine rivers. All this, said McKay, "cannot be done in one year or either in two, as it ought to be carried on, stept by stept with Security and prudence" (A11/5,fo.124d). The first prudent step was taken on schedule in 1791. After a winter march from Albany to Red Lake House, McKay, with three companions, including John Sutherland as second in command, set off further inland on 13 May 1791, reached Portage de l'Isle, went on into the Lake of the Woods area, and was back at Gloucester House by 8 July. One of McKay's tasks had been to suggest suitable post locations for the future trading expansion to the west. When he reported that the strategic position of Portage de l'isle had become even more apparent to him on this visit, a decision to build a house there was made.

In the autumn 1791 packet, Edward Jarvis enclosed two maps (A11/5,fos.173d–174,175d). The first was "The Genl: Map" (37A) (plate 14) which "tho' in most it cannot have pretence to geographical accuracy, will serve to point out the connexion between your interior settlements above Albany, and those above York" (A11/5,fo.173d). The second map (57c) was "a proper appendix, to the general one, and points out more clearly than we knew before, the rout of the Canadians thro' Lake St. Anns to their several posts in the Little North, and the communication of Osnaburgh with Portage de l'isle" (A11/5,fo.175d).[20] The two maps seem to have been prepared in concert. The details on the second, the McKay chart, appear on a smaller scale and in a more generalized form on the general map by McKay and Jarvis. Moreover, both maps appear to have been planned and likely drafted prior to McKay's summer 1791 expedition, except for one small item of information added later.

The general map shows territory extending from the Eastmain, west and north to the Athabasca and Great Slave lakes, into the North Saskatchewan River area, the Assiniboine-Red region, the upper Great Lakes and the Ottawa River down to Montreal. The most detailed configurations are to be found in the Albany and Moose River basins, the Grand Portage to Lake of the Woods, and along the Winnipeg River. These latter elements are central to the purpose of the map, but have been set into the larger framework to give an overview of the company's and Canadians' trade areas. There are six references to McKay; the details included are related closely to McKay's trading career and the policies for expansion he recommended. A 1791 date can be assigned to it because all of the McKay references precede his 1791 and later activities; all of the posts in the Albany area opened to the winter of 1790–1 are included; and none erected in the period 1792–4 are shown. The details of the Slave Lake area are based on one of Peter Pond's maps of 1785, which would have been available to the company by 1790–1. All of the waterways shown west of the bay come from company maps and journals, available, very likely, in copies at Albany. Although it is unsigned, detailed examination of the maps and records shows that both McKay and Jarvis worked on the map, with the drafting likely done by Jarvis, who was already experienced in map drawing. It does not appear from cartographic evidence to have been the work of John Hodgson, who had had little contact with McKay during this time. Nor does the Henley journal make any reference to Hodgson doing any drafting at this period. Probably both the general map and the McKay chart were finished at least in rough during the time between the arrival of McKay and his departure on his first company journey. Because he was very familiar with this region, he had most of the details already in mind, including the 1791 spring route, when he arrived at Albany in 1790. The one item of information which refers to a later action, is the notation, "Mr. Sutherlands House," added to the map in the Portage de l'Isle vicinity. McKay had approved the site on the spring journey and James Sutherland had been chosen to carry out the building of a house later. Jarvis likely entered the note when the final touches were put on it and before it was sealed in the packet, as a reference to an action that was anticipated. This map, undoubtedly the "Genl: Map" enclosed by Jarvis, is significant as the Hudson's Bay Company's most detailed, small-scale, general map of the Canadian interior produced by this date. It was a prime source for Philip Turnor in the compilation of his final manuscript map of 1794, and for Aaron Arrowsmith when he produced his historic map of British America in 1795.

What was the McKay chart like? Jarvis's cryptic comment suggests that it extended at least from Lake Nipigon west through the Rainy Lake-Lake of the Woods area, north to the Albany basin at Osnaburgh, and into the Winnipeg River area, including the Portage de l'Isle site; that is, the area which was the prime goal of the Albany program. Thus, the logical conclusion is that the chart was similar to this limited area as it was shown on the general map, only the chart was at a larger scale and in more detail. Another map prepared about this time, and delineating the same general area is "An Accurate Map of the Territories of the Hudson's Bay Company in North America" (38A) (plate 15).[21] This is not the "McKay Chart" since it is not greatly different in scale from the general map, and there is absolutely no similarity in cartographic style or in the depiction of various features. The Lake Winnipeg area on the "Accurate Map" is oriented east-west; the Red-Assiniboine river system is not shown. Donald McKay was too familiar with this region to admit of either; evidently the "chart" has not survived.

What then of the "Accurate Map"? It has no date or author on its face. It is a fair copy in ink, with colour added. The map would have been prepared after 1786, because it includes Osnaburgh House, established that year. The earliest date for its execution was probably later. On 12 August 1790, while on his journey to Red Lake to the new house site there, James Sutherland recorded that he had named one of the lakes en route, Prince of Wales Lake, that day being the Prince's birthday. On this map, the name is applied to a lake near Red Lake, and the same name is applied to the same lake on the McKay-Jarvis general map, which was sent to London in the autumn of 1791. Sutherland was back at Osnaburgh House from his Red Lake venture on 12 June 1791, bringing his journal with him. Moreover, this map includes the details of the waterways between Lac Seul and Portage de l'Isle and Lake of the Woods, details that were available from McKay's journal of his spring-summer 1791 trip, after his arrival at Gloucester House on 22 July. It is difficult to determine the latest possible drafting date, but another piece of evidence, which may be crucial, is that Martin's Falls House is not named on the map. It was built in the autumn and winter of 1792.

Therefore, the "Accurate Map" was probably drafted in late 1791, using a variety of sources available at Albany Factory. The special purpose of the map may be related to the fact that the river and lake connections of the area beyond Osnaburgh House and west of Lake Nipigon to the Winnipeg River are outlined with a red ink wash. All other interior waters are in blue, while the seashores are in green. The character of the map detail and colours suggest the cartographer wanted to show the major routes discovered and used in the red-marked region, the newest portion of the company's trading area in the south. The details marked in red do not correspond in shape to the same features on the general map. As to the cartography, it can hardly be ascribed to Jarvis, for it is so unlike the general map in character. The two likely contenders are John Hodgson and James Hudson, both of whom had been preparing maps in this district, both of whom were at Albany in the late summer of 1791. Of these two, Hodgson is more likely the author of The "Accurate Map." He had much more cartographic experience than Hudson and had compiled maps earlier.[22]

Fidler Sends Information on the High Plains

While Turnor was on his way home in 1792, Peter Fidler was being reintroduced to the high plains of the west. Buckingham House was under construction in the autumn higher up the North Saskatchewan River when Fidler reached the site, but he did not stay long before striking southwest by horse to the Bow and High river area at the edge of the foothills of the Rockies. After returning to the new post in spring, 1793, he was instructed to survey farther up the North Saskatchewan and to advise on the most eligible site for the next house, beyond the posts of the Canadian traders who had been passing Buckingham bound upstream. The details of this shorter walking trip as far as the Sturgeon River tributary, along with the information Fidler gathered on the southwest plains, were sent to London and used by Turnor on his major map.

Mapping by David Thompson

David Thompson, very disappointed that he was not able to be a member of the Athabasca expedition, wrote of his mishap to the governor and committee. He told them of the observations he had made inland and while convalescing, and said that he had made rough drafts from his observations on the trip from Cumberland to York, via the Knee Lake to Hayes River route. He promised to send the fair copy to them the following year. He also wondered whether, in lieu of the set of clothing which he was to receive in 1791 upon completion of his apprenticeship, the committee would instead ship him certain instruments which he listed (A11/117,fo.54).[23] Naturally, the committee was delighted with Thompson's attitude and activities, and it shipped out all this equipment as a gift, without deducting any clothing cost. Thompson was assured that if he continued to make himself "useful" to the company there would be further "Encouragement"; he was told that the committee looked forward to receiving his map (A5/3,fos.64d,65). Thompson included the promised map, (58^C) in his letter of 19 September 1791, to London (A11/117,fo.109d).[24] At this point, his apprenticeship ended and he was appointed a writer.

In the committee's reply, the map is not mentioned directly, although doubtless it was understood to be part of the reason for the general acknowledgement of his worth, which the letter contained. The map is not mentioned in the catalogues and does not appear to have survived. It may well appear in a reduced form in a later map of 1794. This large map, undated and unsigned, which depicts Thompson's surveys from 1790 through 1794, was also probably drafted by him (56^A) (plate 17). In any event it was completed before ship departure time in the late summer of 1794.[25] In 1793, after his leg had healed, Thompson was sent into the Rat Country, the area north of the Nelson-Saskatchewan route, to examine the lakes and rivers and their interconnections. He continued into the Burntwood River country, an area essentially unknown to the company, then on to the Churchill system as far as Reindeer River. It is not clear whether he ascended this waterway any distance before turning back to spend the autumn and winter, 1793–4, at Buckingham House. Since all of the map's features appear as small-scale generalizations on Turnor's final composite map of 1794, Thompson would have had to complete it by ship departure time from York in early September.

TURNOR'S FINAL 1794 MAP

The information David Thompson entered on his map would not have reached London until ship arrival time in November 1794. It is revealing that on the 1794 Turnor map (57^A) (plate 18), the only area in which extensive erasure markings are found is that where the details of the Thompson map are depicted: the light green ink wash, which covers all land areas on the map, has been erased. There are obvious signs of a previous delineation having been removed and new details having been drawn in. This would suggest that Turnor had completed his 1794 map by the autumn of that year, much of it based on his own surveys. But when he received the new Thompson map in November 1794, he could see that here was information that ought to be added to the map in order to make it as up-to-date as possible. So the erasures were made and the new configuration, based on Thompson's work, drawn in.

The governor and committee received Turnor's large and handsome map by the end of the first two weeks of 1795, for on 14 January they ordered the Secretary to pay Turnor £100 for his services in having explored and surveyed parts of Rupert's Land and for having "laid down the same in a large and accurate map." The company also showed its appreciation by having the watch Turnor had carried for many years "engrav'd on the outside Case" with the company's arms and formally presented to him as "a Mark of the committee's Regard for him." (A1/47,fo.47). Such regard was well deserved as this period of company mapping history may fairly be designated the Turnor period, and his 1794 map deemed one of Canada's greatest cartographic treasures.

9 Mapping to the Columbia and Behind the Eastmain, 1795–1821

> Peter Fidler ... sent home some Maps & Papers which ... convey much curious information respecting North Western Geography.
> Secretary to Alexander Dalrymple, December 1802

The year 1795 is a significant one in the cartographic history of the Hudson's Bay Company and of Canada. In that year Aaron Arrowsmith published "A Map Exhibiting all the New Discoveries in the Interior Parts of North America," largely based on Philip Turnor's final map. After this time, as a result of the company's "liberal communications," geographical information was regularly transferred from company sources to the Arrowsmith drafting tables and hence to other cartographers and the public. As the Arrowsmith family firm became unofficial compiler of regional base maps for the company, their cartographers were allowed access to the records of exploration at Hudson's Bay House itself so that they could continually update maps of the company's trading territory.

During the quarter century to 1821, the kind as well as the number of maps being produced by company servants increased, although the dominant concern was the surveying and mapping of trade routes in the western interior. Several types of mapping were introduced. The lot survey, for example, was introduced by Fidler when he mapped property lines for the Selkirk settlement. In addition, the company instituted a program for mapping its fur trading districts. More than thirty men drafted at least 147 maps, and Peter Fidler, continuing his earlier practice, drew many more segmental sketches. Fidler and James Clouston and Joseph Howse were the pre-eminent maps makers in these years. Fidler ranged widely over the west and north, describing and illustrating the routes he travelled more fully than any of his contemporaries. Clouston, erstwhile teacher, opened the country behind the Eastmain and to the south of Lake Mistassini. Howse, later distinguished student of Indian linguistics, carried the company flag deep into the Rocky Mountains and across into the Columbia River valley. Finally, several maps were produced as a result of the company's assistance to the British Admiralty on its first overland expedition into the Arctic through Rupert's Land.

PETER FIDLER'S CAREER

Upon the retirement of Turnor, Peter Fidler, then only twenty-three years of age, became the most active surveyor and map maker in the company's employ.[1] When Turnor chose Fidler as his assistant for the Athabasca expedition in 1789, Fidler had just arrived in Rupert's Land from his home in Derbyshire, England. At the age of nineteen, he had been hired by the company as a general labourer. Unlike most others employed in this capacity, he was reasonably well educated and had had sufficient mathematical training to pick up practical astronomy quickly. Also unlike most of his fellows, he augmented his practical experience by study, drawing on his own eventually extensive collection of material in mathematics, astronomy, geography, and related subjects. During his surveying and mapping career which lasted until 1820, Fidler drew more maps than any other company cartographer. The period in which he travelled and mapped most extensively began with his inaugural trip with Turnor and lasted until 1811–12 when he took a year's leave-of-absence in Britain. All of the 373 segmental sketches most characteristic of his work were drawn during his travels in these years, but his total map production included several other types. The eight smaller-scale integrated maps of various regions (only three of which have survived), based on a combination of the sketch maps, are one type. The compilations he was to make from the sketches and journals of some earlier company travellers are another. As a district head, he also took part in preparing the regional maps of trading districts that accompanied reports for the governor and committee. He drew several other special purpose maps, including one based on his cadastral survey of the Selkirk settlement. Although he did not prepare coastal and harbour charts in particular, he did show on several maps the coast of Hudson Bay from the Severn River to the Churchill River, based on observations he had taken while walking along the shore. And he prepared two large-scale shoreline sketches that showed the vicinity of Churchill. Most representative of his continuing search for geographical information, were the forty-four maps, some in two or three sections, that resulted from his years of contact with the Indians and with colleagues in the trade. He regularly asked for descriptions and sketches of areas and routes known to them and would then transcribe the information and sketches into his own map journals. In most cases he was scrupulous in identifying his informants and noting the dates the transmission took place.

Fidler's field notes, taken rapidly and continually while in transit, illustrate his surveying and mapping methods and the completeness of his observations. He recorded the results of his observations in his journal and transferred the details into a series of annotated sketches (plates 16 and 21). Along rivers he took compass readings for every change in the angle of the water course, judging distances along each of the segments, both by dead reckoning and by general sight estimation. He plotted shore outlines and indicated island locations. Rapids, water falls, and portages were identified by symbols; the lengths of the portages were noted, and areas with strong currents remarked upon. Along lake shores he did a considerable amount of simple triangulation. Sights were taken to major shore-line features, the distances estimated, and the configuration of the coast and the location of islands noted. He inserted descriptions of the terrain, of vegetation and of distinguishing land marks: "rather bold land," "low flat shoal swampy & wet scrub pine and juniper," and "high pine Id. [Island] a good mark." When crossing the plains he did not draft such sketches but kept a careful log of directions and distance estimations, although he might sketch the outlines of lakes he passed in his wanderings.

Fidler is remarkable not only for the volume of his output, but for its quality. The careful measurements, which provided an expanding network of more precise locations for the map of British America, were testimony that he was an apt pupil of Philip Turnor. His maps were explicit in the conformation of rivers and lakes. His influence is seen in the work of several company map makers, most notably George Taylor Jr, who emulated his field and drafting technique. Although Fidler's maps provide examples of the various styles being used by other explorers and map makers at the time, he was less interested than some of them in the cartographic niceties of colour, typography, title, and cartouche design.

Fidler's Map of the High Plains

The first of Fidler's integrated maps was described as a "Map of his Journey to the Stony Mountains with the River Saskatchewan" (65^C). It was sent to London in 1795 but arrived too late for all the details to appear on Arrowsmith's 1795 map. However, before the 1 January 1795 publication date, Arrowsmith had seen Fidler's 1792–3 journal and had

Map of Swan River and Lake

As described earlier, company traders from York and Albany entered the large region just to the west of the Manitoba lakes after 1790. From Albany, the expansion of trade began when Donald McKay prepared to cut across the Canadians' main trade route. By 1795, the Albany parties had constructed Brandon House at the junction of the Assiniboine River with the Souris and had opened two cabins up the Assiniboine beyond the Qu'Appelle River tributary. In the autumn of 1795, Charles Isham led a York Factory party, which included Peter Fidler, from Somerset House on the Swan River and opened Charlton House on the upper Assiniboine River.[2] The map of the Swan River and Lake district in the archives is not dated nor is the cartographer named, but the date is likely 1795 or 1796, and evidence clearly indicates that it was drawn by Peter Fidler (59[A]).

In his journal Fidler indicated that he had entered the area from Cedar Lake, arriving at Swampy portage and crossing to Lake Winnipegosis on 3 October 1795. He surveyed the full length of the portage, measuring the distance as was his practice, with a rope or cord, one-sixth of a mile in length; he found the crossing to be just over four miles long (E3/2,fo.39d).[3] Fidler remained with the party through the winter, trading out of Charlton House, before he returned to Cumberland House the following spring. His map depicts Gods (Good Spirit) Lake, which he mentioned in his journal, where he also pointed out that both the Red Deer River and Swan River emanated from Eetow wemammis Lake (although this lake is not easily identifiable in the region). He uses the name Red River for the Assiniboine; Thunder Hill, shown as being located near Somerset House, is also mentioned in his journal. The map is very much akin to Fidler's later maps, and it is possible that he drew it at Charlton House in the winter of 1795–6.[4] Charlton and Marlborough houses are both located on it. In the York general letter of 31 May 1797, the company chided Fidler for not having sent any of his journals to London for two years past, i.e. in 1795 and 1796, and for not providing remarks relating to a map or maps which he had transmitted to them. If he did not attend to their request in 1797, "Measures will be taken accordingly," they warned (A6/16,fo.44). The Swan Lake map, which must have been sent to London on the 1796 ship, would surely be one of those referred to for a replica of it appeared in small scale on Arrowsmith's 1796 edition of his new map of the north of the continent.

Mapping in the Beaver River Region

From 1796 to 1799, Fidler was concerned principally with management and trade at Buckingham and Cumberland Houses and very little with surveying and map making.[5] In the summer of 1799 he was transferred to the Beaver River region, an area he had previously only skirted. It included the wooded country lying athwart the Athabasca River from Lac la Biche to Lesser Slave Lake. This secondment lasted for less than a year, but time enough for two houses, Bolsover House at Barren Ground (Meadow) Lake and Greenwich on Red Deer's Lake (Lac La Biche) to be established. Fidler drew fifteen sketches in total, from the time he reached Meadow Lake until he arrived at Buckingham House on his way out, and compiled a map of the "Beaver & Athapescow rivers," based on his sketches and a considerable body of detailed observations and locational calculations (67[C]).[6] Likely, this map included also the area as far as Edmonton and Buckingham houses.

The farthest west Fidler travelled was the junction of the Lesser Slave River with the Athabasca River, a journey he made in early 1800. His purpose was outlined in a letter written the previous autumn to James Bird at Edmonton House (B104/a/1,fo.34d). A party of Canadian traders had left Lac la Biche to go to Slave Indian (Lesser Slave) Lake to erect a post. Fidler had hoped to follow and build beside them, he told Bird, but his plan was foiled because of lack of provisions and the unwillingness of his men to go along. However, in January another opportunity arose when three Canadian traders passed en route to the lake. Fidler decided to follow them with a small party in order to become acquainted with the route as well as to survey it for future reference. Hauling sleighs over the frozen lake, his group reached the bank of the Athabasca River. They then turned upstream to the mouth of the Slave Indian (Lesser Slave) River, which drains Lesser Slave Lake, where a Canadian post had been opened. Fidler did not have time to go farther, but from conversations with the traders and a glimpse of the lake from a vantage point at the top of a tall pine tree on the high bank of the Athabasca River, he was able to piece together a mental map and then draw a sketch of this large body of water and its connecting river lying to the west of his position (72[A]). The party then hurried back to Greenwich House, with Fidler taking time to record the distances covered. He estimated that they were able to make about forty miles per day and that the total distance from the mouth of the Lesser Slave River to Greenwich House was approximately 178 miles, although it would have been somewhat larger if they had not cut across some "points of land." The details of his sojourn in this region appeared later on Arrowsmith's maps.

Fidler on the South Saskatchewan: Indian Maps of the Plains

By the time Fidler had passed down the Saskatchewan to Cumberland House in May 1800, plans had been made to build a post on the grassland plains. It had become necessary for the company to extend its own pemmican collection into the heart of this region because the Indians of this area were not willing to bring it north themselves. It was not a rich fur area, although some furs were obtained, especially fox. No company employees were known to have traversed the South Saskatchewan much beyond South Branch House and the course of the river was unmapped, except for segments of its upper tributaries.[7] Fidler was the suitable choice to carry the post-building plan into effect. As well as being an experienced trader and surveyor, he had lived among the Piegans and spoke their language, and he was familiar with the environment of the high plains. He entered new territory above the abandoned South Branch House in the third week of August 1800 and reached the site chosen for the new post, Chesterfield, adjacent to the confluence of the Red Deer River and the South Saskatchewan. He left the district in the spring of 1802, arriving back at York Fort on 23 June. During the open season of 1801, he had taken the returns to Cumberland and brought back new provisions and trade goods to Chesterfield House; by this time Fidler would have viewed the South Branch of the Saskatchewan on four passages up and down the river and had a fairly detailed knowledge of its channel.[8] Although during his stay at the post he travelled no more than a few miles west, he was visited by hundreds of natives and in the process perfected his technique for asking questions that provided geographically useful answers. He was also able to persuade some Indians to sketch out maps for him on which he recorded the Indian toponymy. Thus, far out in the dry grass lands, he was able to fashion a picture of the high western plains and of the Cordillera.

Fidler transcribed nine maps based on Indian information, one small scale composite map of the far western plains (70C), and one segmental sketch during his two-year sojourn (76A). The nine Indian maps covered an immense area from the upper Missouri River tributaries north into the Red Deer and Battle River region and west past the long rampart of the Rocky Mountains. Four of the nine were essentially similar sketches, based on two maps originally drawn by Ac ko mok ki, or the Feathers, a Blackfoot chief, one on 7 February 1801 (77A) (plate 19), and the other a year later in February (82A). The original 1801 sketch was also transcribed in a copy by Fidler in 1801 (78A), and the 1802 sketch in 1802 (81A). Two similar sketches were prepared from an original sketch by Ak ko wee ak, a Blackfoot Indian in 1802 (83A, 84A). Two more sketches were transcribed from the original material of Ki oo cus or the Little Bear, a Blackfoot chief in 1802 (85A, 86A) (plate 20). A further map used unidentified Indian information, its content indicating that it was also drawn in 1802 (80A).

Fidler's four copies of Ac ko mok ki's original materials are oriented to the west, and all depict the Rocky Mountains extending across the map in a straight line from north to south. Two of the maps show the great plains only to the Rocky Mountain ranges (81A, 82A); the other two extend beyond the mountain line, one to the Pacific coast (77A), and the other (78A), some distance beyond the Rockies. Fidler's two 1801 maps, based on Ac ko mok ki's, depict the country farther to the south than do the 1802 maps, that is, they show the area south of the Big Horn River tributary of the Missouri. The two 1802 maps reach only into the more northern Missouri tributaries or as far south as the "warm water river," probably the Judith River in Montana, north of the Yellowstone River. The two 1801 copies include the courses of many rivers with their Indian place names, the locations and names of thirty-two Indian groups with the numbers of tents in each group, the Indian toponyms for many hill and mountain features, and the distances between Rocky Mountain landmarks expressed as the "Number of Nights the Inds. Sleep in going from one place to the other."

Ac ko wee ak was less ambitious in the coverage and detail of his map than his fellow Blackfoot tribesman. Neither were the details of the Ki oo cus contribution as comprehensive. But the two 1802 copies based on Ki oo cus's sketch give a more accurate picture of the general vegetation and of the terrain forms of this immense land than those derived from the sketches of the two other Indians. What the Ki oo cus map depicts, in addition to river features, is the "woods edge" starting in the north east in the vicinity of the lower course of the South Saskatchewan and middle course of the Battle River, that is, the break between the more wooded park belt and the more open grasslands. This boundary is shown as extending west and southwest to the Rocky Mountain foothills just to the north of where the Red (Bow) River flows out onto the high plains. The Rocky Mountains extend unbroken south to the upper Missouri River, where the mountains trend south and east to form the great bow of the Absaroka Range and Big Horn Mountains, transected by the Red Deer (Yellowstone) River. Across the high plains are traced out various Indian tracks; distances are represented by a symbol for each night's "sleep."

The last transcribed map (80A) in this series is more restricted: it portrays the locations and tent numbers of Indian groups in the Missouri River region, with a few to the west of the mountain line, although it also purports to show some "Spanish settlements" far to the southwest. There is a possibility that this sketch was a trial by Fidler himself to work out a population pattern; his data came from Indian sources, but the map base itself was not necessarily from an Indian original. His transcription showing Indian encampments, population numbers, routes of movement, and natural characteristics other than waterways, was unusual among Hudson's Bay Company maps. Graham's 1774 map of Indian tribal boundaries was the only previous example. Later, when trade districts had been reorganized and district reports and maps expected from the district officers, the committee became more interested in such information.

Elements of some of these maps were integrated with other features of western topography in the final composite map Fidler drew. At Oxford House on 10 July 1802, he wrote a remarkable letter to the governor and committee concerning western geography, which he enclosed with two maps, one a composite map, drawn in 1802 (70^C), and the other (77^A), one of the sketches based on Ac ko mok ki's map.[9] The committee turned the three items over to Aaron Arrowsmith since the members believed that these documents threw "considerable light, on the North Western Geography" (A5/4,fo.104). The committee also informed Alexander Dalrymple and Sir Joseph Banks, President of the Royal Society, that these materials were sent to Arrowsmith's cartographic offices because "these Discoveries should be of sufficient Importance to attract ... their notice" (A5/4,fos.103d,104). Fidler's letter to the governor and committee articulated in a refreshing manner both his understanding of and his speculations about the physical and human geography of this vast region. Apparently, only the copy of the Indian map was returned to the company's office. The disappearance of the composite map was an unfortunate loss because of its significance to the history of western Canadian cartography. Along with the Indian sketch, it was used to flesh out the depiction of this large region by Arrowsmith in his 1802 publication. However, for some reason Arrowsmith did not completely draft the course of the South Saskatchewan River between Chesterfield House and the former site of South Branch House, showing it instead as only as a generalized dashed line. The composite map must have been at a fairly large scale for it was drafted on six sheets and had an extra sheet-and-a-half annexed to it with a depiction of the Missouri River and tributaries, based on Indian information. It would have repeated the details of Fidler's map of his 1792–3 journey, including the plains from the Stony Mountains in the vicinity of the Bow and High Rivers, across to the North Saskatchewan River area at Buckingham House, and added his new information about the South Saskatchewan. He also would have included the "wood's edge," as he found it. Fidler reported that the place names were in Blackfoot, but that he had included English translations when possible. For many years afterwards, Fidler's geographical configurations were accepted as the model for representations of the northern grasslands.

Fider Returns Twice to the Athabasca Region

From the autumn of 1802 when he was thirty-three to the mid-summer of 1806, Peter Fidler passed through trying times in the Athabasca region. The Canadian traders were firmly ensconced in the upper Churchill River system and north of Methy Portage, and the company had to depend on its most stalwart traders to confront the burgeoning competition. The governor and committee were still uncertain as to the best route to take across the Mackenzie watershed. The path ordinarily used led across Methy Portage to the Clearwater and down the Athabasca River; an alternative route, discovered by David Thompson for the company, went through Reindeer and Wollaston lakes to the east end of Lake Athabasca. Thompson had not prepared a map of this route, having refused to send the committee any cartographic information for about two years before his resignation in 1797.[10] In 1807 the company decided to develop this more easterly route. But it had to be carefully assessed, and Peter Fidler was chosen to carry out the surveying and mapping necessary. Provided with an abridged version of Thompson's original journal notes, he left his family at Cumberland House and with a survey party of four embarked on 2 June 1807 to provide the committee with his appraisal of the Reindeer Lake route to Lake Athabasca as compared to that across Methy Portage. In addition, he was to survey the length of the Churchill system from the mouth of the Reindeer River to Hudson Bay. Sixty-three days later on 4 August the party arrived at Churchill Fort. Chief Factor William Auld had high praise for the men and their accomplishment, as, at a later time, the committee did also. Altogether, 147 detailed sketches of the journey were prepared in Fidler's usual style (100^A). They illustrated a journal that recorded distance and direction and perceptive observations on local geography. Fidler also drafted two integrated maps based on this enterprise, one large and one small scale. He reported in a letter to the governor and committee, sent on the 1808 ship, that he had enclosed in a packet "a 12 Sheet Map of the Communication between ... Frog Portage and the Athapescow Lake ... also the Track from the Lower End of Deers River, down thro' the Missinnipe or Churchill River to that Factory" (73ᶜ). He explained that he had added to this configuration the Seal River to Southern Indian Lake track of Stayner and Linklater in 1794 and the 1786 track of Ross and Charles across into the Nelson River area. Fidler also included a small-scale version of this large map with the journal of his expedition (106ᴬ). It is very probable that these two maps were compiled by Fidler during the 1807–8 season when he was master at Swan River House. His sojourn there was reasonably uneventful, and he would have had enough free time to prepare the fair copy of his journal, work out all his computations, plot his courses, and complete the art work. It seems logical to assume that the multi-sheet version was turned over to Aaron Arrowsmith, for the details are inserted on his maps of British North America, at least those after 1811. Although Arrowsmith may have not have returned the multi-sheet map to the company, the smaller version, at least, remained at Hudson's Bay House.

Mapping Lake Winnipeg's Shores

Fidler drafted his last integrated map at Reindeer Lake during the winter and spring of 1810 or in the autumn of that year at Ile a la Crosse (144ᴬ). (In spite of the turmoil there as a result of fierce competition with Canadian traders, he was apparently able to attend to the map's compilation.) It was drafted in three sheets. Two of the sheets showed the Nelson River track inland from York Fort to Cumberland House; on the third Fidler added the contours of Lake Winnipeg. An annotation in pencil by an unknown hand on the Lake Winnipeg sheet reads: "East coast – may be depended upon – but the West not," an accurate assessment of the geography as portrayed on the map. In 1807–8, Fidler had been master at Swan River House, and when he was recalled to York in the early spring of 1808, he decided to take a roundabout route there in order to survey some territory new to him. He was motivated to make this detour by the visit of a private trader, John McDonald, to Swan River in the winter of 1808. McDonald provided information on which Fidler based a map of the Assiniboine River-Red River area (110ᴬ). It is made up of five sketches that together form a picture of the region from the Carrot River junction with the Saskatchewan, the Assiniboine River from its source to its junction with the Red River, and the Qu'Appelle River. This series of sketches must have been used as a road map by Fidler when, from 29 March to 1 July, he followed a circuitous path across to the upper Assiniboine River, down past Brandon House to the Red River junction, and thence to Lake Winnipeg. At this point he began his observations, recording data on the entire east shore of the lake. The eighteen sketches in his journal provide a record as far as Playgreen Lake (107ᴬ). McDonald had also given Fidler data for the Cedar and Dauphin lakes region with the connection to Lake Winnipeg (109ᴬ).

When Fidler reached York Factory from Lake Winnipeg, he was almost immediately sent back to the lake on some unspecified mission. It is probably during this time that he spent several days charting a segment of the west coast of Lake Winnipeg, from Sturgeon Bay into which the St. Martin Lake connection with Lake Manitoba enters, northward. He titled his six sketches, from "Dogs Head to the Grand Rapid" (114ᴬ) (plate 21). All this information was combined with his complete survey of the Nelson River system down to York Fort after leaving Cumberland House on 2 June 1809. This waterway from Playgreen Lake to Flamboro House, along the Nelson, was depicted in thirty-nine sketches (124ᴬ).

Fidler Transcribes Sketches by Indians and Traders During His Travels, 1801–12

Fidler also transcribed a large number of maps based on sketches by natives and by colleagues with whom he was in contact during his travels. Although they were not all of equal significance either to Fidler or the company, individually they offered larger regional outlines or were used to fill in local topographic information. None were finished map products, but rather sketches based on drawings by amateurs or on oral descriptions. Seventeen of the cartographic informants were Indians (three of whom had given Fidler their sketches at Chesterfield House), one was an Inuit, and fourteen were fellow traders. There was hardly a major area of the north west which did not receive some map attention, from the Missouri to the Arctic, from the Rocky Mountains to the west shores of Hudson Bay.

While stationed in the Athabasca region (1806) Fidler obtained information for a sketch which was the earliest company map of part of the transmontane region of the west in present Canadian territory (95ᴬ). It delineates some of the terrain and water courses from the northern bend of the Columbia River south to the Flathead River and lake region, a valuable extension of detail to the north and west of the earlier Indian sketches. This map was based on a draft by Jean Findley who, while working for David Thompson, had cut a trail through the Rocky Mountains in 1806 to establish a suitable supply route for the North West Company.[11] How the original sketch came into Fidler's hands is not known, but he appears to have drafted his version in 1806 either at Nottingham House at Lake Athabasca or after his return to Cumberland House in the summer.

That same year, Fidler produced a sketch based on one by a local Indian, Cha chay pay way ti, which is an excellent example of native mapping (96A) (plate 22). It is in the cartographic style sometimes called "beads on a string": convoluted patterns of river and lake shores are generalized into essentially straight lines joining rounded or elliptical water bodies. It was valuable for relative directions and the locations of lakes one to the other along a connected route. A clearer sense of the area shown is obtained by viewing the map with Cumberland House, near the Saskatchewan River,[12] turned to the left side and Split Lake to the right. The lakes and rivers are depicted between these locations along the Minago, Grass and Nelson River routes. Place names are largely those used by natives; the names of several rivers and lakes have been translated from the Indian originals.

From 1809 to 1811, at Clapham House on Reindeer Lake and then at Ile a la Crosse, although Fidler was absorbed mainly in the exhausting activities of the trade in the face of the fierce competition from the Canadian traders, he managed to expand his conception of the northwest interior and even part of the Arctic coast. From a young Chipewyan native, Cot aw ney yaz zah, at Reindeer Lake on 17 February 1810, he procured a map of the lakes and rivers from the east end of Lake Athabasca, through the Cree River to Cree Lake, and south to the Churchill River (140A). On 4 June 1810, just before embarking for Ile a la Crosse from Clapham House, a native, Ageena (Ageenah), drew the very generalized outlines of the waterway from Wollaston and Reindeer Lakes along the Cochrane River, connecting them through the portage east to the "Seal River Track" (141A). Immediately adjacent to this sketch in Fidler's personal map journal is a map depicting Methy Portage, Peter Pond Lake and Churchill Lake, and the upper Clearwater River system with its several source lakes (142A). This was most likely drawn from information by Ageena, and obtained on the same day. Lastly, a third sketch, again by Ageena, and likely prepared at the same time, illustrates, although inaccurately, the Arctic coast from the mouth of the Mackenzie River east to the Coppermine and across to the north arm of Great Bear Lake (143A).[13]

Most of the sketches drafted by Fidler from his associates' information were of smaller regions.[14] For example, he made two maps based on information provided by John Charles, a former Grey Coat apprentice who had become a trader. The first of these, drafted in 1806, shows a string of lakes extending southeast from Northern Indian Lake on the Churchill system to beyond Pike Lake on which Portland House was situated (97A). The second map was of a lake in the vicinity of Methy Portage (118A). Another trader, William Cook, from York, filled in considerable detail for Fidler about the middle Nelson River region. One sketch, based on his information, showed Split Lake on the Nelson system (131A), and the other was a two-section map (130A) of the country from Split Lake across to Foxe Lake, and from there along the Steel River to the Hill River.

Other maps based on traders' information were of more extensive areas. For example, while at home in Derbyshire on leave in the autumn and winter of 1811–12, Fidler had the time to compile three regional maps from a variety of sources, including some of his own materials. Fidler sent these maps with a covering letter to the commitee in March 1812. None have remained in the company's records; the company likely turned them over to Arrowsmith for his use, as was its custom at the time. One of them was a map of the west shore of Hudson and James bays between Albany and Severn "with the different Inland Communications," which he had compiled from the various materials of his colleagues Thomas Bunn, George Taylor Sr, and William Tomison (79C). He believed he had been able to plot the details fairly accurately, but he did not want to vouch for precise locations. A second map was of the west shore of Hudson Bay between Severn and York with some of the rivers communicating with the interior, particularly the Severn (80C). He used Thomas Thomas's coastal measurements from 1809, corrected by George Taylor Sr's latitude and longitude observations. For the area inland behind Fort Severn he used James Swain's records, partly corrected by Thomas Swain's calculations. The Steel River track was from his own work; he depended on two separate Indian sketches for rivers and lakes falling into the Severn River near the port. Finally, a map, consisting of two and one-half sheets, depicting the Rocky Mountains from Acton House to Howse's House to the south, was based on data Fidler obtained from Canadian sources, including some indirectly from David Thompson (81C).

The Last Phase of Fidler's Mapping Career

Before he sailed for his leave in England, Fidler had been offered an official position as surveyor to the company at an annual salary of £100. The position required that he be willing to survey wherever the company wished, however remote the area. Both the appointment and this stipulation would, no doubt, have suited him earlier, but at this point he had a wife and family from whom he would have had to have been absent for long periods. Also, he had passed his fortieth year, and it seems clear that he had become tired physically and mentally of the tedium of surveying. He waived acceptance of the position in 1812 and asked Chief Factor Auld for an inland post master's position, saying he had decided that it held out more substantial advantages "than any celestical contacts whatever." It was then he was posted to Brandon where he was master for five trading season; he spent one season at the Selkirk settlement, and then three seasons at Fort Dauphin in the Manitoba-Winnipegosis lakes area. Surveying, mapping, and drafting occupied only a minor portion of his working time in this second period of his career. References in letters testify to his declining health over the last several years of his life, and probably his appointment to Fort Dauphin in the relatively quiet Manitoba district was in a way a sinecure. He did not travel far, the longest journey likely being to Norway House with fur returns and the annual packet.

Between the time he arrived at Brandon in 1812 and his death at Fort Dauphin in 1822, only one of the eight maps drafted in this second period of his career, that of the route from Red River to Martin's Falls, resulted directly from a deliberate surveying and mapping project on his part (90C). He drew a further map of Lake Manitoba after another passage through the area – and after receiving more information about the lake (182A). He also assisted with incipient "engineering" works at portages along the Hayes River route to Lake Winnipeg. Two maps from the period immediately after Fidler's return from leave were a result of the company's interest in opening the area between Lake Winnipeg and Hudson and James bays (164A, 165A). Both maps were compilations based on information gathered by James Sutherland and Richard Perkin during their exploration of this region between 1784 and 1790. Fidler, obviously the most experienced draftsman and knowledgeable explorer in Rupert's Land, must have been asked to transpose details from the original sketches and journals onto two similarly scaled projection graticules.

Although he was sent as master to Brandon, his first task after his return to Rupert's Land was to assist in the foundation of the Selkirk settlement. Over the next several years, he became increasingly involved with the struggles of the fledgling colony, having earlier aided a party of colonists to pass inland from York to the Red-River Assiniboine junction. On numerous occasions, he moved back and forth from Brandon House to Red River giving help and advice. In 1814, 1817, and 1818, he laid out lots for different groups of settlers. He and another cartographer, likely Aaron Arrowsmith, drafted maps which were attached to several deeds of conveyance of land to the Earl of Selkirk (148A, 175A). Fidler also provided a map for an affidavit that was concerned with the disturbances at the Selkirk settlement (174A). Members of the colony apparently discussed with Fidler the possibility of his working full time for the colony, as surveyor and in other capacities for which his long experience in the country made him eminently suitable. Nothing came of the idea, perhaps because his services could be called upon when the need arose.

Fidler's last annual report on the Red River District, which he wrote in the late summer of 1819, contained a map, oriented to the south, with the major topographic features of this area shown in considerable detail (181A).[15] The village lots from Fort Douglas to Frog Plain and along the Seine River in the settlement, cart trails south to Fort Daer and west to Fort La Reine, and all the company and North West Company posts operating within the region are set out along the waterways. He used simplified hill symbols to show upland areas such as Pembina Hills and Turtle Mountain.

The last map known to have come from Fidler's pen is dated 1820, during the first year of his last posting at Fort Dauphin (187A). It may have accompanied his first district report. It was his rendition of one or more of the local Indians' sketches, which he copied or compiled for the use of the company's officers. In his characteristically honest way he attributed the map as a "Sketch a la Savage" (plate 23). Although directions, shapes, sizes, and distances are awry and distorted, relative positions could easily be seen. Fort Dauphin, the district centre, the outlying posts, and the limits of the district are all depicted.

Fidler died at Fort Dauphin and was buried there in December 1822. The flavour of this unusual man has been preserved for posterity – by the journals and letters that reveal his keen interest in geographical, astronomical, mathematical, and cartographic subjects, by the library he bequeathed to the citizens of the Red River settlement, by the sketches and maps he drew, by the personal journals of maps and commentary he willed to the company. For various reasons Fidler did not carry on active field work for a longer period, even though the company urged him to do so and was chagrined that he did not. In the second half of his career, his cartographic production did not approach the level or quality of that in the earlier part. He excelled in careful measurement and observation in the field and thus contributed to the more precise locational grid begun by his mentor, Turnor. By virtue of his transmission of the detailed information of his fellows and of the Indians, Fidler made the production of a more accurate map of the north and west possible.

JOSEPH HOWSE ON THE SASKATCHEWAN AND COLUMBIA RIVERS

Joseph Howse, an officer of diverse talents, contributed to the expansion of knowledge of the grassland high plains and led the first company expedition over the Rocky Mountains to the Columbia River system. His career extended over two decades from 1795 when, at twenty-one years of age, he came to York Factory as a writer. Through most of his active career, Howse was stationed at posts on both the north and south branches of the Saskatchewan River. He drafted three maps and collaborated on a fourth with William Auld at York Factory. The three maps, of areas little known or unknown to the company at the time, have not remained in its possession, although new features of this territory appeared on Arrowsmith maps.

Howse's first map, of the South Branch of the Saskatchewan River, was probably drawn in 1809 (78^C). He had been master of Chesterfield House during the 1804–5 trading season and was master of the new Carlton House on the North Branch from 1805 to the summer of 1808. If he went up the South Branch again during this period, a likely possibility, since the committee wanted further exploration of the area, he would have had fresh in his mind the topographical details that appear on the map.[16] Also, in the early summer of 1808, he had joined Peter Fidler for a trip to York Factory. Thus he would have had the opportunity to collect more information on the South Saskatchewan area as well as the chance to get drafting instruction from Fidler.

Almost nothing definite is known of Howse's pioneer journey to the Rocky Mountains in the summer of 1809. His foray up the North Saskatchewan River to search for a suitable pass across the range to the Rocky Mountain Trench was a result of the company's decision to try to move past the mountain front in order to compete there with the North West Company – led by David Thompson. The pass, later named after Howse, was reached, and may have been traversed, but little definite information or a map was forthcoming.[17] His second and final journey in 1810–11 across the mountains is commemorated in Howse's second map, drafted in 1812 (82^C). It seems to have shown the interrelations of the Athabasca, Saskatchewan and Missouri rivers and of the upper Columbia, Kootenay and Flathead rivers, and Flathead Lake. Near the lake, Howse constructed a wintering cabin which appeared on maps for some years afterwards as "Howse's House." He was willing to return to this region later, but warring between the Indians forced the local officers to decide that it was too dangerous, despite good trade prospects.[18] William Auld at York extolled Howse's bravery, leadership, and his "ardent and unquenchable zeal which sets fear & hesitation alike at defiance" (A11/118,fo.24d). He recognized also that Howse was "an elegant scholar" of languages as well as a successful trader.[19] As a result of Auld's enthusiasm, the committee decided to award Howse a gratuity of £150 "for his past & an encouragement for his future exertions" (A1/50,fo.82).

In the early summer of 1812, Howse went to York to wait for the ship to Britain where he was to take his leave. During the waiting period, he assisted Auld in surveying the Nelson River from the estuary to Split Lake and in preparing a detailed six-sheet map series (153^A).[20] The two may have worked together at York on their records and on the map in the month prior to September ship departure, but it is more likely that Howse drew it while he was on furlough in Cirencester, England.[21] The map series is not on a graticule, so one may assume the survey was made with a compass. Features affecting transportation, such as the location of falls and rapids and the lengths of the various portages, are emphasized. Information useful to boats as to the character of the currents, heights of river banks, and the characteristics of the channel are noted: for example, "boats may be handled up or down" and boats "can run down south side loaded."

A sketch of the "Columbia River" has been attributed to Howse and dated 1815 by the company in its catalogue (84^C). It is not based on a further journey to this region as Howse spent his two final seasons in the north in the Saskatchewan and Ile a la Crosse areas, after which he retired from the company. The extent of the map's coverage cannot be determined, but it likely involved some redefining of his former map, as further information became available during his final sojourn in the west. Corroboration for this likelihood is the difference in the river pattern of this region depicted on the 1814 and the 1818–19 editions of Arrowsmith's map of British North America. If his 1812 and 1815 maps provided the basic data for Arrowsmith, then Howse is a significant link in the development of the configuration of the vast region flanking the western ranges of the Cordillera, even though his original maps are not available to the modern scholar.

JAMES CLOUSTON, INLAND FROM THE EASTMAIN

In contrast with Joseph Howse's maps, six of the seven maps which James Clouston drafted of the country inland from the Eastmain have remained in the company's archives. In 1808, at about the age of twenty-one, Clouston had come to Eastmain House as a school teacher.[22] He served there for three years. Clouston was better educated than most company men, especially his fellow Orkneymen; he was attracted to and adept in exploration; he was able to gain the respect and assistance of the native peoples. A one-month winter fishing trip in 1811 demonstrated his predisposition for outdoor activities and observation. In the spring of the same year, he wrote to the committee requesting that a reflecting telescope, a brass sextant, and a case of instruments be forwarded to him. He may have decided by this time that he wished to become a trader rather than remain a teacher, and he may have known that he was to be "tried" inland in the following summer.

The large region inland from the Eastmain coast was essentially unexplored at this date. The harsh environment afforded little local food supply, nor was bark for building canoes available. But more critical than these factors was the lack of major motivation. To the south, from the Lake Mistassini area across to Lake Abitibi, the company was able to maintain a relatively stable trade front with the Canadians. To the east, northeast, and north, the contacts of the company's Eastmain traders with inland natives had strengthened the impression that opportunity for trade was limited. But to gauge the appropriate trade strategy, the company needed to check this impression and learn the annual cycle of activities of the various native groups. The impetus to gain such information had been lacking until Clouston appeared on the scene; his ambition and curiosity sparked the program of exploration.

Clouston first appeared as a map maker in the 1814 district report from Neoskweskau House, where he had been in the summer of 1811 and then full time from 1812. He titled the map: "Ruperts and slude Rivers together with various lakes," noted that he had also shown the "Falls of the Rivers" (presumably rapids and portages), and that the map included only those features for which he had taken distances and bearings (83ᶜ). The committee was delighted to receive this map, particularly since, alone among the post masters, Clouston provided the map the committee had requested be included with district reports. Two years previously the committee had also applauded Clouston's "zeal" in moving into the trading service and requesting transfer to the interior of the Eastmain. It had awarded him £120 as a mark of satisfaction and "an earnest that his future merits will neither be unnoted or unrewarded" (A6/18,fo.97).

Clouston's second map, sent to Governor Vincent at Moose Factory in the late summer of 1816, showed what he had learned during three journeys made to the south of Neoskweskau Post into the upper Rupert, Broadback, and Nottaway river basins in the preceding two open seasons (168ᴬ). The map particularly highlights Lake Mistassini and its environs. In the annual report of his district for 1815–16, Clouston also provided a large-scale sketch of the general site of Neoskweskau settlement, the water line, ponds, swamps, and main terrain features (170ᴬ) (plate 24), and he included as well a plan of the house. The next year, Governor Vincent assured Clouston that this further evidence of his cartographic skill was highly satisfactory to the committee and that the members encouraged him to persevere in such work.

In the same 1817 letter to Clouston, Vincent brought to his attention the "considerable Tract of Land lying between Mistassiny Lake and Whale River" north up the Eastmain and suggested that it should become the object of his attention (B143/a/18,fo.4). In fact, by this time the London committee already had the idea of using some of the rivers north of the Eastmain River to explore what was thought to be valuable hunting grounds far inland towards Labrador. The plan was to have one part of the country after another explored until there was "a distinct idea of all the interior of Eastmain as far as the streights," or at least as far as one could go without running into danger from the Inuit, a plan that amounted to a clear admission of the committee's lack of understanding of the geography of the area (A6/18,fo.270). They recommended that Clouston should try to reach the upper stretch of one of the Whale Rivers (Great or Little), which he should then follow down to the coast. A successful ascent of the Great Whale River had already been made by George Atkinson II, along with his son and another company servant, in the summer of 1816. A rough sketch of the area (87ᶜ), possibly by the son, George Atkinson III, who had been assigned to make a sketch, was redrafted by an anonymous person, probably the father, at Moose Factory before it was sent on to London (169ᴬ). This map indicates that the group had gone slightly farther up the river network than Lac Bienville before turning back.

In the same year, the energetic Clouston had begun a long range program of expansion which reflected his and the company's interest in the region north and east of the Eastmain River region. By the summer of 1817, Nitchequon House had been built on Nichicun Lake in the Big River headwaters, and Clouston had surveyed the route between it and Neoskweskau and had produced his map of the interconnections (173ᴬ). By that time also, he had refined a plan for further action, which included the opening of several posts on the frontier. He also planned to search for a river to the north east that reportedly flowed to Labrador and into the Atlantic. The possibility of a feasible route to the Atlantic watershed was a tempting one.

The second phase of exploration the company plan called for was carried out in the summer of 1818. Atkinson Sr was ordered to follow the course of the Little Whale River and to go as far inland as he could to find out whether there were any Canadian settlements behind the Whale rivers and the Big River. He was provided with a pocket compass and encouraged to keep a journal. The map Atkinson produced is undoubtedly the very large one on several sheets of paper pasted together, executed entirely in pencil and oriented to the west, that shows a route from the Great Whale River mouth north to Richmond Gulf, across to Lac a l'Eau Claire, Lac des Loups Marins, south and west to the Little Whale River, and thence downstream to Hudson Bay (179A). Some elements were based on Indian information. Atkinson's superior, Thomas Alder, was not greatly impressed; he considered the sketch to be superficial, "containing nothing particularly interesting" (B77/a/6,fos.17,17d). He reported that although no Canadian posts were discovered inland, there was an injurious counter-trade from a Hudson's Bay Company post at Nitchequon newly-established by an "enterprising young man." And, lamented the official, "with a few more such persevering characters as Clouston scattered amongst us, the opposition would be rendered equal at least to that of the NW Company" (B135/b/40,fo.3).

Clouston's move beyond Nitchequon had to be delayed until the summer of 1820. Because he could not get a proper replacement in the Waswanipi river and lake area, he was forced to return there from Rupert House until the late summer of 1819 (B19/a/1,fos.10d,15d). He surveyed part of this area in early 1819 and produced a map of the complicated routeway, north through the upper Broadback River network to the lower Rupert River, that he followed back to Rupert House (180A). By March 1820, his party was at Nitchequon preparing to leave for the north. During the month's wait for break-up, Clouston gathered information from Indians as to the pattern of lakes and rivers beyond the post. He decided to strike out for a lake called Caniapiscau, which Indians estimated was about one hundred miles to the northeast "over the height of Country Towards the Labrador Coast" (B143/a/20,fo.16). He thought it possible that the water issuing from Caniapiscau Lake was part of a system flowing eventually to the Atlantic.

The expedition, having departed on 17 April 1820, reached its farthest limit on a high vantage point looking north over the lowest course of the Koksoak River flowing into Ungava Bay. Clouston was forced to conclude that the estuary seen was not in the Atlantic. His group turned west from the Koksoak along an Indian route leading to Richmond Gulf and then down the Eastmain coast to Rupert House. The map drafted by Clouston of this formidable round-trip journey (188A) combined his route in ink and several insertions in pencil which were transposed from a "drawing" made by his Indian guides during the journey (94C). In his journal he analyzed the northern geography, reasoning that there was likely a line of mountains running parallel to the coast of Labrador, from whose highlands large tributary rivers flowed west into the northward-flowing Caniapiscau River. Although this area might be worth examining later, he concluded that initial exploration and trade development should be focused east and south of Nitchequon and similarly of Mistassini, and towards the Gulf of St. Lawrence. Larger events in the development of the company's structure and operations, growing out of the union with the North West Company in 1821, intervened, and for the remaining portion of his career to 1827, Clouston was involved in only local exploration, either on the Eastmain coast, or in the southern part of the region.

The last of Clouston's mapping projects was the assembly of all the detail he had obtained, compiled on four integrated map sheets, set on a one-degree projection grid (195A) (plate 25). There is also some pencil planimetry based on Indian intelligence. In essence, this is an updated and expanded amalgam of Clouston's five previous maps. Some generalization has occurred, although the map is at the same scale as his 1821 draft of the northern journey. Clouston was not a trained surveyor, as his maps readily attest. He made errors in observations of latitude and longitude, as well as in the use of the barometer in obtaining heights. But for that matter, so had Turnor and Fidler, who had more training and experience. But, in concert with his detailed, perceptive, and readable journals, Clouston's maps were easily understood and added critical information about the main travel routes of a relatively sparse native population. His maps revealed the enormous range of territory the natives traversed in their annual cycle of activities. Moreover, his 1825 map, in particular, was used by the Arrowsmiths as the source of the basic river-lake network inland from the Eastmain, a network that remained constant even after later company investigation east of the Caniapiscau River added new features to the map of this vast peninsula. As was the case with Turnor, Thompson, and Fidler, Clouston's full potential for exploration, surveying, and cartography was not exploited by the company. He had only a short career as an explorer – seven years, though he drafted maps for a slightly longer time. Although his most notable contributions concerned the region north of the Eastmain River, most of his trading and travelling career was spent south of it.[23]

MAPPING FUR TRADE DISTRICTS, 1814–1820

The preparation of a series of maps of the company's two departments, northern and southern, and their district divisions was a main cartographic goal in this period. The company also restructured the supply and transport systems. A set of instructions about "reporting procedures" was issued in the 1810–14 period (A6/18,fos.200–6). The committee's expectation was that in the 1814 packets, or as soon afterwards as possible, the governors of the two departments would be able to provide through their district chiefs, district reports, in which prescribed topics would be covered. Other than an assessment of trading opportunities, the company wanted discussion of two aspects. The first was a "Topographical account" of each district, or, in modern terms, a description of the physical environment and of the social geography of the area, dominated, of course, by an estimation of native population and activities. The committee had always wanted information about the native tribes, their populations, and other aspects of the indigenous population that bore upon trade possibilities, but this was their first attempt to assess the entire trading region in these terms. The second was a "Topographical Sketch, or Plan" of each district, showing lakes, rivers, and other physical features of value along with an indication of human occupation patterns, especially in those areas where no accurate surveys had been made. The committee wished to have some idea of the relative situation of places that were frequently mentioned in journals and letters but which did not appear on current maps. It provided the two departmental governors with copies of the largest scale map of the country in order to give them "an outline of the parts which remain to be filled up" (A6/18,fo.206). Later, the company secretary wrote to Arrowsmith for "traced Copies" from his "large Manuscript Map," at the scale of four miles to a degree, of "all those parts of the Company's Territories where it is desirable to gain further Information for completing the Topography of the Country" (85ᶜ) (A1/51,fo.22d). How many copies were made and of what areas is not known, but the order does indicate that the committee was concerned with the extension of exploration and precise surveying. The company also attempted to interest their employees in "laying down Sketches of the Country." To make the task easier for them, it had a set of "plain Mercator Charts divided into 32 Compartments with an Index to a projection" prepared and shipped to Rupert's Land in 1818, to be used by officials there as they saw fit (88ᶜ) (A6/19,fos.63d–64).

This district mapping program was not completely successful. Reports were sent to Britain, but many of the post masters were reticent about engaging in map making as requested. Thomas Alder at Whale River, for example, did not attempt to draw either a map of the area or to detail the latitudinal locations of "the different Points, Bays, Creeks, etc.," since as he said these had been provided for over twenty years by more able hands than his, nor did he write a topographical description: "I don't understand it," he admitted bluntly, "and others who do attempt it are (I believe) not much better" (B77/e/1b,fo.3). The committee recognized that many of the officers were not "capable of making out a distinct & accurate plan" but advised them that this did not matter. The object should be to get information from local persons, even from "the rudest Indian sketch" (A6/18,fo.205).

Maps from eleven districts, drawn between 1814 and 1820, have been located in the archives.[24] The most complex map was of the Severn district, a three-sheet map at medium scale (158ᴬ, 159ᴬ) (plate 26), based on Indian information to a large degree, and drawn by James Swain Sr. Because of the nature of the sources, and because no accurate surveying had occurred in much of the interior area, the map is not based on a graticule, nor is there directional or scale information. But the combined map extends through the Severn, Winisk, and Equan basins, across into the Hayes River network some distance. Robert Kennedy's Lesser Slave Lake district map of 1819–20, again based especially on native sources, has scales which vary from about twenty miles to the inch in the better known areas to seventy-five miles in the region for which the information came largely from natives (183ᴬ). It is oriented to the west, as the normal traveller would approach the lake from the bay. James Clouston's map of the Rupert and Eastmain rivers and district and Fidler's Red River district map have been discussed previously. Finally, Donald Sutherland's sketch of the Berens River district of 1819 is a useful example of the standard type of sketch received (178ᴬ) (plate 27). It is oriented to the east, fronting on the east shore of Lake Winnipeg. Although Sutherland asked the committee to excuse its "Rudeness," it was the most complete rendition of this extensive region prepared to that time. As such, its features made their way directly into Arrowsmith's 1822 edition of his major map.

COASTAL AND HARBOUR CHARTING

Coastal and harbour charting continued during the early decades of the nineteenth century. More up-to-date renditions of immediate off-shore features were always welcome: it was essential for safe harbourage in river estuaries as shifting sand bars quickly altered seabed topography.[25] At least nine coastal or estuarine charts were prepared. Captain Henry Hanwell Sr, who had been in the sea service for thirty-four years and was very well acquainted with the bay by 1803, drafted a chart that year of "the South part of Hudson's Bay"; in reality it is of the most southerly curvature of James Bay from Albany to just north of Eastmain House (89A). (plate 28). Included on the face of the chart is a list of latitudes and longitudes of significant locations and a list of distances between various places. It was the most exact chart of this busy sea basin at this date. For the first time a detailed chart of the soundings of Charlton Sound was also drafted for the company in 1805, as an appendage to a report Hanwell made to the committee on the general situation on Charlton Island after he inspected the area in 1804 (91A). The island had been occupied by a North West Company group in 1803 as part of a more widespread attempt to force its way into the bay. Captain Hanwell apparently encouraged his junior staff to learn hydrographic charting as part of their training. One of them, George Roberts, completed a chart of Hudson Bay which he presented to the company (77C), and in 1808, the committee gave him a gratuity of ten guineas as "Encouragement to his steady perseverance in the improvement of his Nautical Skill & knowledge" (A1/49,fo.58). When the Captain's son, Henry Hanwell Jr, was a member of a company ship whose crew was forced to winter on Charlton Island in 1811–12, he charted some of the same areas his father had (150A).

Three charts of the Nelson-Hayes estuary were produced in 1811, 1812, 1813, apparently each one more exact and acceptable to the committee than the last. The first was an anonymous, quite crude chart attached for some reason to the inside front cover of the Churchill letter book of 1811–12 (149A). It was likely the work of William Auld, who, as noted above, later aided Howse in mapping the Nelson River. The other two charts were the work of William Hillier, a former master in the Royal Navy, who arrived at York Factory on the 1811 ship to conduct a party of Irish and Orkney people to the newly established Red River settlement. Since it was too late in the season to lead such a group over the long distance, it was suggested by Auld that Hillier should survey the Nelson River mouth in 1812 to search for a more advantageous anchorage for company ships and a better location for the factory.

Hillier's first chart did not satisfy the committee (152A). In their 1813 letter to Auld they required that "further investigation must be made by accurate sounding" (A6/18,fo.106). Only then would the members be able to consider the "comparative difficulties of the Navigation of the two Rivers," and to make a decision on the best new site for the factory before any expense was incurred in removing the buildings. Hillier returned to York from Red River on 23 June 1813, and on 13 July set off in the Factory yawl "to sound a channel in the Nelson" (B239/a/120,fo.26). Two days later he was back, having decided on the proper channel. Further instructions in 1813 asked Hiller for data on the spring neap tides, for his opinion on the possibility of ships going up river and lying at anchor near shore instead of lying off at Five Fathom Hole, and on the possibility of a ship wintering in the Nelson if detained over winter. It is likely that Hillier drew a chart on which he inserted the new 1813 soundings and that he personally delivered it to the London committee in 1814 (154A).

MAPS RELATED TO FRANKLIN'S OVERLAND EXPEDITION

Among the several other maps that came into Hudson's Bay House in this period, only one set will be mentioned. This map series was not drafted by company employees, but was rather the product of a collaboration between Thomas Thompson, master of the brig *Wear* of London and Lieutenant John Franklin of the Royal Navy. It was the initial outcome of the co-operation between the company and the Admiralty in the latter's long-lasting geographical investigation of the North American Arctic. In 1819 the first Franklin overland expedition received the logistic support of the company. That summer, Franklin and Thompson carefully surveyed the Nelson River system and part of the Saskatchewan River. Then on the way inland towards the Mackenzie district, Thompson drafted six integrated sheets, based on the previous survey, each sheet covering two degrees of longitude and one degree of latitude (186A). Thomas Thompson also drew a chart of the coastline of the bay from Cape Tatnam to beyond Churchill in 1820 (185A). The involvement of the company with Franklin lasted until 1855 when it ceased to be directly concerned in the endeavour to find the explorer and his party, who had, in fact, perished in 1845.

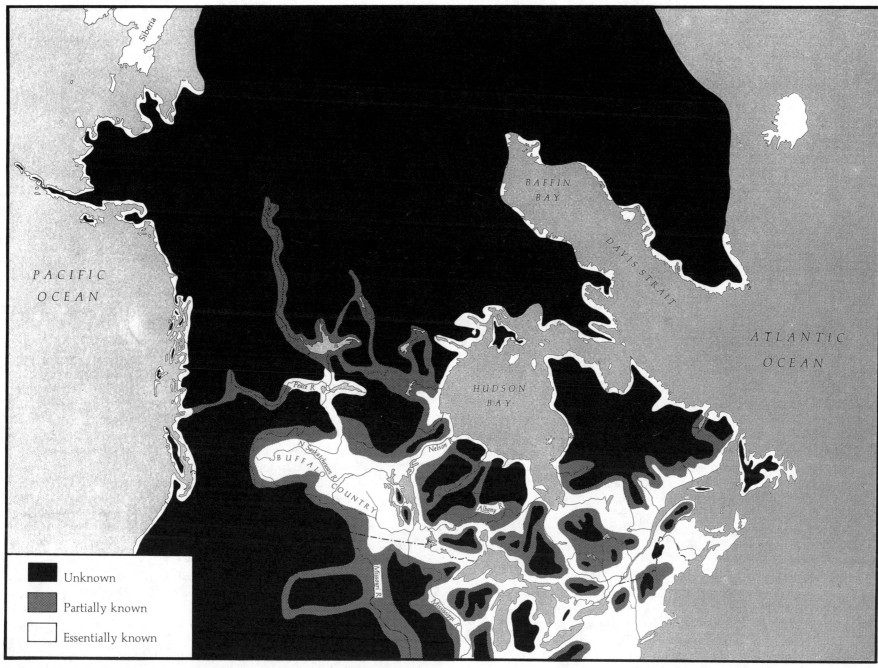

Figure 4 Retreat of the unknown: European knowledge of Canadian territory in 1795

10 Mapping Rupert's Land, the Mackenzie Basin, and the Arctic Shore, 1821–1849

> ... what a field to face the imagination, what a number of ideas rushes in at once, all for the means to investigate a Country so interesting.
> Edward Smith to the Committee, March 1825

The amalgamation of the Hudson's Bay Company and the North West Company in 1821 unleashed new energies, new expansionary forces, and new managerial procedures in the revitalized company over the next quarter century. Operationally, the union spurred diversification in resource use and trade, in land development and property management. Territorially, it turned attention to areas peripheral to the previously central regions of fur trade and settlement – to northern Quebec and Labrador, to the Mackenzie and Yukon river basins, to the near-Arctic, and to the west of the Rocky Mountains. Cartographically, the change the amalgamation wrought was reflected in the increasingly diverse documents which passed through the hands of officials in North America and into Hudson's Bay House in London. About 190 separate maps and 184 segmental sketches were produced. These were still dominated by those showing routes of exploration and by charts of coastal areas, river estuaries, and harbours. But as employees were seconded to the Red River settlement and the Columbia valley to carry out the subdivision and plotting of land holdings, more cadastral surveys entered the collection. Among other special purpose maps were numbered the unusual series of beaver conservation maps, witness to a singular program established on Charlton and a neighbouring island in Hudson Bay. Main freight and passenger routes from Hudson Bay to the interior were re-measured and depicted more precisely. In the far west, horse brigade trails leading from the Thompson and upper Fraser districts to the lower Fraser River were proposed, mapped, and then made operative. This chapter is concerned largely with the territory east of the Rocky Mountains, although mapping efforts in the Yukon, beyond the Mackenzie and Richardson Ranges, which front on the Mackenzie River valley, are also described. Mapping west of the mountains to the Pacific from 1821 until 1849 is the subject of the next chapter.

MAPPING INTERIOR QUEBEC AND LABRADOR

At the time of union in 1821, as noted in the previous chapter, the company withdrew its support of inland exploration and post building from north and northeast of the Eastmain River in Quebec. However, in deliberations a few years later, the committee had to take into account new circumstances, among which were increasing signs of overtrapping in the older trading areas. Under Governor Simpson's active promotion,[1] trade in the southeast towards the watershed of the St. Lawrence and in the unexplored territory south of Ungava Bay between the Koksoak and Caniapiscau rivers and the Labrador coast were developed to provide new sources of supply. At least nineteen maps are known to have resulted from company operations in the interior Quebec-Labrador regions up to 1850.

Simpson's plan called for the erection of a centre for trade development on the lower Koksoak River near the Ungava shore. From there Indians south towards the Moravian missions on the Labrador coast and also up the Caniapiscau could be reached. The purview of the new house would also include the Inuit along the coast. Two maps depict the first stage in the evolution of this trading operation. William Hendry, surgeon and surveyor to the expedition,[2] with George Atkinson II as his assistant, was instructed in 1828 to journey from Richmond Gulf overland to the Koksoak mouth and return to the Little Whale River outlet; the party would be crossing some of the area mapped in 1820 by Atkinson and Clouston.[3] A marine party was then to pick up Hendry and Atkinson at Little Whale and sail north around Cape Diggs and into Ungava Bay. There the group would establish the new post at a chosen site on the Koksoak. Because a seaworthy sloop was not available, the marine part of the plan was abandoned. The two maps show the same details of Hendry and Atkinson's trek to the mouth of the South (Caniapiscau) River, a tributary of the Koksoak, and from there to the mouth of the main river. One depicts in addition the area along the coasts of Hudson Bay, Hudson Strait, and Ungava Bay (215A). Although the less extensive map (214A) (plate 29) has an endorsement on it which states that it was a "Chart ... by Willieam Hendry," the authorship of the more extensive one is less clear. The evidence does point to Hendry even though minor discrepancies in the locations of some places he would have observed do not reflect the usually careful nature of his field work.

Four similar maps represent the next phase of the Ungava project. The venture, which began in the late summer of 1830 and extended to the late summer of 1834, was superintended by Nicol Finlayson, who was aided by Erland Erlandson. The party chose a site, some twenty-seven miles up the Koksoak River, for the new station, Fort Chimo. From there, sorties were made both along the coast and inland. One exploring party reached northeast to the mouth of the George River, and a second was turned back by ice from the northwest coast of Ungava Bay. In an essay up the South River in 1831, a third group reached only as far as the first major waterfall (Limestone) on the river. Another attempt in 1832 under Erlandson succeeded in going only halfway to Swampy Bay River, his original destination where he had hoped to build a cabin; instead South River House was erected on the South River. By 1833, both Finlayson and Erlandson had learned enough from their experiences to become convinced that there would be no value in setting up any more houses in the interior, unless they were located far to the south across the height of land in the St. Lawrence watershed: the possibilities for trade further north seemed minimal.

Under Simpson's direction, Finlayson and Erlandson began to search for Indian guides who could be persuaded to lead Erlandson south from Chimo to Mingan on the north shore of the St. Lawrence. They were unsuccessful until the spring of 1834. Erlandson's party then followed a complicated series of waterways to Lake Petitsikapau where the Indians refused to proceed south in spite of being offered liberal rewards. Instead, they diverted the party to Esquimeaux Bay (Lake Melville-Hamilton Inlet) on the Labrador coast. In a convincing example of by indirection finding direction out, this area became the terminus of a regular series of connections between the Labrador coast and Ungava. Erlandson drew three manuscript maps in ink and one rough sketch in pencil, none of which he signed. Two of the inked maps were sent to London with reports (228A, 229A).[4] In addition to many place-names identifying the lakes and rivers across which they passed, Erlandson noted, in red ink on one map, (228A) the animals he had seen and their fur bearing potential. Another annotation "recommended a post to be established" at Lake Petitsikapau. Erlandson drew the third inked map at a slightly larger scale than the other two and inserted details on a graticule (230A). Most likely it was this map that Finlayson sent along with Erlandson's report to Governor Simpson early in 1835 (B38/b/2,fo.45). The pencil version, much rougher than the others, Erlandson may have kept at Fort Chimo (231A).

This new and arduous overland connection brought to the fore the central problem which the company faced in this region: how to supply Fort Chimo in its remote location by a practical and assured method. Various alternatives were proposed and some tried, but most involved potential danger and great difficulty.[5] By 1837, a novel system had been initiated by John McLean who became master at Fort Chimo in that year. The plan was to connect Chimo to North West River House (Fort Smith) on Lake Melville, which would then act as the Atlantic terminus for the supply route to the north and west. McLean's geographical and cartographic contributions during the pursuit of such a viable supply line were extensive. In at least four journeys across the northeast of the peninsula, he traced out the main lineaments of this vast area, concentrating particularly on the use of the George River-Naskaupi River connection as the major pathway. Unfortunately, none of the individual journeys by McLean was memorialized in separate maps nor did McLean sign any maps. But there are three maps in the archives that depict the routes he followed in a clear fashion. In the spring of 1840, Governor Simpson had requested that McLean produce a "rough chart of the country and coast, shewing the situation of the different rivers, lakes and posts and how they relatively stand by the districts and posts of Rupert's River and Esquimaux Bay" (D4/25,fo.93), and he drew his attention to an Arrowsmith map which might provide useful background. In September 1840, McLean reported to Simpson from Chimo that he was then forwarding the finished chart (D5/15,fo.325d). This 1840 map illustrates the country from Fort Chimo, along the ocean coast to Hamilton Inlet, and inland to the Caniapiscau and Kenogamisi rivers (242A). Significantly, a pecked line is identified as "Mr. Mcleans Route Summer 1840" up the George River and southeast to Lake Melville. Also shown are the other routes south from Ungava to Petitsikapau Lake and from there across to the Labrador coast. The two other maps are generally related, at least in content and date to the 1840 version, although they are on different scales and projections (254A, 255A). These two maps illustrate the situation at the culmination of a long, painful, and costly effort to organize a workable trading system in this part of the company's territory. Together with the 1825 summary map of the west side of the Quebec-Labrador peninsula by James Clouston (195A), discussed earlier in chapter 9, these maps show us the then current state of geographical knowledge for this interior area; they were the foundation upon which the company had to depend when making policy decisions in the fifth decade of the century.

Five maps of two sections of the coastline of Labrador and Ungava were drafted by company personnel during this period along with one

Mapping Rupert's Land

map of the route followed in a journey from the Labrador coast to the St. Lawrence shore. W.H.A. Davies was probably the author of four of them, and it was likely at his instigation that the coast was explored and mapped. As master at Fort Smith he was responsible for all trading activities within the extensive area of Labrador surrounding his post, but he was also especially active in charting the intricate coastline of Hamilton Inlet-Lake Melville and the next inlet to the north, Kaipokok, as well as the intervening hill country. His first map, in 1838, was, as he called it, "a rough sketch ... badly done" of the two inlets (130^C) (B153/b/1,p.20). He sent it to his superiors to give them an idea of the relative positions of the features of the area. This was followed by a more detailed map of Kaipokok Bay, likely drawn in 1840 (243^A).[6] Davies crossed overland from Melville Lake to Kaipokok in late December 1839, and early January 1840, and returned via a different land route. His purpose was to explore the country between the two bays for its prospects as a hunting area, and more specifically to search for a more direct overland path between the inlets in contrast with the normal circuitous coast route. With his report he enclosed a sketch of the two routes (136^C), which it would seem from his description of the land would have emphasized the "abrupt hills" and "numerous swale[s] & shallow lakes," (B153/b/1,p.41). He also described the country as having been ravaged by fires, which gave it a "naked & desolate appearance." Davies's map delineating the inner body of water, Lake Melville, was likely the most useful for the company (244^A). It showed the existing fur trading posts and the outlying summer and winter houses of various traders, indicating in what years these were occupied. Associated with the map is a list of distances from Fort Smith for a number of locations.

The fifth map was a sketch of the route followed by George Duberger, an employee at Fort Smith, who was directed to go from the house on 22 July 1838 to explore the nearest and best canoe route across to Musquaro in the Seigneury of Mingan on the St. Lawrence north shore (131^C) (B153/b/1,p.3). He returned on September 6 with the report that, because of shallow water and boggy and swampy conditions, the shortest track was impracticable as the route for large trade canoes.

The company, after closing the outlying posts to the northeast of the Eastmain and Rupert Rivers, reopened Nitchequon and Caniapiscau posts for a short time in 1834 in an unsuccessful attempt to aid the Ungava operation. To counter the Canadian trade incursions from the St. Lawrence basin, it had also extended a network of posts and increased its interest in the arc of country south of the Rupert River and Lake Mistassini. Interestingly, much of the motivation for these moves came from the Nitchequon post, and especially from two energetic men there, John Spencer, post master, and Thomas Beads, a mixed-blood employee, both of whom enjoyed working on maps. Beads journeyed from Nitchequon cross-country to Tadoussac, which he reached in 1837, an unexpected turn-up since he had headed for Lake Manicouagan. Spencer encouraged him to draw a map of his entire trip (133^C). The original was sent to the district master, Robert Miles, who forwarded it to Governor Simpson (B186/b/37,p.4); a copy accompanied the Rupert River district report for 1837–8 to London (B186/b/36,fo.1d), and another was kept at Nitchequon by Spencer to be used in further map making.

By now, Spencer had discovered that Beads was a competent worker and as concerned as he was with compiling a more integrated picture of the topography of the district. Spencer suggested to his superior, Robert Miles, that the district masters at Abitibi, Timiskaming, and at the posts strung from St. Maurice to Three Rivers, along with those situated in the Moose and Albany regions should be encouraged to do what he and Beads were undertaking, namely district maps, which could then be put together to form "a very handsome Map" (B147/a/5,fo.34). To begin this project Spencer and Beads gathered together in 1839 all the known detail into a general map of the region from Hudson Bay to Ungava and the St. Lawrence River (134^C). He conceded that it was a "very plain" map thus far (B186/b/38,p.39). Miles was pleased with the map, but he cooled Spencer's ardour somewhat regarding a mapping program by "Gentlemen of the adjoining Districts," for as Miles reminded him, not all were so adept with the pen as Spencer was, and few would even attempt to compete with him in this regard (B186/b/39,fo.19).

Fortunately, a copy of the last and most comprehensive map, prepared through the joint effort of Beads and Spencer in 1841, reached London (247^A) (plate 31). This map is a composite one mainly "pencilled out" by Beads, who was then stationed at Caniapiscau House; Spencer made several additions which he had gleaned from Indian reports. In contrast with the first map sent out in 1839, the coverage must have been quite impressive, although Spencer apologized for the blank areas for which they had been unable to obtain information. This map shows the area extending from the Nottaway River basin up the Eastmain coast to Richmond Gulf and across to Fort Chimo. It is especially detailed for the area east of the Caniapiscau River and south to the St. Lawrence River, from Tadoussac downstream towards Sept Iles. The cartographers had indicated with simple terrain symbols the mountainous height of land north of the St. Lawrence, as well as other hill lands. The density of the river and lake network is impressive, even though the scale and orientation of many of the lakes and rivers are somewhat distorted. Some of the larger lakes are unnamed, but Spencer suggested that doubtless in time they would be "christened" (B147/a/8,fo.8d). Two copies of the map were sent to Robert Miles, one for his own use and one for Governor Simpson's. The copy in the archives is likely that sent to the governor, for it has a pencil endorsement: "For Govr Simpson if you please." One copy may have remained with Spencer since, in spite of the lack of official interest, he still pressed for his idea of a mapping program, which could not, he concluded, fail to result in a "valuable Atlas of what has not yet ... appeared on paper, at least to any extended degree of interior delineation" (B147/a/8,fo.8d). During this period the Hudson's Bay Company did not have in its employ in the Eastmain and interior Quebec a cartographer of the calibre of Turnor or Fidler who might have assembled the configurations for the Eastmain which had come in or were then flowing into London from such men as Clouston, Hendry, Erlandson, McLean, Beads, and Spencer. Nor did the committee choose to reintroduce a full-scale district mapping program like that they had promoted in the early nineteenth century – perhaps because that program had not been particularly successful.

MAPPING ACROSS THE MACKENZIE AND RICHARDSON MOUNTAINS

The cessation of the competition between the two major fur trading companies in 1821 paved the way for a more single-minded attack by the Hudson's Bay Company on the peripheral region of the Mackenzie River basin and especially the area west into the Cordillera. The first episode in this assault between 1822 and 1825 upon the Rocky Mountain fastnesses, originated from Fort Simpson at the junction of the Mackenzie and Liard Rivers and Fort St. John on the upper Peace River. Part of the impetus for this expansion was the steadily decreasing supply of valuable furs from many of the established northern districts such as Athabasca and the lower Peace: new sources were urgently required. Furthermore, there was growing pressure from the Russians in the Pacific North West. The influence of these traders was being felt in several large river basins west of the mountains as increasingly fewer native traders from the tribes in the mountainous interior came to the Mackenzie and Peace River posts east of the mountains. One of the tribes often mentioned in company correspondence was the Nahanni, who lived in the northern tributary valleys of the Liard River, especially those of the South Nahanni, whose streams drained the southern ridges of the Mackenzie Mountains. On several occasions, small Indian groups, outfitted to pass the winter in the mountains, had been provided with gifts to entice Nahanni traders to Fort Simpson or Fort Liard. The ploy had not been successful.

The key purpose of the first expeditions from Fort Simpson into this region was to contact the Nahanni hunters and trappers. Neither the first expedition led by Alexander R. McLeod to the mouth of the South Nahanni, nor the second and third ones, led by John McLeod up the South Nahanni, resulted in mapped outlines.[7] The next explorer, and the first to draft maps for this region, was Murdoch McPherson. In 1822, he had assisted in the construction of Fort Liard, situated above the South Nahanni tributary on the Liard River.[8] Indian hunters whom he had sent west to contact Nahanni groups had been unsuccessful, but they had described the country up the "West Branch," that is, the main Liard River, as consisting of a series of three mountain ridges flanking a "stupendously high" fourth range, which McPherson correctly assumed to be the Rocky Mountains (B116/a/1,fo.40). They also spoke about a rich beaver population in the region. Recognizing that discovering the correctness of the reports would require his personal investigation, McPherson set off up the Liard for a month in July 1824. The course he followed – up the Liard River past the East Branch (Fort Nelson River) as far as the forks, where the West Branch (the main Liard River) is joined by the north bank tributary, the Beaver – is indicated on a map he drafted (190A). He investigated this latter river first until he found it to be a lesser stream issuing in a rough passage through very mountainous terrain. Turning back, the party then went west up the main Liard River a short distance to test "the practicability of making progress against that uncommonly strong stream" (B116/a/3,fo.13). They found the going too difficult and retraced their course back to Fort Liard. At the same time, in 1824, McPherson supplied a "McKenzie's River" district map (191A) (plate 32), the coverage of which extended from the west end of Lake Athabasca north to the Mackenzie River delta. Included is the western portion of Great Bear Lake, McLeod's information about the Liard River, McPherson's own 1824 data, and the Peace River to the forks of the Finlay and Parsnip rivers. He traced in the "supposed situation of the Main Ridge Rocky Mountain," which he indicated was based chiefly on Indian information.

McPherson's map, however, did not include the details of a journey Samuel Black had made, between late May and late August 1824, up the Finlay River.[9] Black had travelled from the upper Finlay across the grain of the country to the Turnagain River in the southern Liard basin. He had been disheartened at the end of his journey by the barrenness of the country, which precluded his party supporting itself over a winter. Because of the lateness of the season, he decided not to go on farther north to the Liard River. Although he kept a journal of his travels, he did not provide the company with a map. There were so few local natives and so few beaver that the company, on Black's recommendation, decided further expansion in the Finlay River area would not be worthwhile. Nor did the area Black explored appear to provide a potential route to Pacific-flowing rivers.

By the latter half of the 1820s, the country beyond the crestlines of the Mackenzie and Richardson mountains to the west of the Mackenzie River, had been breached westerly only to a point a few miles up the West Branch. The streams issuing out of the eastern flank of the Mackenzie Mountains were so obstructed by rapids and falls that they were impassable except by Indian canoes. Thus the likely routes would be located around the southern or northern flanks of the mountains. The southern flank of the Mackenzie Mountains was tried first, the entry point being the West Branch of the Liard, which would eventually give the traders access both to the Stikine River and to the Yukon River systems. The man chosen to break the new trail was John McLeod, already familiar with the region through his earlier expeditions in the South Nahanni area and considered to be of "enterprising character" with "persevering habits" (A12/1,fo.389d). Both enterprise and perseverance, especially the latter, were of utmost importance during the four seasons from 1831 through 1834 he spent exploring this route. McLeod's journals show him to have been both a careful field observer and accurate recorder of measurements. His report on the 1831 trek was welcomed as "a plain account of facts and incidents as they occurred" (B200/a/7,p28). McLeod drafted a map (217A) (plate 33) in the same spare but effective manner. The map and his journal provide a stirring and easily comprehensible account of his group's laboured course almost to the source of the Liard – and almost over the continental divide to the Pacific-flowing Stikine River. McLeod was deceived, however, in thinking that he had reached the source of the Liard. In fact, this lay some miles to the northwest of his farthest position, for he had mistakenly turned up the Frances River tributary of the Liard which he had christened the Stewart, instead of continuing up the Liard River.[10] The presumed source lake, which he named Simpson's (Simpson) Lake, lay to the west of the Frances River in a mountain-girt basin. At this lake he left details of the event carved into a topped and stripped pine tree and then turned back, "satisfied that the object of the voyage was fairly accomplished" (B200/a/14,fo.12d).

Although textual sources do not fully detail McLeod's activities over the next two seasons, he was "in the West Branch exploring the Rocky Mountains," during the summers of 1832 and 1833 (B200/a/15,fo.5). The second McLeod map, based on an Indian chart, outlines his fourth and last journey far west of the Mackenzie River (233A). This journey from his post, Fort Halkett, extended over most of the summer of 1834. It fulfilled the Northern Department Council's intent that his mission should be to discover "the Countries Situated on the West side of the Rocky Mountains from the Sources of the ... [West] branch of the Liard River" (B239/k/2.fo.17d) McLeod's party passed up the Dease River tributary to its source, Dease Lake, and, crossing from this body of water over a portage of about fifteen miles, reached the Stikine River. The conclusion reached by one of the company's officers upon reading McLeod's report, observing his map, and noting the successful crossing of the divide was that it must be "but a hop, step and a jump to the Pacific Ocean" from this most westerly position (B200/e/11,fo.2d). The committee immediately decided that a post should be established on the Stikine; Robert Campbell was brought into the Liard region to carry out the decision.[11]

Campbell's map of 1844 illustrates the knowledge gained in a decade of exploration in the region of the West Branch of the Liard (261A). He recommended against a post on the Stikine in view of the Russian trade and hostile Indians in the area. Instead, in 1840 Campbell passed up the Liard to the present Frances River and Frances Lake, moved west across the Pacific drainage divide and reached the Pelly River, which later proved to be a branch of the Yukon River. At first, he was convinced that this river was really the Colvile, a river that the Franklin expedition west of the Mackenzie River delta had identified as flowing into the Arctic Ocean. In 1843, Campbell was back on the Pelly after two seasons at Fort Halkett, where on several occasions he lamented that he had been "as dull and melancholy as ennui can possibly make him" (B85/a/11,fo.5d). He was now under orders to try to reach the mouth of the Colvile River. However, his party travelled downstream only to the mouth of what he considered to be a large southern tributary of this river, which Campbell christened the Lewes.[12] He chose this junction as a site for the eventual construction of a post. The 1844 map of the area demonstrates that company men still considered the Frances River to be the headwaters of the West Branch of the Liard (that is, of the main Liard River). It also shows that the point reached and the location suggested for the first company establishment past the Pacific divide from the Mackenzie region, was far to the northwest, outflanking the enormous mass of the Mackenzie Mountains.

While the Liard River avenue to the Pacific watershed was being explored, the Fort Simpson retinue were also attempting to move west from the lower Mckenzie River. They hoped to be able to pass around the north end of the Mackenzie Mountains and reach the waters of the Colvile River, along with other streams flowing out to the Arctic Ocean.

Although it was suggested that some way might be found to get across from the Mackenzie via one of its lesser tributaries in the vicinity of Fort Good Hope, Indian informants were able to convince the explorers that this would be too difficult.[13] The better approach would be to go to the mouth of the Peel River, to ascend the river, and then to cut west across the Richardson range. The only two maps known to have been drafted to illustrate this aspect of northwestern exploration were one by John Bell and another by his colleague, A.K. Isbester, neither of which is in the collection.[14] Isbester's map centred on the Peel River area and adjacent country and was based on explorations in 1840 (11ᴮ).[15] Bell's map would have depicted what he had learned during his decade-long attempt to reach beyond the Richardson highlands (147ᶜ). After several trials beginning in 1839, he had managed to explore the Peel River basin for nearly two hundred miles from its junction with the Mackenzie. He also discovered, on his third attempt, a route over the mountains for heavily-loaded parties to the Bell River, then on to the Porcupine. In 1845 he reached the point where the Porcupine flows into the Yukon River. Sometime between the Yukon River trip and the early months of 1848, at Campbell's request, Bell sent him the map. However, since Campbell complained that the latitude and longitude of the Porcupine junction with the Yukon was exceedingly incorrect when checked against the graticule on a contemporary Arrowsmith map, he may have decided that the map was not worth forwarding to London (D5/22,fo.162d).

The end of the long struggle to find a way west past the main cordilleran ranges came when Alexander Hunter Murray was sent to open Fort Yukon at the Yukon-Porcupine junction in 1847, and when Campbell, after erecting Fort Selkirk at the confluence of the present Pelly and Yukon rivers in 1848, finally travelled down the Yukon to the new Fort Yukon in 1851. On that expedition Campbell went on to Fort Simpson via the northern route along the Peel and Mackenzie rivers. The Yukon-Pelly River was shown not to be the Colvile, rather it must be a much shorter river flowing north from the Yukon River-Arctic divide into the Beaufort Sea. Where the mouth of the Yukon River was situated was still a mystery.

MAPPING THE SOUTHERN SHORES OF THE ARCTIC OCEANS

Although the basic motive for company activity in this northwestern region was to gain an understanding of its geography, from 1847 on this motive was interwoven with the humanitarian desire to rescue possible survivors of the Franklin Arctic expedition of 1845. Officers and servants of the company, especially in the Mackenzie and Athabasca districts, knew at first hand the kind of hardihood, courage, and daring that was required for anyone exploring across the interior into the Arctic region. No doubt this gave them a special empathy for Franklin and his men and a special incentive to learn the causes, the location, and the events of whatever disaster had befallen them. Certainly, the company's commercial concern in the Arctic area was minimal, although the increase in geographical knowledge that resulted from its assistance in the search for Franklin did prove advantageous later on. The company also gained national and international prestige for its role and played a major part in completing the delineation of the northern coastal outline of North America and some of the island archipelago.

The company's involvement ranged from general and reimbursed assistance to one British exploring party, to full funding, manning, and carrying out of five of its own expeditions. Additionally, it seconded employees to participate in the operations of or to provide logistical management to four other British research groups. Company men produced at least thirteen manuscript maps (likely more) during this period. Some of these maps provided documentation for the compilation of published maps, and several of the originals were used to accompany articles in learned journals. Five of the maps are not in the collection.

Not since Samuel Hearne's epic journey to the mouth of the Coppermine River in 1771 had company men journeyed to the mainland north of the Arctic Circle.[16] In the years between 1819 and 1835, however, the company had assisted indirectly in Arctic exploration by providing shelter, provisions, transport, and on occasion personnel to British expeditions, notably the John Franklin expeditions in 1819 and 1825. During these latter expeditions, the explorers had succeeded in delineating the Arctic coast from the Mackenzie delta west to Return Reef, near the present Prudhoe Bay, and east to Cape Alexander. In 1822, the company proposed an Arctic coastal expedition of its own under the leadership of George Barnston.[17] His party, after sailing as far north as possible along the north west shore of Hudson Bay was "if found practicable" to cut across the land to the Arctic shore and then to "follow the coast ... in order to fall upon the mouth of the Copper Mine River" (A12/1,fo.19d). Although this plan was strongly supported by Governor Simpson, it was fortunate that the cogent criticisms presented against such an impracticable and foolhardy adventure were heeded and that the scheme was abandoned.

The company was finally successful in organizing its own Arctic expedition when Peter Warren Dease and Thomas Simpson set out in 1837 down the Mackenzie River to the coast.[18] According to the company's directive, their general purpose was nothing less than completing the exploration and surveying of the northern shore of the continent. However, the more circumscribed goals were to extend coastal exploration from Return Reef to Point Barrow in the west, and from Cape Alexander to Chantrey Inlet at the mouth of the Back River in the east. Both aims had been brilliantly achieved by the time the party returned in 1839. In their first summer, Simpson and Dease with their party turned west from the Mackenzie delta along the Arctic coast. Eventually Simpson went on shore past the ultimate point reached by Franklin and proceeded west on foot for four days, reaching Point Barrow. His observations were compiled on the first map the expedition produced, a map he appears to have drawn, likely at Fort Confidence on Great Bear Lake, during the late summer and autumn of 1837 (237A) (plate 34).[19] Simpson also wrote a paper concerning this journey, which the governor of the Hudson's Bay Company, J.H. Pelly gave to the Royal Geographical Society. His map was printed along with the paper in the society's journal in 1838.[20] The map includes notes on ice conditions, water depths, sea floor conditions, and some data on vegetation and terrain. A second map, neither dated nor signed, but similar to the first, is on a much smaller scale (238A). Besides differences from the first map due to generalization, there are other variances in coastal features and nomenclature, although such dissimilarities need not imply that the map was not Simpson's. In the summers of 1838 and 1839, Dease and Simpson in two journeys passed Cape Alexander and finally pressed on to the eastern shore of Chantrey Inlet at the mouth of the Castor and Pollux River. While regrouping at their base camp at Great Bear Lake between the 1838 and 1839 journeys, several members of their party explored the eastern shore and peninsula of the lake and also the Dease and Kendall river routes to the northeast. Simpson combined their accumulated information with details of the inward and outward tracks of 1838 and 1839 to the Arctic shore on a map he drafted in 1839 (9B). The map, "Winter Discoveries on Great Bear Lake and Routes through the Barren Lands to the Coppermine River," is housed in the Royal Geographical Society; it is likely one of the original charts presented to the society by the company.

The success of the expedition brought honours, prestige, and financial recompense to both of the chief participants, as well as recognition to the company for its efforts on behalf of science and British imperialism. The two men were given leave for a year in light of the hardships their achievement had entailed. In 1839, at Fort Simpson, Simpson wrote up his journals and prepared several large-scale maps of the journeys, maps he took with him in his luggage on his trip out from Red River towards St. Paul. Tragically, he was murdered or committed suicide en route.[21] His brother, Alexander Simpson, borrowed these maps from the company when they reached London and had John Arrowsmith make copies at a reduced scale. These he used to illustrate the narrative his brother had written – which he expanded and edited.[22] He also expressed his desire to present the originals to the Royal Geographical Society.[23] After some hesitation the governor and committee gave permission for the transfer. The company archives contain two copies of an Arctic chart (246A), which shows the continental coast from the Coppermine River to the furthest point east reached by Dease and Simpson, as well as some of the southern shores of Victoria and King William islands that they had viewed from a distance. These copies would have been drawn after April 1840. The basis for this date is that the letter Dease and Simpson wrote from Fort Simpson on the Mackenzie in October 1839, describing their journeys, reached London in April 1840. A copy of the letter was sent to the Royal Geographical Society, which printed it with an accompanying map in 1841.[24] This printed map and the two manuscript copies are very similar. The two manuscript maps may have been originally drawn by Simpson and sent with the October 1839 letter, but the large-scale, detailed, original maps, that is, those used by Alexander Simpson and sent to the Royal Geographical Society had not reached London by March 1841. On the other hand, the two copies may have been taken from the map that was drafted for the geographical publication.

With the renewal of the British naval search in 1845 for the Northwest Passage under the direction of the by-then Sir John Franklin, the committee gave approval for a new company attempt to complete the investigation and mapping of the Arctic coast. Dr John Rae was chosen as leader of this second Hudson's Bay Company Arctic expedition.[25] The approach on this occasion was one suggested earlier by Governor Simpson, namely the route from Repulse Bay in Hudson Bay westward to the Arctic shore. The hope was for either a water passage or a short land transect. Rae wrote to Simpson that he intended to search Wager Bay and the shore between this inlet and Repulse Bay for a possible opening, but that there was little doubt in his mind that he would be able to find a route through or across from Repulse Bay to the northwest. From July 1846 to August 1847, Rae crossed the isthmus, traced in the shoreline of Committee Bay northwest as far as Lord Mayor Bay and northeast to about eighteen miles from the entrance to Fury and Hecla Strait. There he was forced to turn back because of lack of provisions, but his journal provided much useful data to scientists as well as to geographers, and he obtained all the readings necessary to prepare a map, three versions of which are still in the archives (see below). He had proven that the Gulf of Boothia does not connect with the waterways west of the Boothia Peninsula whose shores Dease and Simpson had reached. He had also shown that this land is not an island, but a great projection of the continental landmass.

Upon the completion of his journey, Rae wrote to Governor Simpson, sending along with his letter "a rough draft on tracing paper of ... [his] discoveries," drawn in pencil (291[A]) (E15/3,fo.39d). The coverage is of the northwest end of Repulse Bay, Rae Isthmus, and the coasts of Committee Bay. Latitude and longitude data, the track followed, and the various dates upon which points were reached, both coming and going, are shown. Later, Rae wrote again to Simpson saying that he was about to take ship from York to London, where he intended to complete the full report and the more precise version of his map, it "being still in a very rough state" (E15/3,fo.43d). There he was entertained by the governor and committee to whom he gave his report. Before leaving for his home, Rae gave his completed manuscript map to John Arrowsmith, who used it as the basis for the printed map appearing on 1 January 1848 in Rae's published narrative.[26] On 13 March 1848, Admiral Beaufort of the Admiralty wrote to the company asking for the loan of Rae's original "drawings and documents from which Mr. Arrowsmith had put in the outline of Dr. Rae's exploring expedition in his map" (A10/24,fos.216–216d). The Admiralty wished to complete the official version of the Gulf of Boothia on its charts.[27] There seems to be every indication that Rae's original map, used by Arrowsmith and Beaufort, is the map endorsed "Rae's Discoveries Admiralty Sketch" (292[A]) (plate 35). The rectified latitude and longitude graticule, in red ink, was likely added over Rae's original pencil grid, either by Arrowsmith when plotting his map, or by an Admiralty Hydrographic Office draftsman when their new chart was being produced. A third copy of this map, a tracing, is extant, but is of rougher draftsmanship.(293[A]).

Even before Dr Rae had returned from the Arctic in 1847, it had become apparent that Franklin and his men of the 1845 expedition were missing. In December 1847, at the request of the Admiralty, the company agreed to allow Rae, then in Britain, to be attached to a search expedition. Both Rae and John Bell, because of their travel experience, were transferred to the party under Sir John Richardson in 1848. The hope was that, after reaching the mouth of the Mackenzie and following the coast east to the Coppermine while looking for any signs of the Franklin party, the search party would cross to Wollaston Peninsula on Victoria Island and examine its coasts. However, the treacherous condition of the ice prevented them from crossing Coronation Gulf to Victoria Island. No map of the full journey was forthcoming, but Rae, in a letter to Governor Simpson in the spring of 1849 enclosed a sketch map (306[A]) (E15/5,fo.33).[28] This was the result of his investigation of the area from the east end of Great Bear Lake, including the Dease River, Dismal Lakes, and the Kendall River to the Coppermine River. The sketch showed, by different symbols, the route followed earlier by Dease and Simpson, and also the Richardson party's projected course chosen for the summer of 1849, when it was to return northeastward to the Arctic Ocean to continue the search.

In the autumn of 1850, the committee assigned Rae to head a company expedition to try again to find evidence regarding the fate of the Franklin expedition. His plan was to carry out the uncompleted portion of the previous journey, namely to reach Victoria Island and to search its shores. From the base camp at Fort Confidence, Great Bear Lake, Rae and two men set out in April 1851 with dogsleds, crossed over the ice to Wollaston Peninsula, Victoria Island, and charted the coast west and north to the farthest point near Cape Back. Since his practice had been to try to keep a map under construction daily during his journeys, Rae was able to include a rough tracing of the Wollaston Peninsula in a letter to Simpson (159[C]) (E15/8,fo.61d). In the second leg of this journey, after a short rest at the base camp on the Coppermine, the party crossed again to Victoria Island farther to the east, then went around its eastern shores north up Victoria Strait. The farthest point of land to be seen from the terminus was given the name Point Pelly, after the governor of the company. On the return trip they skirted the entire south shore of Victoria Island and crossed to the main land at Cape Krusenstern. Dr Rae provided Governor Simpson with two charts, neither of which is found in the archives. One chart showed the route along the south and east coasts of Victoria Island (160[C]) (E15/8,fo.66); the second was a rough chart of the complete campaign (166[C]) (E15/8,fos.77d–78). The earlier sketch of the Wollaston Peninsula and the second map were combined by John Arrowsmith into a map of that part of the Arctic coast examined by Rae from Point Pelly to Point Back. This map was used to illustrate a report in the *Journal of the Royal Geographical Society*.[29]

During the early 1850s, the most intensive sea search ever undertaken was carried out by the British navy and by ships and crews supported by private funds. In 1850, the government had instituted a series of awards for the rescue of all or part of the Franklin party or for reliable information concerning its fate. The Hudson's Bay Company and Dr Rae, in turn, were anxious that the company complete the task with which they had been involved, that of charting the remaining gap in the continental coast, between the Castor and Pollux River, Chantrey Bay, and the more northerly coasts of Boothia Peninsula. Rae had raised this matter in May 1852, when he had offered his services for a further company expedition. The committee accepted his proposal, and by mid-July 1853, he was on his way north of Churchill. He wintered at Repulse Bay near his old site, and in the early spring of 1854, his sledge party crossed Rae Isthmus and ventured west to Chantrey Bay at the mouth of the Castor and Pollux River. By 6 May, he had reached Cape Porter, the northern terminus of his journey. At this point, he was able to determine that King William Land was an island. More significant even than the geographical results of his journey was the momentous discovery of a number of relics of the Franklin party and his meeting with Inuit people who gave him definite reports of the fate of some members of the expedition. At York Factory in late August, Rae completed the writing of his observations and "filled in rough tracings" of the coast. He carried this map back with him to London by autumn ship (184^C) (E15/9,fo.94d),[30] and in due course it was prepared as a finished map by John Arrowsmith to illustrate Rae's report published in the *Journal of the Royal Geographical Society* in 1855.[31] The Franklin expedition relics were passed from Rae to the company, which turned them over to the Admiralty. After some delay, Dr Rae was presented with £10,000, the government award for obtaining the first definite clues to the fate of the Franklin expedition. This was shared with the members of his party. Rae's map and his report were loaned to the Admiralty; the Hydrographic Office inserted the details it provided on the Admiralty Chart of 20 January 1855 (G3/176). These details appeared also on Arrowsmith's own commercial editions. Unfortunately, Rae's original map was not returned to the company.[32]

In order to verify Rae's statements, and it was hoped discover further traces, and even human remains, from the Franklin expedition, the company supported one further search in the summer of 1855. Only one map of a small segment of the route of the journey by the experienced northern traders, James Anderson and James Stewart, was made (190^C). The two men travelled from Great Bear Lake along the Back River to Chantrey Inlet and searched its west shore toward Ogle Point and several off-shore islands. A further few relics were found, and reports were gathered concerning the fate of a boat party from the Franklin expedition which reached this area, but no remains were located.

CADASTRAL SURVEYS EAST OF THE MOUNTAINS

Cadastral surveying was the concern of two reasonably well-trained surveyors and cartographers, William Kempt and George Taylor Jr. Kempt was hired as surveyor for the Red River colony and Sheriff of the District of Assiniboia in 1822. After a short period at York in which he worked for the company,[33] he arrived at Red River and took on several surveying tasks. Two maps of parts of the settlement were drawn, but neither have survived. The first was a plan of the proposed site for a village for retired company servants on the Assiniboine River near the company fort. (95^C).[34] Rather than taking on the task of laying out the lines for the village in the spring, Kempt was instead put to settling the boundaries of existing lots and preparing the second, a full cadastral map of the Red River colony (96^C). Apparently, the plan of the village and the general map of the settlement were turned over to the colony in 1823, for there is a listing of all the lots with their measurements in the company archives (E6/10).[35] Kempt was high in the regard of the governor of Assiniboia, Andrew Bulger, who commended him for his zeal and industry and contrasted the precision of the measurements in his survey with Fidler's earlier and less precise figures. In spite of a promise for an increase in salary and additional preferments, Kempt withdrew from the position in 1824. He was then hired by the company to study the navigation on the route to Red River from York Factory.

George Taylor Jr entered his useful career with the company in 1819 at the age of nineteen. Son of George Taylor Sr, a company employee in Rupert's Land, and a native woman, he entered the marine service like his father and spent most of his early career as master of sloops or schooners at Severn Fort and between that post and York Factory. In addition to the training he had received in the use of navigational instruments and charts, he had been sent to Scotland by his father for several years of schooling. It would seem that he had received some education there in surveying and drafting. He applied this training to the surveying of Severn Fort in 1823 and prepared a plan. After engaging in a variety of surveying and other tasks for the Hudson's Bay Company, Taylor moved, probably in 1834, with his family to the Red River settlement where he acted as surveyor to the colony. Later, after 1840, he became inspector of roads and bridges and superintendent of works. Although preceded in the task by Fidler and Kempt, it was Taylor who undertook the first major cadastral survey of the colony and who produced the two master survey maps of the settlement. Although not strictly its projects, the company was interested in these surveys and the resulting cartography. The first map by Taylor, along with a copy, was drawn at a scale of two inches to the mile; because of the elongated nature of the riverine settlement, it is over five feet long and two feet wide (234^A, 235^A). The lot lines are delimited, numbered, and listed with the owners' names on an accompanying registry list. The map is both undated and unsigned. Taylor's third map, the major settlement map of 1838, is among the most ornate and finished maps in the company's collection (236^A) (plate 36).[36] It is almost a square map, with a complex border, ornamental lettering, and employs a green colour wash on numbered lots to show the ground which had been surveyed. No other cadastral maps are known to have been drafted by Taylor, who died at Red River in 1844, still a relatively young man. It is likely that he had plotted sufficient lots for future purchases, and that his further work lay in the actual setting of the lines on the ground, registering the indentures, and perhaps providing lot plans for the owners.

Several aspects of company-colony business were illustrated on other cadastral plans which were filed in London. Property details were involved in an 1839 sketch of land held by the Catholic church in St. Boniface, along with adjacent holdings largely by Metis inhabitants (241A); and similarly in an 1844 map of the mission's land in the same area (260A). Finally, the company received a query from a member of the British military regarding possibly vacant land near the fort at Red River that might be assigned to army pensioners. The inquirer pointed out that the pensioners might aid in defending the region from possible American hostility. A "rough chart," sent to London in 1848, laid out the sidelines of the occupied lots in the immediate area, along with the location of the company fort, the churches, and the main road through the settlement (300A).

Another form of property surveying and mapping occurred in the summer of 1827 when George Taylor was sent from York to the Red River settlement with the object of taking the company's first observations for the location of the 49th parallel, the southern boundary of the company's territory. The more specific reason for this exercise was the report that American traders had approached within sixty to eighty miles of Brandon House and planned to build a trading post near Turtle Mountain, which the company thought lay wholly north of the forty-ninth parallel. Taylor's route lay up the Assiniboine valley to Brandon House, then south to the west shoulder of the mountain. But more than the terrain posed problems: instruments broke, the weather was cloudy, and skies were obscured by smoke from the grass fires raging in the area. After determining the boundary location both at the west and east ends of the upland, where he carved inscriptions on tree trunks, Taylor then observed for the boundary position of Pembina in the Red River valley. The conclusion reached from this short expedition was that American posts lay south of the line and that the company could not "make good" a charge of trespass against the traders. Taylor drafted two large-scale maps and two copies. They showed the location of the boundary at the west flank of Turtle Mountain (207A, 208A), and the shape of this upland with Taylor's route along its north side from west to east (205A, 206A).

The Hudson's Bay Company was also faced over the years with providing official surveys of its various properties at trading posts, as settlement impinged upon these sites, and in defining the extent and boundaries of its land claims as new colonial governments required official registration of properties which lay within their territories. This was a concern of the province of Canada during this period. For example, in 1842, a dispute over timber licences between the company and a private contractor at the lower end of Lake Timiskaming required land claims to be adjudicated by the Crown Timber office. The land claim details were drawn up by the district master, Angus Cameron, in a local map in correspondence with the Lachine headquarters of the company (253A) (B134/c/52,fo.15). In this case the verdict was in favour of the private lumberman.

There were several further illustrations of this cadastral process. A map was submitted by the company to the Crown Lands Department in 1846 showing "the exact position of the land at Sault Ste. Marie required for the use of the establishment there" (142C). George Barnston, post master at Tadoussac in 1848, made a sketch which posed a problem for a later government land surveyor who was supposed to map the boundaries of the Mille Vaches Seigneury at Portneuf: Barnston's map showed conflicting company-seigneury boundary claims (303A). Finally, a map sent by George Gladman, post master in 1848 at Tadoussac, illustrated fur reserve land boundaries at that settlement. The map made it clear that there might be friction with persons the company believed to be trespassers. It also showed the location of a block of land which Gladman suggested should be purchased in order to prevent future interference in the company's fisheries at Tadoussac (307A).

MAPPING THE MAIN ACCESS ROUTES TO THE INTERIOR

The struggle to establish the Selkirk colony necessitated a simultaneous exertion by the company to review and refashion the access route inward from York Factory to the junction of the Red and Assiniboine rivers. The ultimate requirement was a more efficient method of transporting an increased volume of goods and, it was hoped, the export of produce such as flax, wool, and tallow. The absolute requisite was a shift from the canoe to wooden boats and other transport systems. William Kempt was employed in solving this transportation problem in 1824 after leaving Red River and before departing for Britain where he resumed his practice as a surveyor. He was asked to see what improvements could be made in the river and lake route from York to Norway House. He was provided with a boat, equipment, and men, and was instructed to undertake minor projects like clearing portages. The summer of 1824 witnessed a considerable improvement in the navigation of the south track, according to Governor Simpson, who wrote to the committee that, nonetheless, the cost of the summer's work would apparently not amount to more than one hundred pounds. A regional map from Red River to York and two local maps resulted from Kempt's efforts. The regional map is a line sketch, without extensive detail, of the Hayes River track to Lake Winnipeg, continuing south through the lake and river to the Red-Assiniboine junction (192A). The most interesting features of the map are the delineation, in black and red pecked lines, of the "Present track of the Boat in Coasting the Lake," and the "Route to be taken by a decked Boat."

Among other locations on the regional map were both White Fall Portage and Trout Fall. These two sites were areas improved by the crew, and large-scale maps were drafted of each. The sketch of Trout Falls depicts the physical characteristics of the site, with proposals for simple river channel clearing, which would allow for safer "running" down the river and hauling up (193A).[37] The White Fall Portage map is much more ornate, is in colour, has brush hachures, and shows proposed and completed engineering works (200A) (plate 37). This portage is located on the Hayes River between present Robinson and Logan lakes and consists of a series of four falls, which interrupt the river course over a distance of approximately one mile. Kempt's crew cleaned up the portage, built a "bridge" over a swampy area and rebuilt a second one over a side stream.[38] On the map he outlined a fascinating scheme whereby water from upstream would be held back by embankments in a reservoir. When desired, the water could be let out below the lowest waterfall through a series of three small locks. This was a pioneering proposal for any waterway in the west and north of the continent. But it was impracticable as it could be used only by boats going downstream unless a pumping system were installed for upstream movement. Kempt also prepared a copy of the portage map (108C). The original map and the copy, dated 1825, were likely completed by Kempt after his return to Britain.

George Taylor Jr was involved in both precise measurement and mapping of the major waterways, and he, like Kempt, also worked on what might be called route improvement. When he was sent inland in 1827 to make the first company boundary delimitation of Rupert's Land at Turtle Mountain, he made a very careful plot of the track from York to Red River via the Hayes River and Oxford Lake. During a winter at Red River, he made a series of latitude and longitude observations and drafted a map of his inland journey to the settlement and southwest to and from Turtle Mountain (209A). Taylor's method was to keep a log book from mile 0 at York Fort, making a record of compass directions and distances from point to point. Occasionally he observed for latitude and compass variation. The details were recorded on rough sketches (204A), each section proceeding from the previous page and previous sketch; the left hand side of each page was the north point. He later transformed the rough sketches into a series of fifty-five finished maps of the journey in a separate book (203A) (plate 38). This was similar to Fidler's field survey method, though Taylor's final maps were much more polished. The maps were valuable indicators of both specific distances and total mileages along the waterways between the posts. After he carried out the 1828 colonial census, he and his wife set out from Norway House to perform the same type of meticulous survey of the waterways leading up the Saskatchewan River and across to Jasper House. He mapped this route in the form of seventy-eight segmental sketches both in rough and in finished form (211A). One of the sketches noted the birth place of the Taylor's first child, and his proud astronomical observation at this event on the bank of the Athabasca River reads, "*Mary Thomas Taylor*, Born Oct. 12, 1828. Lat 54°10'7"N"!

The most innovative aspect of the route improvement program was the concept of extending the transport routine from the open water season through the winter by means of the construction of a winter road system in the Oxford Lake to Playgreen Lake section of the track.[39] The scheme was one which used the frozen surface of rivers and lakes, with cleared and blazed paths between water bodies and across points of land; these same paths could be used as portages during the summer. Dogs, horses, oxen, or even, as was unrealistically suggested, reindeer could be used to move cargo. From 1812 to 1826 the winter road program suffered from setbacks of various kinds, but, although the system proved impractical and the Red River settlement incapable of producing the exports required to cover the costs, the Northern Department Council and the London committee were willing to maintain the program and to introduce new elements. Thus by 1828, further impetus was given to the scheme, and Taylor was brought onto the team in 1829 as surveyor, cartographer, and eventually foreman, responsible for the rebuilding of many sections of the largely land-route road. His task was to determine the locations of places and to combine astronomical observation with measurements along the route to try to find a more suitable path between the upper Hayes River near Lake Winepegucies (Molson Lake) and Lake Winnipeg. Holy Lake, on which Oxford House was located, a lake he considered had been poorly delineated previously, was the central focus of the route. His first map, drafted in May 1830, of the upper course inland from Oxford House, was based on the detailed examination and measurement of existing and potential alternatives to the route (127C).

Taylor's second winter road map is an excellent example of the clean-cut properties of his drafting style and shows his predilection for an ornate lettering style, a recognizable feature of his cartography (226A) (plate 39). No visual or written scale or directional symbol is included on the map, likely because this information was included on the first one. This map depicts the route of the old road and of the newly-completed portions. Included are the six "stages" along the road, where small dwellings and stables were being constructed. Taylor noted vegetative features, such as "Good Wood on this Portage and thickly Set," and "Small Pines and Swampy." South of the winter road is a depiction of the normal track along the Hill River and Knee Lake to Oxford House on Holy Lake, with the main portages.

SPECIAL PURPOSE MAPS

Several regional topographic maps came into the company's office in London from employees who compiled them at the committee's request when it required a better understanding of certain local problems. Two of these maps are of considerable interest, both having been sent to London in 1827. Taylor had been inland from Severn Fort in 1822–3 in charge of a party looking for wood of different types and qualities for the making of planks and for use as pickets and firewood. His training and experience in surveying and mapping, travelling inland, and in evaluating timber properties were then put to use in the spring of 1827 when he was sent from York Factory on a circuitous journey up the Nelson River, across to the Fox and Steel river area, and back to the post. His goals were to ascertain the boundaries of the lands "belonging to the Indians of the place," to indicate the locations of beaver lodges and of any free traders' posts within that territory, to make a map of the area, and apparently to evaluate wood resources in the local region above York (B239/a/135,fo.11d). He kept close compass control of his route, took seven sets of observations for latitude, and drew twenty-one segmental sketches of lakes and rivers on the way (202A). He also asked several Indians to draw sketches of their locales; one, a Sturgeon River Indian, drew a sketch of the river and a lake, a map Taylor found of value (118C). Taylor also drew a composite map of his whole route on which he plotted the locations and numbers of beaver lodges seen en route as well as the locations of fur trading posts (116C). It is likely that he also indicated the character of the forests for at the end of his written report he offered a page of "physical geography," emphasizing the extent and quality of the woods (B239a/138,fo.8d).

The second special purpose map of particular interest shows part of the Lake Huron area (210A) (plate 40). It is undated and unsigned. The evidence proves it to be the work of John McBean who was in charge of the Lake Huron District. His headquarters were at La Cloche post in the North Channel of Lake Huron. On 26 August 1827, he sent a sketch of "the communication from Lake Huron to the height of land" to Governor Simpson (B109/b/1,fo.18). The reason for the production of this map was mainly to show what McBean considered the best route to be followed in the provisioning of the Timiskaming district, which lay northeast byond Lake Nipissing. Earlier, Simpson had suggested that this district be supplied from Sault Ste. Marie along the North Channel and through Lake Nipissing. McBean was of the opinion that there was a better and more direct route using a river-lake track, from just east of La Cloche, that led northeast via Lake Panache to Wanapitei Lake and thence through Temagami Lake to Timiskaming Lake. A second purpose of the map was to convince Simpson that the Great Lakes, with La Cloche as the supply point, and a route using the Spanish River over the height of land to Lakes Mattagami and Kenogamissi in the upper Moose system, would be easier to use as a provisioning route for this region than that coming in from James Bay via the Moose and Mattagami Rivers. He relied on a "half-breed in the Company's employ" and other local Indian people for much of his information (D5/2,fo.251). He did not succeed in convincing the governor.

Other individual maps illustrated additional aspects of the company's activities and interests. When William Swanson reported on his investigation of the murder of a company employee, his wife, and six Indians by local natives near Hannah Bay, he added a sketch showing the route he and a party took from Moose Factory to the area (219A) (B135/a/138). An anonymous plan (221A) (plate 41) of Lachine village in 1832 indicated not only all the buildings in this settlement, but especially three properties the governor and committee were being encouraged by Governor Simpson and the local company officials to buy. One would serve as the permanent home of Governor Simpson, another as a depot, and the third as headquarters office. These proposed properties were measured in relation to their distances from the pier and entrance to the Lachine Canal, which was to be the main artery for company boats and canoes. The map, in black ink with colour washes, would have been invaluable to the committee because its clarity and detail allowed the final choice to be made in London.

Regional maps which portrayed the principal waterways of several areas of the near west, still relatively unknown to the company, were prepared by several employees. The Winisk River watershed and adjacent districts, especially of Severn River tributaries, with their main interconnections were the subject of two other maps. One of these was drafted by John Work in the year 1824 or 1825 (194A). Work had been based at Severn Fort and later at Island Lake post from 1815 to 1823, and had been inland in this larger region on several occasions. He had been sent to investigate the trade prospects of the Winisk Lake vicinity and had set up a small outpost cabin there. In 1833, George Barnston led a party from Albany to open a new house, Fort Concord, on the upper Winisk River. Because a dispute arose over the effect of this post on Severn's trade, Concord was closed in 1834. Barnston drafted a map in that year of the Winisk country and also of the Equan and Attawapiskat River systems (232A). Offshore, Akimiski Island appears, somewhat distorted and wrongly oriented.

MAPS OF A BEAVER CONSERVATION PROGRAM

Another cartographic innovation of this period was the inventory and mapping procedures initiated by Robert Miles, master of the Rupert River District in 1839, as part of a beaver-conservation program. This was a response to over-trapping in the older, established territories of the Eastmain. In the spring of 1836, Miles wrote a private letter to Governor Simpson in which the concept of preservation was presented (D4/22,fo.65d). Simpson approved of the idea and reported in the spring of 1837 that the interest of the governor and committee in London had been "excited," and that he had written to masters of the districts and posts of Timiskaming, Abitibi, and New Brunswick. Young beaver were to be gathered in the spring of 1838 for transportation to Rupert's House and then to Charlton Island in the spring of 1839. Typical of Simpson were the detailed instructions he sent to Miles in February 1838 (D4/23,fos.105,105d). First, Miles should have the island "surveyed" thoroughly in the summer of 1838, with the necessary water and food resources indicated and the habitat the animals required noted. Then several trusted Indians should be sent to clear out the otters, which could destroy the new inhabitants. Miles reported to Simpson that, as of August 1838, one of the oldest and reliable coast Indians, who had originally suggested the idea to him, was already on Charlton looking over "every nook and corner of the Island." On 12 February 1839, this man, Cauc-chi-chenis, and his family, and another man, Kennewap, were sent back to Charlton, with provisions, to trap otters and to make a survey. In order that they would be as exact as possible in their mapping, they had been provided with a sheet of cartridge paper and pencils. They returned to Rupert's House in mid-April 1839, and gave Miles "a very satisfactory chart laid down by them as they proceeded" (135C). The map had sixty-one lakes marked on it, all of which the Indians considered desirable as beaver habitats. In addition, many lakes were not included which could readily be made habitable. Miles sketched a pen and ink version of the pencil original (251A); the Indians later made some additions in red ink to this copy. The map was copied by Henry Connolly (240A), and this copy was sent to Governor Simpson and eventually reached London.

A map of Ministickwattam Island was begun in the summer of 1842 by Nabowisho, a Nottaway River Indian, who was sent by Miles to survey this island also for conservation purposes. Some ninety-two lakes and rivulets were charted. The Indians' pencil sketch was inked over by Miles, as he had done previously on the Charlton map, but his work was not completed until the summer of 1843, and the map then sent to London (256A) (plate 42). Further reviews of the situation on the islands were made in 1844 and 1845–6, and several further maps drawn, purportedly showing the beaver census on these occasions (268A, 275A, 276A). However, the program did not come up to the expectations of Simpson and the committee. In spite of a series of very favourable reports on beaver counts made by Indians over the years, including the years up to 1851, during which Indian hunters were sent in to winter, the hunts were very disappointing. In 1852, for example, authority had been given to harvest five thousand skins, but only eight hundred were trapped. Simpson accused Joseph Gladman, who had become responsible for the program, of reporting inflated counts. In 1858, only 264 beaver were harvested. After this the conservation program was allowed to lapse. Altogether, the project resulted in eight maps, six of which are still in the archives.

Soon after the mid-nineteenth century, company traders ceased extensive exploration in Rupert's Land and along the Arctic shore. Except for a few maps portraying routes of exploration in the Mackenzie-Yukon region, cartography of the lands east of the mountains was concerned with property delineation. West of the mountains it was a different story as will become clear in the next chapter.

11 Mapping West of the Mountains, 1821 – 1849

> ... send to us ... Information respecting the coast and any navigable Rivers which fall into the Sea to the North of the Columbia, also a general description of the Country, numbers and Conditions of the Natives ...
> Committee to Governor Simpson June 1824

After 1821 and its merger with the North West Company, the Hudson's Bay Company's domain extended from Ungava and the Labrador coast in the east to the Pacific estuaries and offshore islands on the west, along the breadth of the Arctic continental shoreline north, and into the mountain-framed basins of Utah and California south. In the first phase of the company's occupation of its newly absorbed western domain, company servants mapped on land and at sea; they sketched the routes they followed as they explored the vast mountains, plateaus, basins, and river systems of the new territory; they surveyed settlements and post locations; they charted the coastal areas and potential harbour sites. Annual trapping expeditions extended their contacts with the country as far southward as northern California and nearly to Great Salt Lake. Moreover, traders probed the mountainous interior of British Columbia and the Yukon to discover less arduous and more efficient routes for the carriage of trade goods and returns. Finally, a new entrepot for the western trade, Fort Victoria, was sited on a fine, natural harbour on the south shore of Vancouver Island, an essential move when the 1846 boundary treaty between the United States and Britain brought the company's occupation of the Columbia district to an end.

At least twenty-two men were involved in this exploring, map making, and surveying; a number of them were part of the company's "inheritance" from the North West Company; five were sea captains; only one surveyor had actually received training. It is impossible to determine exactly how many maps were drawn, but there were about eighty, only half of which are still in the company archives. As in previous periods, coastal and harbour charts were prominent among the cartographic records; one-quarter of the maps produced between 1821 and 1849 were such charts. Property surveys are also an important part of the record for this period.

MAPPING THE COLUMBIA DISTRICT

Alexander Ross was one of the cartographers the Hudson's Bay Company inherited. His first known map, dated 1821, appears to have been the only map drafted in the Columbia region during the time of the negotiation of the agreement for union (7B).[1] It represented fairly the state of general knowledge the company had about its new territories. The original map would probably not have been available to the London committee, since it was kept by Ross. But he would no doubt have shown it to the Columbia District officers and probably to Governor Simpson, who could have gained a useful impression of the district's topography from it. The map accentuated details of the areas Ross knew best, the lower Okanagan area, the middle Columbia valley, and the Snake River basin. Although there was some inexactness in locations and distances, to which he freely admitted in his annotations, the map was a worthy successor of those David Thompson had made of this region for the North West Company. This one cannot be considered a true company map, since Ross was not an employee at the time he constructed it.

Three other maps of the Cordilleran region, which might have been useful, were available in the decade of the twenties. One was a map of the "main ridge of the Rocky Mountains" (98C). No author was noted in the archival catalogue entry, where the only reference to the map in company records is found. The two most obvious possibilities are Ross and Archibald McDonald, a former employee of Lord Selkirk in Rupert's Land, who was also in the Cordilleran area in 1824–5. Ross knew far more of the region than McDonald as he had already delineated this highland ridge in his 1821 map. And in 1825 he had produced the second map that showed the region, a three-sheet map of the Snake Country, which would have included the Rocky Mountain area (101C).[2] It is unlikely that he would have drawn yet another map of the main ridge of the Rockies at this time. The third map, drafted on five connected sheets, which could have been helpful to company officers, was one drawn by McDonald in 1824–5 of the "Country lying West of the Rocky Mountains" (99C). McDonald had been sent by Governor Simpson to the Columbia region in 1821 to provide a full account of the situation in the district. Simpson himself went west in 1824, and once there, was dissatisfied with the available maps. He asked McDonald to compile a map for him from the various sources at hand in the Columbia district. Simpson apologized to the committee because it was not as well "finished" as he could have wished, but McDonald had had only minimal exposure to cartography while preparing in London for his service with Lord Selkirk. The projection base was taken from David Thompson's map of the area, and the various details fitted into it.[3] There is no record indicating the extent of the map's coverage, but since Simpson was concerned with all company activities west of the mountains, it may have included details of the Columbia basin, New Caledonia, and whatever was known of the area called the Snake Country.

MAPPING THE SNAKE COUNTRY

Snake Country was an ambiguous geographical term used by the company to designate a territory extending far past the confines of the Snake river and its tributaries. It was used as well to refer, in an indefinite manner, to certain Indian tribes living in an area that included parts of present day Oregon, Idaho, Wyoming, and Nevada. The company considered Snake Country to be a kind of march area, which would insulate the Columbia and New Caledonia regions to the north from the incursions of the American fur traders. Although the company had exclusive rights of British trade in the Pacific Northwest, it had no right to prejudice or exclude American citizens from the region. In its reorganization plans, the company stressed the importance of exhausting the fur resources of the Snake Country by encouraging overtrapping. As a result of this strategy, which meant annual trapping and trading forays into the area, the American traders would not find it profitable to enter into serious competition. It was also hoped that the scheme would hinder an American advance towards the Columbia River, which company officers conjectured might become the line of the international boundary in future negotiations with the United States.

The prevailing geographical theory underlying the company's marketing plan was that somewhere in the heart of the Snake Country lay mountains that were a common height of land. There, from a source of lake or lakes, the major rivers of the west and south west would flow. The hope was that a river system existed to link the Snake Country to an accessible harbour on the Pacific shore, or that the system would connect north to the estuarine area of the Columbia River. Although in the beginning, exploration was a secondary purpose of the annual fur trapping and trading expeditions into the area, it gradually assumed primary significance, especially for the excursions after 1824 that were led by the forceful Peter Skene Ogden.

In all, Peter Skene Ogden led six Snake Country expeditions for the Hudson's Bay Company from late 1824 until 1830.[4] In the process of exploration, Ogden crossed and recrossed an enormous range of territory from the Columbia River, the Snake River junction, and the Flathead area south, eventually to touch upon San Francisco Bay and the Gulf of California, and then east into the vicinity of Great Salt Lake and almost to the Wyoming border. Unfortunately, Ogden's cartographic ability did not match his geographical interest or general observational skills. He was not driven, as Peter Fidler had been, to measure all the nuances of course changes, or to observe locations exactly.

Ogden's first expedition in 1824, with William Kittson, clerk and trader, as second in command, started from Flathead Post on the Flathead River and wended south, criss-crossing the mountain ranges of eastern Idaho and western Montana, over the upper Snake River valley to within a short distance of the eastern shore of Great Salt Lake. The intention had been that Ogden would then strike west to search for a river, the Umpqua, which flowed into the Pacific in Oregon Territory. This river, which was supposed to be a main east-west water route to the ocean, in reality is just a short one, not a Columbia-like passage across the grain of the country. However, although Ogden believed he was only ten days' march from the upper Umpqua, he did not go west of the middle Snake River valley (in the vicinity of present day Boise) because of a number of desertions among the free men in his party to an American trapping group. Instead, he went down the Snake and across to Fort Nez Perces. Ogden turned his notes and observations over to Kittson and asked him to combine this data with his own to make a map (D4/19,fo.82). Kittson did so, on two sheets, drawn at twenty miles to the inch, and sent it in to Dr John McLoughlin, chief factor of the Columbia district at Fort Vancouver, with his own journal of events (199^A) (plate 43).[5] Another map of the first expedition was drafted by Archibald McDonald in late 1825 when McLoughlin brought Ogden's journal to Fort Vancouver. McDonald would have finished the map before he left Fort Vancouver on 7 January 1826 for his new assignment at Fort Kamloops. It is doubtful that he had Kittson's journal or map with him since only Ogden's material is mentioned. Apparently, McDonald drew the map and a copy (104^C), for McLoughlin sent two maps to Governor Simpson, suggesting that one be kept at York and the other be sent on to London. The latter would have accompanied, and helped the committee understand, the "communication sent ... on the subject of the Snake Country" (D/4/6,fo.50).

Ogden's second expedition in a sense was a continuation of the first journey, for, with less than three weeks' rest, he started off again on 21 November 1825 with a smaller party. His plan was to join up with another company group, led by the trappers and traders Finan McDonald and Thomas McKay, and head south to the Klamath Lake region of southern Oregon to trap beaver in an area where the McDonald and McKay party had already been earlier the same year. Ogden hoped to sort out the perplexities of the river patterns in this more southerly area. Instead, after the two groups united, the decision was made not to search for the Klamath because the season was late, and because they did not know enough about the supply of provisions to the south. The united group travelled through the upper Snake basin, then turned back west to the Willamette River valley and eventually to Fort Vancouver. They had learned little about the lakes and rivers to the south of the Willamette. At Fort Vancouver, Ogden wrote his journal of the trip and apparently drew a map of the route they had followed (109^C), although the letter from McLoughlin to the committee does not say so concretely (B223/b/2,fo.21d). Two copies of the map were apparently prepared and sent on two different ships to Britain (B223/b/2,fo.29). Neither remained in the collection.

The map Ogden drew to illustrate the 1825–6 report on the Snake Country is also lost (110^C) (B202/e/2,fos.1,1d,2). In this report, on four occasions, he refers to a "chart": once when he describes the Snake River and some eighteen tributary streams, once when he indicates the former location of a Pacific Fur Company post, once when he discusses exploring the course of the Snake River (and gives the location of the grave of one of the party), and finally when referring to the salt lake which he said he had seen in the spring of 1825. In this report he claimed to have seen almost all of the Snake Country, and one may suppose therefore that he tried in his map to portray this very large district.

Ogden's claim would have been something of an exaggeration until the end of the third and fourth expeditions. In the third journey, in 1826–7, he attained the goal of the previous expedition, namely, Klamath Lake in southern Oregon. The exact path cannot be traced because Ogden's map of the trip cannot be found, even though it was received in London and catalogued (113^C).[6] The fourth journey, in 1828, involved a transect of the upper Snake River's eastern tributaries and of the Blue Mountains west of the river. No map is mentioned nor has one been located.

On the other hand the map of the fifth expedition in 1828–9 has survived (216^A). On it, Ogden traced his journey south from Nez Perces House (Walla Walla) to Malheur Lake, over the divide to the Humboldt River and Carson Sink in Nevada, and north of Great Salt Lake in the Bear River valley. A northern tier of basin and range country south of the Snake River basin was thereby added to the sum of geographical knowledge. Although it has considerable distortion in direction and distance, amounting to a 90° rotation in the Salt Lake area, it was a significant, early map of this immense territory.[7]

Finally in 1829–30, Ogden headed on his final expedition south. Regrettably there are no cartographic documents to record a spectacular feat of exploration, since his notes were lost in an accident on the Columbia River during his return trip. Following up on his previous expedition, he went south to Carson Sink in central Nevada and over to the Colorado River. He trailed this river's course down to the Gulf of California, then retraced his steps for many miles before heading west across the Mojave Desert and through the mountains into the Great Valley of California. His route was then north, with a side trip to San Francisco, up the Sacramento River valley, northeast across the high plains of central Oregon, eventually to the Columbia River and Fort Vancouver.[8]

As part of the program to deplete the country south of the Columbia of its furs, and in concert with the Snake Country expeditions, several parties were ordered south down the coast from Fort Vancouver. Reports of good beaver country in the basins of several "large Rivers" had filtered north; these reports and the hope of discovering such useful east-west routeways were all the incentives required. Neither expectation was fulfilled: only small fur returns resulted, and only insignificant rivers were found. Alexander Roderick McLeod was chosen to lead two of these expeditions in 1826 and 1826–7. The parties crossed all of the Pacific-bound streams, including the Umpqua and the Coquille, which were two of the main river crossings through the mountains, as far south as the Rogue River. In the same expedition, they investigated both sides of the Coast Range. Only one map resulted; McLeod drew it to accompany his journal of the first journey, which lasted from May to August 1826 (112^C).[9] The journal and map were forwarded to London by McLoughlin (B223/b/2,fo.23).

CHARTING PACIFIC COAST RIVERS AND HARBOURS

The Pacific coastline had been generally delineated, using Spanish, British, and some American surveys. International settlement of the rights of passage to the coastal waters of this seaboard in the early 1820's allowed the company to develop plans for a coasting trade and for the opening of trading posts north of the Columbia River. What was then required was the better charting of specific harbours, of major water channels, and the introduction of company-owned vessels to carry on both trading and charting functions. By 1824, the committee considered that the way was clear for a maritime venture up the coast. They obtained a vessel in Britain, the *William and Ann*, appointed Henry Hanwell Jr, son of one of the company's veteran sea captains, Henry Hanwell Sr, to his first post as captain and dispatched him to the Columbia River in the spring of 1825. Hanwell was assigned two general tasks: to "ascertain if there were any good harbours on the coast" to the north, and to discover "whether a beneficial trade may not be carried on with the natives" (A6/21,fo.11). Specifically, he was to search for a good roadstead or harbour at Portland Inlet, to enquire after the entrance to Simpson's (Nass) River, to go to Sitka if possible, and to examine the entrance to the Fraser River. This pioneer coasting expedition was a failure; little information of significance to the company resulted. Hanwell called in at only a very few points, keeping well off-shore (B223/a/1,fo.3). Even when the ship was at anchorage, he refused to go very far from the ship himself or to permit the ship's company much contact with the natives. He was, it seems, very fearful of hostile Indians (B223/a/1,fo.18d; D4/7,fo.205d). Hanwell did prepare three coastal or river mouth charts: of the "Columbia River (105^C)," of "Observatory Inlet (106^C)," and of the "Gulf of Georgia" (107^C). It is difficult to judge whether the Observatory Inlet chart went much beyond the first few miles, that is, just past the entrances to Portland Canal and Nass Inlet. The title of the third chart suggests altogether too extensive a purview. The course of the ship was given in a journal of the journey (B223/a/1), the return leg being in the Gulf of Georgia, southern section, and not around the entire gulf as the map title would suggest. Their course was to the Point Roberts vicinity, then through the San Juan Islands. All three maps were used by the Arrowsmith firm to compile commercial maps of the coastal area.

The undecided political fate of the country lying north of the Columbia River prompted Governor Simpson to support the development of a new interior supply line to New Caledonia and the interior Columbia area using the Fraser River, with a new company marine depot being situated somewhere near the mouth of the river. The lower Fraser valley had not been examined since Simon Fraser's descent in 1808. James McMillan was chosen to lead a group to investigate and report on the navigable quality of the entrance and lower part of the river, on the route of communication from the Columbia mouth to the Fraser delta, on the physical and human geography of the lower valley, and on the possibility of opening "a friendly intercourse" with the native peoples (D4/7,fo.72). Included in the expedition, which set out in late autumn, 1824, were John Work, Thomas McKay, and Francis (François) An-nance, all of whom produced maps.[10] The first map accompanied Captain McMillan's report; Work co-operated with McMillan on the report but prepared the map by himself (100^C). Although missing from the archives, the catalogue indicates that the map depicted the Pacific coast north to the Fraser River, with the route taken by the expedition from Fort George, near the mouth of the Columbia, to Puget Sound, along its shores to Boundary Bay, and across into the Fraser above its delta. Work had come down river through the delta on his return. He and McMillan reported that, as far as the party had traced its course, the Fraser was navigable. They were deceived, however, by informants who claimed that the middle courses of the Fraser and of the Thompson rivers as far as Kamloops were not barred by dangerous rapids or falls – an optimistic but seriously flawed claim. Annance had also delivered on his promise to prepare a map based on his observations of the countryside (102^C). So had McKay, who handed in a map of the "N.W. Coast of America" which may have shown area extending beyond the Fraser River and down the coast from the Columbia's mouth (103^C).

Other ships and other captains soon came into the coastal service of the company. Among them was Aemilius Simpson, a former Royal Navy lieutenant, who was a relative of the governor. The young Simpson was engaged in early 1826 as surveyor and hydrographer on the Pacific coast when these services were needed and as commander of a company vessel. His ship was the new vessel *Cadboro*, which did not arrive from England to anchor off Fort George until June 1827. As soon as Simpson reached the west coast, he was engaged in both marine and mapping tasks. Moreover, it was soon clear that, given his intelligence, interest, and enthusiasm, he could usefully investigate the area's potential for the fur trade. Unfortunately, Simpson's promising career was cut short by his sudden death in 1831 at Fort Simpson.

It is impossible to determine the number of charts for which Simpson was responsible during his brief career: the catalogue entries are too vague. He drafted at least eight, only one of which is still available. His first cartographic assignment – and his first cruise on his new ship in 1827 – was to make a survey and chart of the Columbia River from Cape Disappointment to Fort Vancouver (114^C). In that same summer Simpson began his maiden sea voyage as captain of the *Cadboro* when he sailed the ship from Fort Vancouver into the Fraser River. The ship was to act as a defence for the party erecting Fort Langley, the initial company post north of the Columbia. He was up and down the Fraser on several occasions, and during these trips charted the river; his chart was catalogued as having been received in 1829 in London (124^C). Simpson's third chart, and the one still extant, is a sketch of New Dungeness on the coast of the Olympic Peninsula, in the Strait of Juan de Fuca (213ᴬ). The map is less of a hydrographic chart than a diagram of the coastal area. It includes the locations of anchorages and local features relevant to an unfortunate matter then at hand, a search for the Indians who had been responsible for the deaths of five company employees.

The last chart that Simpson is known to have prepared from his own observations was one of the entrance to the Nass River in Portland Inlet (125ᶜ). The various bases for this chart extended as far back as Hanwell's initial voyage, which had included the search for Simpson's (Nass) River. Altogether, Lieutenant Simpson was on Nass River expeditions in 1829, 1830 and again in 1831, when Fort Simpson was established at the mouth of the Nass (B223/c/1,fo.26). He arrived on 28 August 1829 and proceeded to survey from the ship's boats. Simpson and his crew found a "good broad channel," and went on "to view a position which from its appearance seem'd best calculated for erecting an Establishment" (B223/c/1,fo.26). He handed the chart of the river, as far as it had been sounded, to Dr McLoughlin. Apparently, Simpson improved the details on this chart in 1830 (B223/c/1,fo.25d; A12/1,fo.378d).

In the catalogue there is a list of twelve charts of the northwest coast.[11] Simpson's name is noted with four of these which can be dated 1828 (120ᶜ, 121ᶜ, 122ᶜ, 123ᶜ). Several of the twelve were drawn from original charts made by other persons, as accompanying notes show. It may be that Simpson used such sources for the production of his four charts. He may have added features and corrections to the four if, indeed, he had visited their locations, but this cannot be verified since they are no longer in the archives.

The ten extant charts in the cartographic record of the Hudson's Bay Company for the northwest Pacific coastline during the years from 1830 to 1850 do not reflect an intensity of maritime activity nor an increasing knowledge of the shore configuration and offshore navigation hazards in these formidable waters. The charts are concerned rather with the characteristics of various harbours. That four of them depict the dangerous passage across the mouth of the Columbia River testifies to the company's growing concern with the problems in using this river as the site for the chief company post in this region or as a route to the interior (218ᴬ, 239ᴬ, 248ᴬ, 305ᴬ). In addition, the area was politically unstable because the river was potentially a border line. Five of the ten charts were drafted in 1831 and 1832 by Thomas Sinclair (218ᴬ, 222ᴬ, 223ᴬ, 224ᴬ, 225ᴬ), a former York Factory sloopmaster and first mate on the *Cadboro* under Simpson, whom he succeeded in 1831.

MAPPING FARM OPERATIONS

In 1825, the company had declared its intention to affirm British sovereignty more concretely north of the Columbia River by establishing a new post on its north bank, some miles up river from Fort George, the post on the south bank. The new post, Fort Vancouver, would serve as the company headquarters. This political-commercial decision also reflected the need to provide a regional food supply in order to cut the cost of feeding employees – and animals. Alluvial terraces around the site provided a natural base for the production of timothy and clover for hay, and of barley, wheat, peas, beans, and potatoes. In the early 1830s, a farming operation was attached to another new fort at Nisqually on Puget Sound, and later a farm specializing in cattle and sheep husbandry was opened in the Cowlitz valley, on the trail between the two forts. Political considerations led to the separation of these agricultural operations from the fur trade, and in 1839 the Puget's Sound Agricultural Company was floated under a related but separate board of directors. The existing farms at Nisqually and Cowlitz were transferred to the agricultural company's control. Slowly, a handful of settlers were introduced into leaseholds, but for a number of decades much of the work was done by the new company's employees.

There were several maps drafted in the early years of the agricultural company's operation, which were helpful to both the Hudson's Bay and Puget's Sound Company officials. The earliest map, made with a copy, was drawn under Governor Simpson's direction, as a result of a tour of inspection that he made in 1841. It may well be that these were produced by Adolphus Lee Lewes (son of the long-standing company official John Lewes), who had been trained in land surveying. Simpson passed across the country between Nisqually and Cowlitz twice during the autumn of that year, coming from Fort Vancouver on a return journey north to Russian Alaska accompanied by Chief Factor James Douglas. The sketch is in the form of an annotated depiction of the terrain, watercourses, and vegetation, and emphasizes the number of grassy plains to be passed along the trackway (249ᴬ).[12] One copy was sent on 25 November 1841 to the London committee (D4/10,fo.23d), and the other to the Puget Sound Agricultural Company directors (D4/59,p.117). A second map, entitled "Sketch of the Prairie Land about Nisqually" and forthcoming at the same time, was made under Douglas's direction and as a result of the tour he made into that region (137ᶜ). He was accompanied by Captain William H. McNeill, a captain of one of the company ships. McNeill would have taken Simpson and Douglas to and from Sitka.

Later William F. Tolmie, the chief at the agricultural establishments, reported that Douglas "had a plat of the country made out" (F12/4,fo.30).[13]

In the same year, 1841, Simpson appended a plan of the Cowlitz farm in a report to the agricultural company (250A). It was the first of a number, drawn each year, perhaps by William Tolmie, with a diagram of the fields showing acreages and types of crops grown, as well as building locations (257A, 258A, 262A, 274A, 141C). Chief Factor McLoughlin sent to London such a map that he had received from Tolmie. It showed what had been sown in 1843 and what was to be sown in various fields in the next spring seeding (258A) (plate 44).[14]

Between the years 1844 and 1846, several maps were drawn depicting the characteristics of Fort Vancouver and the surrounding agricultural community. The area was then at the height of its development as a British settlement – before it was infiltrated by American settlers and encumbered with an American army garrison after the signing of the Oregon Treaty in 1846. Three maps portray the ravages of a devastating fire, which swept to within three hundred feet of the fort stockade on 27 September 1844. H.N. Peers drew a map that vividly depicted the farthest extent of the fire and the fire lines ploughed up or burned to act as fire-breaks (264A).[15] Lewes was likely the draftsman of a larger-scale map which, in effect, put a magnifying glass on the area of the fort, showing vividly the fire front enveloping several barns, a house, and part of the orchard, just to the rear of the fort (267A). Two other maps of about the 1846 period illustrate the environs and actual detail of the fort (277A, 278A). The most artistic is that by Richard Covington, which includes a large-scale settlement plan of the fort and immediate village and, at a smaller scale, the cultivated, pasture, and wooded lands.

BRITISH SECRET MILITARY MAPPING: THE VAVASOUR AND WARRE REPORT

A suite of twelve maps in the collection is closely associated with the Hudson's Bay Company, even though they were not drawn by employees. Nine of these were of locales west of the mountains. They were prepared by two men engaged in British government service during the years 1845 and 1846 and presented to the company with an extensive report. Their existence was due to the increasing political tension between British and American interests on the plains and also west of the mountains in Oregon Territory. In these regions a surge of American frontiersmen and settlers had increased the company's alarm that the international boundary on the grasslands would be ignored by the Americans. Even more, the company feared that the boundary west of the mountains, in dispute since 1818, might be pushed to the "54°40' North" position exploited by the Democratic party as a campaign slogan during the 1844 American presidential election, rather than being along the Columbia River or even as an extension of the forty-ninth parallel to the ocean coast. Governor Simpson and the leaders of the company called for British military aid in the event of hostilities and asked as a preventative measure for an assessment of the situation by military representatives. Lieutenants Mervin Vavasour of the Royal Engineers and Henry Warre of the 14th Regiment, stationed in Lower Canada, were chosen to make the long excursion west. Vavasour was an engineer and able to survey and make drafts, and Warre was an accomplished artist, adept at sketching. They were to make recommendations for the protection of British interests, to assess the feasibility of transporting troops overland to the Pacific coast, to advise on possible fortifications on the Columbia River and the feasibility of protecting the various forts against hostile action. They journeyed as if they were privileged young tourists, interested in travel, hunting, and fishing, examining the fauna and flora of the various regions, and admiring the scenery by mapping and sketching. Their military identities and true purposes were kept secret.

Vavasour prepared plans of the establishments and immediate locales of forts Ellice (284A), Carlton (285A), Pitt (145C), and Edmonton on the plains (286A) (plate 45), Colvile (269A), Nez Perces (Walla Walla) (270A), Fort Vancouver in the Columbia area (271A), and Fort Victoria on Vancouver Island (280A). He made a map of the country along their route between Fort Vancouver and Fort Nisqually (272A), an "Eye Sketch of the Plains &c about Nisqually" (279A), and a chart of "Cammusan" (Victoria) harbour (280A), on which he inserted his Victoria plan.[16] He also drew a rough chart of the Columbia River downstream from the fort, showing soundings in front of Fort Vancouver and at the mouth of the Willamette River and some of the changed positions of sandbars at the river mouth (281A). Since they had been asked to advise on the possibility of fortifying the approach to the river, Vavasour provided maps of Cape Disappointment on the north side (283A) and Tongue Point on the south shore (282A), with proposals for fortifications. A separate chart of the entrance to the Columbia was also drafted, but it is not in the archives (144C), nor is the plan of Oregon City (143C). By the time they were on their return journey, the governments of Great Britain and the United States, retreating from possible confrontation on the Pacific shore, had concluded the Oregon Treaty, by which terms the forty-ninth parallel was chosen as the boundary to the coast, and Vancouver Island remained British. The British had also been able to secure the rights of the Hudson's Bay and Puget's Sound companies in land and property which had been lawfully acquired in the area south of the border; if the United States required the property, the two companies would receive proper compensation. The company also had the right of navigation on the Columbia River until its licence for exclusive trade ended.

MAPPING LAND CLAIMS IN THE COLUMBIA DISTRICT

After the boundary treaty was signed in 1846, the directors of both the Hudson's Bay Company and the Puget's Sound Agricultural Company were faced with the problem of clarifying land ownership and property boundaries in a territory which would be eventually occupied by American officials, military forces, and settlers. Instructions were quickly despatched to the west coast: the companies were to claim that land which they considered to have been in their possession before the treaty, that is, land which was cultivated, in pasture for cattle, horses and sheep, or occupied by buildings. It was recommended that employees and retired servants and officers should have areas assigned to them and agreements drawn up to the effect that the land would revert to the companies upon their deaths. The properties were to be surveyed, and the lines laid down on the ground; the properties were to be registered with local authorities, and the areas were to be mapped so that these maps could be used as references in later discussions. In addition to these actions, both sets of officials asked the officers in the American west to prepare detailed estimates of the current value of lands, farms, and other property in case of future indemnity negotiations.

The first map of such claims from the Columbia District was a rough sketch in 1846 of some of the Nisqually claims in the vicinity of the fort, including sections assigned to Tolmie, John Work, and others (287[A]). The general outline of woodland and prairie areas was included. On 13 October 1847, five maps of Nisqually area were completed and sent to London by James Douglas.[17] It seems certain that the cartographer was Tolmie. The first map was a sketch of the Nisqually fort complex and main farm, with a detailed list of the use of the various buildings (294[A]) (plate 46). Another was of the pasture land adjoining the shepherd's station at Nisqually (298[A]) (plate 47); the other three were likewise of separate sheep and cattle stations (295[A], 296[A], 297[A]). All were drafted by the same hand and show the details of the terrain, vegetation, watercourses, and farm facilities. Three other typical cadastral maps were added to the files during these years, delineating various individual claims for lot registration purposes (288[A], 289[A], 302[A]).

MAPPING NEW CALEDONIA AND HORSE BRIGADE TRAILS

In spite of the activity of the Hudson's Bay Company's employees in New Caledonia since 1821, there was little cartographic depiction of the wide-ranging movements of these men across the mountain, plateau, and river-carved terrain of this region in the earlier decades of this period. In the spring of 1826, on his first journey out of Fort Kamloops, Archibald McDonald travelled to Nicola Lake, to Shuswap Lake, and south to the Columbia River, via Okanagan country,[18] with the annual horse brigade carrying New Caledonia returns. In the autumn, he had returned from a trip with the chief of the upper Okanagan Indians to the Thompson-Fraser River junction, and had drawn a sketch of the Thompson and Okanagan area, which he sent out to Fort Vancouver (111[C]). McDonald's next map, a sketch of the "Thompson's" River district, was drawn in 1827 to illustrate his district report of that year (201[A]) (plate 48). In his report he gave a list of the names of all the regional tribes, their chiefs and principal men, and their population counts. On the map he has drawn the tribal boundaries in various coloured inks and designated the tribes in red ink. McDonald had reported that he had taken observations for latitude and longitude at the Fraser-Thompson junction in 1826 (B97/a/2,fo.29). One can note on the extant map that he has situated the junction about one degree too far west. Okanagan Lake is correct for longitude but is displaced about fifty minutes too far south in latitude. This map of the region was an outstanding complement to the few existing maps which, whether manuscript or published, were available to the company's officers, since it added to and corrected the lineaments of this large interior region as they had been drawn previously.

Company men became reasonably active over the next several decades in drawing maps of New Caledonia, especially those which detailed the search for a new route for the supply and fur brigades. Because this had to lie within British territory, north of the newly-defined border, it meant the eventual abandonment of the Columbia River route and of the headquarters depot at Fort Vancouver. An early proposal for supplying New Caledonia by the northern Pacific rivers, rather than by the very long Columbia route was illustrated on an 1833 map, drafted by Simon McGillivray, chief trader at Stewart's Lake (227[A]). In the late 1820s, a suggestion had been forwarded to London that the river draining Babine

Lake be used to connect with the sea. This river, the Babine, was thought to enter the Pacific at Nass, or at least near Nass, although it is actually a tributary of the Skeena River. When the exploring party found the upper Babine River not navigable, McGillivray traced a practicable horse brigade trail west from Babine Lake over the Babine range, leading to Simpson's (Bulkley) River, and thence to the Skeena. The "rough chart" of his route, included in his report to George Simpson, shows the great arches of the mountain ranges flanking Simpson's River and the "unexplored Babine River," along with the route of the high summer track from which McGillivray claimed he could see the whole course of the upper Babine River. Because he did not proceed to the ocean, his superiors considered his expedition a failure, although he believed that he had proved there was a reasonable route for a future supply system.

Further searches for an alternate brigade trail to and from New Caledonia began in 1846 as the Oregon Treaty came into effect, although consideration of the problem and the planning for its solution, as discussed above, began much earlier. The board of management of the Western District of the company chose Alexander C. Anderson, an experienced officer, then at Fort Alexandria, to continue the search. Anderson had previously been stationed at Fort George and Fort Vancouver, Kootenay Post and Nisqually in the south, but also at Fort McLoughlin and Fraser Lake Post in New Caledonia before his appointment at Fort Alexandria. Since he knew first hand the difficulties of the long horse brigade tracks and boat supply routes that had to be taken to reach Fort Vancouver, he was well-suited to the task. In February 1845, Anderson had written a letter of "Suggestions" for the exploration of a new route, with an accompanying map (273^A) (plate 49) (B5/z/1). His idea for the new track, based on Indian information "and other sources," was a route from present Lillooet on the Fraser through Seton, Anderson, and Harrison lakes, and the Lillooet River to the lower Fraser River above Fort Langley. The sketch map, with the proposed "horse Road" dotted in red ink contains some mistakes. The first is the existence of a very long "Harrison's" River, which drains from the northwest to the Fraser above Langley. This is the present Lillooet River, but the map lacks the two largest lakes in this chain of waterways, that is, Harrison and Lillooet lakes. Moreover, Anderson has his river entering above what would be the great bend of the Fraser at Hope, rather than downstream from it. The northern portion is much more realistic. The northern terminus of this proposed route (at Lillooet) he estimated to be distant a "Horse portage to Alexandria 7 days loaded," and "to Kamloops 4 days loaded."

In 1846, Anderson reconnoitred the route west of the Fraser, which he had outlined previously. After a stop-over at Fort Langley, he went back to Kamloops via a course he negotiated from the mouth of the Coquihalla River, at the great bend of the Fraser River, over the mountains and interior plateau surface northeast to the Thompson River valley. Assessment of this route by Anderson and the board of management immediately afterwards indicated two areas of concern. One was that the most difficult section of the road, from the Fraser River to the upland plateau, would need considerable clearing and levelling and likely alterations in course, and the second was that the depth of snow and its duration would make part of the road impassable to brigades until about mid-July. Therefore, supply brigades would be delayed two weeks longer in their arrival back at Thompson's River than they would be by the Columbia passage. For this reason, and because the members of the board believed the natural difficulties to be too great, and the "communication … at best … tedious and dangerous," they decided not to use the Harrison Lake route (B223/b/35,fo.37d).

Alexander Anderson enjoyed drawing maps, it would seem. He produced a number of them while employed by the Hudson's Bay Company and also after he resigned from his position and moved to Fort Victoria. However, it is interesting to note that almost none of those drawn while he was an employee of the company remained in the files. The map which he drafted of the entire 1846 trip is a case in point. He prepared at least two copies, and possibly three, which he said he flattered himself would be found to be "tolerably accurate" (D5/19,fo.288). One copy was sent to Governor Simpson for his inspection (146^C). A second map would have been the original from which Anderson said he had made a copy for Simpson. Logic would suggest that either this original was kept by Anderson for his later use, or that he sent it to the board of management along with his report (B97/a/3,fos.1–16). However, no mention is made in the records of a map accompanying the report. But there is a map, signed by Anderson, which he entitled "Original sketch of exploration between 1846 and 1849," which he would undoubtedly have completed in the latter year (13^B).[19] This could be Anderson's original 1846 map, or his copy of it, to which he added further details of his attempts to find more advantageous trails in the interior in the years up to and including 1849. It covers the area from Fort Langley to Kamloops Lake, including the Harrison Lake country, and south on the interior uplands to the Similkameen River.

In the 1847 season, the board of management of the Western District, convinced that there must be a more suitable way of reaching the interior than that discovered by Anderson, agreed to the exploration of an alternative route that had been reported by James M. Yale, the factor at Fort Langley (B223/b/36,fo.12d). This route, which Yale had pieced together on the basis of Indian information, involved a passage up the Fraser River past the Coquihalla River mouth and then portaging around the Hell's Gate Canyon rapids, which lay within the Fraser canyon. Anderson was ordered to find a horse trail using the Fraser valley, thereby avoiding the mountainous barrier of the first route. He carried out this assignment in May and June 1847. The trail decided upon was one that, after a portage on the west side of the Hell's Gate, crossed the river above the rapids at Kequeloose and then climbed over the high shoulder of the east side of the Fraser canyon into the Anderson (Bridge) River. From there it passed up onto the plateau to Nicola Lake and the Thompson River. The map of his "Travels ... indicating the proposed route" was drawn, Anderson reported, and sent to Governor Simpson in February 1848 (151C) (D5/21,fo.294d). It did not remain in the files of the company, if it ever reached Hudson's Bay House. However, a reduced-scale version of this map appears as a segment of the composite 1846–9 map prepared by Anderson (13B). Despite some advantages over his first route, Anderson unconditionally condemned the proposed scheme. The worst feature was the difficulty of transporting two to three hundred horses in scows across a treacherous, fast-flowing river. Nevertheless, the board of management ordered the interior brigades to use the road. It took just one return trip, however, for the danger and inconvenience of this route to be impressed forcefully upon the officers concerned with inland transport. Immediately, Henry Peers, one of several young clerks who had been taking part in the brigade venture, was sent off to build Fort Hope, a post at the mouth of the Coquihalla. Following this, he was to leave the fort in order to recheck and, it was hoped, to revise Anderson's 1846 return trail, and to verify an Indian report that this route could be shortened.[20] Peers drafted two sketches "to give a better idea of the country than any description" (153C). The maps were probably concerned only with the section of the trail that had been altered. A great deal of work was done on the road to get it ready for the 1849 brigade.

In 1849 a sketch of another brigade trail appeared, but its author is not known (154C). During the autumn of that year, Eden Colvile, the governor of Rupert's Land, visited from Britain and travelled down this trail to Fort Hope. In a letter from Fort Victoria he described the manner of his journey to Governor Pelly in London. To aid the governor and committee in comprehending the topography of the southern part of the region covered, Douglas provided a map of the route from Fort Hope to the upland meadows on the plateau, a distance of forty-eight miles, incorporating "a nearly correct delineation of the vallies, and mountain ranges" (A11/72,fo.174). Douglas would most likely have had Anderson's and Peers's maps upon which to base his useful map.

MAPPING THE FORT VICTORIA AREA

For many years after the founding of Fort Vancouver as the chief depot of the western trade, some ninety miles up the Columbia River, there had been increasing criticism of its situation: it was too far inland in view of the increasing marine orientation of the company; the dangerous bars and shoals at the river mouth made easy access impossible; the time taken by ships to reach the fort upstream, or to sail down to the open ocean, restricted trade; it was inconveniently located for the purpose of provisioning the northern coastal posts. Finally, there was also a growing consciousness that Fort Vancouver was on a politically untenable frontier and that eventually British control of the area would be lost. The committee consequently instructed Chief Factor McLoughlin to have the coast farther north examined for a site for a more suitably located depot. From 1837 at least to 1842, this search was a continuous element in western operations. In 1837, Roderick Finlayson scrutinized the intricate waters of Puget Sound, but McLoughlin reported unfavourably on it as a possibility. During this same season, Captain McNeill was instructed to examine the east and southeast coast of Vancouver Island on his return journey from Fort Simpson. He brought back favourable comments on the character of several inlets on the south coast. In spite of the committee's desire for more specific examination of this locale, it was not until 1842, after Governor Simpson arrived in the Columbia district and concluded that the Vancouver Island area should be the goal for resettlement, that James Douglas was sent north, accompanied by Adolphus Lee Lewes, as surveyor and map maker, and a small party, to report on a specific site for the fort.

Douglas and Lewes nominated Camosack (Camoosan, Camosun) Inlet as the most suitable harbourage, and proposed a site on its southeast shore. The detail of the inlet, which extends from Cadboro Bay on the east to the west end of Portage Inlet, is shown on a four inch to the mile map, the "Ground Plan of Portion of Vancouvers Island Selected for New Establishment" (252A) (plate 50). This coloured land-use map delineates the areas of woods and forest, plains, wet marshes, rocks, hills, and lakes. Lewes drafted the map at Fort Vancouver upon their return, and added an inset map of a "Selected" part of the island at six times the larger scale. In contrast with the rest of Vancouver Island, this small segment was more lightly forested and offered the possibility of farming on a small scale on some of the "plains." The map and report were accepted, and the decision made to establish the new western depot at this site. This map is, therefore the first cartographic document of Fort Victoria and the future colony of Vancouver's Island.[21]

A second map, a chart of Comoosan harbour, was prepared in the summer of 1842 by James Scarborough, captain of the *Cadboro*, the ship which had been used to transport Douglas and his survey group (139^C). In November 1843, McLoughlin in an official letter drew attention to a chart of "Comoosan" on which the situation of Fort Victoria was depicted. Since Douglas and Lewes in 1842 had indicated this site as their choice of location for the future fort on the east shore of the harbour, their map would have been available to Scarborough as the basis for his harbour chart.

After 1843, when Victoria was founded, until 1849, almost no surveying was done and little in the way of maps produced. The company had exclusive rights to trade and was still first and foremost a trading company, with minimal interest in the influx of settlers and the sale of land. Only one sketch is available in the collection regarding land ownership; it outlined the boundaries of the claims of several settlers to blocks of land in the vicinity of the fort (290^A). The sketch was not dated, but it would appear that it was drawn about 1846. Land-claim mapping did not begin on Vancouver Island until 1849 with the arrival of a part-time surveyor. It did not become securely established until 1851 with the employment of a full-time surveyor and the organization of a combined company and colonial office and cadastral and land registry systems.

12 *Pemberton and the Colony of Vancouver's Island, 1849–1859*

The Governor and Committee are ... gratified by the favorable account ... of Pemberton's industry and Exertions in making surveys and also by his ... Map of the Country.
    Committee to Douglas, January 1852

Cartographic activity by Hudson's Bay Company employees was concentrated in the districts west of the mountains from 1849 to 1859.[1] Over seventy-five percent of the approximately 215 maps prepared in this decade were of Vancouver Island and the western part of the Georgia Strait, off the eastern shore of the island. Moreover, unlike earlier years, the bulk of these maps, seventy percent, was significantly cadastral in nature, dealing with land claims, settlement plans, or with company mining operations. The second largest group of maps were maps of exploration and regional maps; twenty-two percent of this second group were topographic maps of the newly-created land districts on southern Vancouver Island. The exploration, surveying, and mapping program which centred on Vancouver Island was the responsibility largely of Joseph Despard Pemberton, who was appointed to be jointly the surveyor for the Hudson's Bay Company and for the colonial administration of the new colony of Vancouver Island.

By 1848 it had become obvious to the governor and committee that Vancouver Island possessed such importance for the future of the company's operations west of the mountains that they petitioned for and received from the Crown a grant of proprietorship of the island.[2] In the meantime, the British government had decided to establish a colony on the island. In June 1849, James Douglas arrived at Fort Victoria, by then the chief depot for the trade west of the mountains. He was not only chief factor and senior member of the board of management of the Western District of the company, but also provisional governor of the new colony of "Vancouver's Island." The Hudson's Bay Company thus found itself, against its commercial disposition and experience, responsible for all colonial activities and development.

On earlier occasions the committee had issued orders through Governor Simpson that Vancouver Island be surveyed to ascertain its resources and potential. Simpson, considering William Tolmie to be well qualified for the job, had sent notice west in June 1848 to have him so assigned, if he could be spared (D4/37,fo.126). By October 1848, however, Simpson had received notice from London that a qualified person was being chosen in England, and he relayed this message to the west coast, indicating that surveying should be delayed until the London appointee arrived.

WALTER COLQUHOUN GRANT, FIRST COMPANY-COLONIAL SURVEYOR, VANCOUVER ISLAND

In November, Governor Simpson received the news that a contract had been signed with a young former army officer, Walter Colquhoun Grant, and that he was to arrive accompanied by twelve to twenty other persons as labourers and settlers. Grant was to be both the first "settler" on the island and as well as the first company-colonial "surveyor" in the northwest. Governor Pelly affirmed that Grant was "quite a gentleman ... peculiarly qualified for the work he had to perform" (A8/6,fo.106), and that he "brought ... [with him] certificates of qualification for a survey" (D5/26,fo.307d). But not long after Grant's arrival, it became manifest that the committee had chosen unwisely. The members had neither the experience necessary for identifying the requirements of a trained land surveyor nor any understanding of what was involved in the task they would assign. Moreover, they made a mistake in allowing someone to fill two exacting roles: chief surveyor and principal settler.

When Grant reached Fort Victoria, he decided to establish himself and his, as it turned out, eight settlers at Sooke on the Strait of Juan de Fuca, some considerable distance south and west of the fort. Although he had suggested before leaving England that he should first make a "general survey of the Island," his proposal was rejected (A11/72,fo.90). The committee's instructions were much more specific. He was to set out the boundaries of all the lands in the vicinity of Fort Victoria that had been in the occupation of the fur trade in 1846 and to designate the other areas local company officials considered expedient to "reserve for the fur trade for cultivation or otherwise." A sub-grant of land should be selected for the Puget's Sound Agricultural Company to occupy near the fort. He was also to survey and make a plan of his own property.

He was instructed to send original sketches or tracings to the committee at every opportunity, as its members wished to have information on lands to be kept or to be disposed of, as soon as possible (A6/28,fos.96d,97). Simpson had widened the scope of the project even before Grant arrived when he advised the board of management to send out a qualified officer to traverse the island "in every direction" making a geographical examination of its physical, economic, and human geography (D4/39,fo.100). Coastal surveying was to be left aside in the hope that British naval ships would do it.

Disillusionment with Grant set in early. In September 1849, Douglas reported that, although he had arrived in June, no surveys had been undertaken. Douglas recommended that another surveyor should be attached to the fort, at the disposal of the company officers. For the position proposed, he suggested Adolphus Lee Lewes, who had, of course, done the first surveying in the Victoria area with Douglas in 1842. This suggestion was not accepted by Simpson. By late October, Grant had still produced no map, but he had taken time to prepare a paper on the physical environment of the southeast of Vancouver Island, a paper which appeared later in the *Journal of the Royal Geographical Society*.[3] He excused his lack of progress in surveying by saying that his instruments had arrived late and that some of them had been broken. By early December, the committee was complaining that the lack of local maps was holding up land sales (A6/29,fo.90d). This was impressed upon Grant, who was ordered to spend part of his time on the mapping enterprise. By mid-February 1850, he had still made little headway but could report that he had "done surveys of the plains of Matchousin and Syusum[4] ... with the intermediate country" (A10/28,fo.273).

Grant declined to take any of his salary for the surveying already done, or "what little ... [he was] likely to be able to do" (A10/28,fo.273). He promised to forward tracings as soon as he could. On March 25, he sent his resignation as surveyor to Douglas, again blaming his lack of accomplishment on delayed or broken instruments, adding that he had not had sufficient field aid and noting that his home area was considerably distant from the fort. He alleged in addition that there was some lack of official support and advice from local officers. However, he did include a "sketch of the south coast of the island" (308A). This first map, the committee concluded, was "of little or no use" (A6/28,fo.160). Grant drew it on six sheets of tracing paper. The extent was essentially the same as the earlier 1842 map, but it was at a much larger scale. He added roads and trails, details of cultivated land and the fort region, as well as of settlement beyond the fort. He did not indicate any of the specific cadastral information the committee particularly wanted.

Simpson and Douglas were by then in agreement in their assessment of Grant's habits, character, and services as a surveyor. Since they could no longer trust him to satisfy the requirements, Simpson sent an urgent request to Douglas to hire a "duly qualified person"; he did not think that Lewes measured up to the position, and he recommended that it be a Canadian. Douglas had by then enquired as to the availability of competent persons in the Columbia area, but the fees charged by practitioners were higher than the company was willing to pay. Douglas was in a difficult position, and, though Grant had resigned, he made an arrangement to pay him $10 per diem to work exclusively on measuring and plotting the boundaries of the Fur Trade Reserve. Douglas confirmed that Grant worked "pretty steadily" until early September.

The final fruit of Grant's labour, a map of the Fur Trade Reserve was received in London in the autumn of 1850 (155C). It was, however, incomplete, because his crew had laid out only the eastern, western, and northern boundaries. They had used Nankuan Hill at the head of Victoria Harbour, Mt. Douglas, Mt. Tolmie, and Gonzales Hill as the centres for the triangulation grid, from which offset lines were taken (BCA,EEG76M) Douglas confessed that the map would be of little use to the committee beyond giving a general idea of the extent of the reserve. He sent them Grant's promise that he would complete the map and forward it later to London (A11/72,fo.322). But Grant left Victoria suddenly for the Sandwich Islands (Hawaii), from which he sent a long letter to London, partly of explanation, of justification, and, to a degree, of promises to complete the task. He told the committee that he had sent from Honolulu a "copy on tracing paper of the survey already made ... with a small portion of the Southern part of the coast etc. roughly filled in" (156C) (A11/62,fo.596). This had been forwarded, he said, through a British person who was taking it with him via vessel to Hong Kong and overland to London! He promised to complete his assignment upon his return to Vancouver Island. This he did not do, although he was again in the region, engaged in various commercial enterprises. Disappointed by the essential waste of almost two seasons of potential surveying and mapping, Simpson was authorized by the committee to search for a new man in Canada and to send him west of the mountains in the spring of 1851.

JOSEPH DESPARD PEMBERTON, COMPANY-COLONIAL SURVEYOR OF VANCOUVER ISLAND, 1851–1859

In the winter of 1850–1, the governor and committee were still expecting a replacement for Grant to be found in Canada. At their meeting of 18 December 1850, they had declined "at present" an offer from a young man in England, Joseph Despard Pemberton, for his services as a surveyor and engineer (A5/17,p.65). There must have been a change of mind soon after for, on 1 January 1851, Douglas was informed that a competent surveyor was to be sent from Britain (A6/29,fo.28), and on 22 January, Pemberton was confirmed as the new surveyor and cartographer for the company and the colony (A1/67,p.45). His first contract was for three years, with a salary of £400 per annum, to commence at the date of his arrival at Fort Vancouver (A1/67,p.46). When he was hired, Pemberton was twenty-nine years old, unmarried, and the Professor of Practical Surveying and Engineering and Second Master at the Royal Agricultural College, Cirencester, England.[5] Previously, he had been a railway engineer in Ireland and England, having apprenticed with the principal engineer of the Midland Railway of Ireland.[6] During his apprenticeship and employment he would have become familiar with the nature and use of topographic maps, particularly because his supervisor had formerly been a member of the Ordnance Survey of Ireland. Moreover, Pemberton would have become a thoroughly experienced surveyor both because of his training and because of his position as a teacher at the Royal Agricultural College. The curriculum of his third year surveying and practical engineering course of lectures and practicals for agricultural students included the use of most types of instruments, measurement and allotment of lands, surveying as applied to tracts of country "at home and in the Colonies," the drawing of cross-sections, mapping, contents of plans, and cartography, including the character and use of drafting instruments.[7] The committee made a more careful examination of Pemberton than it had previously of Grant, and it had hired him as a full-time surveyor and cartographer so there would be no distractions.

The task that faced Pemberton was described in the committee's letter of instructions (A/120,fos.7d–8d). He was furnished with available surveys (probably maps) of Vancouver Island to aid his understanding of the region. He was to undertake surveys which would "form the ground work" on which an accurate map of the island was to be "constructed" (A6/9,fo.37). The object of the surveys and the resulting maps was the eventual colonization of the island. He was, therefore, to report on topographic features, geological formations, the nature and quality of soils, timber stands, vegetation, and other characteristics related to the potential for settlement. He was to take charge of the registration of all land grants and sales and of all subsequent transfers of ownership. Pemberton was to begin his surveying with the area around Fort Victoria, that is, the Fur Trade Reserve and other areas westward from this settlement. Douglas had already concluded arrangements with the local native peoples for the rights to these areas. When similar rights had been obtained, he was to survey at Fort Rupert in the northeastern part of Vancouver Island where coal had been located. Before he left London, arrangements had been made to have an experienced surveyor-engineer choose a suitable pupil who would become an articled apprentice and assistant to Pemberton when he arrived on the island in 1851. The choice fell on Benjamin W. Pearse, already an engineering apprentice in Britain, who carried surveying instruments, paper, and other supplies with him when he left Britain. Pemberton, commencing his company-colonial career in Victoria on 24 June 1851, remained in this position until 31 May 1859. He served as colonial surveyor for a short time thereafter and was then appointed as the first surveyor-general of the Colony of Vancouver Island. He retired from this position in 1864.

Pemberton's cartographic output is difficult to isolate from that of his associates during the period he was company-colonial surveyor, since he did not sign maps nor did those who drafted maps under his direction. He established a small "surveying department" in Fort Victoria, in which there was a drafting area; the cartographic principles followed were those based on his knowledge of map design practices in force in Britain during the earlier decades of the nineteenth century.[8] The surveying department was a joint company-colonial office until 31 May 1859, after which the Hudson's Bay Company withdrew from its operation, and it came under the control of Pemberton as colonial surveyor and then surveyor-general. Pearse was his student and assistant in engineering, which included surveying and mapping, land sales and land registry. Over the years the department had a varying-sized field crew.[9] Pemberton found that he preferred Metis as labourers because they were accustomed to working in the difficult field conditions, were used to rough camps, and could hunt for game to aid in the provisioning of the field kitchen. He hired the largest number of Metis during the 1851–3 field seasons. In the 1856–8 seasons more of the labourers came from an English, Scottish, and American labour force that had entered the colony. In 1858, with an enormous surge of settlers, and a boom in land sales due to the mainland gold rush, at least six surveyors were at work at various times to assist in the hectic pace of surveying lots, preparing survey plans, and drawing settlement maps. Three men, in particular, are noted in the records for their map production: surveyor and architect, Hermann Otto Tiedemann, and Robert Homphrey and Richard Covington, who also aided in surveying.

There were at least fifty-six maps of diverse types, and at a minimum, ninety indenture plats prepared by or under Pemberton's direction. They concerned southeast Vancouver Island, Nanaimo, and the intervening coasts and offshore islands. Several indicate transects between the east and west coasts of the island. Although over forty percent of the first group of maps is not in the company archives, many of them are found today in the British Columbia Surveys and Land Records Office. The maps may be separated into six categories. The largest category contains cadastral plats and maps, a not unexpected statistic since this was the major purpose of the surveying department. This group includes the indenture maps of lots sold to individuals, or held by the company or Puget's Sound Agricultural Company, and the "government," that is, those lots held for public use. Many are topographic maps covering the eight land districts of the southern part of the colony and Nanaimo, at scales ranging from one to six inches to the mile. There are coastal maps from the Victoria area north beyond Nanaimo harbour. The maps became increasingly accurate over the years and were used as the basis of published maps by John Arrowsmith. Of great historical value are the first three town plans of Victoria, produced in 1852, 1855, and 1859. Several of this group are maps that illustrate Pemberton's expeditions across the island. Finally, some were special purpose maps, such as those dealing with water supplies and ground levels.

Pemberton was under pressure to proceed with his work quickly. He had to organize a mapping program that was related to the needs of the colony, that took into account the local environment, and that suited the facilities, equipment, and the labour supply at hand. Although he was familiar with the concept of a regional surveying system, he had no personal experience in new territories. In these pressing circumstances, he turned to the advice and examples of other colonial surveyors and colonial specialists, as outlined in their recent books, notably books dealing with surveying and colonization in New Zealand and Australia.[10] Pemberton inclined to the view of these other men that before becoming involved with large-scale cadastral surveys, one needed to lay out a triangulation grid, prepare topographic maps of the central region, map land use while assessing the areas suitable for agriculture, lay out major roads, and choose a main townsite. Other than these books, he had to rely mainly on his own counsel. Although reality forced him to undertake topographic and cadastral activities simultaneously more than he would have wished, he was able to outline a systematic program of action and to maintain its coherence. He had two masters, company and colony, both under the direction of one man, James Douglas. As a company employee his first and urgent responsibility was to aid its officers in identifying and demarcating the Fur Trade Reserve, which was to include those lands in the possession of the company and used for its own purposes previous to the Oregon Treaty of 1846. He was also to outline the property assigned to the Puget's Sound Agricultural Company. Concurrently, as a colonial servant his prime obligation was to design a cadastral system and an acceptable parish or district structure and to develop a satisfactory surveying and registration procedure such that existing and future land owners could have their lots measured, registered, and the indentures and indenture plats quickly supplied. But, to satisfy these requirements, he had to organize simultaneously a topographical survey and a mapping program for the area that could receive settlers within the reasonably near future. The section and lot boundaries would be delimited on these maps for local and home country use. Further, there was an insistent demand from the company in London for the exploration of the south and east coast of the island, the nearby island groups, and the interior of the colony. These explorations were meant to provide corrections and additions to existing maps, which could be transcribed by commercial cartographers into new and published maps. In pell mell fashion, other chores were imposed upon Pemberton; some were obligations of his office; some were requests that took advantage of his engineering training. The company and colony quickly found Pemberton to be not only able, but willing and interested in all facets of their requirements; he became all things to all people during the first years of colonization of the island.

THE SUMMER AND AUTUMN ACTIVITY, 1851

Pemberton's first summer, 1851, was a hectic but successful one. In spite of fires breaking out during a dry period, and not having all the necessary instruments at hand, he completed a basic reconnaissance of the local area, established a triangulation network for the Victoria and Esquimalt areas, defined the boundaries of the first established district, which was the Fur Trade Reserve at Fort Victoria, and surveyed part of the second established district, Esquimalt.[11] The draftsmen were able to prepare their first map (14^B),[12] to a six-inch scale, and to prepare several tracings or copies (310^A, 311^A, 312^A). The original of the map apparently remained at Fort Victoria for later use. The tracings, said by Pemberton to be "unavoidably rough," were submitted with a report to Douglas, who transmitted them to London with the comment that the map was "very correct" (A11/73,fo.61). The map and tracings displayed the boundaries of the reserve, the lots sold to officers and servants, the course of the proposed road to Metchosin and Sooke, land available for immediate cultivation, symbols for pine forest (areas defined as unimproveable rock and swamp), and the site of a proposed portage from Esquimalt Harbour to the long inlet (Portage) extending northwest from Victoria Harbour.[13] The committee greeted the three-sheet tracings with satisfaction, although it chided Pemberton for not having noted those lands that had been in company possession in 1846. Clarification was requested.

The autumn of 1851 was just as productive. Before the winter rains, the crew had extended the trigonometric grid and prepared a topographical survey north of Victoria up the east coast of the Saanich Peninsula. In places the forest was so thick they had to use the tops of the highest hills and climb the tallest fir trees to sight over longer distances. Pemberton had trekked on foot from the head of Esquimalt Harbour down its west side to Pedder and Becher bays, small inlets along the Juan de Fuca shores. Moreover, the crew laid out the lots of the settlers and company employees and set out the townsite around the fort. The town survey notes provided the data for the initial pencil draft of a forthcoming town plan. Pemberton had already satisfied the committee's request for specific cadastral information by drafting three sketches showing three blocks of land of 1300, 1144 and 640 acres in Victoria District, under tillage or in cattle range, and marked as Fur Trade Reserve land (161^C). An amended copy (167^C) of sketch number 1 (310^A) was sent to London the following summer. The first two of many indenture lot maps were prepared and also sent to London. The plan of lot 1, section VI, the property of Governor Douglas, on the south side of James Bay, was the usual simple black-line cadastral plan (313^A). On the other hand, the plat of lot 2, Section I, comprising property inland from Clover Point on the Strait of Juan de Fuca, was a coloured, large-scale topographic map, with lot lines overdrawn in black (314^A) (plate 51).

THE 1852 WORKING SEASON

The long spring and summer of 1852 provided a working season that Pemberton used to good advantage. Field parties continued basic triangulation and topographic surveying in two directions, west and south in the Esquimalt area, and north and east in Saanich. Enough work had been completed in March for a fourth topographic sheet to be drafted and despatched to Britain (319^A). It was of the peninsular area north of Mount Douglas, north to Cowitchin (Cowichan) Head, and several miles inland from the coast. The draftsmen had also prepared a revision of the topographic map of the Oak and Cadboro bays area of Victoria District, in which some more precise instrument work had been undertaken (318^A). London was advised to transfer the new detail to the sheets already provided. As part of his regional analysis, Pemberton had been put to the job of examining fresh water supplies for the local region. His investigation resulted in two maps, one of smaller and one of larger scale, both dealing with water sources. The first, compiled in March 1852, was "Roughly Sketched" of the Saanich Peninsula and the Victoria and Esquimalt districts, the purpose of which was to show the region's "water supplies," that is the rivers, streams, lakes, and major marshes which could provide drinking water for settlers, for domestic herds and flocks, and sites for the water mills (321A). Drawn at the scale of one inch to one mile, it was in accordance with Pemberton's decision to set up a system of map scales ranging from six inches for best quality areas, four, two, and one inch for less important regions and for areas least likely to be settled in the near future. The second map was based on a field examination of the land north of the fort, to the east of Rock Bay, where there were a series of ravines leading down from a "gravel hill or bank" (320A). On this hill were several springs that could be tapped by wells to supply the fort. Pemberton measured the elevations and the ravine gradients; on the map he personally noted the levels and the way in which a water-line could be constructed southwest to the Victoria townsite.

By the time of the opening of the field season in 1852, the drafting room had finished its map of the Victoria townsite, which had been in draft form the previous autumn. Two copies have been noted, one being labelled by Pemberton as No. 1, "A Plan of the Town of Victoria Shewing Proposed Improvements," and which was kept in the fort surveying department office (15B). The second, an exact copy with the same title, was sent to London (323A) (plate 52). This is the first town plan of Victoria. It is oriented to the east, and delineates the fort with interior buildings, the company dock, a proposed pier, storehouses, all other buildings on occupied properties, and the lots and streets planned by that date.

The company had instructed Pemberton as to its views on the use of the Esquimalt district. The land was to be laid out principally for the Puget's Sound Agricultural Company, although the company wished to retain command over Esquimalt Bay until the British government had decided whether it wished to establish a naval station there. He was asked to advise on the best location in the district for the principal town of the colony, and on a suitable harbour for company jetties. Pemberton readied a report on the district, based on previous assessment, a re-survey with better instruments, and close examination of the shores of Esquimalt Bay. With the report was a map of the "Country Around Esquimalt Harbour," on which the available land for agricultural purposes was separated by red lines from the hilly and therefore uncultivable land to the west and north (322A). The map included estimates of the amount of the different types – "open, pine, rocky, and muck-swamp" – of land. In a letter to Deputy-Governor Colvile (A6/120,fos.64–65d), Pemberton had inserted a sketch map (with a copy for Governor Douglas) of the southeast of the island, on which he marked an area situated northwest of the head of Esquimalt harbour (168C). This he recommended as the most suitable site for a principal settlement. He also marked the location of Village Bay on the east side of the harbour, recommending it for company wharves. In the spring of 1852, he forwarded a map of the Puget's Sound company's agricultural reserve at Esquimalt, with boundaries delineated as "pink lines" (169C) (F12/2,fo.431).

Already Pemberton had noted that "office" work was taking more time than anticipated (A6/120,fo.44). In addition to general administration, he was responsible for lot sales and land registration. He felt it necessary also to ask for certain instruments that he had not contemplated needing earlier. These included prismatic compasses, a simple, strong, and light portable theodolite, and a small sextant; all were to have strong cases with carrying straps. He asked that all the bright brass instruments be japanned to reduce the pilfering by natives attracted to the bright surfaces.

In the late summer of 1852 Pemberton travelled by canoe with Governor Douglas north along the east coast to Wentuhuysen Inlet (Nanaimo); it was his first journey beyond the south east corner of the colony. He had several purposes: to map the Nanaimo area and prepare a geographical report; to make a general geological study of surface coal deposits there; to begin a series of observations on the coast to define its character and mark exact locations; to begin an examination of the Gulf Islands and to map Haro Strait with the intent of correcting published maps. Douglas reported that the Nanaimo region was "one vast coal field" (B226/b/7,fo.7d). Pemberton's very thorough report demonstrated both his excellent scientific and practical training and his high level of skill in illustration, particularly in making cross-sections and sketches. His cartographic expertise may be observed in the three resulting maps. The most extensive was a four inch to the mile topographic map depicting a number of offshore islands as well as area extending inland to the higher hill-lands (327A). The offshore soundings were obtained from one of the company's ship captains. To establish the framework, he observed the astronomical location of the site of the fort at

Nanaimo harbour and sighted on three mountain peaks seen from there.¹⁴ The second map was less extensive and less informative, concentrating on the coast (328ᴬ). The third was a similar map, but on it he plotted the coal seam outcroppings on the main shore and on nearby Newcastle and Protection islands (329ᴬ).

Upon his return to Fort Victoria from Nanaimo in 1852, Pemberton prepared a map of the coast from Victoria to Nanaimo (170ᶜ). Part of Georgia Strait was based on maps by captains Vancouver and Kellett, the accuracy of which Pemberton decried. Nanaimo's location, he declared, was off true position by about thirteen miles. He confessed that his map also was inexact because his chronometers were in unsatisfactory mechanical condition. A sextant was used to obtain locations on Finlayson Arm and a compass for much of the remaining coast. Later, he borrowed a ship's chronometer and went up coast as far as Nanaimo;¹⁵ he also obtained better soundings from Captain Charles E. Stuart. His revised map was sent to London (172ᶜ). But he also had to indicate his regret that his borrowed timepiece could not be relied upon. He repeated a request for three small portable chronometers which could be used to rate one against the other for greater accuracy (A6/120,fos.53,56d,58). In the meantime he borrowed three of them, rated them over a month, revisited the coast in early spring 1853, and then submitted a "tolerably correct" outline of about one hundred miles of coast from Victoria to Valdez Inlet, north of Nanaimo, based on over five precise determinations of latitude and longitude, with intermediate points fixed by compass and distance estimation (353ᴬ).

THE 1853 SEASON: DELIMITING THE BOUNDARIES OF THE EIGHT SOUTHERN LAND DISTRICTS

The 1853 season was spent largely in the vicinity of Victoria, but Pemberton was on several occasions in Nanaimo and once travelled across Georgia Strait to Bellingham Bay and the San Juan Islands. With the arrival of better instruments, the surveying department spent a good deal of time resurveying many lot lines and trying to complete work in Esquimalt, Metchosin, and Sooke. The work was also aided by the hiring in June of William Newton as a draftsman. Newton, who soon impressed Pemberton with his industry and ability, had been with the Puget's Sound Agricultural Company as assistant to the bailiff until he transferred to the Hudson's Bay Company. No fewer than twelve maps were produced in the surveying department that year, and at least five of these may be classed as topographic sheets, displaying the general terrain and vegetative characteristics, hydrographic features, and settlement patterns. One was the revision of the first topographic sheet encompassing Victoria and Esquimalt land districts, with 1852 and 1853 amendments, but drawn as one sheet, on a scale reduced to four inches to one mile. It depicts the Fur Trade Reserve, with all surveyed lots shown by fine, red lines, the lots reserved for the company, and several kept for future public use were indicated (354ᴬ). A second, the first topographic map of Metchosin District, was accompanied by a report on land use and soil quality for agricultural purposes (356ᴬ). On the map, available land was delimited with a dotted line, land already under the plough was shown, and the Puget's Sound Agricultural Company's Langford Farm was outlined. In the same year a similar type of map of Sooke District was completed (357ᴬ). Two new Nanaimo maps were produced as a result of two field trips to the settlement. One was a revision of the map of the "Country Round Nanaimo Harbour" (358ᴬ); and the other was an amendment of this revised map, with additional company land holdings shown on it (177ᶜ). Altogether 6650 acres had been assigned to the company there. The small group doing the work at Nanaimo had time enough to range, cut, and partly measure about ten miles of property boundaries.

A major enterprise completed in late 1853 was a map encompassing the entire southeast of Vancouver Island. It was prepared on four sheets at a two-inch scale, and showed territory extending from Sooke to "Cowitchin," including the Saanich Peninsula and Finlayson Arm (179ᶜ). On it were shown for the first time the boundaries of the eight land districts established by Pemberton: Victoria, Esquimalt, Metchosin, Sooke, Lake, Highland, South and North Saanich. In the accompanying report he described the basic characteristics and potential of each district. He was anxious to learn of the safe arrival of this map project for, as he said, it had "cost a greater expenditure of time and labour" than any which the committee had yet received (A11/74,fo.455). Pemberton explained to the committee his working method for the survey of the region. First, he laid out base lines in the Sooke, Victoria, and Saanich areas. These were carefully measured and their magnetic bearings ascertained. From the bases he set out a network of triangles carried across the hill-tops by observations to and from flag-poles tied to the tops of pine trees on the summits. From these fixed points, the positions of valleys and streams were determined, sometimes by observations, sometimes by chain and compass, but more often by stretching a long rope from one fixed point to another. The seacoast and inlets were carefully surveyed in canoes, and the surveyed outlines connected with the other bases.

Two maps concerned the Esquimalt area. The survey crew had managed to range the lines of the four Puget's Sound Agricultural Company farms, each of about six hundred acres; three were on the Esquimalt peninsula and the fourth lay northwest of the head of Esquimalt Harbour. These were on a map of the peninsula, which may well have been largely Newton's handiwork (355^A). Pemberton also sent to the committee a map of the Village Bay area on the peninsula on the east side of Esquimalt Harbour, to illustrate his choice of a site for a company boat slip (176^C). He supported this choice by noting that this bay could be easily connected by road to Fort Victoria; consequently he was designing the plan of a bridge over Portage Inlet at a narrow, rocky gorge site.

By 1853, the northern interior of Vancouver Island was better known than the southern interior. In 1849, John Work had examined Quatsino sound on the northwest coast, and Captain McNeill in 1850 had extended the search, hoping that it might become a coal harbour and possible export route for coal from Fort Rupert. This coal centre had been the starting point for a transect by John Muir to Quatsino.[16] Other men crossed to that area in early 1852. In addition, another small group found their way from Fort Rupert up the Nimpkish River and over to Nootka Sound.[17] They returned by coasting the entire northern projection of the island to Fort Rupert. No maps resulted from these journeys, but three maps were produced by Boyd Gilmour, the chief of coal operations at Fort Rupert. He had made an investigation of the island coast from the Nimpkish River north past Beaver Harbour and the fort to determine the value of this area as a coal field. Two were sketches of coal operations at two sites in the district, which were to be sent to a mining engineer in Edinburgh for his technical comments (324^A, 325^A). The last map was a regional survey of the geological outcrops along the neighbouring coast, or, as he spelled it, a "skitch of the Mettles" (326^A) (plate 53). Gilmour hoped that it would "serve as an explanation to those who have not seen the country" and would certainly show "how the thing [the mining potential] actually stands" (A11/73,fos.587,587d). In fact, the Fort Rupert coal operation had closed, and the miners were moved south to Nanaimo.

There are five other maps that relate to the interest of the company in regional coal resources. In 1853 Pemberton took a five-day canoe trip to Bellingham Bay on the mainland coast, stopping at some of the San Juan Islands. The information gained as to the American coal resources in the Georgia Strait region was passed on in a report and a map, sent to Deputy Governor Colvile (178^C) (A11/74,fos.416–417). The company was provided with another chart of this coal operation by Joseph W. McKay, at the time the company's master at Nanaimo (182^C). The third map, obtained by Governor Douglas, was by a member of the United States Corps of Engineers (365^A). The governor added to his collection of maps of the coal industry with a sketch Gilmour made of Newcastle Island and Nanaimo to indicate coal measures and suggestions for the best location of a mine shaft (181^C). McKay sent a map showing the "Works and Improvements effected at Nanaimo" (364^A). On it were indicated several pits, coal outcroppings, various workshops, and dwellings of the settlement and the shipping wharf. The company fort was on a knoll at the southeast corner of the site, commanding the line of buildings. Douglas complimented McKay because he considered it to be a correct plan which gave "a perfect idea of the place and improvements" (B226/b/7,fo.104).

UPDATING MAPS OF THE VICTORIA, ESQUIMALT, AND NANAIMO REGIONS, 1854–55

The 1854 season was an extremely busy one. It did not involve as much travelling but was concentrated on surveying lots in the main districts, on managing the sales and registration office, and on drafting. Pemberton's contract was renewed, with the strong support of James Douglas and the expressed satisfaction of the governor and committee. Pearse and Newton were also continued. Cartographic output was not as voluminous as in the previous season, and a number of the maps that were produced were revised and updated versions of earlier ones. Such was a map of the southeast coast of the island extending from Victoria north to Cape Mudge with the offshore islands (185^C). Pemberton considered the map to be imperfect although he had no hesitation in stating that it was the best map yet provided for the region. In compiling it he had compared his data with those of several company ship captains. There were also two revised topographic sheets for the Sooke District (373^A) and for the Victoria and Esquimalt districts (374^A), with updated topography and newly surveyed allotments. More detailed maps of the four developing Puget's Sound Agricultural Company farms were provided (375^A, 376^A). These farms were the Constance Cove, Craigflower, and Viewfield operations on the Esquimalt Peninsula, and the Esquimalt or Langford farm northwest of the harbour. Further illustrations of the activities in the Nanaimo area were added to the London files this year. Two similar maps of Colvile Town (Nanaimo),[18] with the workings, proposed mine shafts, coal outcrops, and all of the industrial and residential buildings were provided (370^A, 371^A). Pemberton announced that his department had completed surveying and laying out the company's land purchase at Nanaimo, for which a sketch was produced, and deeds had been prepared (372^A) (A11/75,fo.7A). Boyd Gilmour, the company overseer of the coal works, just before resigning, drew another map of the Newcastle Island operations in the autumn, with new proposals for expansion (385ᴬ) (A11/75,fos.417–418). Douglas, who believed that Gilmour knew "just enough of Physical Geography, to confuse himself, and everyone about him," did not press him to remain (B226/b/14,fo.37). Because of his inability and indolence, Douglas pointed out, he had not discovered a useful seam of coal either at Nanaimo or earlier at Fort Rupert.

George Robinson, the mine agent who replaced Gilmour at Nanaimo, was a much more qualified and experienced person. He kept Chief Factor Douglas apprised of the developing mining operations by means of frequent reports and maps, which detailed changing conditions over short periods. Five maps were produced in 1855 (406ᴬ, 407ᴬ, 408ᴬ, 409ᴬ, 410ᴬ), and others in each of the following years to 1858 (417ᴬ, 435ᴬ, 436ᴬ, 463ᴬ) (plate 54).[19] They were large scale maps concerned not only with surface conditions, but interestingly with the location and extension of the underground galleries. Shaded portions on the maps indicated where the coal had been entirely extracted.

While Pemberton was on leave in Britain in 1855, Pearse was acting surveyor, assisted by five labourers in the field season. Two useful maps, as well as a number of indenture maps were produced in this period.[20] One was a revised, updated tracing on four sheets of the 1853 map of the "S.E. Corner of Vancouver's Island from Soke to Cowitchin including the Saanich country and Inlet" (387ᴬ). This map was prepared by Fr. Augt, about whom no further information has been noted. The second map was more significant, since it was a new town plan of Victoria (25ᴮ).[21] The map is based on the first plan of 1852, with more recent details added, such as the new church, and an "Indian Reserve of 10 acres."

ACROSS THE ISLAND TO BARKLEY SOUND, 1856

Upon his return in December 1855, Pemberton was again plunged into his regular duties as surveyor, land sales agent, registrar of lands, arbitrator of land disputes, director of the drafting room, and engineer involved in road building and other public works such as bridge, school, and church construction. In 1856, he was elected as a member for Victoria to the new colonial legislature. Other than cadastral maps prepared by the office, the most unusual mapping project was the cross-island expedition carried out in October and November 1856. Pemberton's object was to examine Nitinat Inlet on the west coast of the island, which was reputed to have exposed coal deposits, and also to make careful observations of the coast for his general island mapping project (A11/76a,fo.397). He did not get as far as this inlet, but reached Alberni Canal and Barkley Sound. In fact, he had been preceded to the canal in the spring by a small party led by Adam Horne of the company's Colvile Town colliery. Captain Stuart, the factor at Nanaimo, relayed this news with a report and "a very interesting sketch" of the route (191ᶜ) (B226/b/12,fo.125d). Pemberton's small party also started from Colvile Town, and proceeded north to the mouth of the Qualicum River, turned inland up the river valley, following an Indian trail to Horne Lake, and passed over the central mountain ridge to the valley at the head of Alberni Canal. They examined the south end of Great Central Lake, taking soundings there as they had in Horne Lake. After following Alberni Canal and Barkley Sound shores to Cape Beale they turned south to Port San Juan. After his return to Fort Victoria, Pemberton drafted an original map (192ᶜ), and made a tracing (413ᴬ). He observed to Douglas that the map should be regarded only as a reconnaissance effort since he had taken only nine principal observations of latitude and longitude, the intervening detail being filled in by taking compass bearings and estimating distances.

ACROSS THE ISLAND TO NITINAT INLET, 1857

In the winter of 1857, Pemberton went north to Fort Rupert to fulfil a request of the committee that a plan be prepared, and a financial evaluation made, of the fort and its accoutrements, the wood and coal yards, the gardens, the wharves, and two small bridges (A11/76a,fo.666). Both tasks were accomplished (422ᴬ, 423ᴬ).[22] Similarly two copies of a map and an evaluation of the company's premises at Victoria were provided (424ᴬ, 425ᴬ) (A11/76a,fo.694,700). The map is based on the 1855 town plan but has only company building sites depicted. Pemberton, Pearse, and their crew were increasingly busy in the southeast region of the island as the inflow of settlers grew. The demand for land was such that they had to extend allotments into Saanich and the Cowichan valley. For this and other purposes, Douglas and Pemberton went north into the Cowichan area in September 1857. Pemberton was sent on from there across to the west coast. Because of a fear that the Cowichan Indians might be hostile, a royal navy lieutenant, two marines, and two seamen volunteers accompanied him. The fears were groundless, however. Because only a few hunters were attached to the party, it moved quickly. Pemberton carried the principal instruments including the chronometer himself. Unfortunately, the going was made rough by the massive trees that covered the heavily timbered land and by the slipperiness of the fallen and rotting tree trunks. Pemberton fell several times and so damaged his instruments that he had to forsake astronomical observations and rely mainly on his compass. His map was, therefore, only a compass sketch. At least two map copies were made and sent to London (426A, 427A) (plate 55). The map indicates the course along the Cowichan River to Cowichan Lake, and across the central mountains to the long narrow-mouthed Nitinat Inlet. The party returned to Victoria by walking southeast along the shore of Juan de Fuca Strait.

THE FRASER RIVER GOLD STRIKE AFFECTS VICTORIA

The flow of settlers became a torrent in 1858 as the news of gold strikes on the Fraser River spread. Victoria was crowded with new arrivals, many of whom decided to emigrate and settle or to purchase lots for speculative purposes. By September, the surveying department had been forced to hire five more surveyors, most of them from San Francisco. The draftsmen were not able to prepare any other maps beyond the normal indenture plats for land sales. The eighty indenture maps now located in the company archives are, it is clear, only a portion of the total that had been prepared by the draftsmen over the years. The temporal order of their arrival at Hudson's Bay House over this time indicates the pattern of land sales outward from Victoria settlement. An example of the normal indenture map is that of Lot 26, Sect. 1, Esquimalt District, which is the Esquimalt or Langford farm of the Puget's Sound Agricultural Company (378A) (plate 56). A further cadastral map, related to the Puget's Sound company activities, reached London in 1858 (462A). Alexander G. Dallas had been sent to Victoria to deal with problems associated with this company.[23] One of his tasks was to try to lease several large areas (five hundred to a thousand acres) of rough and forested land in Esquimalt for sheep pasturage and to arrange further pasturage rights on all adjoining forested lands unlikely to be purchased for some time. The map indicated the boundaries of all the agricultural company's farms as well as a number of waterfront lots in Esquimalt, which it hoped to sell at high price. Because the general character of land use was also depicted, the map implicitly suggests the location and type of property being sought in leases.

THE COMPANY CLOSES THE SURVEYING DEPARTMENT

By the autumn of 1858, Pemberton was receiving instructions concerning the future of the company's role, and therefore of the surveying department, in the colony. The mapping facility was to cease operation as a company enterprise on 31 May 1859, the date on which the Hudson's Bay Company's role as proprietor of the colony would cease. He was instructed to arrange for the closing of the surveying department and to give notice to all workers in the office. Since his personal contract ended as of 16 December 1858, his offer to stay on until the later date to ensure that the transition ran smoothly was accepted. The last major map produced under Pemberton's direction as company-colonial surveyor and cartographer was the Victoria District map of 1859, in effect the third town plan produced during this period (30B).[24] The chief draftsman for the project was Hermann Tiedemann, who started the work in late 1858, but did not complete it until the close of the succeeding year. There were several maps of properties within the town of Victoria prepared, these being of utility to show the locations of colonial office buildings and the main streets of the centre of the settlement (469A, 477A).

PEMBERTON'S ACCOMPLISHMENTS

The governor and committee had been faced in 1849 with a colonization-settlement scheme for which they were unprepared. Although their first appointment for the responsible post of surveyor was an unhappy one, in their second they were indeed fortunate. During the eight years Joseph Despard Pemberton was an employee of the Hudson's Bay Company, he more than lived up to their expectations. He was the best-trained surveyor and cartographer employed by the company over the years of its operations in the north and west of the continent. He had designed and produced, at large and medium scales, detailed and exact cadastral maps as well as a topographic series depicting natural environment and settlement features on a precisely surveyed triangulation grid. From personal observation of many prominent coastal and terrestrial locations, he had developed a graticule on which to organize the small-scale maps he drew of both land and water areas. Moreover, he had been responsible for the design, organization, and nomenclature of the administrative land districts of the colony, the registration procedures for land holdings, and the pattern of roads and properties that became the skeletal structure of further settlement.

13 Exploration and Mapping in the Northwest and the Establishment of Company Land Claims, 1849–1859

> You should endeavour to make a rough chart of the route you pursue, taking the distances & courses by compass ... make due allowance for the variation of the Compass, ... when traced on a map we occasionally find rivers followed far out into the Arctic Sea, while they are made to cross other streams and mountain ranges in a marvellous manner.
>
> Sir George Simpson to James Anderson,
> August 1852

Although cartography related to colonial affairs and land claims predominated in this period, a significant number of maps depicting the routes followed by company servants exploring the northwest of the continent were also produced. New geographical outlines were traced in the Yukon and Mackenzie regions, and the archipelagic character of the Queen Charlotte Islands was disclosed. The company became more closely involved in making and mapping claims to its property in Rupert's Land and on the western mainland and in fighting a form of rear guard action in the old Columbia district against the imposition of American land control up to the forty-ninth parallel.

MAPPING OF EXPLORATION IN THE NORTHWEST

The Hudson's Bay Company was attracted to the Queen Charlotte Islands by the reports of gold deposits there, although some geographical inquisitiveness was also a factor in the attraction. This archipelago had been thought of as one large triangular-shaped island until 1850. At that date, Pierre Legace, a servant from Fort Simpson on the northwest coast, discovered the water passage from Skidegate on the east coast to Englefield Bay on the west and reported that Indians claimed still other channels existed. Further visits by John Work and captains Charles Stuart and William McNeill confirmed the existence of gold. The three men organized both a gold mining enterprise and the gathering of the precious metal by local Indians. In 1851, George M. Nutt, second officer on a company vessel, prepared a chart of Mitchell Harbour on Englefield Bay, which was sent to Governor Douglas by John Work (163^C). In the same letter Work included another chart, a sketch of "Skiddigates" passage, which Stuart, the captain of the ship *Beaver* had drawn (164^C) (A11/73,fo.181d). Stuart probably also drafted the "improved map of Queen Charlottes Island" that was transmitted to London in 1853 (183^C). Not only coasts and harbours, but also the character of the interior were displayed. Point Rose's position was located more accurately, at least five miles south of its position on contemporary charts. Two company ships had been wrecked on this spit; Governor Douglas warned the company secretary that the location error should be corrected on commercial charts (B226/b/11,fo.7).

A great deal of new geographical information on the northwest interior was entered on maps in this period. In the summer of 1852, Governor Simpson had requested from James Anderson, master of Mackenzie District, who was stationed at Fort Simpson, a description of the country lying west of the mountain front – even a "rough chart" would help (D4/45,fo.59). Anderson in his turn had several months earlier chided A.R. Peers at Fort McPherson for not sending a map of the routes involved over the Richardson range from the Peel to the Rat River (B200/b/26,p.30). In January of 1852, Anderson had asked Robert Campbell for a map of the Yukon River, as well as of the Lewes, Porcupine, and Rat rivers and of the "bordering country" (B200/b/26,p.23). The previous summer, Campbell had established the identity and interconnections of these rivers. He had descended the Yukon from Fort Selkirk to Fort Yukon, and had then taken his annual returns to Fort Simpson via the usual supply route to Fort McPherson (D5/31,fo.267). It would seem that none of these requests or admonitions resulted in any maps being prepared. Alexander H. Murray, the Fort Yukon master, did give Anderson "much valuable and interesting information" about the Peel, Bell, and Porcupine river regions (B200/b/26,p.11), and what is more, compiled a map which was carried to Peers at Fort McPherson who, Anderson thought, could make best use of it (175^C).

A most informative map of the Northwest, a sketch of the Alaska peninsula showing the Yukon River and most of its tributaries and the major mountain ranges, which also delineated the territories of eleven of the tribes around the fort, was obtained from William Hardisty at Fort Yukon in 1853 by Governor Simpson (366A) (plate 57). Hardisty, a knowledgeable and intelligent officer, who had been living "upwards of thirty years ... among Indians," had shown a keen interest "in their character and disposition" (B200/b/32,fos.24,25). Drawing on his experiences at Frances Lake, Fort Selkirk, and Fort Yukon, he had made a special attempt at Fort Yukon to open communication with more distant tribes and to assess the possibilities of trade with them. Hardisty enclosed a "rough sketch of the Youcon river and surrounding country, to show the position of the several tribes around the Fort," with the report he sent to Governor Simpson (D5/38,fo.75d). In this report, Hardisty advised against a proposal to close out both Fort Selkirk and Fort Yukon and to move the latter post higher up the river. His arguments against the change hinged on the negative effects of such a move on the company's present and potential trading relationships.

Having a reasonable knowledge of the territory west of the Mackenzie River, the company began to focus attention on the little known area east of the river and north of Great Bear Lake, a region untouched by company explorations directed toward the Arctic coast and the Coppermine valley. James Anderson at Fort Simpson began to gather information by sending a series of questions to post masters in the Mackenzie district instructing them to question natives coming from the area. Roderick McFarlane at Fort Good Hope was particularly successful in eliciting comments (B200/b/31,pp.61–67). Anderson, in 1855, wrote a description which he "considered as approaching correctness" that he had "extracted from a mass of contradictory statements and possible exaggerations" (B200/b/31,p.66). With this he sent his own rough sketch of the Beghulatesse or Inconnue River and adjacent country (412A) (plate 58). The map depicts the course of the Mackenzie River to its mouth, a very generalized Arctic Ocean shore east to the Coppermine River, Great Bear Lake, and the lower course of the Coppermine River. Anderson then outlines what he postulates are the major tributaries and associated lakes of the, as yet undiscovered, "Unknown River."[1]

Anderson was again involved in mapping in 1856 when he took part in the preparation of a composite map of the northwest of the continent (D7/1,fo.238b). The history of this joint map began with Sir John Richardson, the Arctic explorer, who had sent a map (likely a printed one) to Alexander Murray, master at Fort Yukon. In addition to the data on this map, Murray had also provided Anderson with information about the Yukon river system. The chief factor then had Bernard Ross, an employee at Fort Simpson, trace out the combined features on a composite map (419A). As well as more local details, the map showed the location of the Colvile River, flowing to the Arctic, and the supposed outlet of the Yukon River on the south shore of Norton Sound, Alaska. Anderson's ultimate purpose in having the map made was to buttress his comments on the proposals for altering the posts and trade network in the Northwest.

MAPPING COMPANY LAND CLAIMS WEST OF THE MOUNTAINS

Continued encroachment on land the Hudson's Bay Company's claimed as its property was the motivation for the drafting of a number of maps of the west coast mainland both north and south of the international boundary. In 1858, Alexander Dallas was concerned with the situation which the company was facing at two posts, Fort Hope and Fort Yale, as new settlements were expanding adjacent to company forts. In a report he sent to Governor Berens in London, he included a map that indicated how the street and lot pattern of the Fort Hope settlement crowded close to the walls of the fort and even infringed on the company's horse paddock (464A). Furthermore, lots on land claimed by the company were being sold. A similar map of Fort Yale showed the location of the company's store and its claimed land reserve, but the new town plan completely ignored this private property and the lot lines (465A) (plate 59). Three further maps drawn in 1859, one of Fort Hope (198C), and two related to Fort Yale (196C, 197C), are also illustrative of these vexatious land claim problems.

By 1849, the reasonably amicable relationship of both the fur company and the agricultural company with American settlers and officials had changed to an increasingly testy one as large numbers of people pushed north of the Columbia to squat on or near lands claimed by these enterprises. In 1851, the Puget's Sound Agricultural Company alleged that there were twenty-eight trespasses; in 1853, it claimed fifty more.[2] The committee and the board of directors of the Puget's Sound company ordered local officials to warn offenders orally in a reasonable and pleasant manner, but, if this failed to produce the desired response, they were to give trespassers written notices. Copies of notices were to be filed for future action.

A broader issue in regard to land claims was brought to the company's attention by the proposal of John B. Preston, surveyor general of the American Oregon Territory, to have his staff lay out and survey townships as well as investigate all private land claims. William Tolmie, the agricultural company's officer at Nisqually, was directed by his superiors to give the Oregon surveyor general a full description of its claims along with a "topographical outline," that is, a map. The map of the Nisqually claim, prepared in 1851 by Edward Huggins, a clerk at Nisqually, was likely the original kept at that post (317A). A copy was sent to Preston and described to him in a letter from Tolmie as a "colored map of the tract of [land] occupied by the said Company for agricultural and stock-farming purposes since the year 1840" (165C) (B151/b/1,fo.50). Preston recommended that a surveyor should be hired by the agricultural company to lay out the property in proper manner. This was acted upon in the spring of 1852, but the American surveyor, who was hired by the Puget's Sound company, did not finish the job because he was threatened with personal violence by a group of American squatters. Huggins finished the task himself and drew up a map of the Nisqually claim with the surveyed boundaries (173C).[3] A copy was made, probably by Huggins, which Peers handed in to the surveyor general's office by August 1852 (16B).[4] Peers also gave in a copy of another original map prepared at Nisqually by an employee, an "approximate survey of land claimed by the Puget's Sound Agricultural Company on the north bank of the Cowlitz River" (174C, 17B).[5]

In the spring of 1855, James Tilton, surveyor general of the newly established Washington Territory, asked Tolmie for another map of the agricultural company's claims, to assist territorial surveyors in determining the intersections of newly laid township lines in relation to these claims.[6] A map of the Nisqually claim was sent to him in 1855. It was likely a copy of the 1852 map (189C). Dugald Mactavish, the Hudson's Bay Company chief factor at Fort Vancouver, was sent a similar request, but he refused to provide a map on the grounds that this action would not have been authorized by the governor and committee.[7] Actually, a map of this type existed then as Adolphus Lee Lewes had been put to such an exercise in 1852. His production had displayed the fort and surrounding region, including other company buildings, settlers' farmsteads, enclosed and unenclosed cultivated land, cattle pastures, and so on (352A). In addition, that year John Ballenden, in charge of the Columbia District, was to have similar maps prepared of Hudson's Bay Company property claims at the forts of Colvile, Nez Perces, and Okanogan. Alexander C. Anderson, post master at Okanogan, surveyed and mapped his post (351A); he was to have done the same for the other interior forts, but no resulting maps are now in evidence.

By 1859, a joint American and British Land Claims Commission had taken the matter of British land claims in Washington Territory in hand and were discussing financial compensation for lands relinquished by British subjects. The Hudson's Bay Company and the Puget's Sound Agricultural Company were to supply documents and maps to be used later in taking testimony. Several maps in the archives bear on this important issue. A large-scale "Map of Fort Vancouver and U.S. Military Post with Town Environs" was drafted by Richard Covington (471A). This carefully drawn and complete visual inventory of the settlement in this area was submitted by the company as part of its evidence later. Similarly submitted in evidence by the company was an American map, printed in 1848 by J.C. Fremont. Charles Preuss, the cartographer, had drawn part of the "Arro Archipelago" (the San Juan Islands) in an enlarged scale inset. This map was included because it indicated that the San Juan Islands were British territory. In 1853, Governor Douglas had had a tracing of this 1848 map made in Victoria and sent to London to illustrate his discussion with the committee on the political geography of the Strait of Georgia region (368A) (A11/74,fos.424-427d).[8]

Several maps of the canyon region of the Fraser River valley were not associated directly with land claims, but were related to company interests in gold rush activities there. Governor Douglas had a map drawn of "Fraser's River" from the Thompson junction down to Hope, which "gives a good idea of the country" (195C) (B226/b/15,fo.38d). He sent this "Gold Fields Map," as it has been called, to the committee in February 1858 as a convenient reference when it received reports of the hectic activities on that narrow river defile (A11/76b,fos.960d–961d). Although the original is not in the collection, it was very likely the basis for the map reproduced and printed by John Arrowsmith in July 1858 (G3/19). This was "Presented to both Houses of Parliament by Command of Her Majesty." If so, both original and printed versions were oriented to the west; they displayed the Shuswap (Thompson) River, Couteau Country where gold had been found, and a "Succession of Cataracts and Rapids" all the way to forts Yale and Hope. A further map, a detailed reconnaissance sketch resulting from an examination of the Fraser River Valley from Fort Hope to Fort Yale by James Yale, master at Fort Langley, includes the names of the various gravel bars on the river where gold panning might be profitable (29B). Finally, Alexander Dallas sketched a rough plan of the region to show the location of newly established Fort Berens (Lillooet) on the Fraser relative to other regional posts and in relation to the Harrison-Anderson-Seton Lake route (199C). The Western District was considering using this route as the new brigade trail to replace the Fort Hope to Fort Kamloops trail.

MAPPING COMPANY LAND CLAIMS EAST OF THE MOUNTAINS

In addition to the flow of maps of Vancouver Island during this decade, the company was vitally concerned with its property rights and claims to land elsewhere on the continent.⁹ One group of maps was drawn by a young surveyor, Alexander McDonald, who was under contract to the company in the upper Great Lakes area from 1854 to 1856. In July 1854, the company's claim of twelve hundred acres at Sault Ste Marie was confirmed by the Lands Office of the Province of Canada. McDonald staked these out on the ground, but recommended alterations to the boundaries which would have been of advantage to the company. He drafted not only a plat of the required survey (186C), but also a second map to illustrate his proposals and make clear his reasons for recommending the changes (187C). This map was sent to the Lands Office for adjudication. McDonald surveyed other posts with a small crew and prepared survey plats, which also reached the Lands Office.¹⁰ There officials declared that, except for Nipigon, each survey map showed deviations from instructions they had given him. Governor Simpson replied in explanation that local peculiarities had made it difficult for McDonald to run lines exactly as directed, and in some cases there were clashes concerned with reserve boundaries claimed by Indians. Simpson asked for acceptance of these minor deviations as the company did not wish to have the further expense of maintaining the survey crew in the field, if patents had not been received (D4/52,fo.10d).

Another land problem erupted in 1857 in the valley of the Gatineau River, a tributary of the Ottawa River, when the company claimed that the Roman Catholic church had been squatting on its property at the Desert River Junction. Archibald McNaughton, the company master at Buckingham, produced two maps to explain the matter (411A, 437A) (plate 60). A third map, a copy of the official survey delineating its adverse decision, was sent by one of the Crown Land Office's officials, E.P. Tache, with notification that the church's right was upheld, based on possession at that time (438A). Moreover, said Tache, both parties were really squatters on Crown land, and even though the Hudson's Bay Company had been there for a longer time, it had no right to deny the other party the lands it had improved (D5/44,fos.24–25).

SPECIAL PURPOSE MAPS

Other maps were sent to London dealing with various topics in this period. In 1859 and 1860, Captain J.F. Trivett, of the company steamship *Labouchere*, indicated on four small-scale charts the routes of the *Labouchere* and of another vessel, the *Princess Royal*, between Victoria and London (472A, 473A, 474A, 480A). He also made a larger scale map which showed the complicated route through the Strait of Magellan and the Inner Passage of the Chilean coast (466A). Other cartography included, in 1858, a coastal map of the Eastmain (467A), and a chart of Little Whale River mouth, both likely drafted by Alexander McDonald, who had earlier surveyed Upper Great Lakes posts for the company (468A). These maps had an unusual purpose; the company was investigating the possibility of establishing a porpoise fishery on the Eastmain.

Investigating a Country So Interesting

14 Mapping Company Land Claims and Exploring Inland Routes, 1859–1870

> We learn ... that a strong desire is being manifested ... to pre-empt portions of the Company's claim – Unless our claims in British Columbia are soon defined, a repetition of Oregon annoyances regarding land may, in one form or another, be certainly looked for.
> John Work to Thomas Fraser, February 1860

The Hudson's Bay Company and the Colony of Vancouver's Island had been fortunate in having the services of Joseph Despard Pemberton, a well-trained professional, at an important juncture in their histories. In expanding and reconstituting its business north of the newly defined international border on the Pacific littoral, the company had assumed the novel role of colonial agent with all the attendant rights and privileges, including the cadastral and cartographic responsibilities, which that role entailed. Within a few busy years' time, mapping for purposes of the fur trade gave way as the company became a more urban enterprise, involved in land speculation and different kinds of commerce. This change in orientation was increasingly reflected in the maps produced by company employees or received by the company during the course of business throughout its vast area of operation in the latter half of the nineteenth century.

Of approximately 110 maps received in Hudson's Bay House in London during this decade, over three-quarters concerned property matters. These were maps confirming and delineating the property of the company or of its closely associated agricultural company, maps relating to the selling off of land parcels, maps providing a basis for arbitrating financial compensation for land and establishment losses in Washington Territory, and maps for transferring some reserve land in Victoria to the public domain. During this period, the company did not have an official surveyor-cartographer in its employ, and extensive exploration by servants had essentially run its course. There were, however, expeditions to find a more efficient and economic supply route from the Pacific across the Coast Range in northern British Columbia and to look for more accessible passes from the valleys of the east through the Rocky Mountains. The company was also involved in an overland telegraph project across the plains and Cordillera. The map files reflect both these enterprises. Ninety-seven of the maps produced in this decade were related to the area west of the mountains and Vancouver Island; seventy-five of this group pertained to the island, and sixty-one of these were concerned with the Victoria-Esquimalt region in particular.

MAPPING COMPANY PROPERTY AND LAND TRANSFERS IN VICTORIA-ESQUIMALT

After 1 June 1859 when the Colony of Vancouver's Island was transferred to the administration of the colonial governor, the former company-colonial surveying department became strictly a government office. Pemberton was colonial surveyor and Pearse his assistant, until they were appointed surveyor general and assistant for the colony. The company maintained a relationship with the surveyor's office and hired some of its employees, as well as other men, to undertake surveying and drafting tasks when required. There was a considerable turnover of property owned by the company in the central area of Victoria. Some of it was occupied by or about to be built upon by the government. Company and colonial officials began to challenge one another over ownership of certain land parcels and over compensation to be paid for the lots that both sides agreed were company property. Alexander Dallas was placed in charge of the Hudson's Bay Company's transfer of operations. To aid the governor and committee in understanding the situation, he sent them in early 1860 a ground plan of the company's holdings north of Fort Victoria (477A). This showed the location of former temporary government offices and the new buildings for the police barracks. Accompanying it was Dallas's estimate of the land's value. Over the next several years, this property would rise greatly in value because of increased demand.

One bitter controversy in these early years concerned property rights on the south side of the James Bay Inlet of Victoria Harbour. In 1861, the company obtained a map, prepared under Governor James Douglas's direction, which purported to show all those lots sold by the company between 1859 and 1861 (482A) (A11/80,fo.627). It included a section that the governor considered to be "a government reserve." The lots, he said, lay within the limits of the public park, and one section (lot z) of the government reserve had been sold by the company to a local person,

Mr. Lowenburg. When this gentleman had tried to enclose the land, he had been stopped by colonial officials, then arrested and arraigned for trespass. The case was taken up by the company and referred to London. The company insisted that this plot of land was its property and that the government reserve itself belonged to the company (B226/b/20,p.213). Furthermore, it claimed that the colonial government, which had erected buildings on the lot, was an illegal occupant or squatter. To buttress its arguments, the local company office had a small map prepared, which it sent to the committee, showing details of the location of lot z (201C). This controversy continued until its disposition as part of a general settling of land ownership claims over the next several years.

In preparation for this general settlement, the company hired a private surveyor-draftsman, F.W. Green, in 1861, to draw up another map of the area around James Bay (483A). This map, on linen, depicted lot lines and specific properties in colour; the "government reserve" was in pink, the Lowenburg parcel (z) in green, and other properties in other colours. By February 1862, an agreement was concluded in London between the company and the government on the conveyancing of certain blocks of land to government control in Victoria. To facilitate this transfer, four sketch maps outlining the extent and boundaries of these lots, as the colonial secretary's office in Victoria perceived them, were given by the secretary to Dugald Mactavish, the chief factor at Victoria, in July 1862 (A11/78,fo.475). He returned the maps to the secretary remarking that with one or two exceptions "they are not such as ... the Government can expect the Hudson's Bay Company to acknowledge as being correct" (A11/78,fo.476d). At the same time, Mactavish prepared a set of six maps delimiting these same blocks of land as the company comprehended them, accompanied by written explanations of the boundaries. Six tracings of these were made and sent with copies of the explanations to the colonial secretary's office in Victoria. A further set of six tracings was sent to London in August with explanations of the negotiations and the discrepancies noted between the two versions.[1] The six maps depict: the lots for the harbour master's office (490A); the police barracks and prison (491A); the post office (492A); the public park and school reserve (493A), the church reserve, parsonage and cemetery (203C);[2] and government offices (494A). The colonial secretary replied to Mactavish in August, querying only one of the maps, the first, dealing with the location of the harbour master's lot (A11/78,fo.477).

Lying between James Bay and the Strait of Juan de Fuca was Beckley Farm, which originally belonged to the company. As part of the joint agreement of February 1862, the company was to choose, within eighteen months, the fifty acres of the farm that it wanted to keep. The remainder would be in government control. The farm had been surveyed before 24 March 1861 by the company, and a map of this survey completed in the autumn of 1862 (204C). The map proved to be so incorrect that all the surveying and drafting had to be redone. The committee requested that this be carried out as rapidly as possible so that the company could choose the most desirable lots. Green was given a contract for this re-survey, and his map was available in early 1863 (206C).[3] It delimited eight lots chosen for the company, totalling 48.77 acres, along with the one-and-one-quarter acre farm homestead. The remaining 60.54 acres, together with the entire waterfront along the Strait of Juan de Fuca, reverted to the Crown. At least four other copies of this map were made for various purposes, only one remaining in company hands. This copy is one of two sent to the committee (504A).[4]

In 1865, to explain the land situation in Victoria more clearly to the committee, William Tolmie had Hermann Tiedemann prepare a set of tracings based on earlier maps. Tracing 1 encompassed most of the town of Victoria; it showed the block of 1212 acres that the company owned where land transfers were taking place (538A).[5] Tracing 2 focused on the land assigned to Governor Douglas on the south side of James Bay (539A).[6] Tracing 3 was taken from a small segment of the 1859 town plan, to define the "Indian Reserve" of ten acres, next to Douglas's property (542A).[7] Tracing 4, also from the 1859 plan, illustrated the land holdings along the south shore of James Bay, particularly the Douglas block and the government reserve (544A).[8] The last tracing, number 5, was a composite map showing the various changes in the boundaries of the governor's and the government's land between 1855 and 1862 (546A).[9]

Although most of the differences between the company and the government had been negotiated, the Lowenburg "affair" was still pending. Dallas insisted that it represented an important matter of principle. If the company allowed its "Indenture" to be set aside there might be a "dozen similar attacks made" (A10/62,fo.150d). In 1865, after his retirement, Dallas prepared a map intended to enlarge upon the company's position in selling the lot to Lowenburg (548A).[10] The company finally prevailed; the colonial secretary withdrew the government's claim to the lot in 1868, and Lowenburg was left in possession (A11/83,fo.630).

The final map to be noted in the series of documents associated with company and government property claims in the Victoria district is a Tiedemann map of 1866 (553A).[11] This was drawn for the company at a large scale and in sixteen segments. It provided a definitive delimitation of land ownership, specifically of the lands (in green) reconveyed to the Crown, and of those lots (in pink) remaining as the property of the company. Boundary lines and measurements based on official records were drawn in red ink; ground measurement boundaries were in black.

Maps of the Victoria district were concerned largely with the property adjustments being made between the government and company's Fur Trade Department. In the Esquimalt area the larger number of maps reflected the Puget's Sound Agricultural Company's attempts to subdivide its farms into marketable lots at appropriate times. The first map in this period of any of these properties is of Esquimalt Farm (Langford Farm) (476A). It is especially useful since it is the only large-scale farm map showing the characteristics of these farming enterprises. The homestead, consisting of main house, labourers' houses, bake house, storehouse, dairy, cow shed, cattle sheds, implement shed, and granary, is set in the midst of cultivated fields, pastures, and woodlots. In 1863, the directors of the agricultural company asked Tolmie, its agent, to begin to organize the surveying and subdivision of the 593-acre Viewfield Farm, at the east end of Esquimalt peninsula. They considered it absolutely essential that careful work be done in order "to prevent vexatious discussions or disputes hereafter" (F11/2,p.188). A map should be made also to avoid "endless trouble and annoyance" (F26/1,fo.315d). The lots were to be under five acres in size, and roads through the farm were to be planned. In early 1865, Tiedemann at Tolmie's direction, completed a plan of the farm (213C). A tracing of this first land speculation map reached the directors in the autumn of that year (537A). It is drawn with black and red ink lines, with colour washes, particularly green, indicating the extent of wooded lands. Tiedemann had named several of the lateral streets after officials of the agricultural company.

The directors advised Tolmie to have Constance Cove Farm surveyed as quietly and unobtrusively as possible so that news of the subdivision would not prejudice the sales of Viewfield lots. Craigflower Farm subdivision should be delayed in the hope that developments in the Constance Cove area might enhance the value of Craigflower land later (F11/2,p.220).[12] Tolmie was able to convince the directors that it would be best to complete the Constance Cove subdivison and part of that at Craigflower immediately (F11/5,fo.297). Tiedemann was put to both tasks. He was instructed to have his map also note a property of about five acres in size fronting on the most sheltered cove on the east shore of Esquimalt Harbour where deep water adjoined the land. This was the location the Hudson's Bay Company officers considered best suited for the company's future use (214C).[13]

In the summer of 1863, the governor and committee requested the Western Board of Management to advise them on the suitability of the company moving its regional headquarters to the Constance Cove locale. The board concluded that the company must enlarge its business in order to retain its commercial position; such an enlarged concern would need space in Esquimalt for a new centre of operations. It recommended that a warehouse, wharf, and sheds be built, with plans for additions as required, and that a sixteen-and-three-quarter-acre block at Constance Cove be acquired. Additional site advantages at the cove were suggested: outward bound ships from Puget's Sound were already being admitted into Esquimalt Harbour without cost; outgoing lumber ships would come to this site for seamen and stores if inducements were offered; British naval personnel would be customers; and Indians travelling to Victoria from the west shore and Strait of Juan de Fuca would call in on passing and would sell their furs and fish oil if prices were right (A11/80,fo.550). Finally, the headquarters of the western department of the company could be moved there at the appropriate time. Tiedemann, in 1868, before he fled to San Francisco to escape his creditors (F11/5,fos.464d,465), was put to the task of making two copies of a map of the land selected by the board (570A). This map, on four sheets, was an enlargement of the area as shown on the 1866 map. There were about 528 yards of waterfront, and the lot was bounded on the rear by Admiral's Road (A11/83,fos.366,366d).

The directors quickly notified Tolmie that the acreage was too large, and that they would grant permission for only eight acres. By this time, June 1868, the company's wharf was completed, and a stone warehouse was nearly built at the water's edge adjoining the wharf. A large wooden, two-storey sales shop had been started, as had a forge and an oil store. Ten acres had been cleared and a passable road constructed. All of this had cost the company nearly $14,000. Nonetheless, the directors of the agricultural company would finally allow the sale of only 5.6 acres. It was likely Tiedemann who drafted the site ground plan that was to be sent to the committee – a plan with the improvements marked on it (216C). A similar plan was also sent to the board of directors of the Puget's Sound Agricultural Company (217C).

MAPPING COMPANY PROPERTY AND LAND TRANSFERS
IN NANAIMO (COLVILE TOWN) AND FORT RUPERT

This decade to 1870 also witnessed the relinquishment of the Hudson's Bay Company's affiliation with the coal mining operation that it had established at Nanaimo. Here, the holdings of the company amounted to 6193 acres, mostly on the mainland, but the property also included the islands of Newcastle, Douglas (Protection), and Cameron. In 1860, Governor Douglas was in the process of having a townsite laid out on the government reserve at Nanaimo, which lay north of the Millstone River. The company already had a townsite (Colvile Town) laid out on the small peninsula where the largest share of the coal mining facilities was located. But by this date the old workings were almost exhausted, the shipping wharf had recently collapsed, and the coal tramway had been abandoned. This state of affairs is made clear by a map of the Nanaimo District (478A).[14] The governor and the committee and the western board had been contemplating the company's future in the coal operation for some time before 1861 and 1862 when negotiations took place between the company and a new enterprise, the Vancouver Coal Mining Company Limited, regarding the sale of part of the Hudson's Bay Company's property and of its coal operations to the coal mining company. The chief entrepreneur of this new business was Charles Nicol, who had been the last mine manager for the company. Dallas, who was in Britain on leave in the summer of 1861, took along several maps of the area to show the London officials. One was a large-scale plan of Nanaimo, with an inset of the larger region indicating all the property owned by the company in this area (485A).[15] The more localized map of the Colvile Town peninsula and nearby mainland, locates roads, all property lines of lots already sold to individuals and institutions, the various buildings, coal works, and the saw mill. In 1861, a rough map showing the land involved in the sale was drafted and attached to the draft agreement of sale to Nicol (486A). By the end of the next year, the sale and transfer of the company's coal holdings at Nanaimo was essentially completed. Another map, showing the land owned by the Hudson's Bay Company, was attached to the indenture, ready for the final transfer (495A).[16]

In the summer of 1861, the colonial government was assigned two lots in the Colvile Town waterfront for a harbour master's facility, in consideration of the company being allowed to alter the outlines of the Indian reserve to best suit its "views and wishes (A11/80,fo.267d). A sketch plan, of part of the central area of the community indicates the location of the transferred lots (488A).[17] In 1861 or in one of the two following years, a small map of the section of lots bordering the Indian reserve in the government townsite was drawn to show this transaction (489A).

Although the sale to the coal mining company involved almost all of the company's property in Nanaimo, several lots were excluded since they had been sold to other parties.[18] Two cadastral maps, almost the same in character, were drafted in the spring of 1862 to show the details of these previous land sales (501A, 502A). Two other plots were disposed of also, one to the Wesleyan Methodists, and one to the Episcopalians. All of this alienated land sold prior to the final coal company transfer, appears on a map of 3 February 1863, for the use of the company in arranging final papers for the sale (507A). Three further maps, with an extra copy of one of them, were associated with the legal technicalities of the sale. One map, dated 25 August 1863, of the peninsular area, was attached to the draft reconveyance of land and hereditaments at Nanaimo (508A). A second, of 28 August 1863, showed the larger area surveyed and plotted by the company, in which the property released by deed to the new coal company appeared (509A). Finally, in 1867, a further map, with a copy, of all land around Commercial Harbour, Nanaimo, which was to be transferred to the coal company was attached to a draft reconveyance document (561A). The Hudson's Bay Company had cut its tie to this up-island community after a decade of close commercial and industrial interest.

At the same time that the company was sorting out its land ownership arrangements down island in the Victoria region and retreating from Nanaimo, its officials were defining their property at Fort Rupert, the other main company settlement on the island. The surveying department had first examined, measured, evaluated, and mapped the company establishment in 1857 (422A). P.N. Compton who was attached to the fort at that time, was put to the task of drawing a map of the details of the one hundred acre lot (510A) (plate 61) that had been granted to the company by the colonial government in October 1863. This drawing was to accompany the certificate of pre-emption for this land.

MAPPING COMPANY PROPERTY ON THE MAINLAND OF THE COLONY OF BRITISH COLUMBIA

Land adjustments were also being made between the company and the new colony of British Columbia on the mainland. The company had found what it considered to be encroachments; surveys showed other ownership lines thrust across its territory at several locations. Some years of negotiation preceded final agreements. The colonial government requested that it be furnished with duplicate maps "exhibiting the situation and extent of the claims in question" at all posts, and the company complied (A11/77,fo.586).[19] At Hope, official town plan lines had impinged upon the grounds of the fort, its last remaining hay pasture having been enclosed and appropriated by the government. Chief Factor Mactavish at Victoria had a map drawn in 1862, a so-called "diagram" of Hope, based on the government's official plan of the community (205C). This map indicated the company's understanding of the land agreement ratified that year in London. The company received two blocks at the fort and a five-acre surrounding lot, both of which were marked on the map. One town plan of the district of Hope indicating the street grid in relation to the position of the fort was also drawn and was of general utility (33^B).

At Yale the situation had become disagreeable; encroachments continued, a street had been laid across the staked reserve, and the post master, Ovide Allard, had been given notice in June 1860 to have the company's buildings moved off the street line within four months. Alexander Dallas wrote in late 1860 to Governor Douglas of the colony of British Columbia, enclosing a map of Yale, on which he showed the company's land claim "as laid down by the Governor" some years earlier, that is, when Douglas was still an officer of the company and governor of Vancouver's Island (200^C) (A11/77,fo.756). The map also located the company's holdings in the settlement at the date of the correspondence for comparative purposes.

The situation at Fort Langley on the Fraser River was even more complicated. In 1858, James Douglas, as chief factor, had hired a surveyor, John Gastineau, to survey the land claims there; the survey was to include "the grounds about the Fort, say from the plain above to one mile below the establishment" (B226/b/16,fo.85d). He was to lay out lots and then was to survey the remainder of the company lands at Langley. Gastineau did not carry out his contract. Instead, he spent his time, on company pay, surveying public land. At Langley the government had also sold a lot on the bank of the Fraser River where the company had a wharf and salmon shed. It was mid-1862 before the controversy was settled and the situation clarified. The Hudson's Bay Company was made legal owner of the site of the fort, about two acres, and land around it amounting to two hundred acres; the lot on the bank of the Fraser was restored; and the company's ownership of the five hundred-acre Langley Farm was confirmed, with an option allowing it to purchase up to fifteen hundred acres more (B226/b/23,fos.916d.917d). In 1863, the Western Board of Management decided to promote Langley over New Westminster as the principal exchange point overseas. For this reason a map of Langley, with the various company holdings located, was taken to London (209^C). In 1864, title deeds were received for the Langley, Hope, and Yale properties.

By 1867, Langley Farm had shown great progress in its operations: the cattle from Nisqually had been moved there; the horses, mules, and bullocks employed in inland transport were wintered there. Tolmie called it the finest dairy and pork producing farm on the coast. The farm supplied the steamer *Enterprise* with products. It also provided Fort Yale with grain and nearly all the hay used to feed company wagon teams and pack trains passing through. In 1867, William Newton, the former draftsman at Victoria, prepared a map of the farm (563^A), and a copy (564^A), to show internal structure, field use, and principal improvements made since 1864.

Other ground plans of three interior posts and associated land claims were made. Fort Shepherd was built in 1856 on the west bank of the Columbia River opposite the mouth of the Pend Oreille River, just north of the international boundary. It was connected by a direct trail from Fort Langley, whereby traders could avoid the heavy import duties applied to goods taken down the Columbia. In 1864, Tolmie made an inspection tour of interior territory, visiting this fort and preparing a sketch, which indicates a U-shaped set of buildings sited on a high western terrace of the Columbia, well above the spring high water mark (534^A) (plate 62). The map of Quesnel post was drawn in 1869 (573^A), and of the third interior post, Barkerville, in 1870 (572^A).

MAPPING THE LAND CLAIMS CONTROVERSY IN WASHINGTON TERRITORY

The last phases of the land claims controversy in Washington Territory came to a climax with the appointment of the British-American Indemnity Commission, which began work in 1864 and carried on until 1869. At Nisqually, Edward Huggins obtained certified copies of maps of land claims of the Hudson's Bay and Puget's Sound companies from Surveyor General Tilton's office at Olympia, Washington Territory. These and other maps were used to take attested declarations from the companies' officials in Victoria as to the extent and character of their properties in the United States. Copies were sent to London to Governor Henry Berens to use in discussions with British authorities. In preparation for the commission hearings, a map of the northwest of the United States was drafted in 1863 or 1864, on which all of the Hudson's Bay and Puget's Sound companies' claims were carefully located (511^A). Other maps were drawn, one of Fort Vancouver showing company property taken over by the U.S. army (557^A), and two of Champoeg on the Willamette River in Oregon marking property purchased by Chief Factor McLoughlin (565^A, 566^A). Both maps were attested for legal purposes in 1867.

MAPPING LAND CLAIMS EAST OF THE MOUNTAINS

Two cadastral maps from across the continent helped to bring to a close the discussion of company land claims and other property matters. James

Anderson, who had been moved from the Mackenzie district to look after company interests on the St. Lawrence, sent a map of the Mingan post on the north shore of the river to Hudson's Bay House in London in 1860 (479A) (plate 63). Governor Simpson had asked his advice on moving the post from the mainland to nearby Mingan island. Anderson disapproved of the move, partly because, as the map shows, there was no proper location there – the inner island shore has three hundred-foot cliffs, while the east end is low and sandy. There was no drinking water, no boats could anchor there, and thus no customers would arrive there. Moreover, the company store was already located close to the Indian encampment and the church. He advised the governor to take a lease on land up and down river for several miles. Among other maps was one of the lots sold off to a private individual near the fort at Red River in Assiniboia (535A), and one of company property in the centre of Georgetown on the east bank of the Red River in Minnesota Territory in the United States (559A).

SURVEY MAPS FOR A PROPOSED TELEGRAPH ROUTE

The last group of maps to be examined for this period in the Hudson's Bay Company cartographic history are, in appearance, typical of maps showing exploration, drafted from the journals of persons employed by the company to examine sections of New Caledonia or the great plains. Their purpose, however, was not to provide maps useful to the fur trade so much as to suggest potential transport and communication routes. Twelve maps are involved.

In the early 1860s, the Duke of Newcastle of the British Colonial Office requested that the Hudson's Bay Company consider granting a right-of-way across the plains through Rupert's Land for a public road and a telegraph line. The company could hardly refuse such a request in light of the prevailing political climate, both in Britain and British North America, even though it considered the proposal an attempt to break the company's monopoly by introducing new entrepreneurs into their charter territory. At worst, if the company refused, the committee feared the government would have a pretext to remove this territory from its control. The company decided to avoid these problems by taking on the telegraph enterprise itself. In 1863, it sent Edward Watkins to Red River to report on the general situation on the plains, on the prospects for settlement, and on the conditions prevailing for the commencement of such a project. When he returned a positive report, the committee hired Dr John Rae, of earlier Arctic exploration fame, to superintend a route-finding expedition. Before he set out, Rae provided his superiors with his assessment of the factors weighing most heavily in route selection (A10/58,fos.162,162d,163,163d). He recommended that the line should use a northerly pass through the Rocky Mountains so that the route on the plains would be through the parkbelt – through fertile, wooded land, rather than the drier south. Also, the Indian tribes there were not as "troublesome" as those to the south. The line should pass through, or be as near as possible to existing company posts in the Saskatchewan valley. It should be located as near as practicable to rivers and lakes en route, because timber for telegraph poles grew near their banks. Also, as the best soil was located there, stations and farms could be developed. Rae was asked also to examine the area for gold deposits and coal resources that would be of economic benefit to the company. A young geologist from Toronto, A.W. Schwieger, hired as an assistant, received instruction from Sir William Logan of the new Geological Survey on the best method of searching for gold and other important products (E15/13,fo.7d). As it turned out, his principal tasks for the company were to determine how many telegraph poles would be needed, where they would be situated, and where to obtain fill to use at various low points along the route.

After further advice from Alexander Dallas, who had become governor of Rupert's Land in 1862, the small party left Fort Garry on 25 June 1864, and thirty-four days later reached Fort Edmonton. They had gone by way of Fort Ellice, a more northerly route than they had planned, but the more southerly course would have taken them through excessively burned-over woodlands. Reasonably abundant timber was available along their course west of the Assiniboine River to Fort Pelly. After passing through the wooded Touchwood Hills, they had to traverse the driest part of the road with little timber along the way to Fort Carlton. From there Rae followed the north side of the North Saskatchewan River, with Schwieger investigating the south side so they could select the better route. They agreed that the north would offer greater advantages. Rae sketched their routes on a rough working map, which also showed proposed alternative roads between forts Garry and Edmonton (210C).[20]

Twenty-two days after their departure from Fort Edmonton, following the regular pack-horse track through Jasper House, they reached Tete Jaune Cache (Yellowhead Pass). Rae completed another rough map of this section, while camped at the pass (211C).[21] Then he parted company with his assistant, purchased two Indian canoes, and descended the Fraser River to Fort George and Fort Alexandria. Although he had hoped to discover a shorter, more practicable route from Williams Creek to Yellowhead across country, it was too late in the season to try. Instead he went on to Fort Victoria and there drafted a general sketch of the proposed route from Red River to Victoria, a total distance of eighteen hundred miles (212^C). With his written proposal, he sent a memorandum in which he noted that a cross-country survey, which he had been unable to do, should be undertaken (A11/79,fo.567).[22]

Schwieger and his small crew left Tete Jaune Cache for Fort Edmonton. From this post they went down river by boat to Carlton House to determine distances from point to point. Eventually they reached Toronto where Schwieger prepared a full report and a map of the survey across the plains (A10/60,fo.101). The map was undoubtedly "The Working Plan of the proposed Telegraph Line" (536^A) (plate 64), which laid out the outward and homeward bound journeys. There are notes and tables of distances also. His report dealt with terrain problems, number of poles required, tree resources available for poles, distances of pole haulage to the route, rock fill available, and soil conditions for agriculture. He included notes on the building of a companion road, the practicability of a railway line, and coal and gold deposits. Also listed were the dangers from fire, from falling timber, and from Indians who could destroy the line. Interestingly, he suggested that poplar plantations could be grown to ensure a constant pole supply, and he considered that it would be necessary to bring in a labour supply rather than rely on the local population.

The Western Board of Management hired the experienced traveller Joseph W. McKay to find the best telegraph line route from Tete Jaune Cache to the middle Fraser River. His first trial, in the early spring of 1865, took him from Fort Kamloops along the North Thompson River.[23] From this river, just past the twelve mile gorge, he struck northeast across the Summit Lake Divide to the Canoe River in the Rocky Mountain Trench. Crossing this river, he passed to the Cranberry River, which is an upper Fraser River tributary, and continued along it to Tete Jaune Cache. McKay believed that this would be the best route for a future telegraph line, but no mention of a sketch of the route was made in his report. He did map the next stage, producing a map that shows his route from the Yellowhead pass, down the Fraser to a west bank tributary he christened Telegraph Creek (550^A). Some distance up this creek, he climbed away from it into the upper course of the Shuswap River, followed this back to the Fraser downstream, which he followed to Fort Alexandria. After a little rest, McKay left this fort ready to try a route east to Tete Jaune Cache from the Fraser River, as suggested by Dr Rae. His party had an extremely difficult trek, during which one of them succumbed to "Camp Fever" (A11/80,fo.139). The route began at Williams Creek, went on to the upper headwaters of the Bear River, to Otter Lake, the Swamp River, and eventually to Canoe River. Thus, McKay was again on his earlier spring route to Cranberry River, and from thence he proceeded to Tete Jaune Cache. Although this way was shorter, in his report, he expressed his preference for the North Thompson-Summit Lake route. The report included his map of his final travels on which there is also the Telegraph Creek to Shuswap River detail (551^A) (plate 65).

MAPS OF WATER TRANSPORT IN INTERIOR BRITISH COLUMBIA

As part of the search for reasonable cross-country transport in central New Caledonia, there are two maps showing an approach by means of steamers on some of the main rivers and lakes of this area. In the summer of 1865, the company's western board sent Captain William A. Mouat of the *Enterprise* to examine the feasibility of using steamships on the Thompson River and Shuswap Lake. With a small boat and four Indian aides Mouat proceeded from Fort Kamloops, sounding and mapping the course up the Thompson and through the main body of Shuswap Lake to the company post at the end of the northern arm. Returning, he passed the fort and travelled west downstream to Kamloops Lake and to its western end at Savona. His report was favourable, and he worked with William Newton as draftsman on his map (552^A). This map clearly illustrates the water depths, the depths at low water (a significant factor), river bars, speed of current, the winds, Indian fishing weirs as impediments to navigation, the location of timber along the shore for logs, boards, and firewood, and the locations of harbours for laying by during storms.

Having accepted Mouat's favourable report, the company organized the building of the steamboat *Marten* in 1866 on the lakes to carry supplies, private freight, passengers, and government mail. To provide data for insurance underwriters in Britain, the chief factor at Victoria obtained from the Commissioner of Lands and Works of British Columbia an 1866 official map of the Kamloops and Shuswap region, which had been drafted by J.B. Launders of the surveyor general's office (560^A) (plate 66). It is on a larger scale than Mouat's map and extends farther to the east, showing the trail from the northeast arm of Shuswap Lake to the Columbia River and also a considerable length of this waterway.

Both the use of the steamer on the river and lake system and the increased traffic on this Columbia trail affected transport planning in two ways. One was the inevitable desire to extend steamer use farther up rivers flowing into these waterways, and the second was the hope that more efficient overland trails could be found, linking the waterways to contiguous regions, particularly the Columbia. In the summer of 1867, James Bissett was ordered inland to examine further a number of routes. He was first to determine how far up Eagle Creek, on the south arm of Shuswap Lake, a steamer could run and whether a useful wagon or mule trail could be built over Eagle Pass to the Columbia (A11/81,fo.578). Bissett found out that the steamer could be run up the creek for ten to twelve miles at high water and four to five miles at low water and that the route over the pass was feasible. William Newton was pressed into service in preparing a map of this region (567A). The company urged the government to build a road, but, in spite of promises, it had not done so by mid-1870.

Pressing on with his transport studies, Bissett examined the North Thompson River and its west bank tributary, the Clearwater, as a possible steamer route to supply the Quesnel area and the Cariboo mining district. Following the Clearwater River to Clearwater and Hobson lakes, he crossed over a low pass into a southeast arm of Quesnel Lake. His conclusion was that his method and route were not satisfactory for larger vessels. Newton again supplied a map with his report (568A) (A11/82,fos.629–638). It showed the extent to which the North Thompson and the Clearwater could be used by a steamer such as the *Marten*.

Finally, Bissett returned to Shuswap Lake where his final task was to investigate the shipping possibilities of the lower course of the Spilamacheen River, which drains into the southern end of the lake's southern arm. His conclusions were that a boat could go about twenty to twenty-two miles at high water and about half that at low. The study also included a run up the Salmon Arm of the lake and into the Salmon River. There is no conclusive evidence that a map was prepared (A11/82,fos.589–590).

MAPPING ROUTES INLAND FROM THE NORTH WEST PACIFIC COAST

While these various interior investigations were in process, further attempts to use one or more of the inlets and river valleys as a route to northern New Caledonia were being made. At various times the Skeena, Stikine, and Nass rivers had been promoted as possible entrances. By 1863, a private concern was involved in building a road from Bute Inlet to the junction of the Quesnel and Fraser rivers. The Western Board of Management became engaged in formulating various approaches, and, for this, Tolmie prepared a summary of the navigation attempts to that time. Light-draft stern-wheelers could reach from 130 to 180 miles up the Stikine River and up the Nass for fifty miles. The Skeena was navigable for the eighty miles of its length that had been explored. By 1866, the board had agreed to support an attempt at inland travel from as far up the Skeena as proved navigable over to Lac des Français (François Lake) in the Nechako River basin. The hope was that New Caledonia could be supplied by this route in the spring of 1867. The explorer was William Manson, an experienced traveller in the company's employ, but he was not able to find a suitable pass, being stopped by snow at higher levels (A11/80,fo.4; A11/83,fo.420). However, local Indians spoke of a shorter route from an inlet south of Kitimat Channel into Francois Lake. After wintering at Fort Simpson, Manson went down coast in the spring of 1867, entered that channel, and from there tried to get inland from Kidala Arm up the river of that name. He was overcome with snow blindness but learned that this way led only to high mountain snowfields (A11/82,fo.62). In July, however, he tried Gardner Canal and Kemano River, and eventually he was successful in finding a way upriver through a series of lakes to Francois Lake (A11/82,fos.591–593d). While returning via the same route, he broke his leg. As a result he was unable to carry out the other assigned task of investigating a trail from Dean Channel northeast to Fraser Lake in the Nechako Basin. All of the information he had gleaned was used by William Newton to make a map of the country lying between Grenville Canal and François Lake (569A). This last known work by Newton was, like his previous manuscripts, of high quality, illustrating the terrain and water features of these regions in a very readable manner; it provides an excellent example of the ink and colour procedures used by Pemberton and the other cartographers in the former surveying department at Victoria. The report by Manson, along with the map, made it clear to the western board that the Kemano-Francois Lake route could be traversed only for a few short months because of snow, and that there were falls and rapids in the associated river connections in the François Lake and Fraser Lake region (A11/82,fos.591–593d). Other routes had to be sought.

In the summer of 1870, Manson was sent back to the northwest coast for two further expeditions. The first, which he completed quickly, was to take the route up the Stikine River and northeast to Dease Lake, the western end of the Liard River and Dease Lake connection made many years previously by Mackenzie River district traders. Manson's report included a map, but unfortunately it has not remained in the collection (220C) (A11/85,fos.441–443d). Given the short period the route could be used each year and the high cost of preparing this route for full transport – it would have required a river steamer and crew, two scows, pack train and crew, and other elements – this approach was deemed impracticable by the board. In his second and final attempt, Manson examined the Portland Canal as a potential road inland. After travelling some thirty miles into British territory from the head of the canal, he discovered that the glaciers of the Coast Range thwarted further advance. No map resulted from his journey.

Although the company did not build the telegraph line after Rae finished his investigation, the information, especially that provided by McKay's maps of the river valleys and mountain passes from the Fraser River to the Rocky Mountain front, was available to the company's western officials for later use. Also, although Manson was not successful in finding a satisfactory supply route from the northwest coast across the mountains into the interior of New Caledonia, the complicated coastal and cordilleran pattern had been greatly clarified for company managers by 1870. And the upper Thompson and North Thompson rivers, Shuswap Lake, and the links to the upper Columbia River were revealed and mapped, preparatory to further activities there. This period of company mapping to 1870, largely concentrated in the Pacific region (see figure 5), and especially concerned with tidying up property claims, was followed by a virtual cessation of surveying and mapping. The Hudson's Bay Company no longer had the need or the incentive for its own employees to engage in cartographic endeavours.

Afterword

In 1859, the Hudson's Bay Company lost its monopoly and sole right of trade in the vast regions lying to the west and north west of its older charter grant, Rupert's Land. And, with the deed of surrender and with the transfer of Rupert's Land to the Dominion of Canada in 1869–70, the company also ceased to govern its former domain. However, it retained its posts and stations with small blocks of adjoining land and had a percentage of land sections reserved for it in the fertile belt of the western plains. A land commissioner was appointed in the company to administer these properties. But few manuscript maps were drafted over the following decades by company employees. Most of the remaining manuscript cartographic records are detailed property surveys prepared by professional surveyors hired when the occasion demanded.

In contrast, over the two centuries, 1670 to 1870, the number and diversity of maps, along with the records, indicate that maps had been of considerable utility in a number of ways. They were not drawn especially for the use and guidance of company servants in the field, since these men developed their own particular knowledge of local scenes and their own mental maps of more extended regions. The maps were drafted mainly for their value as a basis for decision making by the committee members, the governor, and the deputy governor in London.

A long-term mapping policy was never codified by the officers of the company, nor was a position of chief surveyor and cartographer established at Hudson's Bay House. The Arrowsmith map firm acted unofficially and partially in this capacity. The company established only two official full-time surveying and cartographic positions over these years, that of inland surveyor, held by Turnor after 1778, and that of company-colonial surveyor of Vancouver Island, retained by Grant and then by Pemberton, from 1849 to 1859. Fidler was offered the opportunity to be surveyor but rejected it. Such appointments were designed to meet specific short-term needs and not meant to be long-term positions. Turnor's task was to construct a base map of known territories by locating the major waterways accurately and by fitting other accumulated geographical data into this framework. Pemberton (and Grant) were hired when the company found itself in the novel role of colonial land manager; when this function ceased, Pemberton's contract was not renewed.

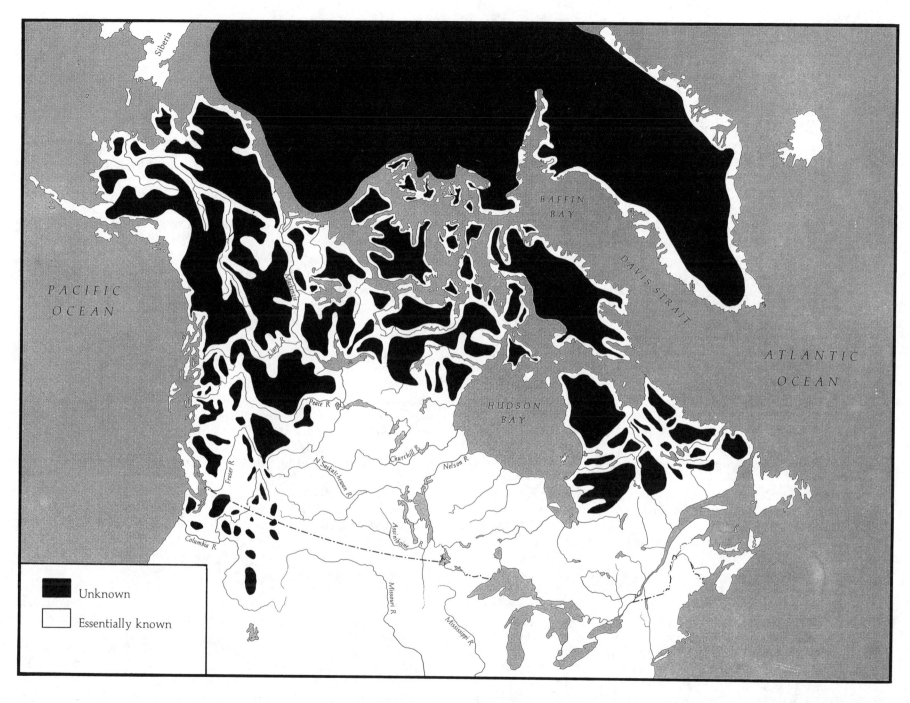

Figure 5 Retreat of the unknown: European knowledge of Canadian territory in 1870

The company lost David Thompson, a young man of obvious abilities and notable enthusiasm, from its ranks because the committee was unwilling or unable to recognize his potential for exploration, surveying, and cartography, and did not encourage him just at that crucial period when he was completing his apprenticeship. Of course, the Hudson's Bay Company was not a mapping agency, but a trading enterprise whose executives discovered that maps and charts would contribute significantly to their activities.

This study has been a survey of manuscript maps associated with the Hudson's Bay Company in its initial two centuries of operation. Only a portion of these documents and a few of the mapmakers involved have been discussed in the text; some of these were mentioned only in passing. Some themes have been touched upon lightly. Elaboration upon all the maps and the full array of cartographers would have resulted in an encyclopedia. In the catalogues, and in appendices seven through ten, readers are able to observe quickly which maps and men were discussed in the text and which were only noted. Readers may have recognized that a number of topics would merit further analysis or have conceived of others that were not touched upon in this study at all. I suggest below a variety of subjects (several of which I am working on at present) that would be interesting fields for enquiry. The list of suggestions is not intended to be comprehensive, but rather indicative of the possibilities.

In the first place some mapmakers and their maps warrant greater attention. Andrew Graham is better known for his *Observations on Hudson Bay* than for the maps ascribed to him. He was much more a naturalist than a cartographer. However, he had considerable geographical orientation as his book and his two major maps attest (21A, 24A). An excellent project would be an evaluation of the role of Graham as an early interpreter of the human and physical geography (especially biogeography) of the country to the west of Hudson and James bays through an analysis of his maps and treatise.

The joint map of Edward Jarvis and Donald McKay (37A) needs greater attention paid to it; as previously noted, it is related to maps 57C and 38A. The map was a useful, although fortuitous, collaboration between these two men. It was also helpful that John Hodgson was available locally to prepare a regional map, which included the details of the missing McKay chart. Jarvis possessed greater drafting experience, while McKay had more personal knowledge of the spatial details they depicted. Since their interpretation of the geography of the interior was also a crucial source for Philip Turnor's compilation of his last great map of the west in 1794 (57A), it would be instructive to delve further into this collaboration, into the careers of these two men, and into the geographical and cartographic characteristics of this influential document. Nor has Turnor's major map yet been evaluated for its full geographical content, for the sources of its detail, or for its cartographic correctness. It is worthy of such appraisal.

James Clouston, one of the better educated and energetic company servants, was also a trade strategist. The knowledge gained during his short but active career is expressed to a major extent in his final, comprehensive 1825 map (195A), which depicts the routes he and several confreres followed. A valuable research topic, centred upon this map, would be to compare the spatial concepts of the interior Eastman and the Labrador interior held by the public, the company, and Clouston during this period of trade expansion.

Finally, although George Taylor Jr did not produce an integrated map based on the information gleaned during his inland surveying career, as did Turnor and Clouston, he did draft several maps at various scales, of an assortment of types, all prepared in a meticulous manner. Taylor has not been the subject of a separate study, but his career deserves further illumination. The son of one of the company's British employees and a native woman, he was educated abroad. His employment represented a positive advancement in the company's personnel practices. Taylor's story is suggestive of another research area: an examination of the father and son combinations as well as of various extended families employed by the company could be instructive.

Many native people aided the Hudson's Bay Company with map sketches and by providing geographical information to the fur traders, notably Peter Fidler, as has been noted in the text. But their fuller, distinctive role in assisting the company's advance across the north of the continent by transmitting geographical data and area concepts has yet to be expounded.

Although the inland wintering policy of the Hudson's Bay Company did not result in many maps, the written and oral reports of Kelsey, Norton, Stewart, Henday, Hearne, and a number of other servants were raw material for their superiors' understanding of the inland country and for the formulation of trade policy. The elucidation of such material as a study of the evolution of geographical concepts needs to be undertaken.

The education of charity hospital apprentices as a background to their employment with the company I have discussed in a previous study. And to some degree the roles of John Charles, George Donald, John Hodgson, and David Thompson have been described in this study.

However, the later careers of all the Blue Coat and Grey Coat boys have not been examined; I am now completing such an examination.

Two studies of relationships between the Hudson's Bay Company and other commercial companies and public institutions have considerable potential. The first would be to examine further the nearly century-long role of the Arrowsmith cartographic firm in map preparation on behalf of the company, and that of the company as a ready and up-to-date source of compilation material for new Arrowsmith North American maps. Although I have written elsewhere about Governor Wegg's relationship with the Royal Society, an expanded study of this company-society collaboration could be worthwhile. Moreover, when the Royal Geographical Society was formed in the mid-nineteenth century and assumed the scientific and geographical function formerly held by the Royal Society, the company transferred its ties to this new body. Some company men became fellows of the geographical society, were honoured with the award of medals, and presented papers on scientific endeavours, especially those concerned with exploration.

Although I have not been concerned with the plans of fur posts, except where they depicted larger sites in some detail, the map archives of the company are replete with such documents. A study of the fur trading post as a settlement form and as the progenitor of larger urban centres could be undertaken as an aspect of the historical urban geography of the nation. There are many other post plans, based on surveys made after 1870, now in the HBCA, which would be valuable for such research.

Three further projects concerning the colony of Vancouver Island could be advantageous; the background for these has been described in this book. The first is a perusal of the maps and plans of Vancouver Island prepared by company-colonial draftsmen to 1859, and by later colonial survey office draftsmen up to the formation of the colony of British Columbia in 1866. These documents are held in the surveyor general's office in Victoria. This study should include analysis of the Land Records Office holdings and could be pursued in several ways; as an archival study, as an investigation of cartographic drafting procedures, or as a review of the organization and procedures of the survey department's drawing room and of the subsequent surveyor general's office after 1859. The second project, and one on which I am currently working, is a full study of Joseph Despard Pemberton's legacy to British Columbia, and to southern Vancouver Island, in particular. My study is concerned with his education and work experience, and his decisions as chief surveyor, cartographer, and director of land policies for the colony to 1859. Another research project would be a study of the history of urban mapping in Victoria and Nanaimo, from fort site to expanding town, based on materials in the HBCA, in Land Records, Victoria, and in the British Columbia Archives. Such an analysis of historical urban geography combined with the history of cartography has not yet been attempted.

Although many maps were lost from the Hudson's Bay Company records, those that remain in the company archives constitute one of Canada's major research resources and the primary group of cartographic documents of the fur trade in North America. The premise of this study has been that the exploration, mensuration, and mapping of the northern part of America is a vital topic in the history of the Hudson's Bay Company, of the fur trade, and of Canada.

PART THREE

Plates

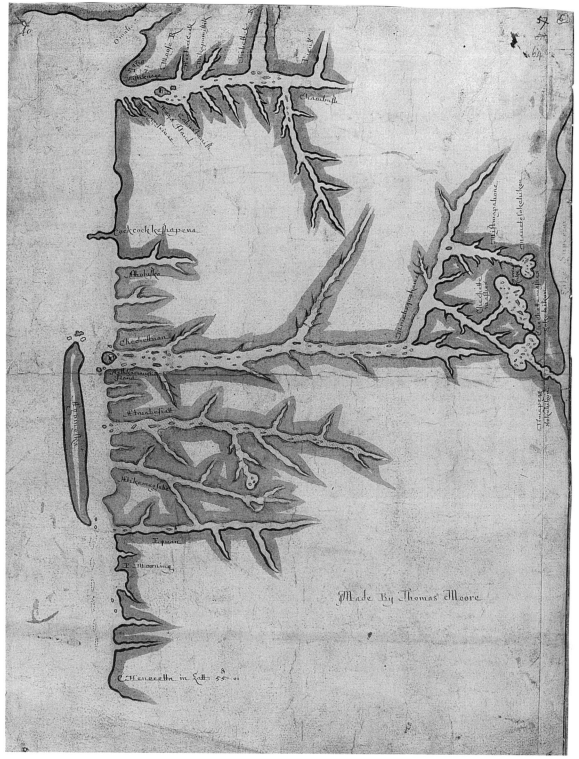

Plates

A SELECTION OF MANUSCRIPT MAPS, CHARTS, SKETCHES, AND PLANS PREPARED FOR THE HUDSON'S BAY COMPANY, 1670–1870

The maps, charts, sketches, and plans in this representative selection are used by permission of the Hudson's Bay Company Archives unless otherwise indicated. The captions for the plates include a title, date of drafting, the cartographer, and the Catalogue A or Catalogue B number. When the date of the drafting of a map or the cartographer's name have been postulated on the basis of firm evidence but are not certainly known, the name and date are given in parentheses. The titles assigned to the plates identify the *geographical* locations being portrayed. They are intended to make it possible for readers to compare how HBC map makers depicted any given area with that same area as delineated in the modern location maps found at the front of this book. Names of locations in the captions follow usage contemporary with the map; when appropriate, modern equivalents are given in parentheses. For an exact transcription of titles as printed in contemporary hands on the maps or for titles supplied and adapted from HBCA documents, the reader should consult the map catalogues.

1 Western Shore of James Bay. (1678). Thomas Moore. Used by permission of the British Museum. 1[B]

Plates

2 Hudsons Bay, James Bay, and Hudsons Straits. 1709. Samuel Thornton. 1[A]

3 North America Showing Hudson's Bay and Straights. 1748. Richard Seale (engraver). 8[A]

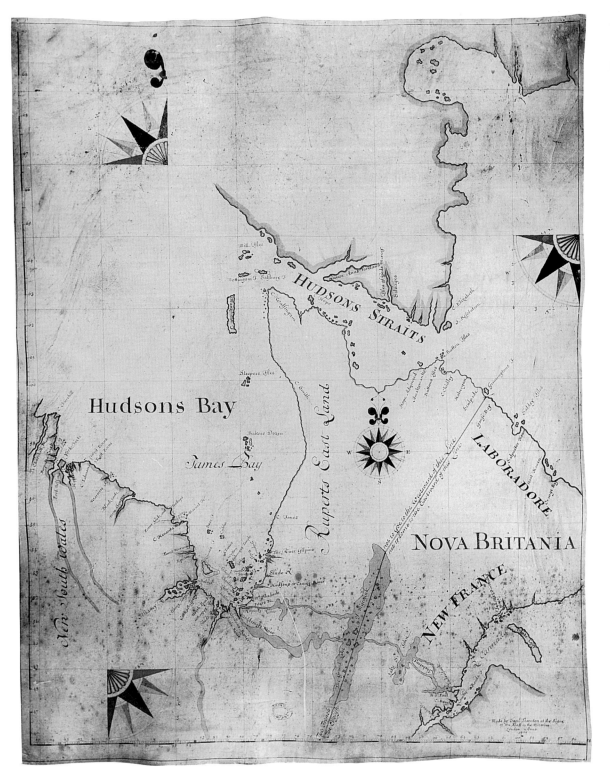

2

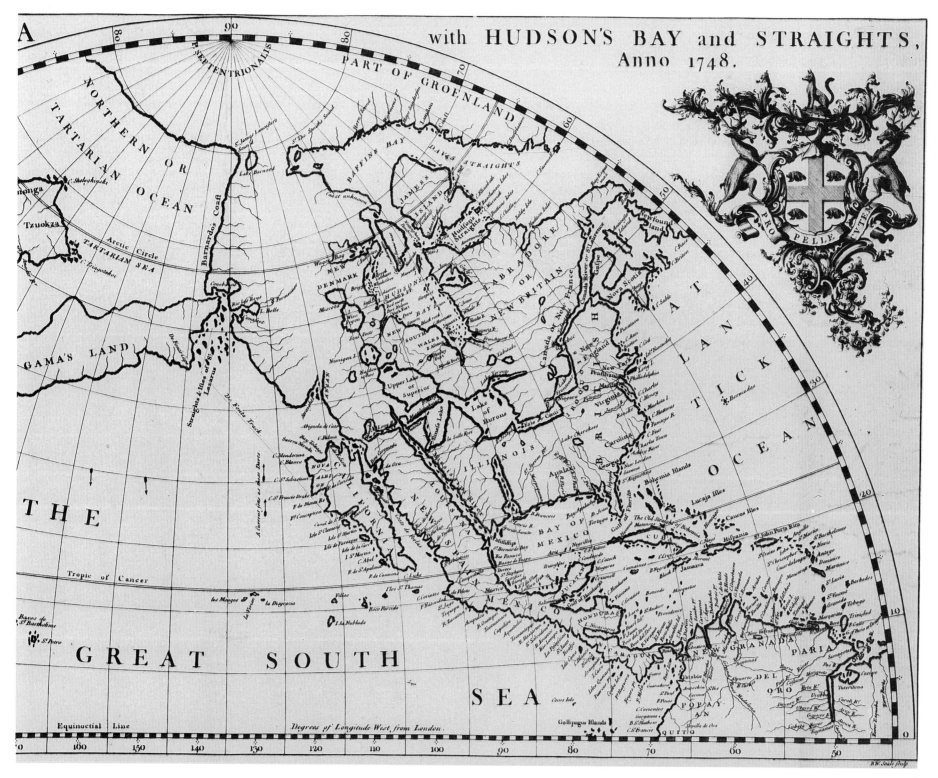

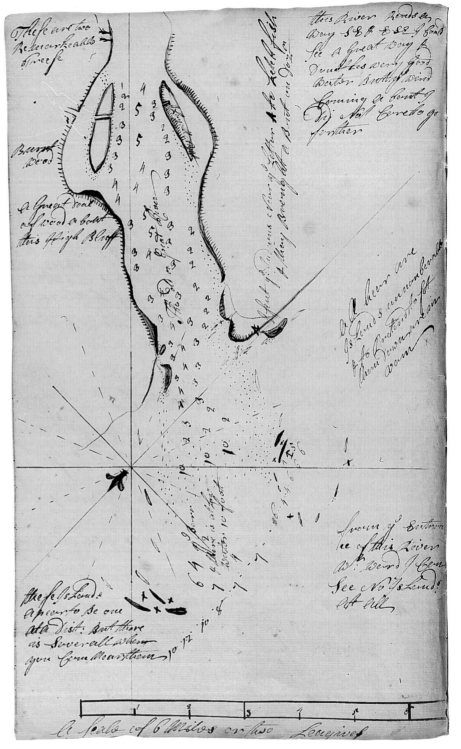

4 The Old or Great River (Fort George River). 1744. Thomas Mitchell. 4[A]

5 Artiwinipeck or Richmond Gulf (Lac Guillaume Delisle). 1749. William Coats. 9[A]

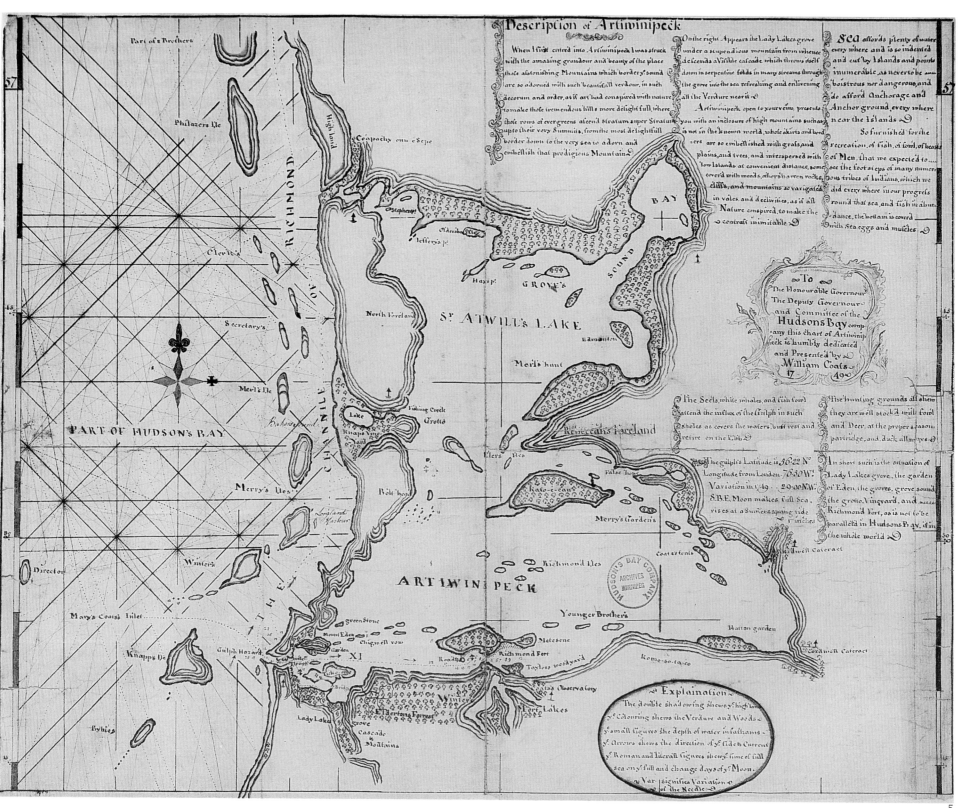

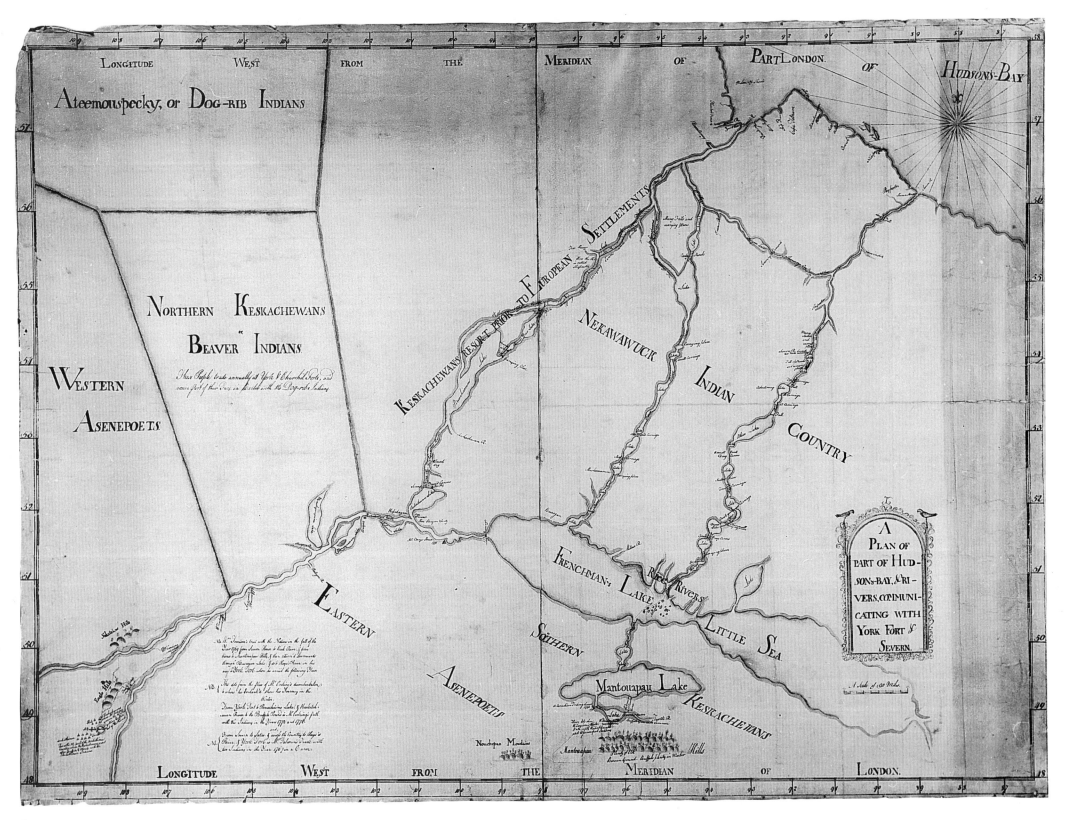

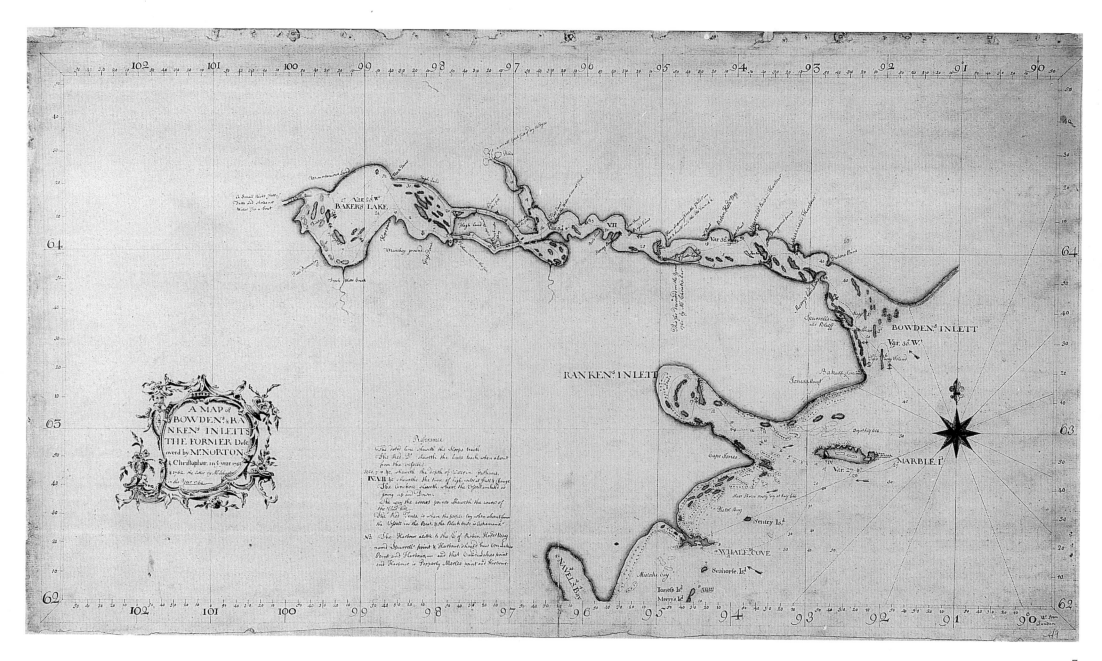

6 Southwestern Hudson's Bay and Rivers Leading to York Fort and Severn House, and Identifying Territories of Native Peoples. (1774). (Andrew Graham). 23[A]

7 Bowden's (Chesterfield) Inlet and Ranken's (Rankin) Inlet, on the West Coast of Hudsons Bay. 1765 or later. (Samuel Hearne). 18[A]

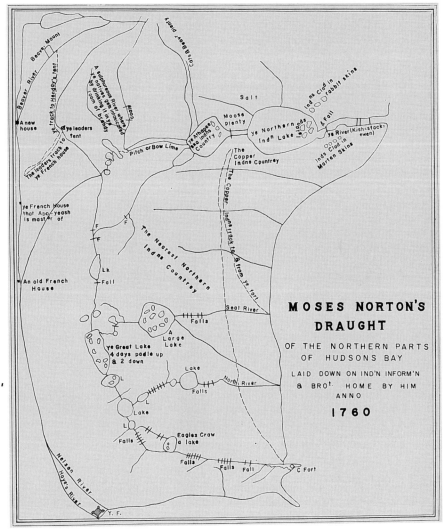

8a Transcription of plate 8 by R. Ruggles from Warkentin and Ruggles, *Manitoba Historical Atlas*.

8 The Northern Parts of Hudsons Bay. ca. 1760. Moses Norton. 16[A]

9 Country Inland from Prince of Wales Fort at the Mouth of the Churchill River. 1772. Samuel Hearne. 20[A]

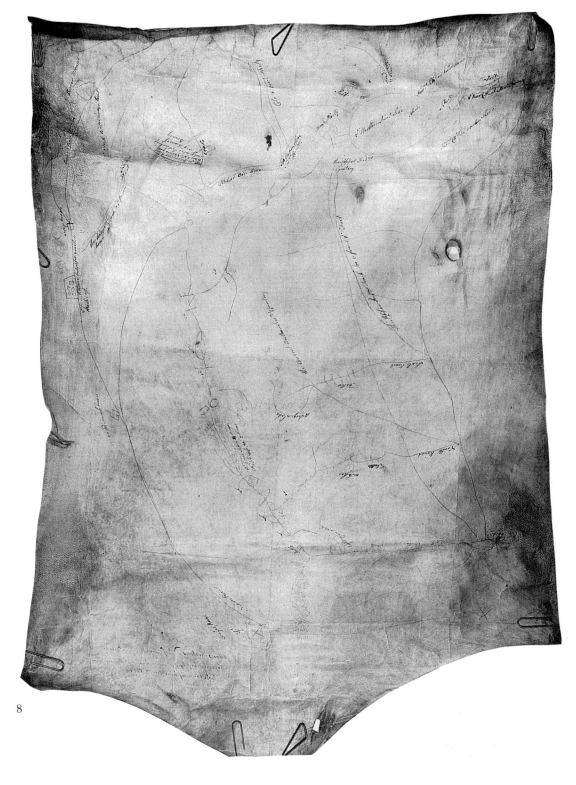

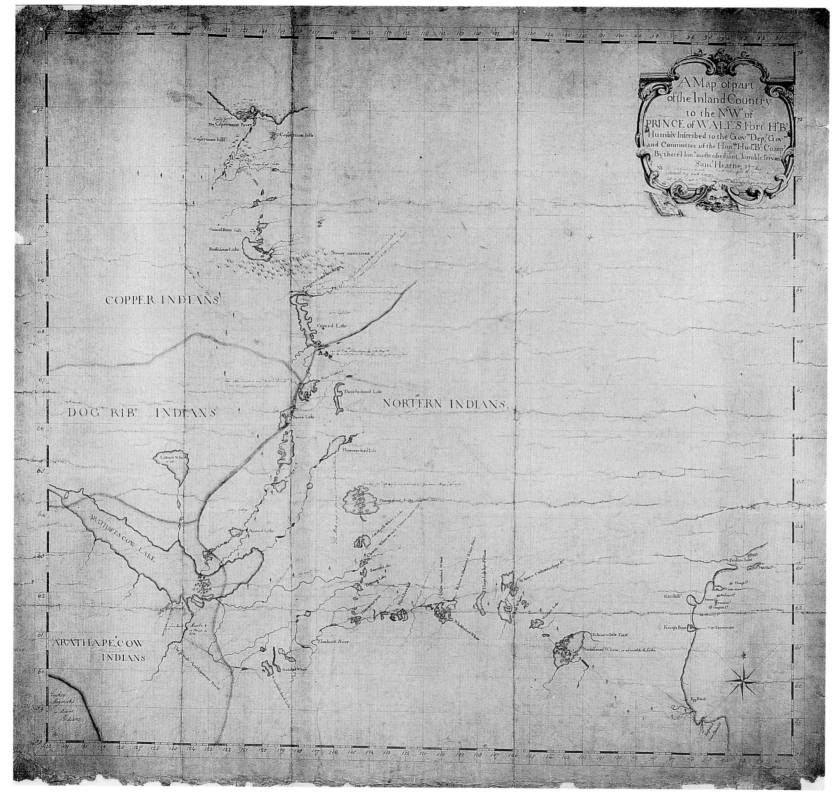

10 The Coppermine River. (1772). Samuel Hearne.
Used by permission of the James Ford Bell Collection, University of Minnesota. 4[B]

11 The Albany River from Albany Fort at Hudson Bay to Gloucester House on Upashewa Lake. 1780. Philip Turnor. 28[A]

12 Rivers and Lakes Southwest of York Fort at the Mouth of the Nelson River and Churchill Fort at the Mouth of the Churchill River. 1787–8. Philip Turnor. 34[A]

13 Rivers and Lakes Communicating with the Albany River above Henley House. (1789). (John Hodgson). 36[A]

14 Western Shore of Hudson and James Bays and Interior Rivers and Lakes Particularly Above Albany Fort. (1791). (Edward Jarvis and Donald McKay). 37[A]

15 Rivers and Lakes West and South of Hudson's and James's Bays to Lake Superior and Lake Winipeg. (1791). (John Hodgson). 38[A]

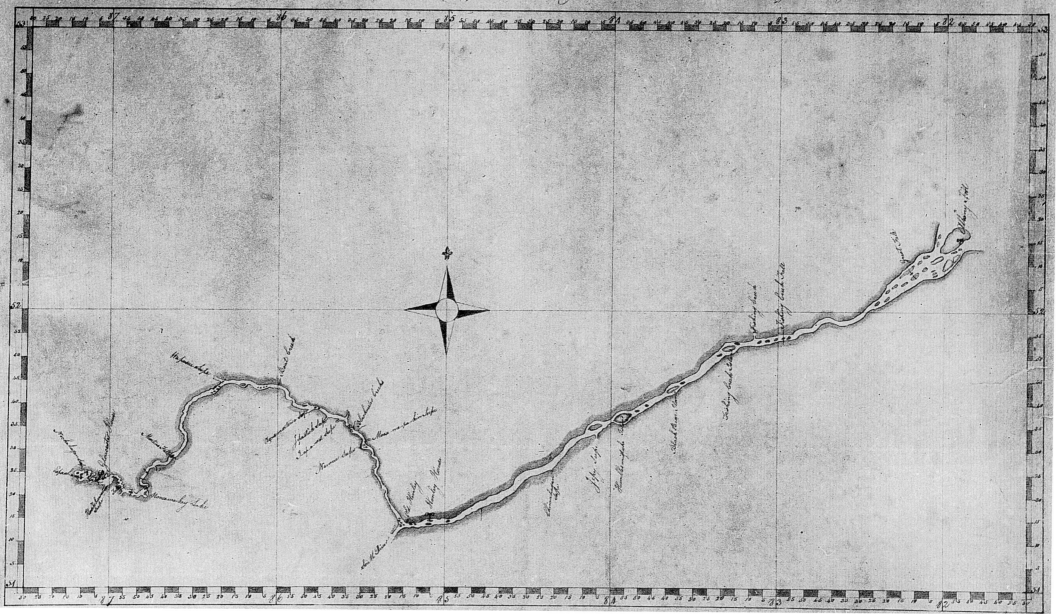

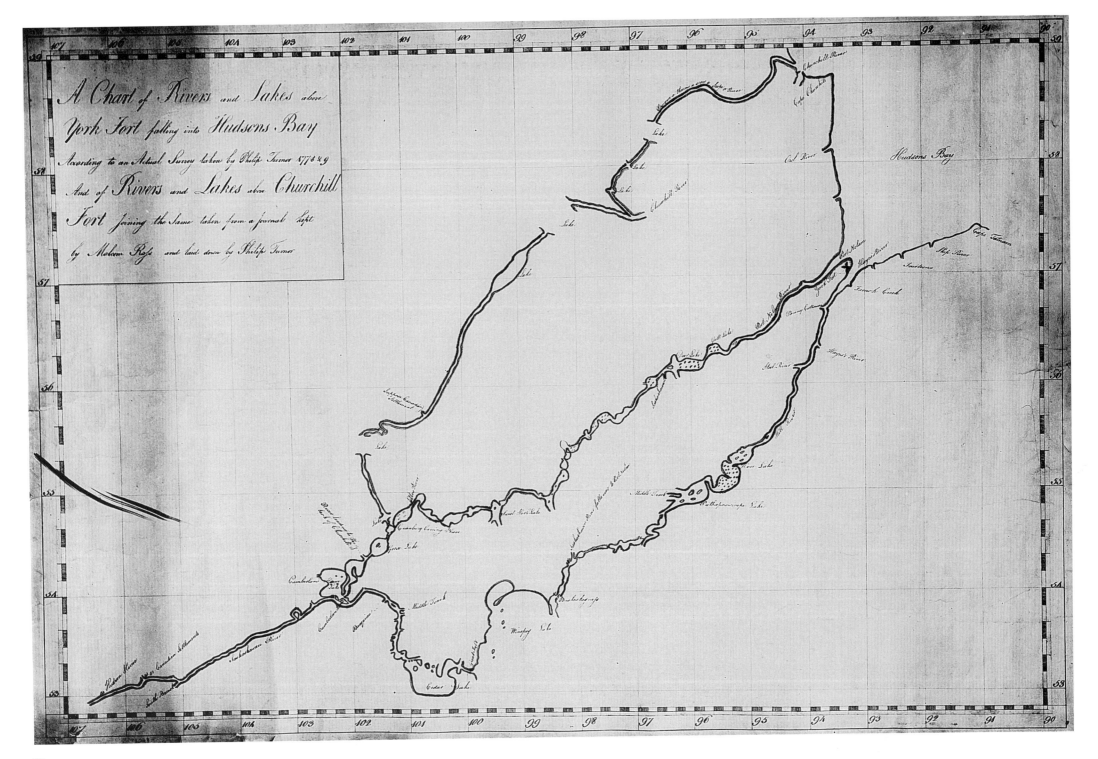

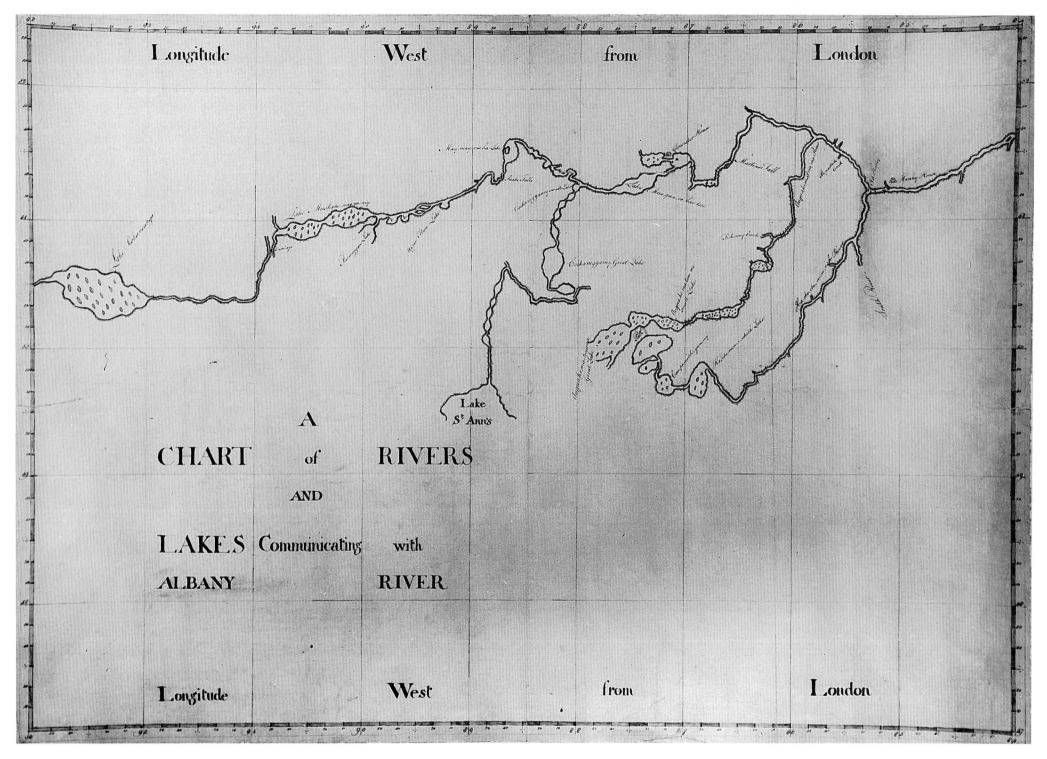

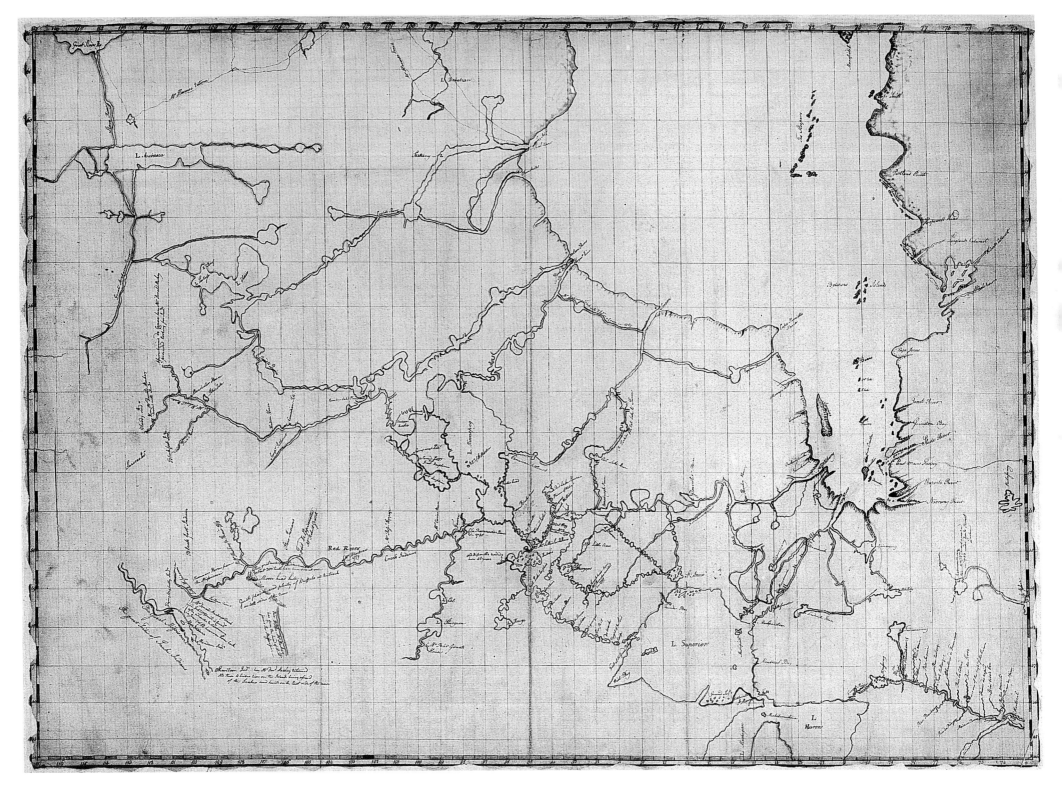

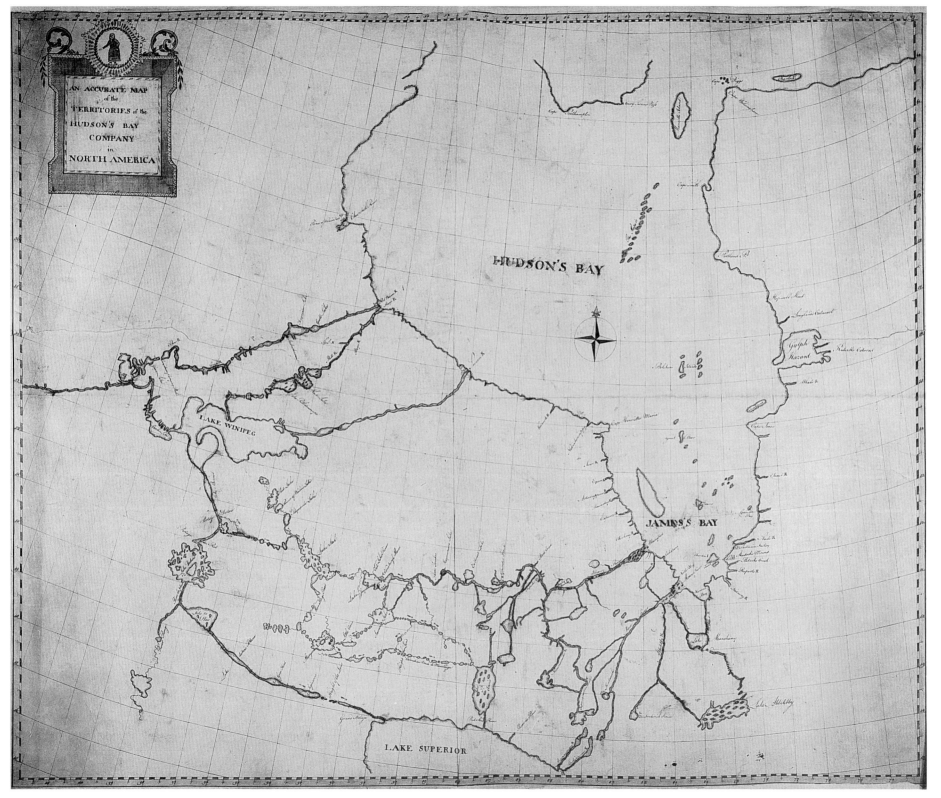

140

Plates

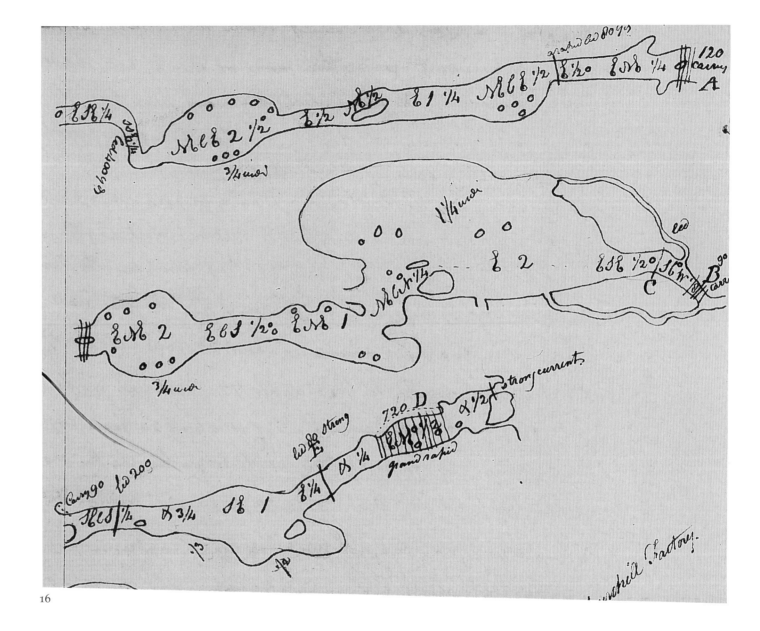

16

16 Sketch Maps of the Churchill River Illustrating a Journey from Île à la Crosse to Cumberland House. 1792. Peter Fidler. 46^A

17 Part of a Map of Nelson and Hayes River and Connections through Lake Winnipeg, Showing an Area above Split Lake on the Nelson River. (1794). (David Thompson). 56^A

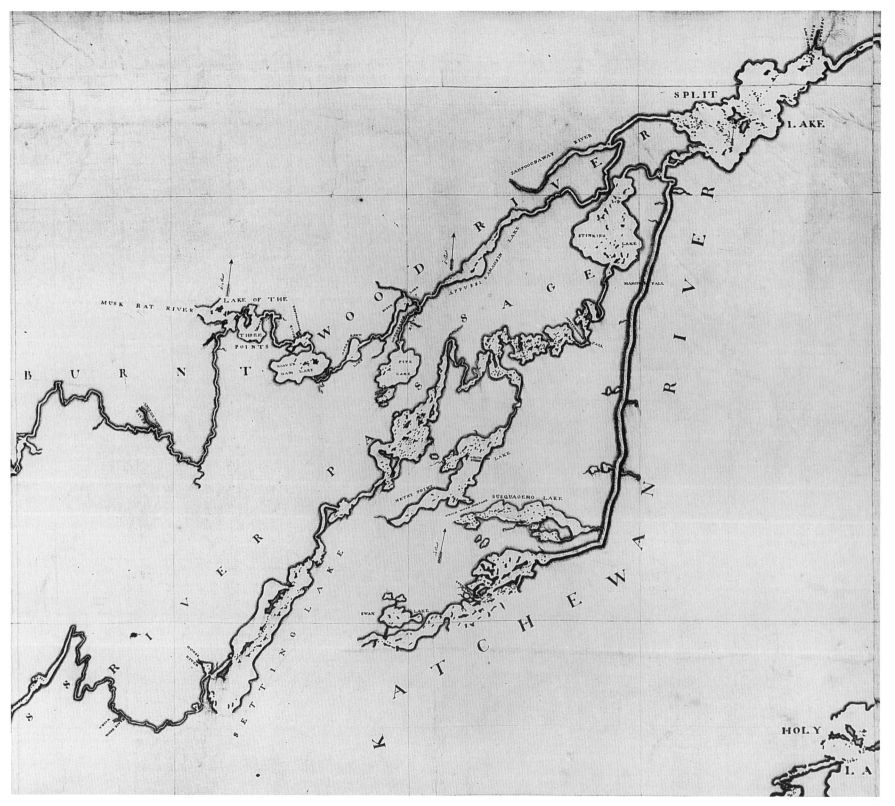

18 Title Cartouche for a Map of Northern North America. 1794. Philip Turnor. 57[A]

19 Ac ko mok ki's Map of the Indian Tribes Living on Both Sides of the Rocky Mountain in the Upper Missouri River Area. 1801. Peter Fidler. 77[A]

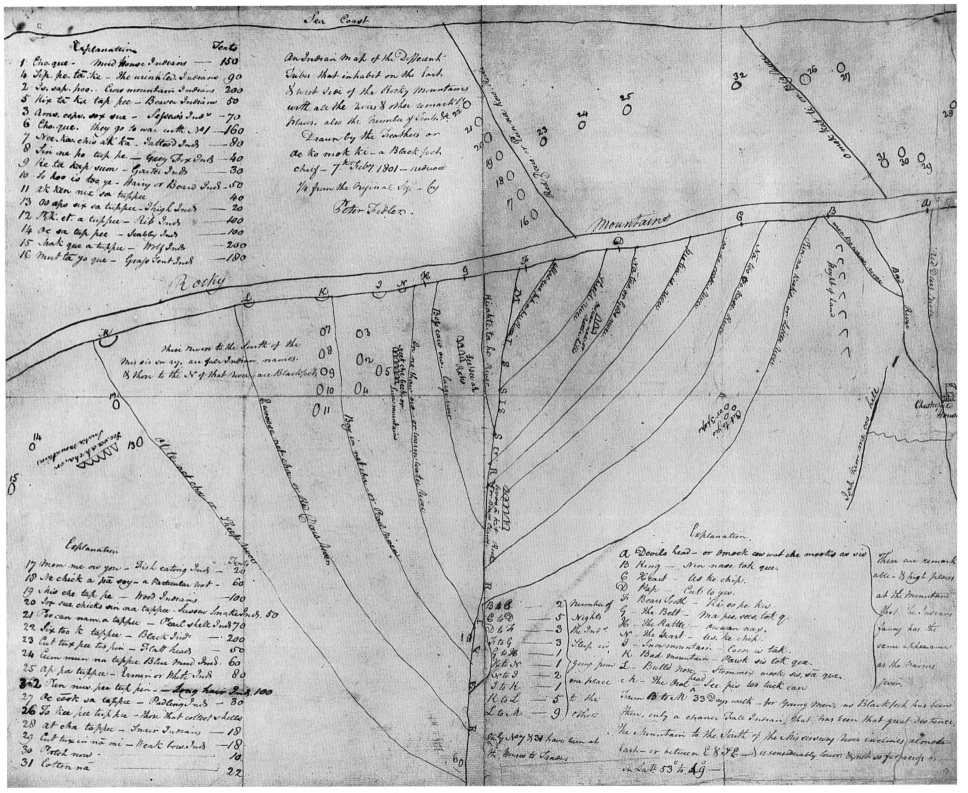

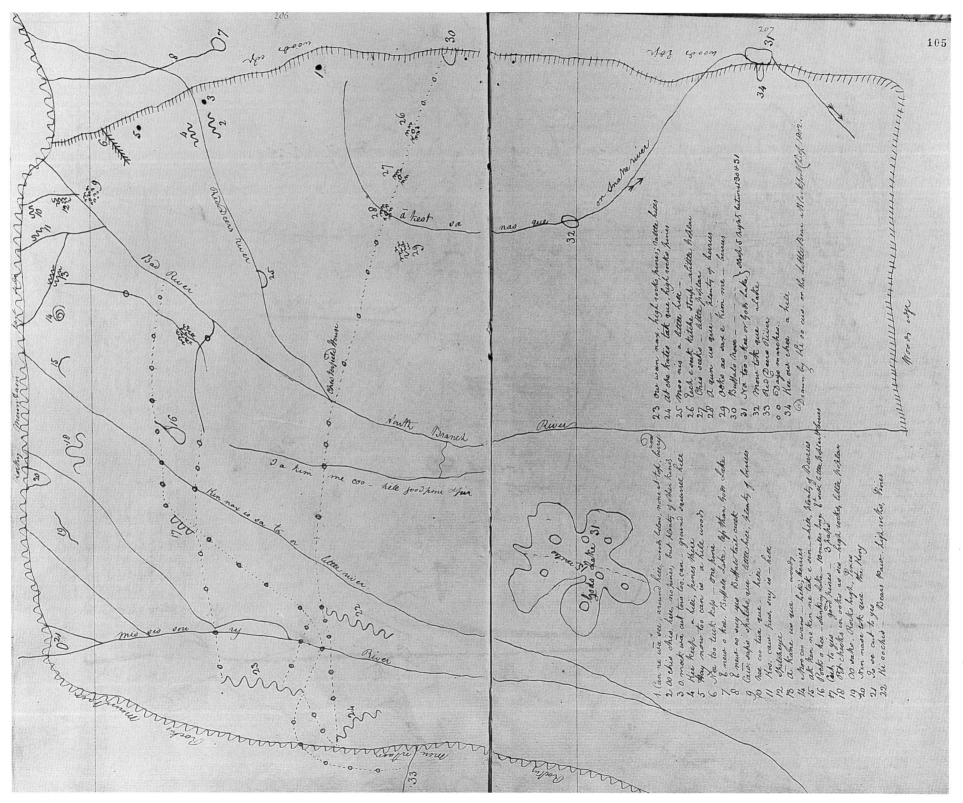

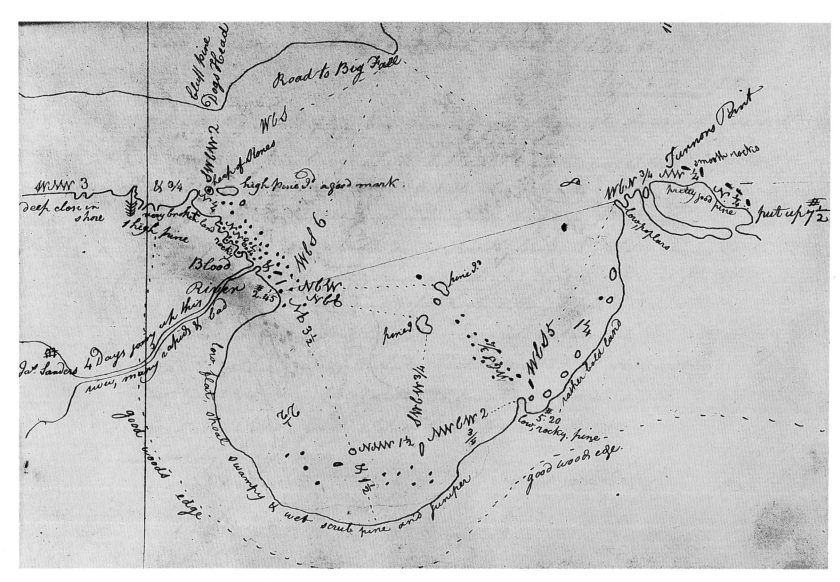

20 Ki oo cus's Map of Area from the Red Deer River South to the Missouri River. 1802. Peter Fidler. 86[A]

21 Sketch Map of the West Side of Lake Winnipeg at Dogs Head. (1808). Peter Fidler. 114[A]

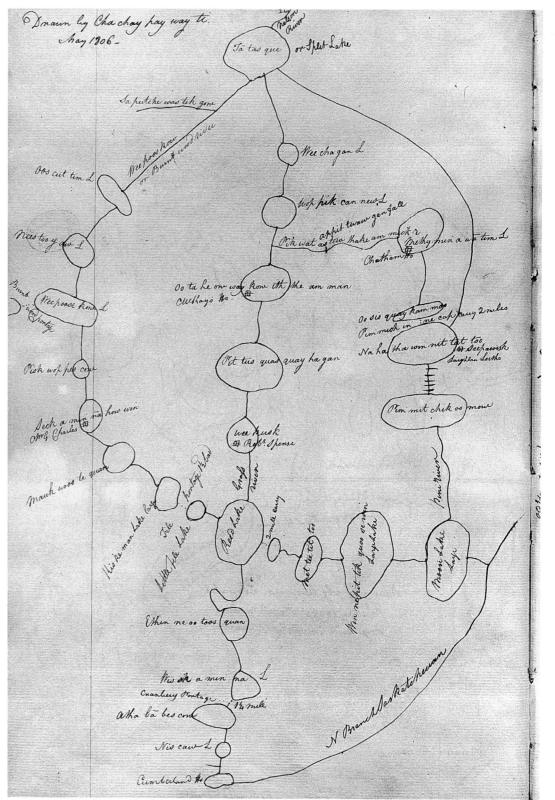

22 Cha chay pay way ti's Sketch of the Area Southwest from Split Lake to Cumberland House. 1806. Peter Fidler. 96[A]

23 Manetoba (Manitoba) District Based on an Indian Sketch. 1820. Peter Fidler. 187[A]

24 Site of Neoskweskau Fort and Surrounding Area. 1816. James Clouston. 170[A]

25 Inland from the Eastmain, Extending from the Koksoak River and Gulph of Richmond South to Rupert's House. (1825). James Clouston. 195[A]

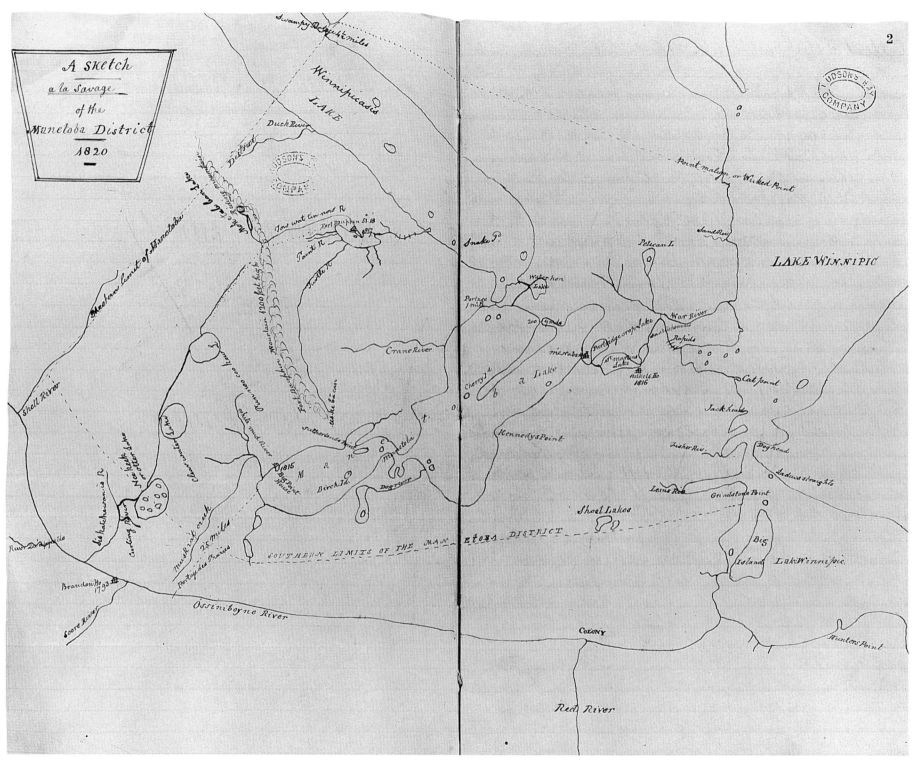

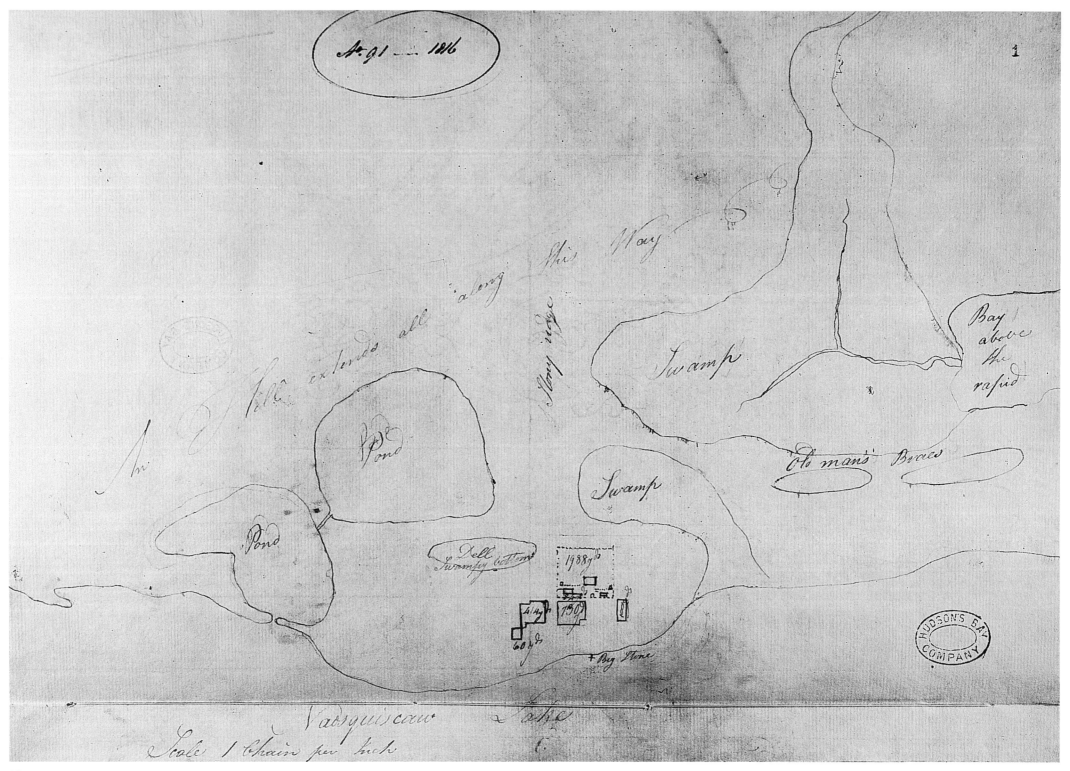

Plates

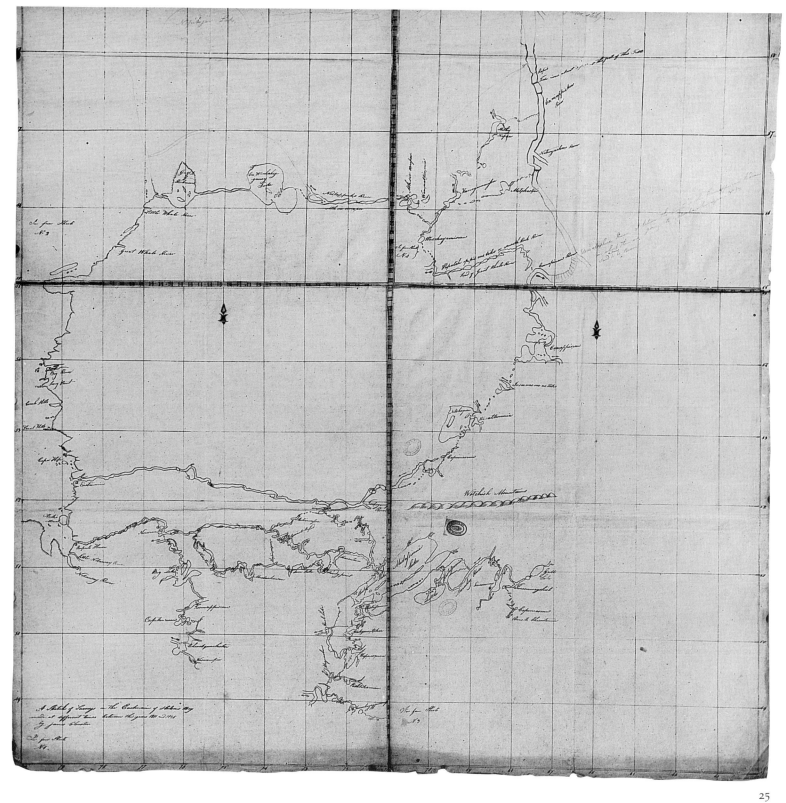

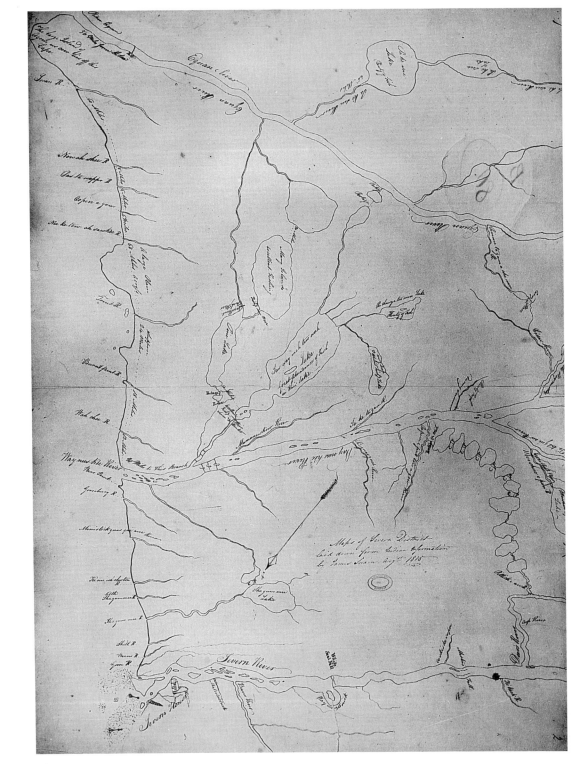

26 Severn District Based on Indian Information. 1815. James Swain Sr. 158ᴬ

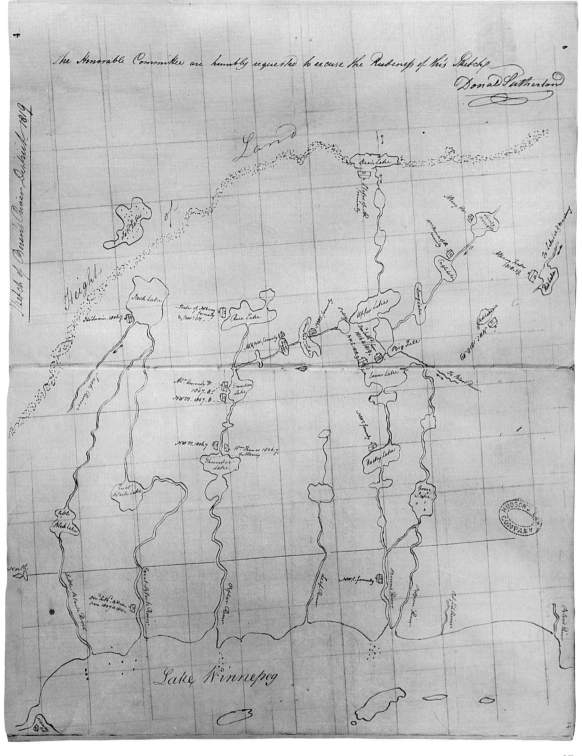

27 Berens River District, East of Lake Winnipeg. 1819. Donald Sutherland. 178[A]

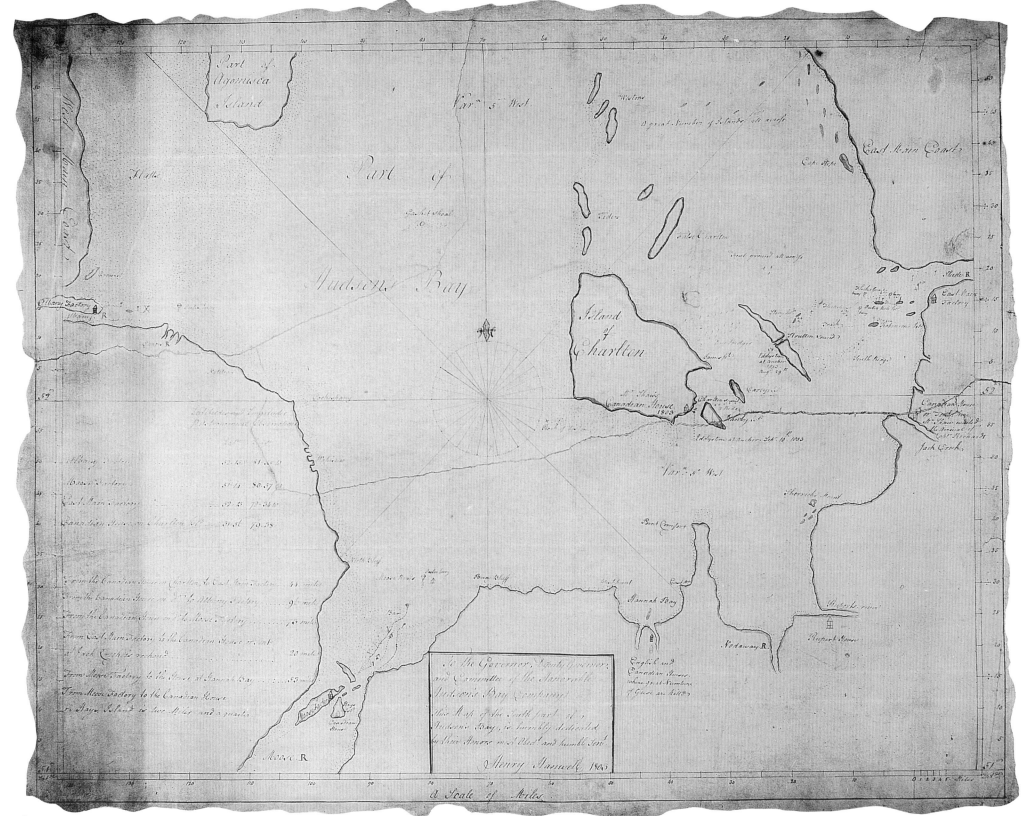

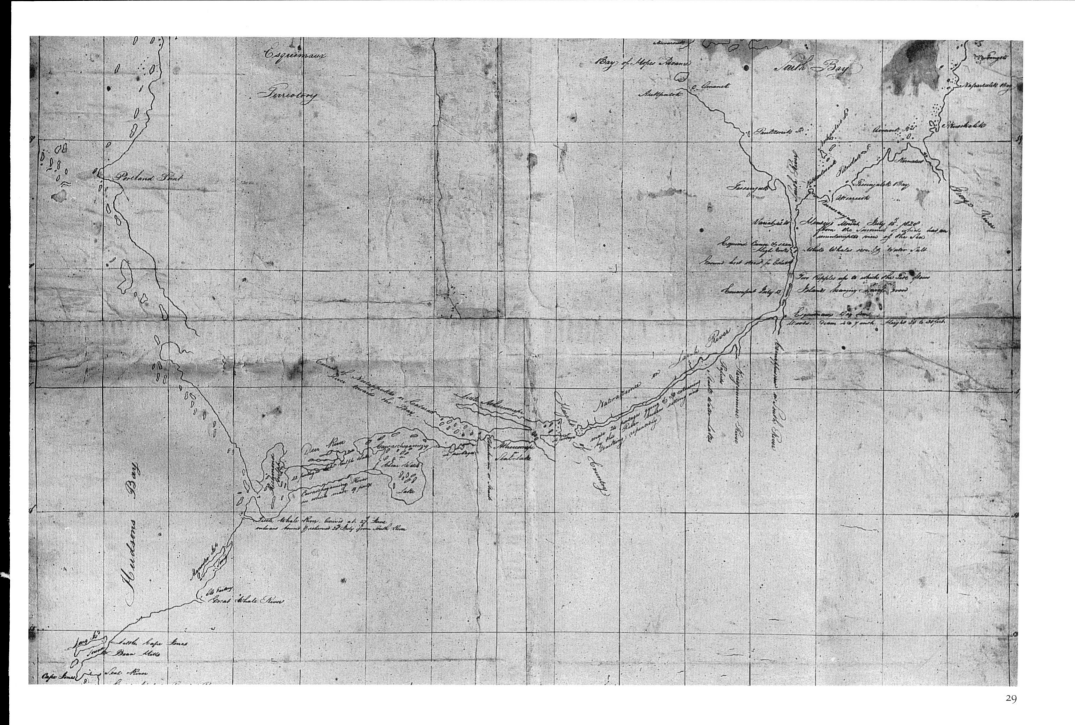

28 The South Part of James Bay. 1803. Henry Hanwell Sr. 89^A

29 The Route of the Inland Expedition from Richmond Gulph to South (Ungava) Bay. 1828. William Hendry. 214^A

154

Plates

30 The Route from Fort Chimo to Esquimeaux Bay (Hamilton Inlet) Labrador. (1834). (Erland Erlandson). 230[A]

31 Area Extending from the Eastmain to the St. Lawrence River. 1841. Thomas Beads and John Spencer. 247[A]

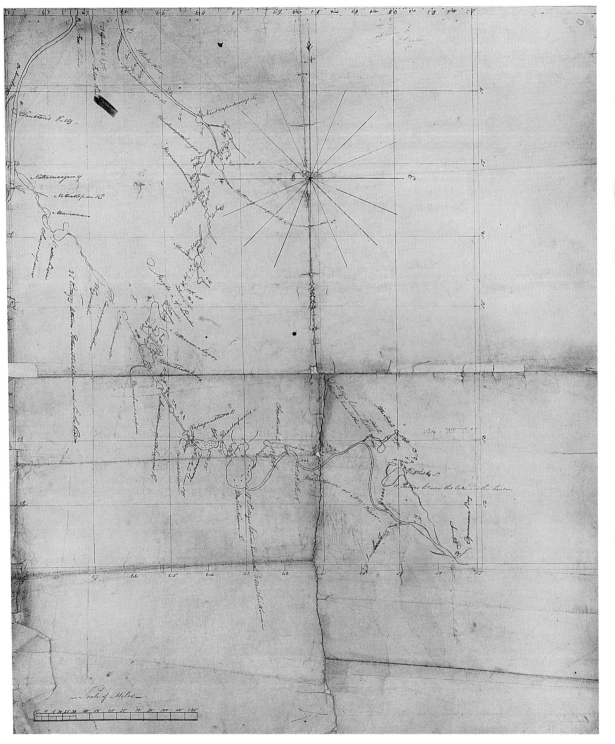

30

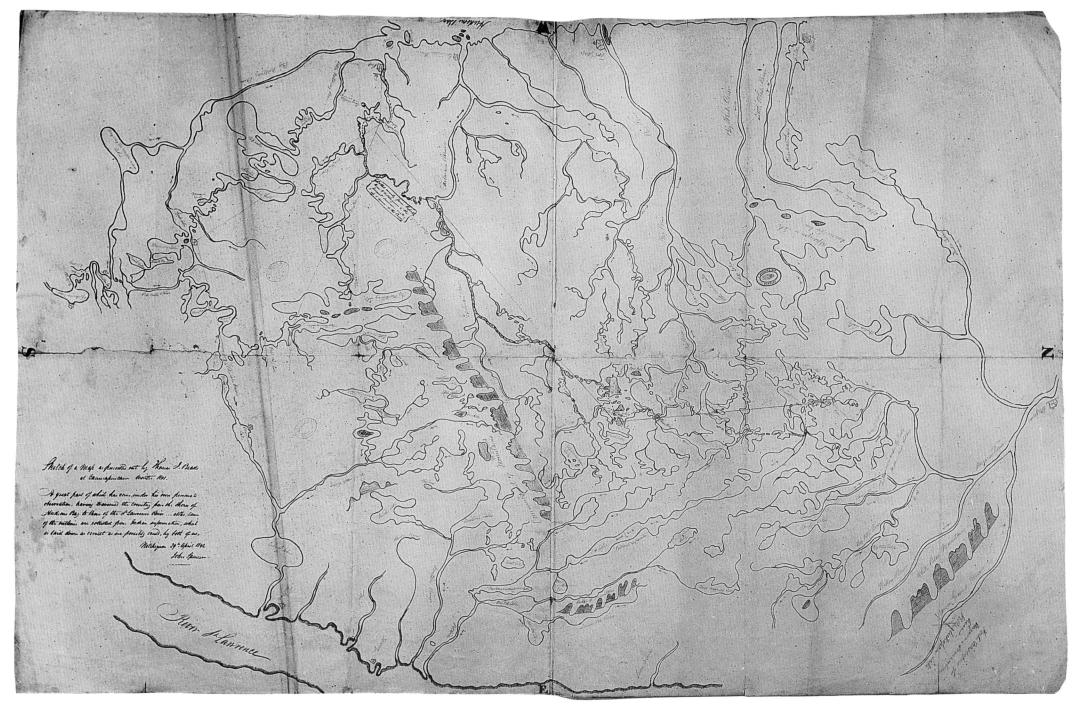

32 The McKenzie's (Mackenzie) River Department Exhibiting the Supposed Situation of the Main Ridge of the Rocky Mountains. 1824. (Murdoch McPherson). 191[A]

33 Expedition up the Liard River to the Frances River. 1831. John McLeod. 217[A]

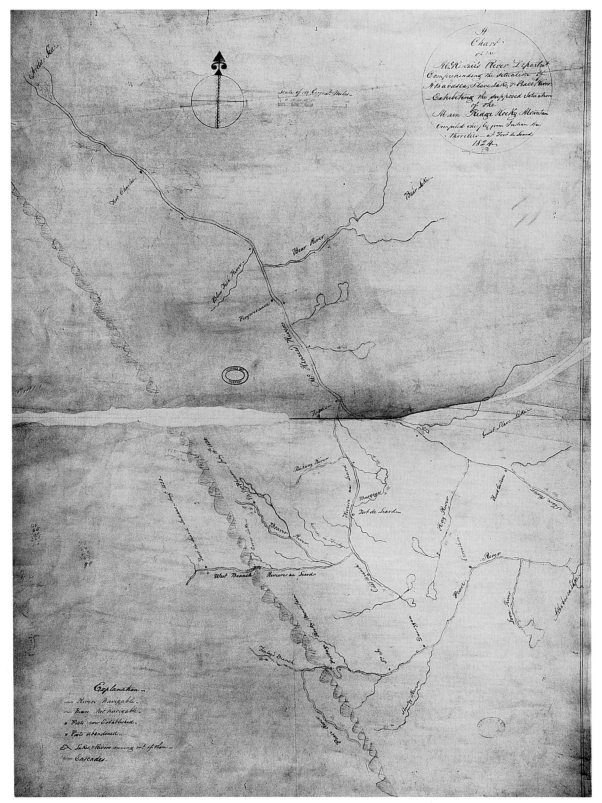

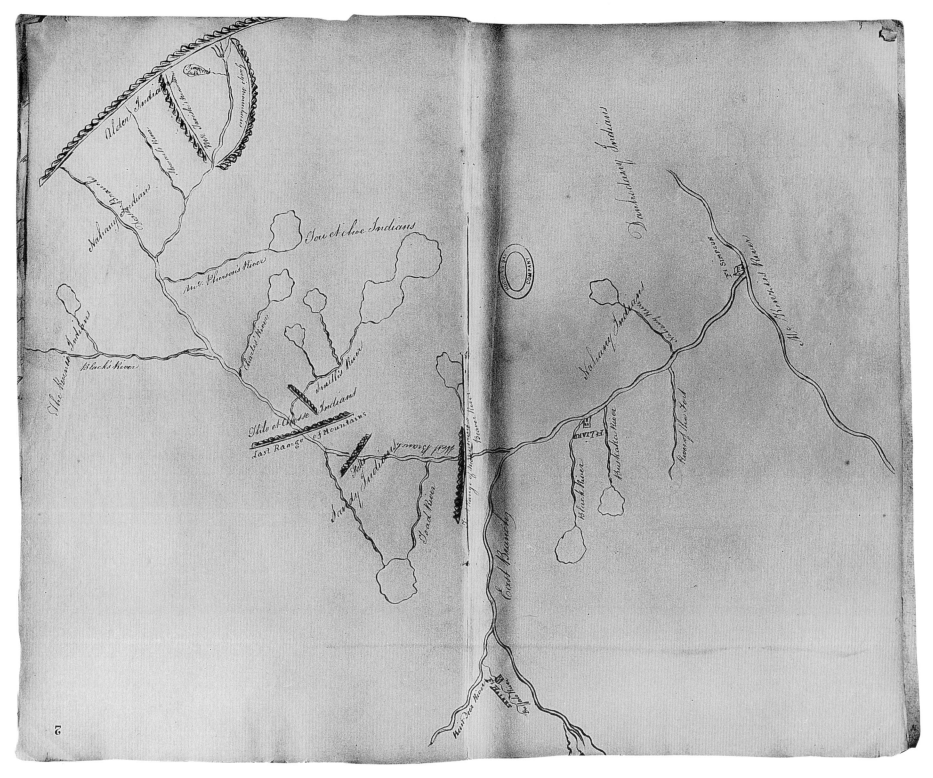

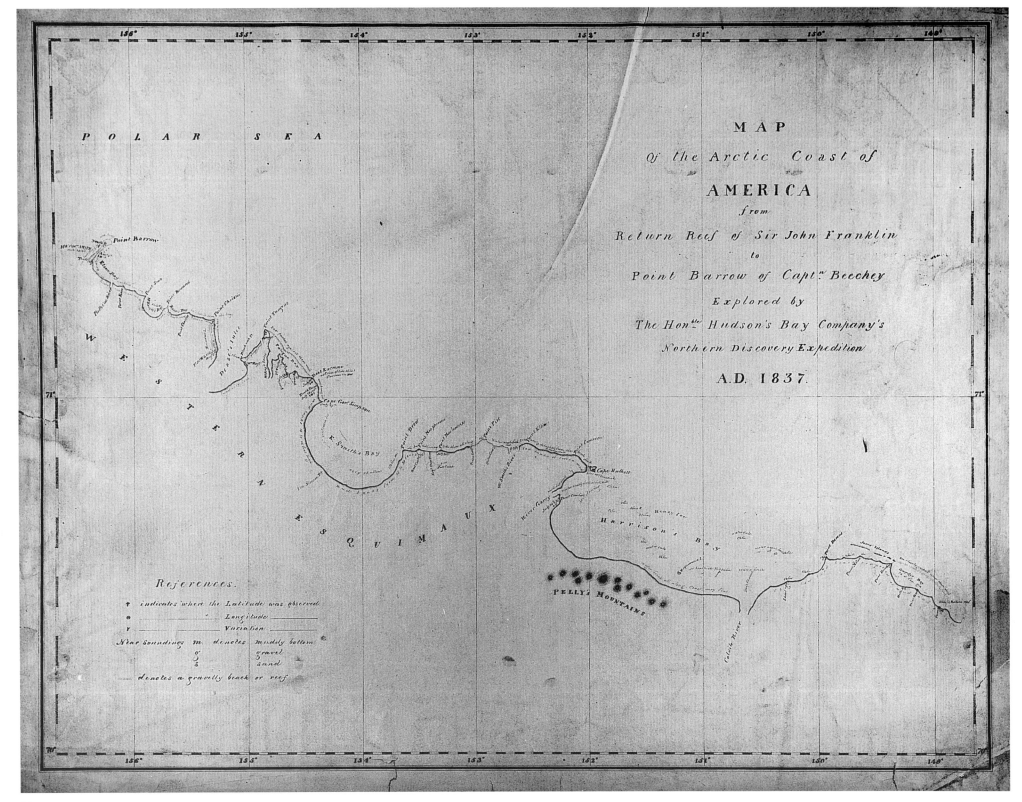

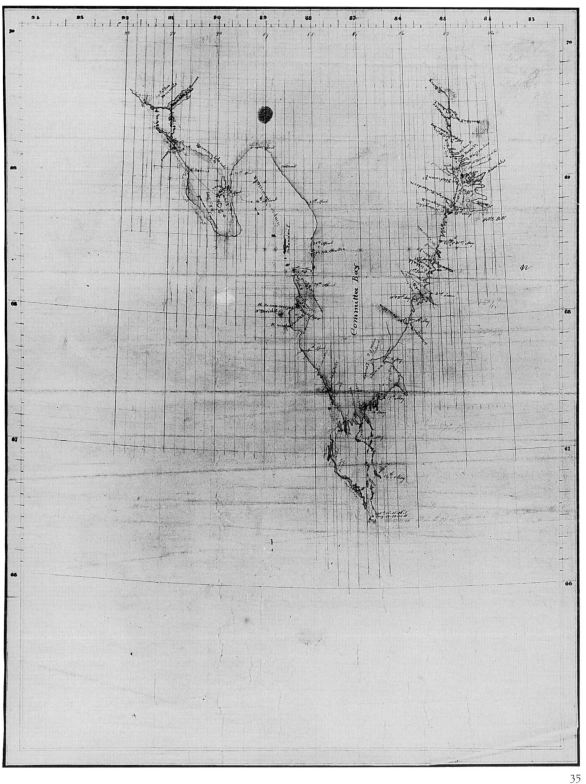

34 The Arctic Coast from Return Reef West to Point Barrow. 1837. (Thomas Simpson and Peter W. Dease). 237[A]

35 Committee Bay on the Arctic Coast West of Melville Peninsula. (1847). (John Rae) 292[A]

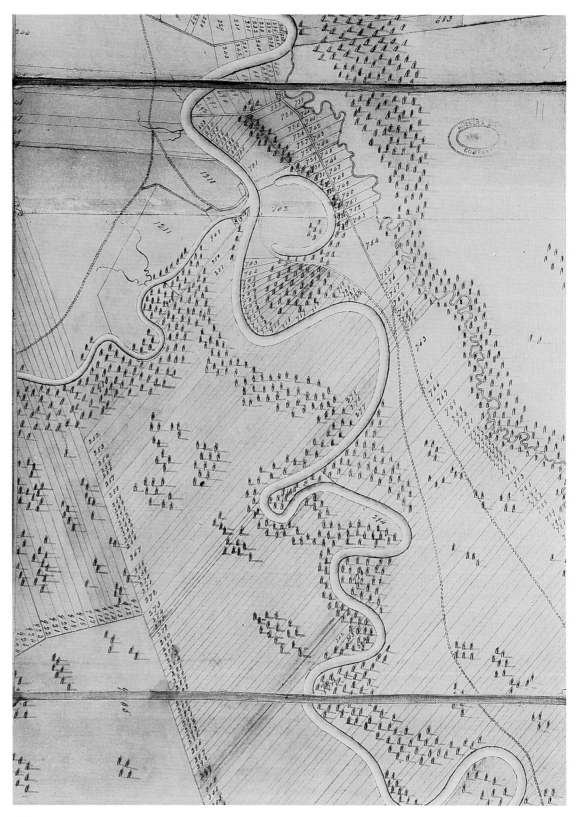

36 Part of a Plan of the Red River Colony at the Junction of the Red and Assiniboine River. 1836–38. George Taylor Jr. 236A

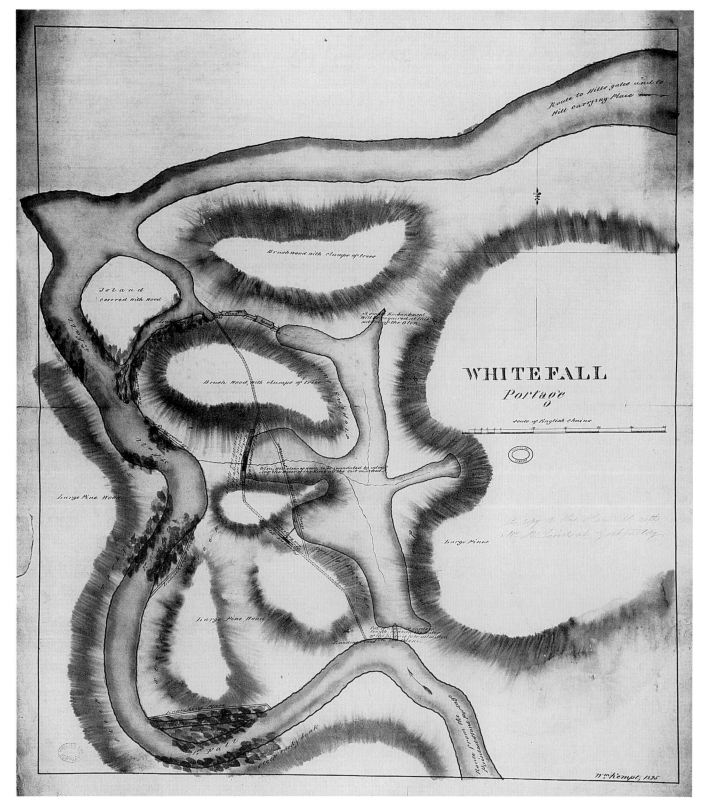

37 White Falls Portage. 1825. William Kempt. 200[A]

38 Sketch of Cumberland House Area, Illustrating a Journey from Norway House to Jasper House. (1828). George Taylor Jr. 211^A

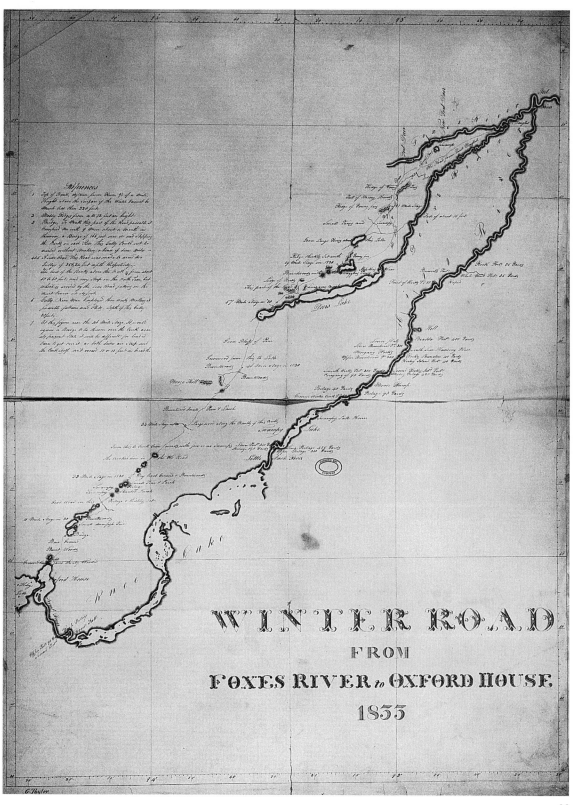

39 Winter Road from Foxes (Fox) River to Oxford House on Oxford Lake. Hayes River. 1833. George Taylor Jr. 226ᴀ

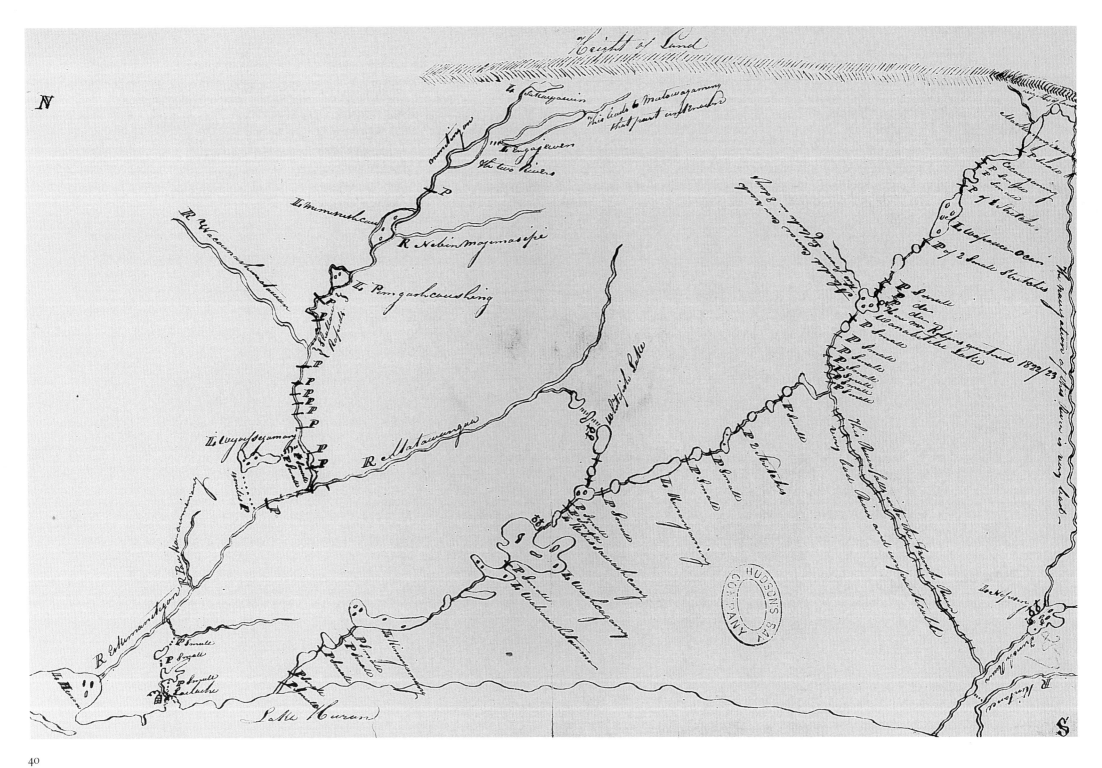

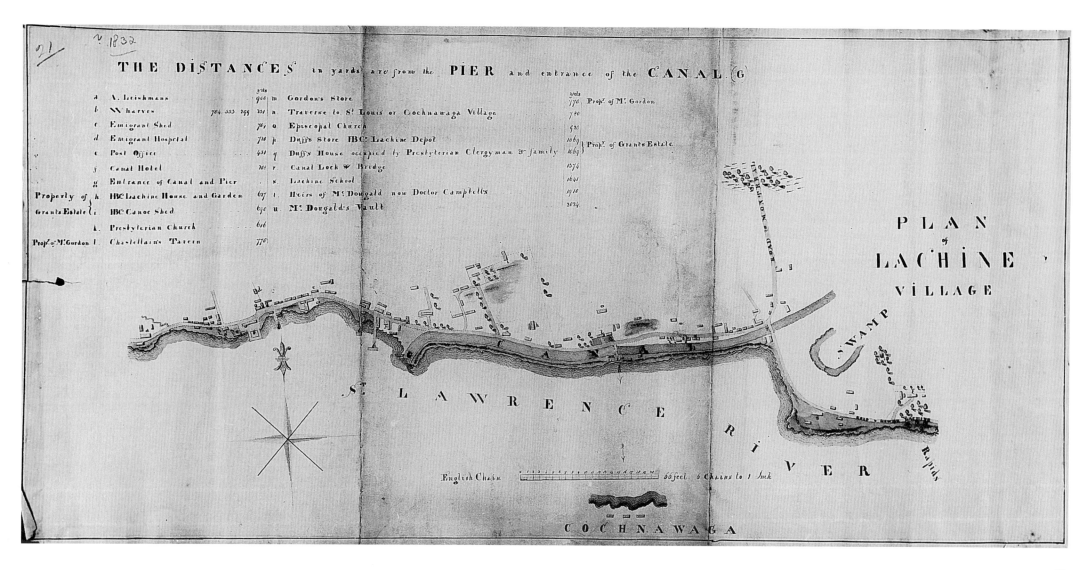

40 River Communication between La Cloche Post, Lake Huron, and Lake Temiskaming. (1827). (John McBean). 210[A]

41 Plan of Lachine Village. (ca. 1832). Cartographer unknown. 221[A]

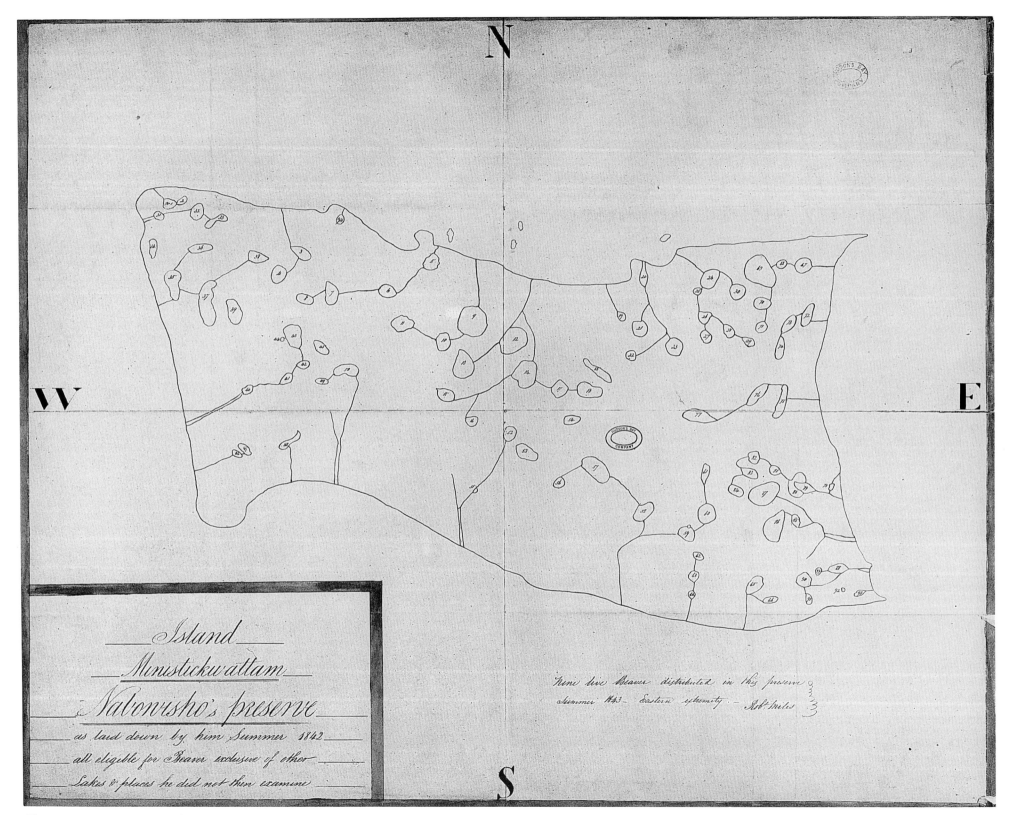

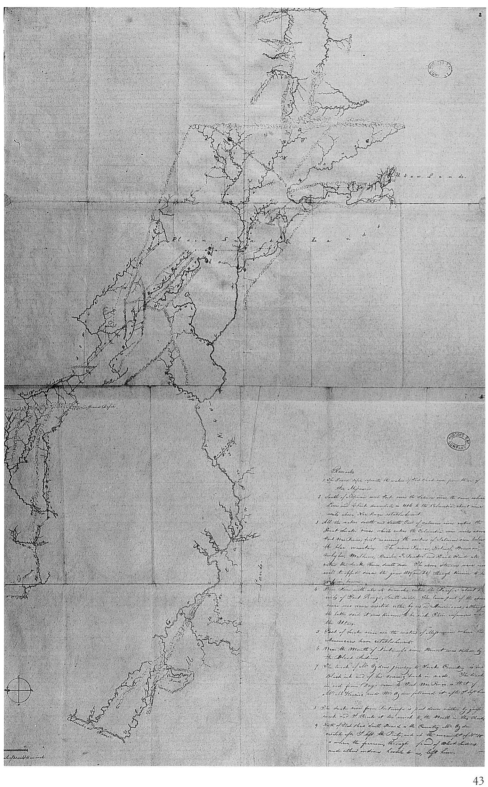

42 Nabowisho's Beaver Preserve, Ministickwattam Island, James Bay. (1843). Nabowisho (with Robert Miles). 256[A]

43 The Snake River Country. (1825). (William Kittson). 199[A]

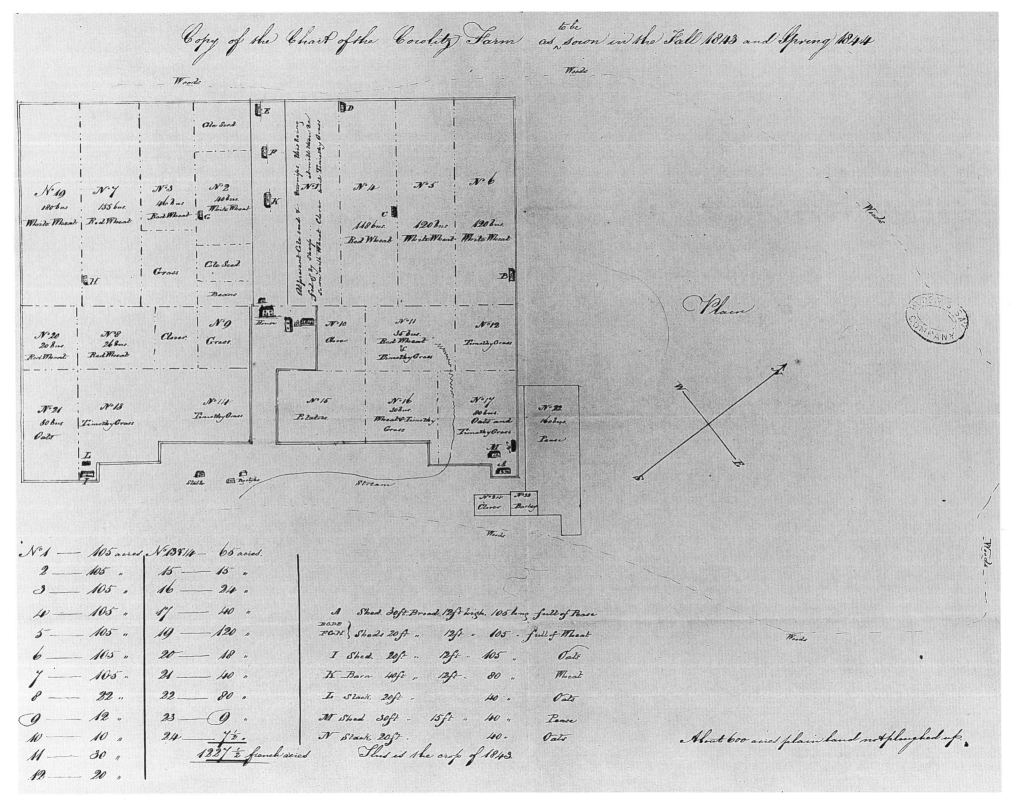

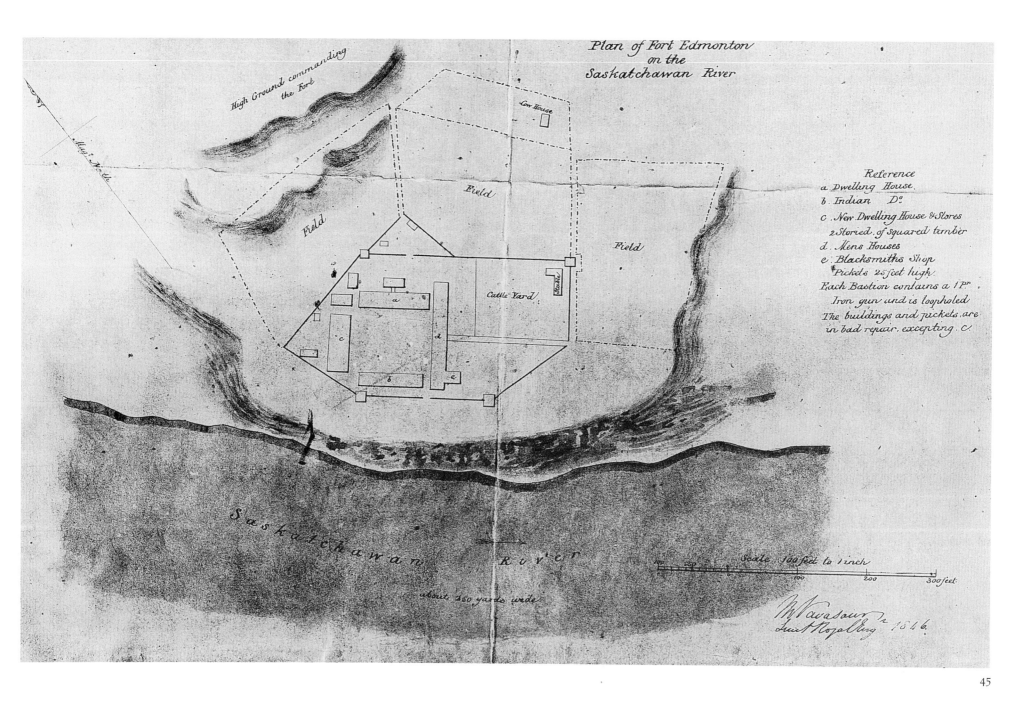

44 The Cowlitz Farm, Puget's Sound Agricultural Company, Washington Territory. 1843. Cartographer unknown. 258[A]

45 Fort Edmonton on the Saskatchewan River. 1846. Mervin Vavasour. 286[A]

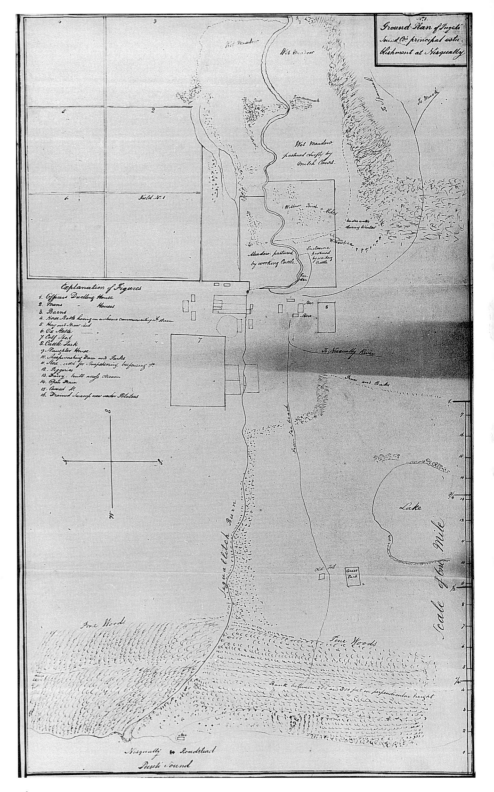

46 The Puget's Sound Agricultural Company's Establishment at Nisqually, Washington Territory. (1847). (William F. Tolmie). 294[A]

47 Pasture Land Adjoining Shepherd's Station at "Sastuk", Nisqually, Washington Territory. (1847). (William F. Tolmie). 298[A]

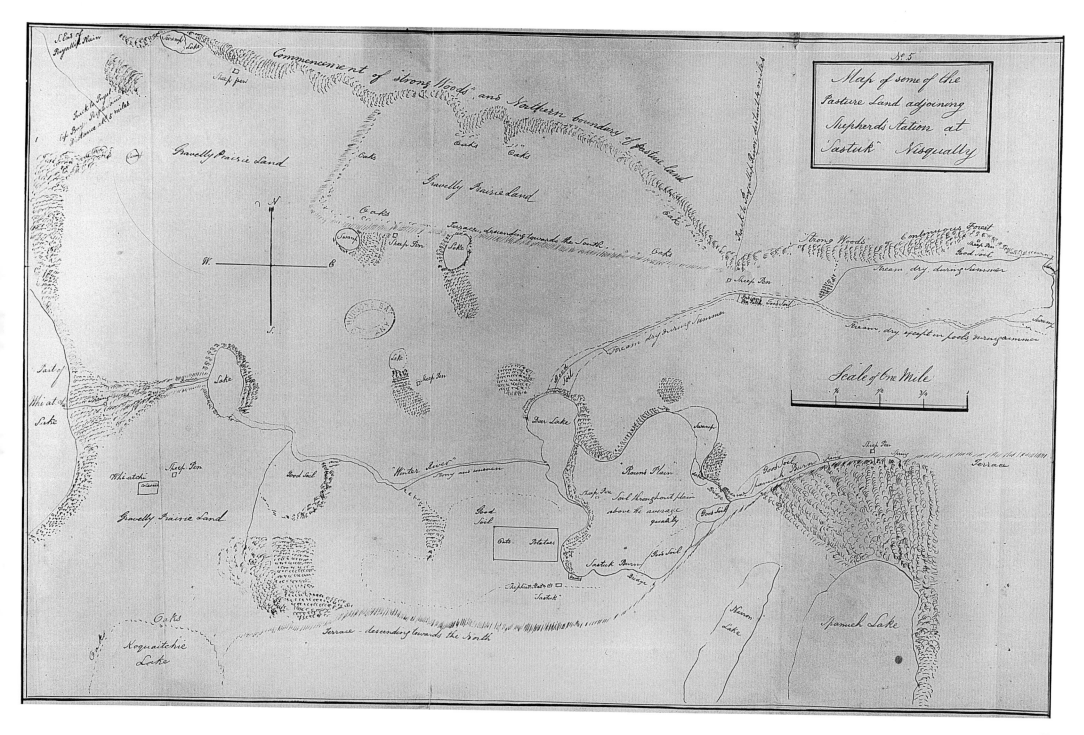

48 Thompson's River District. 1827.
Archibald McDonald. 201[A]

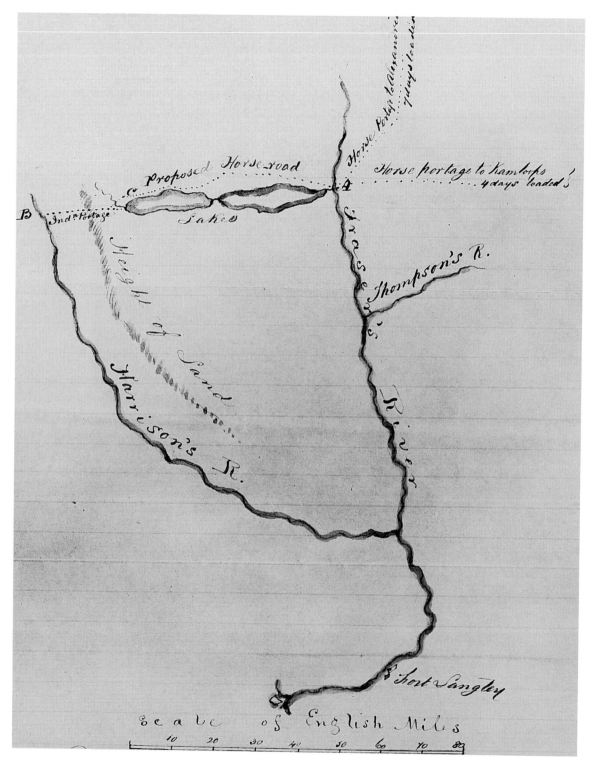

49 A New Route Proposed for the Horse Brigade from New Caledonia via the Harrison River (Harrison Lake) to Fort Langley. 1845. Alexander C. Anderson. 273[A]

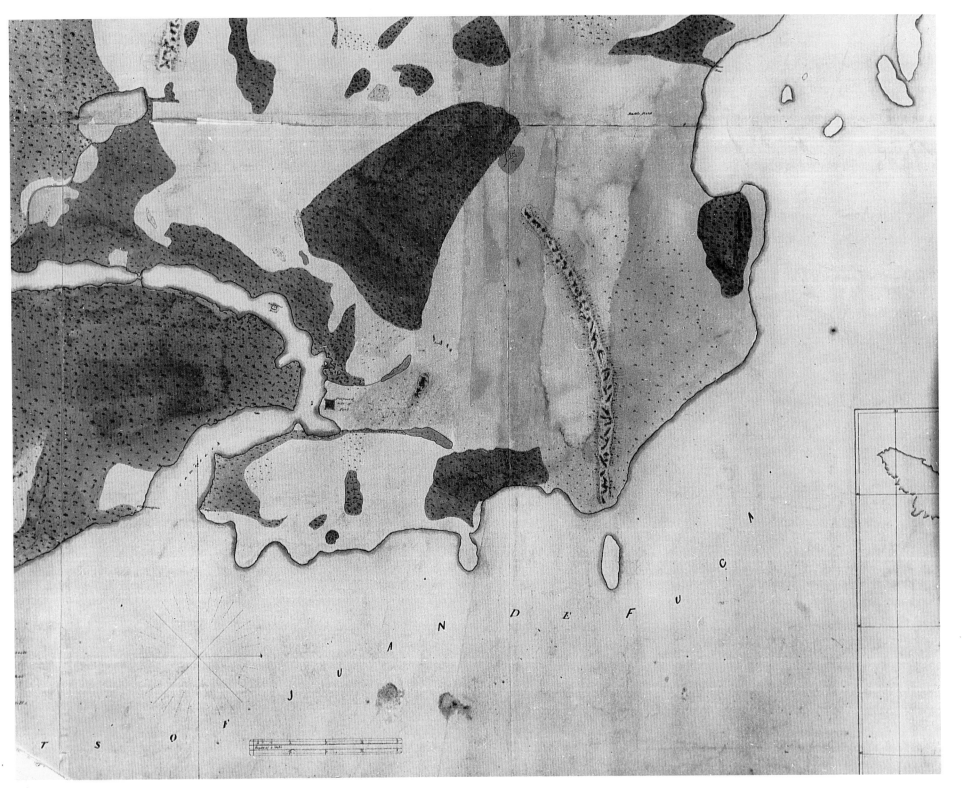

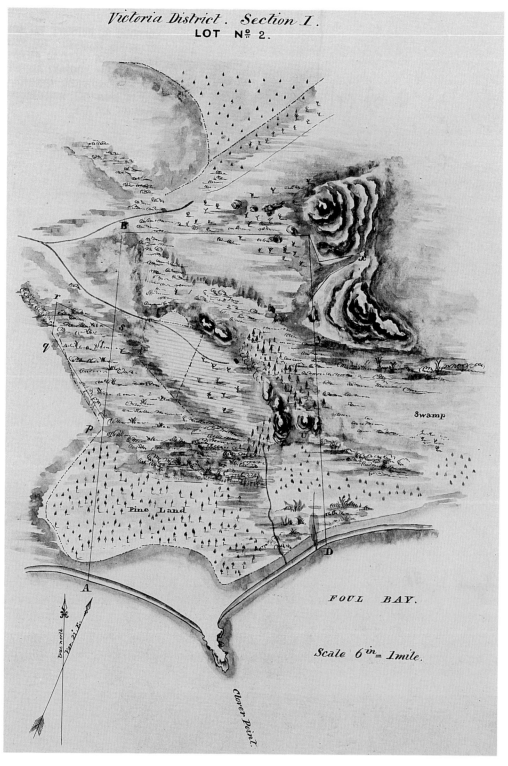

50 Plan of the Portion of Vancouver's Island Selected for the New Establishment (Fort Victoria). 1842. James Douglas and Adolphus Lee Lewes. 252[A]

51 Victoria District, Foul Bay-Clover Point Area, Section I, Lot no. 2. (1851 or 1852). (Joseph D. Pemberton). 314[A]

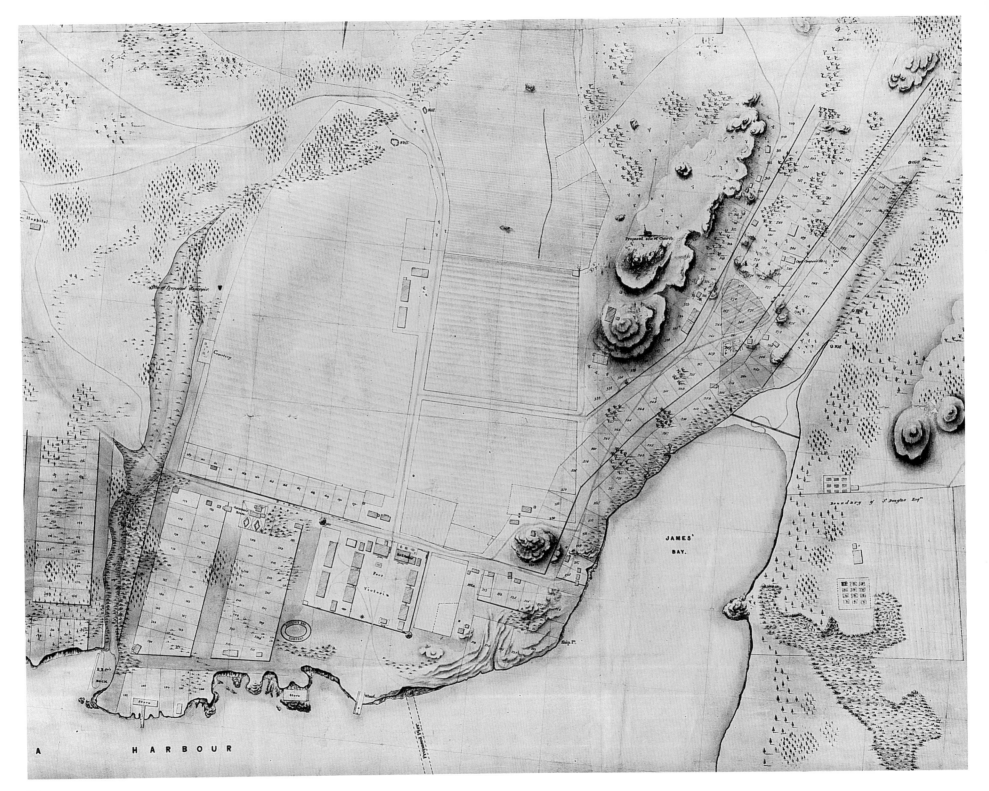

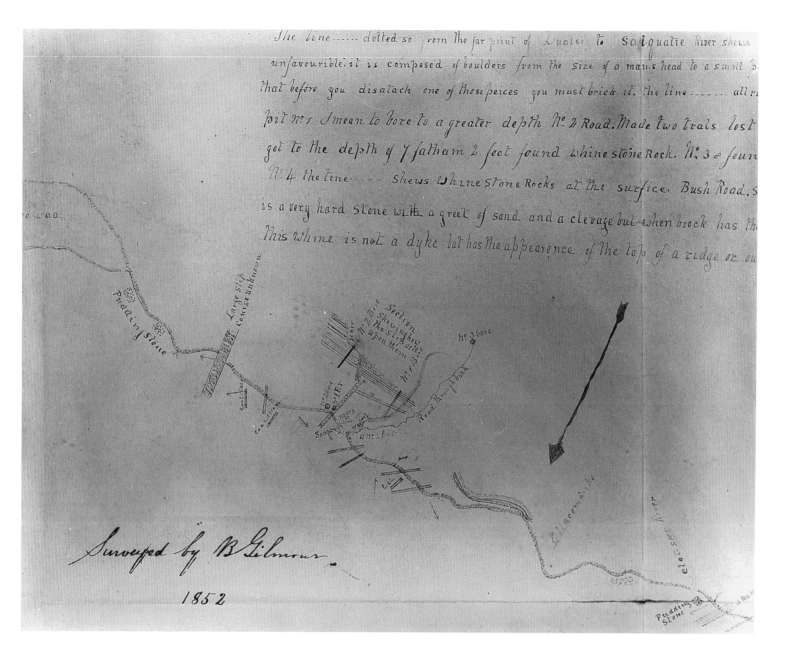

52 The Town of Victoria, Showing Proposed Improvements. (1852). (Joseph D. Pemberton). Used by permission of Surveys and Land Records, Victoria. 323[A]

53 Part of the Shore of Queen Charlotte Strait, Vancouver's Island, Showing Mineral Deposits. 1852. Boyd Gilmour. 326[A]

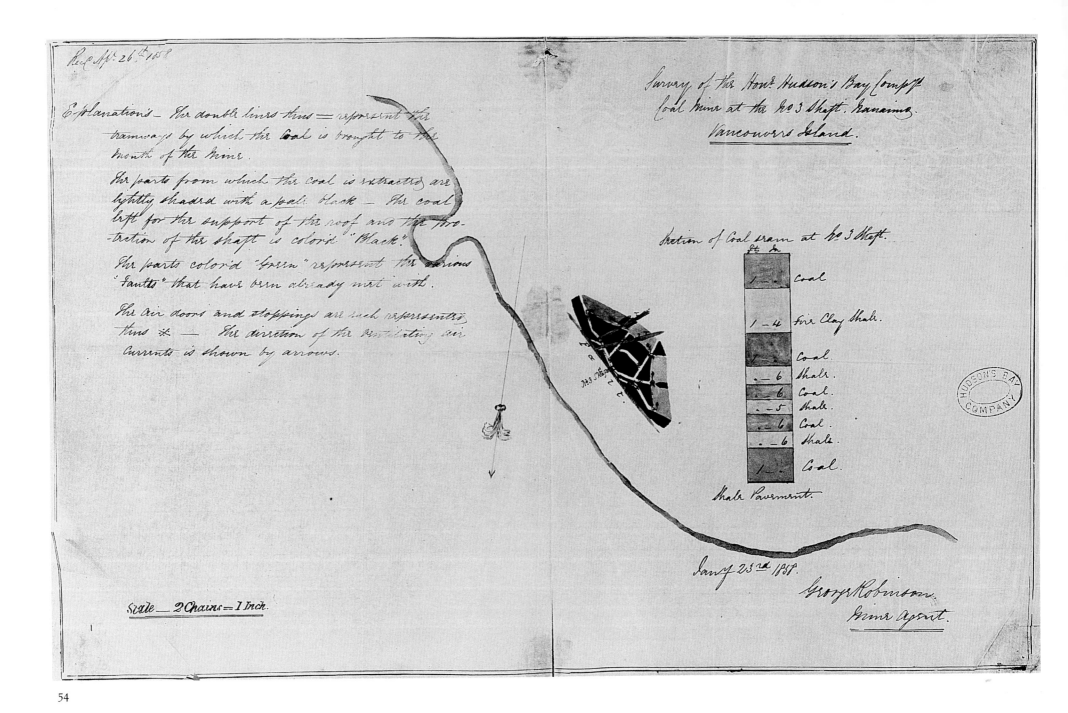

54 The Hudson's Bay Company's Coal Mine at No. 3 Shaft, Nanaimo, Vancouver's Island. 1858. George Robinson. 463^A

55 Journey of Exploration Through the Cowetchin (Cowichan) Valley, Vancouver's Island. 1857. Joseph D. Pemberton. 427^A

56 Esquimalt District, the Esquimalt or Langford Farm. (1854). (Joseph D. Pemberton). 378^A

57 Yukon District Showing Positions of Various Indian Tribes. 1853. William L. Hardisty. 366^A

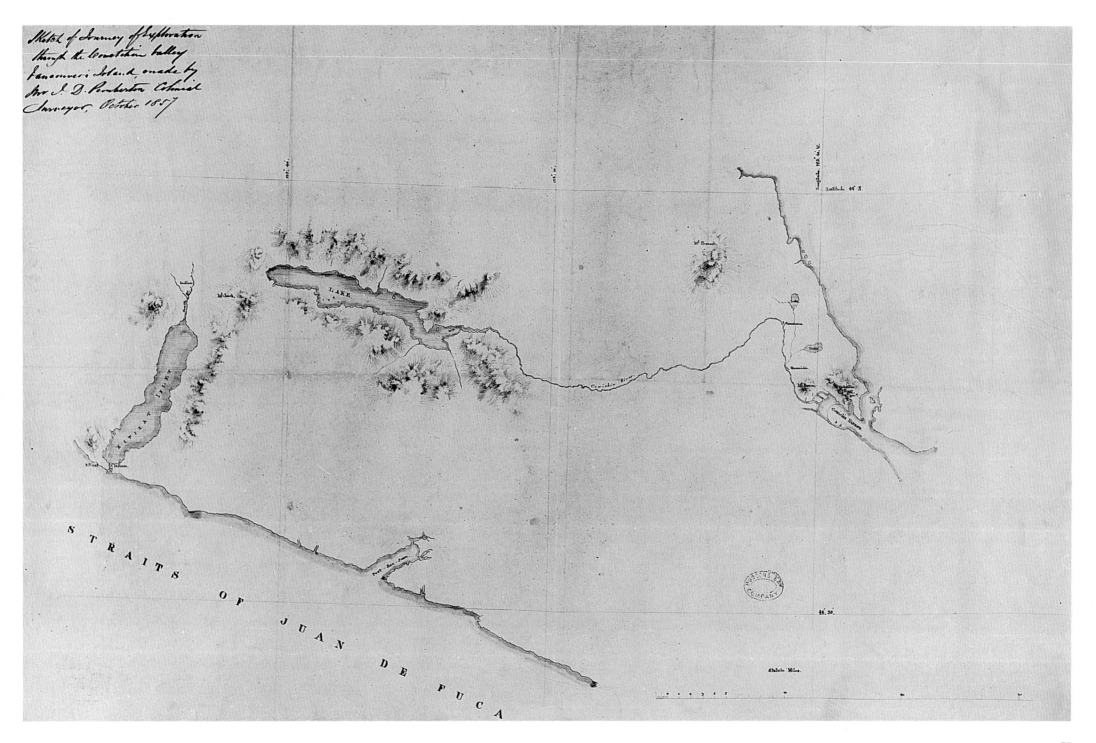

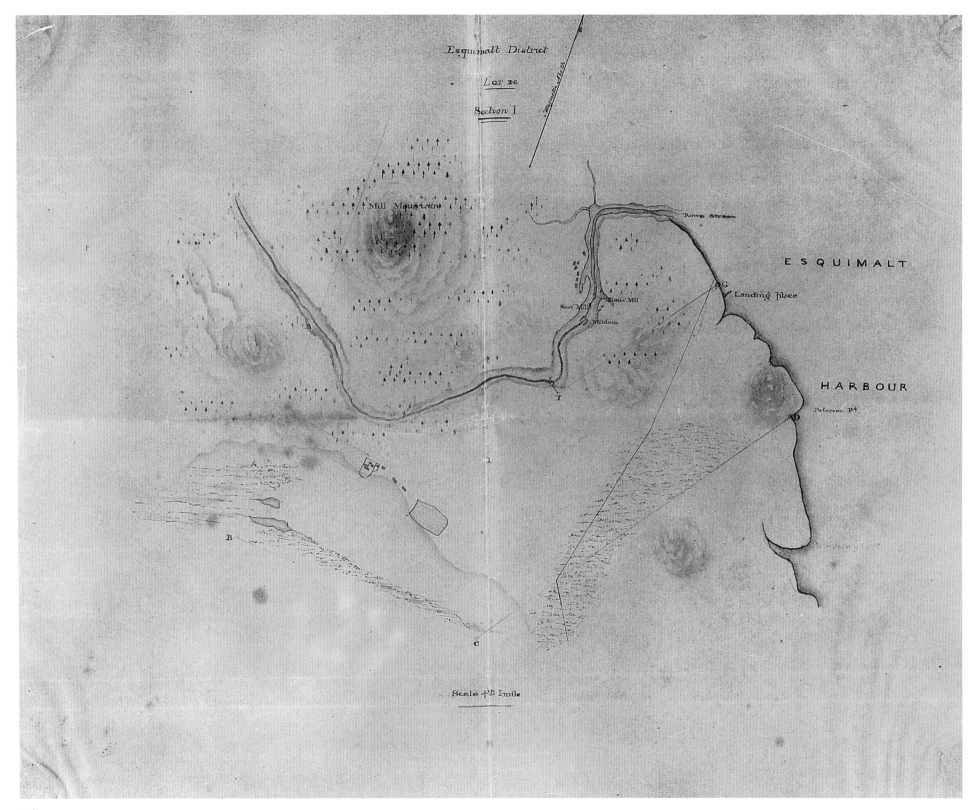

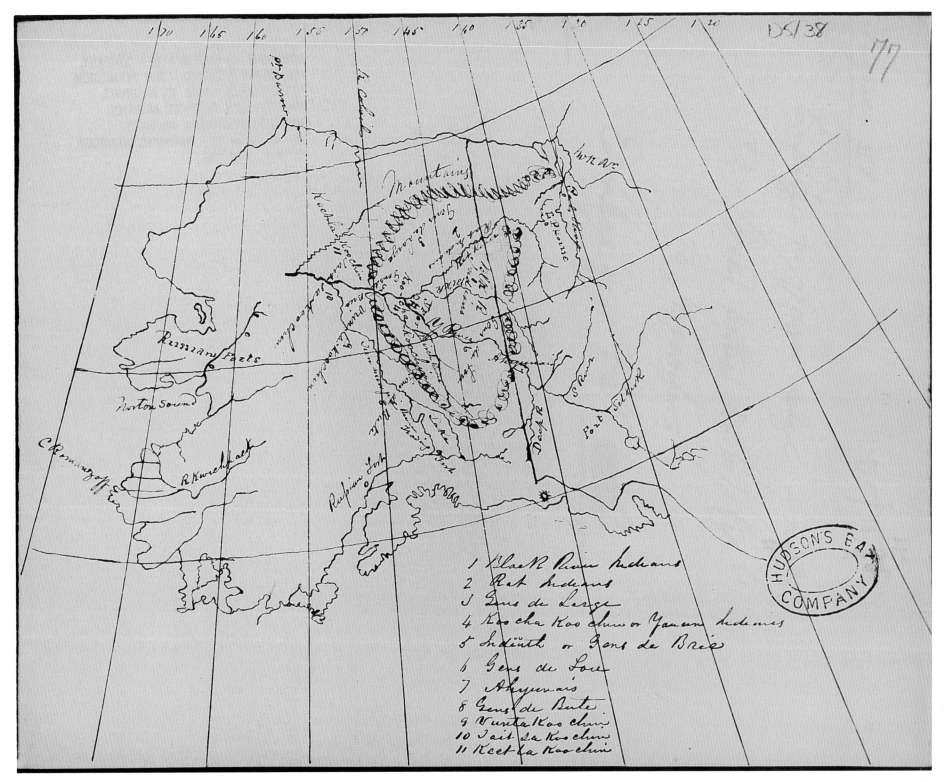

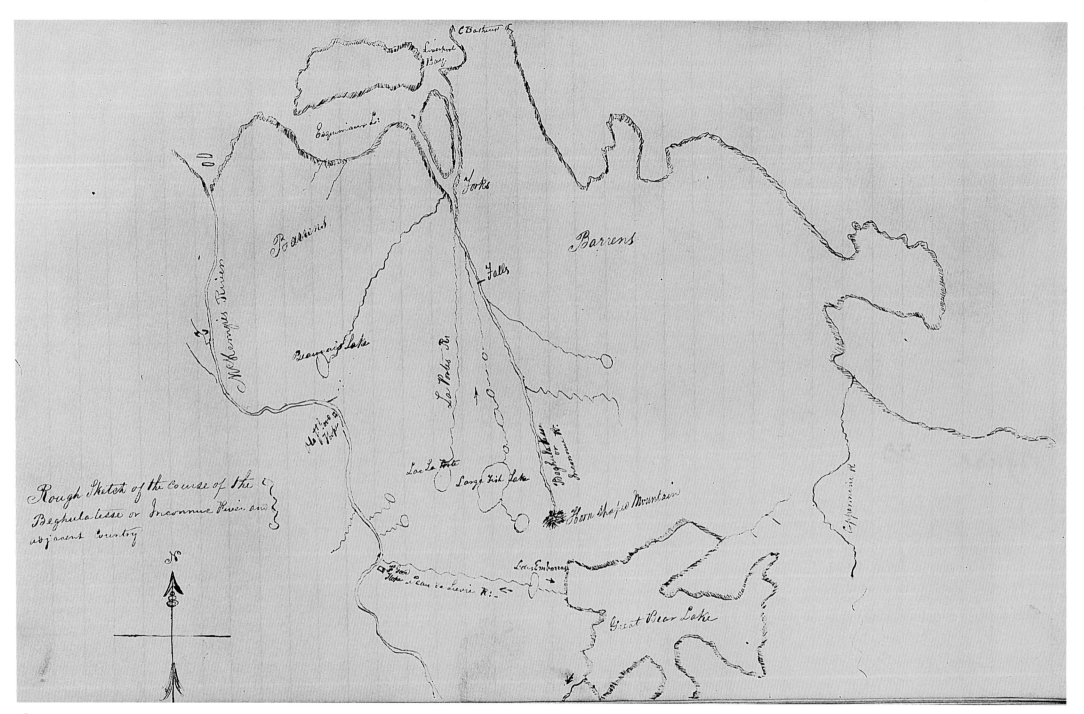

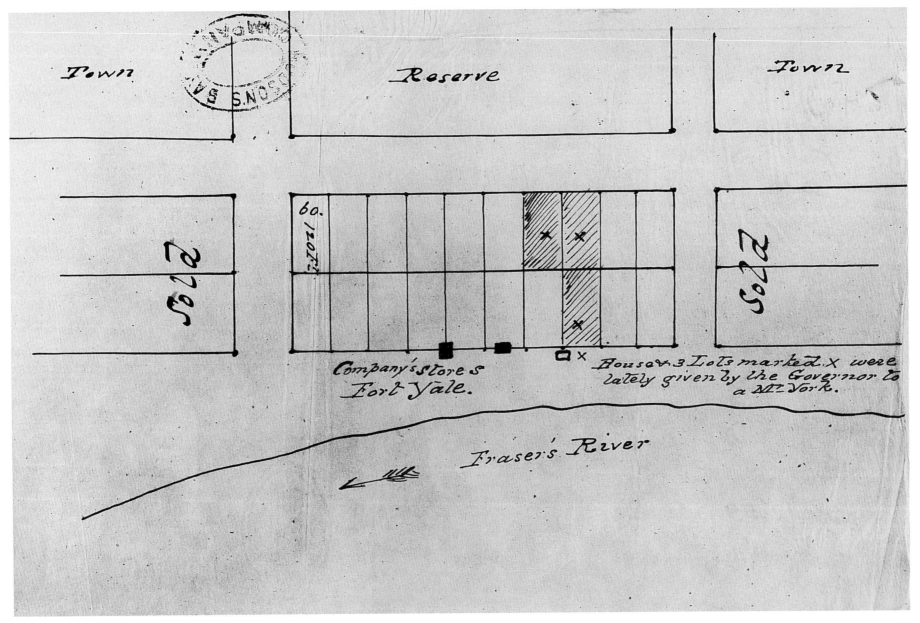

58 The Course of the Beghulatesse or Inconnue (Anderson) River, Northwest Territories. 1855. James Anderson (A). 412[A]

59 Town Lots at Fort Yale, Colony of British Columbia. 1858. (Alexander G. Dallas). 465[A]

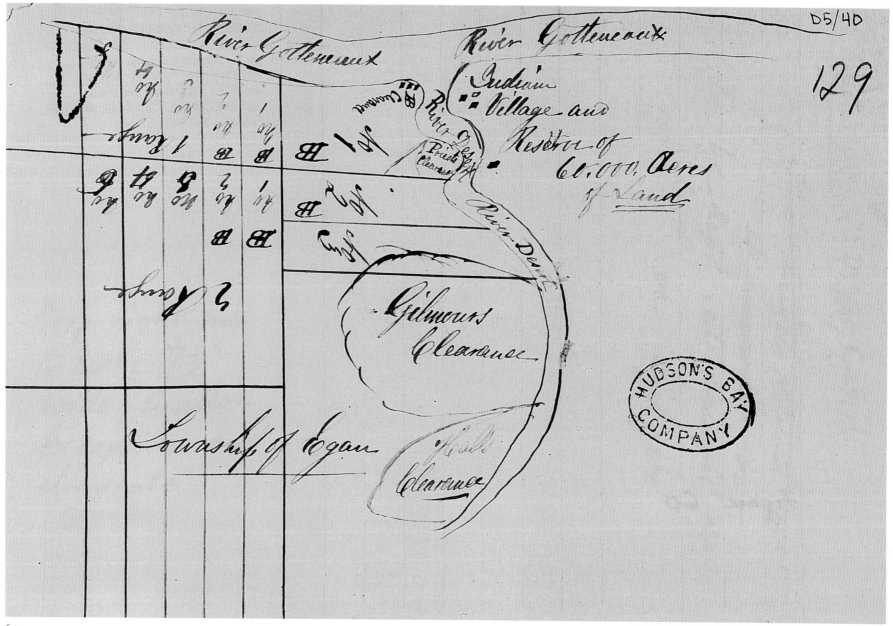

60

60 Desert River, Junction with the Gatineau River, Lands Surveyed for the Hudson's Bay Company. 1855. Archibald McNaughton. 411[A]

61 One Hundred Acres of Land Preempted at Fort Rupert, at Beaver Harbour, Vancouver's Island. 1863. P.N. Compton. 510[A]

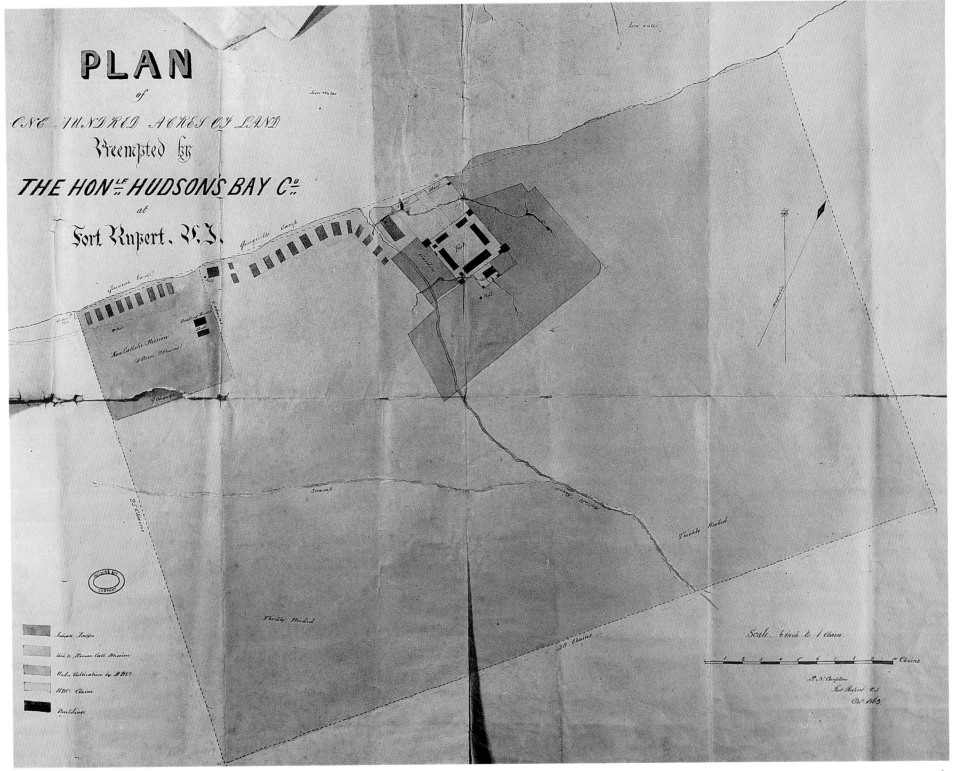

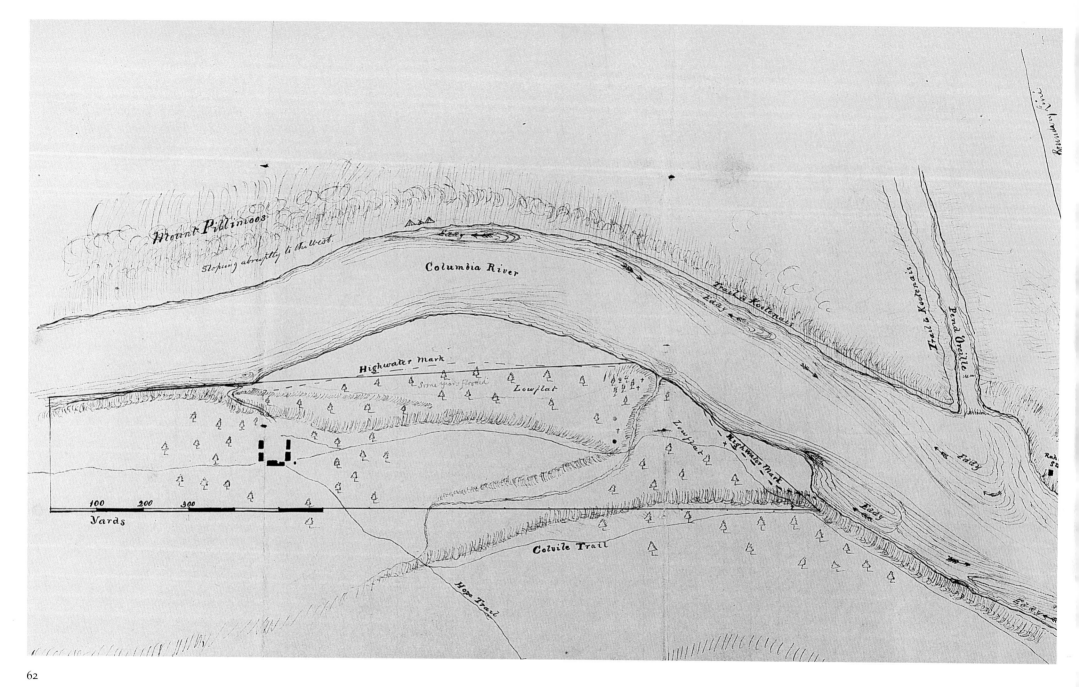

62 Fort Shepherd on the Columbia River, near the Entrance of the Pend'Oreille River. (1864). (William F. Tolmie). 534[A]

63 Mingan River Mouth, Mingan Island, Mingan Post on the St. Lawrence River. (1860). (James Anderson (A)). 479[A]

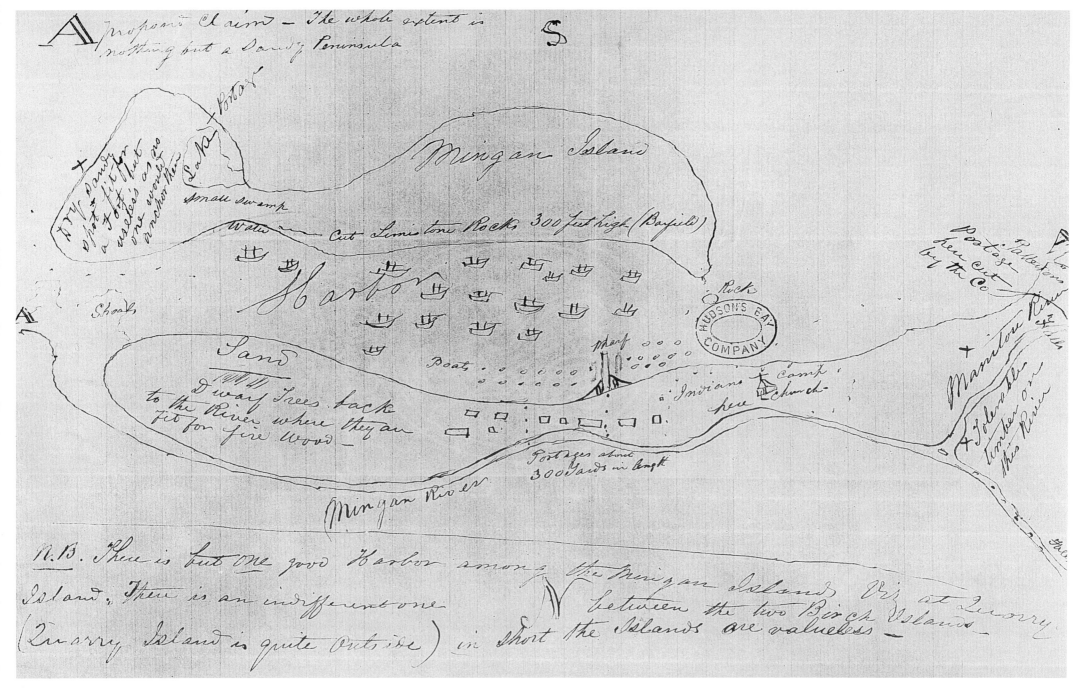

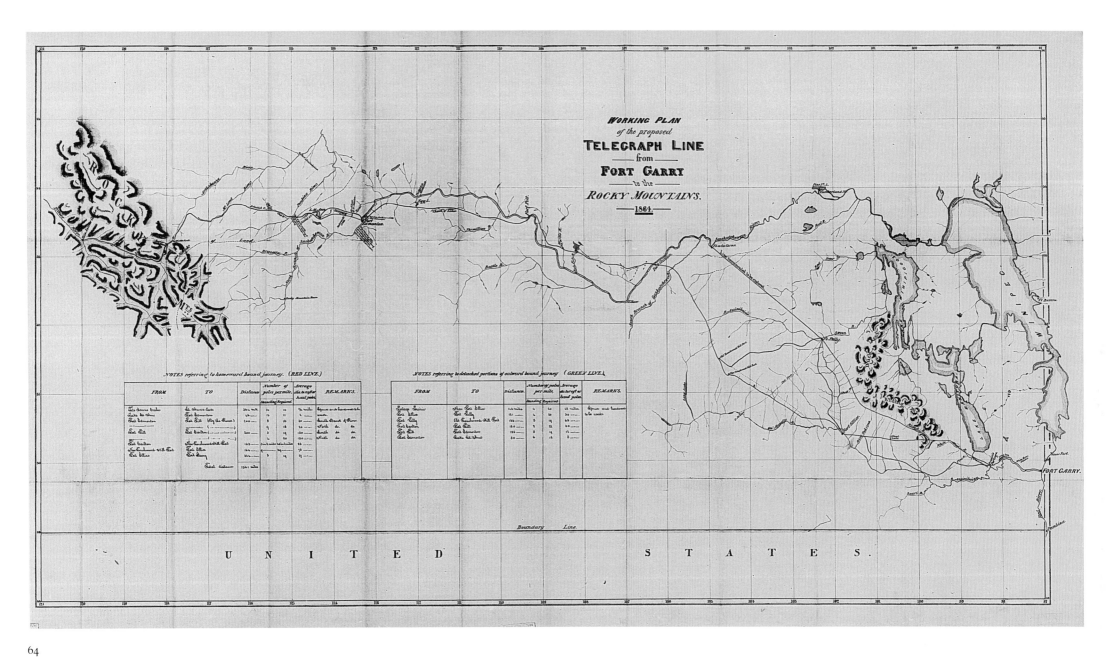

64 Proposed Telegraph Line from Fort Garry (Winnipeg) to the Rocky Mountains at Tete Jaune Cache (Yellowhead Pass). 1864. (A.W. Schwieger). 536[A]

65 The Country Between Fraser's River (near Williams Lake) and Tete Jaune Cache (Yellowhead Pass). 1865. Joseph W. McKay. 551[A]

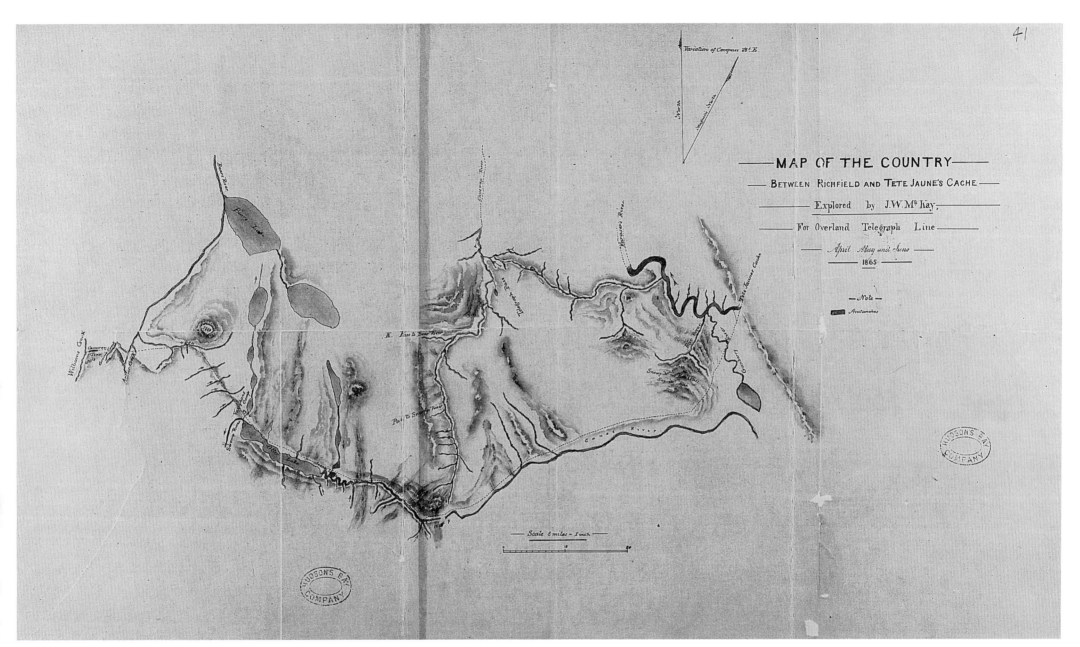

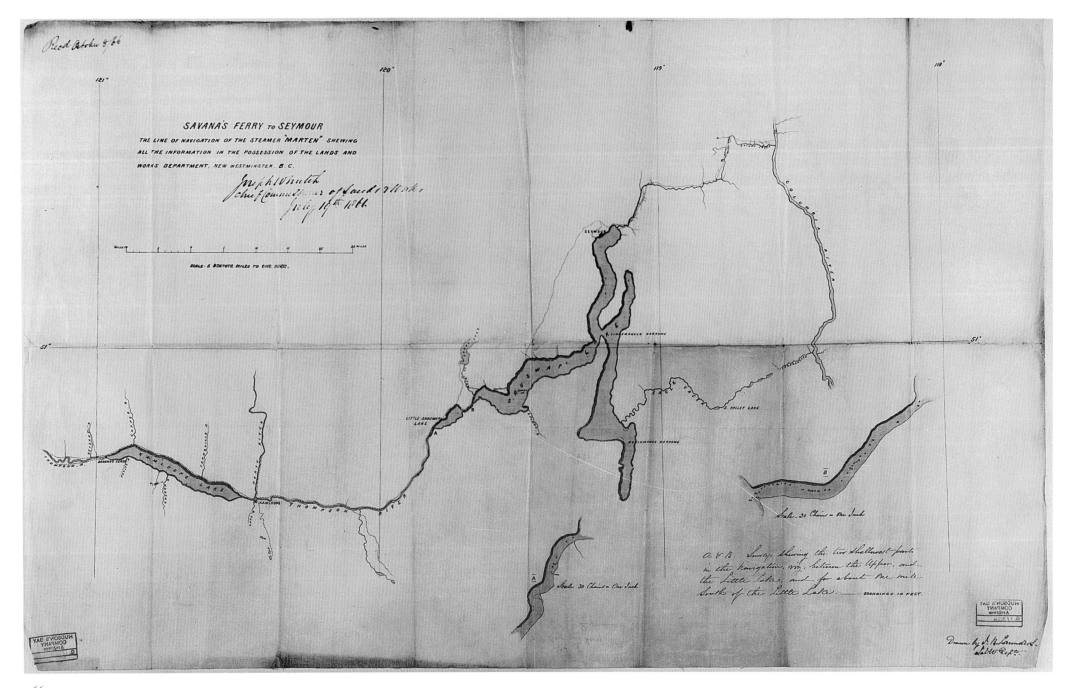

66 The Line of Navigation of the Steamship *Marten* from Savana's Ferry (Savona) on Kamloops Lake to Seymour on Shuswap Lake. 1866. J.B. Launders. 560[A]

PART FOUR
Catalogues

Catalogue A

Manuscript Maps (1670 – 1870) in the Hudson's Bay Company Archives

Maps in this catalogue are referred to in the text by numbers with a superscript A. The text column in the catalogue records plate numbers (in italic) and the text page(s) on which a map is discussed. The reference column gives the HBCA manuscript catalogue number. Dimensions are given as vertical x horizontal. The printing conventions used in this catalogue are:

1. A title or description printed in italic reproduces as nearly as possible the inscription on a map. Example: 8[A] *A MAP of NORTH AMERICA with HUDSON'S BAY and STRAIGHTS, Anno 1748. R.W. Seale Sculp.*
2. A title or description printed in roman and enclosed in quotation marks is derived from other HBC archival sources. Example: 6[A] "a Plain of The Gulph call:d by ye Natives Wenipegg."
3. A title or description printed in roman without quotation marks has been assigned by the author for the purposes of the catalogue. Example: 7[A] Map of part of the Nelson River.
4. Where appropriate, the author has assigned titles or descriptions that include inscriptions as in 1. above or quotations as in 2. above. Examples: 1[A] Hudson Bay and Straits: *Made by Saml: Thornton at the Signe of the Platt in the Minories London Anno: 1709*; 17[A] "The Chesterfield by M. Norton": chart of a journey of discovery to the north of Churchill in the *Churchill* sloop. Some assigned titles may include both quotations and italics. Example: 189[A] *Peak View September 18th. 1824*, showing relative positions of the sources of rivers from "the top of a very lofty peak situated between the sources of River Malade & Salmon River."
5. Titles or parts of titles taken from the reverse side of maps are identified by the word "endorsed." Example: 16[A] Endorsed: "Moses Nortons Drt. of the Northern Parts of Hudsons Bay laid dwn on Indn. Information & brot. Home by him anno 1760."
6. When the date of the drafting of a map or the cartographer's name have been postulated on the basis of firm evidence but are not certainly known, the name and date are given in parentheses. The designation c.u. for "cartographer unknown" is used when the evidence does not permit even an informed guess regarding the identity of the cartographer.
7. As would be expected, some maps in this catalogue were not constructed to a particular scale. The nature of others, most noticeably the sketch maps of Peter Fidler, preclude indicating size as well.
8. In some entries, it is indicated that two or more sheets were used to make up a map or sketch or that two or more copies of a particular map were drawn. If, in these entries, only one scale or size is given, this applies to all the sheets or copies mentioned. For an example, see 8[A]. Where sizes and scales of the sheets or copies differ, these are shown for each sheet or copy. For an example, see 144[A].

NO.	TITLE OR DESCRIPTION	DATE	CARTOGRAPHER	REFERENCE	SCALE/SIZE	TEXT
1[A]	Hudson Bay and Straits: *Made by Saml: Thornton at the Signe of the Platt in the Minories London Anno: 1709*	1709	Samuel Thornton	G2/1; G2/2	1 inch: ca. 56 miles 80.5 x 64.5 cm	28,2, 274n27
2[A]	Sketch of rivers between Prince of Wales' Fort and the "Norther most Coper Mind," giving Indian names	(post–1719)	James Knight	G1/19	1 inch: ca. 40 miles 52 x 66.5 cm	30–1, 277n25
3[A]	Entrance of the Moose River, Moose Factory	(1740)	(George Howy)	G1/117	1 inch: 1 mile 76.3 x 61 cm	35
4[A]	Chart of "The old or Great River" to illustrate the *Eastmain* sloop log entries for 13, 14 and 15 July 1744	1744	Thomas Mitchell	B59/a/9, fo.4d	1 inch:1 mile 32.3 x 19.7 cm	34,4

NO.	TITLE OR DESCRIPTION	DATE	CARTOGRAPHER	REFERENCE	SCALE/SIZE	TEXT
5A	"A Plain of ye White Whale River which Layeth in Latt:d 55° = 35' North & about 02° = 00' Ey Longitude from Cape Jones"	1744	Thomas Mitchell	B59/a/9, fo.8d	3 inches:1 mile 32.3 x 19.7 cm	34
6A	"a Plain of The Gulph call:d by ye Natives Wenipegg"	1744	Thomas Mitchell	B59/a/9, fo.11	1 inch:ca. 3 miles 30.8 x 38.7 cm	34
7A	Map of part of the Nelson River	(1746)	(Joseph Robson)	G1/106 – G1/107	4 inches:1 mile 90 x 143 cm	36
8A	A MAP of NORTH AMERICA with HUDSON'S BAY and STRAIGHTS, Anno 1748. R.W. Seale Sculp.	1748	Richard William Seale, Engraver	G4/20a(7 copies) G4/20b(1 copy) G3/126(1 copy)	1 inch:ca. 200 miles 43 x 82.5 cm	33,3,261
9A	To The Honourable Governour The Deputy Governour and Committee of the Hudsons Bay company this chart of Arti-winipeck is humbly dedicated and Presented by William Coats 1749	1749	William Coats	G1/18	1 inch:ca. 3.3 miles 47 x 59 cm	34–5,5, 277n15
10A	"A Chart of the Entrance of Little Whale River & Gulf Hazard into the Artuirnipeg … Coates"	(1749)	(William Coats)	G1/15	1 inch:ca. 1.6 miles 48 x 59.7 cm	34–5
11A	Map of "Sr. Atwel's Lake or Artiwinepeck" (Richmond Gulf)	(1749–1750)	(Williams Coats)	G1/16	1 inch:2 miles 54 x 64.5 cm	34–5
12A	Map of Hudson Bay and Straits, Labrador, Newfoundland, and New England	(1750)	(James Mynd)	G2/6	1 inch:ca.56 miles 96.5 x 82 cm	
13A	"A Map from Diggs to Little Whale River. Coates"	(1751)	(William Coats)	G1/14	1 inch:ca.20 miles 63.5 x 51 cm	34–5
14A	"A Map of Artuirnipeg or Gulf of Richmond … W. Coates"	(1751)	(William Coats)	G1/17	1 inch:ca. 3.3 miles 52 x 64.7 cm	34–5
15A	"a Ruff Draft of ye River so far as it Concerns ye fishery in my future Intentions"	1759	John Potts	B182/a/11, fo.26	1 inch:ca. 240 feet 11.7 x 19.7 cm	34–5
16A	Endorsed: "Moses Nortons Drt. of the Northern Parts of Hudsons Bay laid dwn on Indn. Information & brot. Home by him anno 1760"	ca.1760	Moses Norton	G2/8	1 inch:ca. 30 miles, but scale varies 64.7 x 88.7 cm	41–2,8
17A	"The Chesterfield by M. Norton": chart of a journey of discovery to the north of Churchill in the *Churchill* sloop	(1762)	(Moses Norton and William Christopher)	G1/161	1 inch:ca. 8.4 miles 48.5 x 62.3 cm	41
18A	A MAP of BOWDEN,s & RANKEN,s INLETT'S. THE FORMER Descovered by Mr, NORTON & Christophar, in ye year 1761 & 1762, the latter by Mr. Johnston in the year 1764	(1765 or later, possibly 1774)	(Samuel Hearne)	G2/9	1 inch:ca. 12 miles 40.7 x 73 cm	42,7

NO.	TITLE OR DESCRIPTION	DATE	CARTOGRAPHER	REFERENCE	SCALE/SIZE	TEXT
19A	Endorsed: "Captain Mea'to'na'bee & I'dot'ly'a'zees, Draught. CR"	(1767–1768)	(Moses Norton)	G2/27	1 inch:ca. 20 miles 69.3 x 139.5	42–3
20A	*A Map of part of the Inland Country to the Nh Wt of PRINCE of WALES'S Fort Hs; By, Humbly Inscribed to the Govnr Depy, Govnr and Committee of the Honble, Hudns, By Compy By their Honrs, moste obediant humble servant.Sam,l Hearne; 1772*	1772	Samuel Hearne	G2/10	1 inch:ca. 30 miles 76.7 x 82.5 cm	43–4,9
21A	*A PLAN of Part of HUDSON'S-BAY & RIVERS communicating With the PRINCIPAL SETTLEMENTS*	(1772)	(Andrew Graham)	G2/15	1 inch:60 miles 54.7 x 76.2 cm	38,40, 120
22A	A sketch of Hayes's River Mouth. Endorsed: "14 No² HB Hase River entrance"	(1772–1774)	(Andrew Graham)	G2/16	1 inch:ca. 0.5 mile 49.5 x 61.5 cm	40
23A	*A Plan Of Part Of Hudson's-Bay, & Rivers, Communicating with York Fort & Severn*	(1774)	(Andrew Graham)	G2/17	1 inch:40 miles 75 x 101.5 cm	38,40–1, 64,6
24A	*A Chart of Hudson's Bay and Straits*	(1774)	(Andrew Graham)	G2/14	1 inch:90 miles 53.5 x 74 cm	40, 120
25A	*A MAP of some of the principal LAKES RIVER'S &c leading from YF to Basquiaw; Humbly inscribed to the Governor, Depty Governor, and Committee of the Honble, HUDSON'S BAY COMPANY, by their Honour's most obedient and very humble servent, S. Hearne*	(1775)	Samuel Hearne	G1/20	1 inch:ca. 15 miles 55.3 x 77 cm	44–5
26A	Various marginal sketches illustrating "A Journal of a Journey from Henley House to Meshippicoot Bay in Lake Superior," 26 May 1776 to 1 July 1776	1776	Edward Jarvis	B86/a/29, fos.16d,17, 19d,20d,21, 21d,22d,25		46
27A	*A CHART of Rivers and Lakes Falling into HUDSONS BAY According to a Survey taken in the years 1778 & 9 By Philip Turnor* (two copies)	1779	Philip Turnor	G1/21, G1/22	1 inch:ca. 20 miles 55.3 x 76.3 cm	20,50, 279n14
28A	A Chart of Rivers and Lakes between Albany Fort and Gloucester House as taken in the year 1780 by Philip Turnor	1780	Philip Turnor	G1/23	1 inch:ca. 9.5 miles 47.7 x 75 cm	50–1, 54,11
29A	*To the GOVERNOR The Deputy Governor-And COMMITTEE of The HONOURABLE-HUDSONS BAY COMPANY. This CHART of HUDSONS-STRAITS, is most humbly Dedicated and Presented by Their Honours most obedient & faithfull servant Iohn Marley. 1781*	1781	John Marley	G2/31	1 inch:ca. 11 miles 96.5 x 137 cm	49, 278n1, 280n25

Manuscript Maps in the Hudson's Bay Company Archives

NO.	TITLE OR DESCRIPTION	DATE	CARTOGRAPHER	REFERENCE	SCALE/SIZE	TEXT
30A	*Chart of part of Hudsons Bay and Rivers and Lakes falling into it by Philip Turnor*	(1782–3)	Philip Turnor	G1/1	1 inch:ca. 17 miles 50.4 x 73 cm	21,50–1
31A	*A Draught from Lake St Joseph to Lake Saul by James Sutherland*	1786	James Sutherland	B78/a/14, fo.1	1 inch:ca. 11 miles 51 x 37.5 cm	55–6
32A	*A Chart of the NE Coast of Hudsons Bay. by George Donald*	1786	George Donald	A11/45, fos.51d–52	1 inch:ca. 36 miles 47 x 37 cm	280n25
33A	*A PLAN of the Banks of the River in Front of YORK FORT Hudson Bay with the LAND and WOODS a short regular distance around. Also of the Proposed Improvement. Humbly submitted, to The Governor Deputy Governor and Committee, of the Honorable Hudson Bay Company: by their very Dutiful, and faithful Servant, Joseph Colen*	1786–7	Joseph Colen	G1/111	1 inch:ca. 140 feet 54 x 75.5 cm	
34A	*A Chart of Rivers and Lakes above York Fort falling into Hudsons Bay According to an Actual Survey taken by Philip Turnor 1778 &9 And of Rivers and Lakes above Churchill Fort Joining the Same taken from a Journal kept by Malcolm Ross and laid down by Philip Turnor*	1787–8	Philip Turnor	G2/11	1 inch:ca. 20 miles 63.5 x 93.5 cm	51–2, 12, 279n14
35A	*A Chart of the Rivers & Lakes from Henley House to Lake La Puew, Jas H: Sculpt;*	1778	James Hudson	B86/a/42, fo.2	1 inch: ca. 10 miles 38 x 50.7 cm	56
36A	*A CHART of RIVERS AND LAKES Communicating with ALBANY RIVER*	(1789)	(John Hodgson)	G1/2	1 inch:ca. 10 miles 53.3 x 75.5 cm	56,13, 279n17
37A	*"A Map of Hudsons Bay and interior Westerly particularly above Albany 1791"*	(1791)	(Edward Jarvis and Donald McKay)	G1/13	1 inch:ca. 65 miles 54 x 75 cm	57–8, 120,14
38A	*AN ACCURATE MAP of the TERRITORIES of the HUDSON'S BAY COMPANY in NORTH AMERICA*	(1791)	(John Hodgson)	G2/28	1 inch:ca. 70 miles 75 x 97.7 cm	58–9, 120,15
39A	*Eight sketch maps from Methye Portage to McLeod post on Clearwater River*	1791	Peter Fidler	E3/1, fos.4–8d		53
40A	*Two sketch maps of the west end of Lake Athabasca*	1791	Peter Fidler	E3/1, fos.10d,11		53
41A	*Two sketch maps at the mouth of the Slave River, Great Slave Lake*	1791	Peter Fidler	E3/1, fos.15d,17		53
42A	*Shew-ditha-da's description of Great Slave Lake, Northern Indian Lake, and Esquimay River to Hudson Bay*	1791	Philip Turnor	B9/a/3,p.83	1 inch:ca. 175 miles 7.5 x 17.8 cm	53

NO.	TITLE OR DESCRIPTION	DATE	CARTOGRAPHER	REFERENCE	SCALE/SIZE	TEXT
43A	Sixteen sketch maps of most of the shoreline of Lake Athabasca	1791	Peter Fidler	E3/1, fos.20–24		53
44A	Six sketch maps of the south shore of Great Slave Lake	1791	Peter Fidler	E3/1, fos.30,31,33		53
45A	Eight sketch maps from Clearwater River to Ile a La Crosse 1792	1792	Peter Fidler	E3/1, fos.50–55		54
46A	Fifty-three sketch maps from Ile a la Crosse to Cumberland House	1792	Peter Fidler	E3/1, fos.56–58		54,16
47A	*Chart of Lakes and Rivers in North America by Philip Turnor those Shaded are from Actual Survey's the others from Canadian and Indian information*	1792	Philip Turnor	G2/13	1 inch:ca. 40 miles 67.3 x 100.3 cm	53–4
48A	Twenty-seven sketch maps from York Factory, along Hill River, Knee Lake, and Playgreen Lake to Lake Winnipeg	1792	Peter Fidler	E3/1, fos.70–80d		61–2
49A	Five sketch maps of the north shore of Lake Winnipeg to mouth of the Saskatchewan River	1792	Peter Fidler	E3/1, fos.81–82		61–2
50A	Nine sketch maps from Grand Rapids, Saskatchewan River to Cumberland House	1792	Peter Fidler	E3/1, fos.83–87d		61–2
51A	*A,Draught Between Osnaburgh and red Lake by,John,Best 1792–*	1792	John Best	B177/a/2, fo.1d	1 inch:ca. 15 miles 31.7 x 19.7 cm	57
52A	Five sketch maps from area between Buckingham House and the Rocky Mountains	1792–3	Peter Fidler	E3/2, fos.3d,4,8,34		61–2
53A	Sketch map of "Egg Lake" (Lac Sante) and Saskatchewan River	1793	Peter Fidler	E3/2, fo.36d	1 inch:ca. 2 miles 9 x 12.7 cm	61–2
54A	Two sketch maps of Lower Hayes River above York Factory	1793	Peter Fidler	E3/1, fo.98		61
55A	*A draught of a Journey from Osnaburgh house to Sturgeon Lake By Jacob Corrigal 1794*	1794	Jacob Corrigal	G1/24	1 inch:ca.3.6 miles 59.7 x 38.7 cm	57
56A	Nelson River, Hayes River, and communications through Lake Winnipeg to, and along, Saskatchewan River, including part of Churchill River	(1794)	(David Thompson)	G2/18	1 inch:ca. 8 miles 118 x 227.3 cm	59–60, 17

Manuscript Maps in the Hudson's Bay Company Archives

NO.	TITLE OR DESCRIPTION	DATE	CARTOGRAPHER	REFERENCE		TEXT
57A	To the Honourable the Governor, Deputy Governor, And Committee of the Hudson's Bay Company This Map of Hudson's Bay and the Rivers and Lakes Between the Atlantick and Pacifick Oceans Is most humbly Inscribed By their most obedient & dutiful Servant, Philip Turnor Tomkins scripsit Foster Lane, 1794 E. Days del. Draughtsman to His R.H. The Duke of York	1794	Philip Turnor	G2/32	1 inch:ca. 28 miles 193 x 259 cm	49–50, 53–4, 59–60, 120,18
58A	Seven sketch maps, Cedar Lake, Lake Winnipegosis, Swan Lake, and Lac de l'eau Claire	1795	Peter Fidler	E3/2, fos.39d,40, 41,42,45d		62
59A	A map of Swan River, Swan Lake area	1795 or 6	(Peter Fidler)	G2/19	1 inch:ca. 8 miles 45.7 x 69.8 cm	62
60A	Rough sketch of part of a lake, from Wegg's House Journal	(1795–6)	(William Sinclair)	B228/a/1, fo.2	19 x 32.3 cm	
61A	Sketch map of rivers and lakes in the Churchill area	(1795–6)	(William Sinclair)	B228/a/1, fo.1	1 inch:ca. 35 miles 19 x 30.5 cm	
62A	Sketch map of part of *Cumberland House Lake 22nd April 1798* showing position of Cumberland House and old Cumberland House	1798	Peter Fidler	E3/2, fo.103	21 x 24.2 cm	62, 280n5
63A	*A Sketch of the Track from Nelson House, to Harpurs House in 3 point Lake with the Burnt Wood River & Lake – in a small Canoe 1798 by Mr. Flew* (five sketch maps)	1798	Peter Fidler	E3/3, fo.48d	36.8 x 24.1 cm	62, 280n5
64A	Sketch map showing position of Bolsover House on Barren Ground Lake	1799	Peter Fidler	B104/a/1, fo.7d	1 inch:ca. 2.5 miles 3.8 x 7.7 cm	62
65A	Sketch map showing position of Bolsover House on Barren Ground Lake	1799	Peter Fidler	E3/2, fo.50	1 inch:ca. 1.6 mile 5 x 15.3 cm	62
66A	Sketch map of small lake in upper Beaver River valley, on route west to Red Deer's Lake (Lac la Biche)	1799	Peter Fidler	E3/2, fo.56	2.5 x 3.8 cm	62
67A	Sketch map of lakes, rivers, and portages from Beaver River to Red Deer's Lake (Lac la Biche)	1799	Peter Fidler	E3/2, fo.56	5.7 x 17.8 cm	62
68A	Sketch map of Red Deer's Lake (Lac la Biche)	1800	Peter Fidler	B104/a/1, fo.36d	1 inch:ca. 3.1 miles 14.6 x 19 cm	62
69A	Two sketch maps of Red Deer's Lake (Lac la Biche) with the location of Greenwich House	1800	Peter Fidler	E3/2, fo.56d	1 inch:ca. 2.5 miles 22.8 x 22.8 cm	62

NO.	TITLE OR DESCRIPTION	DATE	CARTOGRAPHER	REFERENCE	SCALE/SIZE	TEXT
70^A	Sketch map of Red Deer's Lake (Lac la Biche) to Athapescow (Athabasca) River	1800	Peter Fidler	E3/2, fo.56d	1 inch:ca. 20 miles 5.7 x 8.3 cm	62
71^A	Sketch map of meanders in Red Deer's River, with portage	1800	Peter Fidler	E3/2, fo.57	1 inch:ca. 0.3 mile 2.5 x 1.8 cm	62
72^A	Sketch map of Little Slave (Lesser Slave) Lake, east to Athapescow (Athabasca) River	1800	Peter Fidler	E3/2, fo.59d	1 inch:ca. 45 miles 3.3 x 10.2 cm	62
73^A	Sketch map of Little Slave (Lesser Slave) Lake	1800	Peter Fidler	B104/a/1, fo.39	1 inch:ca. 17 miles 15.3 x 12.7 cm	62
74^A	Sketch map of Little Slave (Lesser Slave) Lake	1800	Peter Fidler	E3/2, fo.59d	1 inch:ca. 14 miles 8.3 x 12 cm	62
75^A	Three sketch maps of lakes, rivers, and portages from the Beaver River to the North Saskatchewan River at Buckingham House	1800	Peter Fidler	E3/2, fos.61,61d,62		62
76^A	Hills on north side of the South Saskatchewan River	1800	Peter Fidler	E3/2, fo.65d	1.8 x 6.8 cm	63
77^A	An Indian map of the Different Tribes that inhabit on the East & west side of the Rocky Mountains with all the rivers & other remarkable places, also the Number of tents. &c. Drawn by the Feathers or ac ko mok ki-a Blackfoot chief- 7th Feby. 1801 – reduced ¼ from the Original Size – by Peter Fidler	1801	Peter Fidler	G1/25	37.2 x 47 cm	63–4,19
78^A	Copy of an Indian map *Drawn by the Feathers or ac ko mok ki a Blackfoot chief 7 Feby. 1801.*	1801	Peter Fidler	E3/2, fos.106d–107	37.5 x 48.3 cm	63
79^A	Untitled map of Churchill River "By Mr Stayner 1801" from Churchill Factory to Frog Portage	1801	Peter Fidler	B42/b/46, fo.39d	1 inch:ca. 19 miles 38.1 x 54.6 cm	
80^A	Indian map showing location of Indian tents in upper Missouri River region and west of the Rocky Mountains	1802	Peter Fidler	E3/2, fos.105d–106	37.5 x 48.3	63–4
81^A	Indian map drawn originally by the Feathers or Ac ko mok ki, a Blackfoot chief	1802	Peter Fidler	B39/a/2, fo.93	20 x 32 cm	63
82^A	Indian map *Drawn by the Feathers or ak ko mok ki a Black foot Chief 1802.*	1802	Peter Fidler	E3/2, fo.104	24.1 x 37.5 cm	63
83^A	Indian map drawn originally by Ak ko wee ak, a Blackfoot Indian	1802	Peter Fidler	B39/a/2, fo.92d	19 x 32 cm	63

Manuscript Maps in the Hudson's Bay Company Archives

NO.	TITLE OR DESCRIPTION	DATE	CARTOGRAPHER	REFERENCE	SCALE/SIZE	TEXT
84A	Indian map *Drawn by ak ko wee ak a Black Foot Indian 1802*	1802	Peter Fidler	E3/2, fo.103d	24.1 x 37.5 cm	63
85A	Indian map drawn originally by Ki oo cus or the Little Bear, a Blackfoot Chief	1802	Peter Fidler	B39/a/2, fos.85d–86	32 x 40 cm	63
86A	Indian map *Drawn by Ki oo cus – or the Little Bear a Blackfoot Chief 1802.*	1802	Peter Fidler	E3/2, fos.104d–105	48.3 x 37.5 cm	63,20
87A	*Plan of Eastmain Plantation*	(post 1802)	James Russell	G1/97	1 inch:120 feet 34.6 x 51.6 cm	
88A	Northwest end of Lake Athabasca showing route taken out of the lake from Nottingham House by Thomas Swain, 16 September 1803	1803	Thomas Swain	B41/a/2, fo.12d	1 inch:ca. 5 miles 12.7 x 19.3 cm	
89A	*To the Governor, Deputy Governor, and Committee of the Honourable Hudson's Bay Company This Map of the South part of Hudson's Bay, is humbly dedicated by their Honors most Obedt. and humble Servt. Henry Hanwell 1803*	1803	Henry Hanwell Sr	G2/20	1 inch:5 miles 65.5 x 83 cm	72,28
90A	Salt deposits thirty miles up Salt River from its confluence with the Slave River	1803–5	Peter Fidler	B39/a/5b, (inside front cover)	1 inch:ca. 8 miles 17.8 x 20.3 cm	64
91A	*Part of Charlton Island*	1805	Henry Hanwell Sr	A39/1, fo.46	1 inch:1 mile 32.5 x 39.6 cm	72
92A	PART OF CHARLTON ISLAND SUPPOSED TO BE ABOUT 24 LEAGUES IN CIRCUMFERENCE	(1805)	(James Winter Lake)	A1/220, fo.60	1 inch:1 mile 31.5 x 40.5 cm	
	Attached map: Southern part of Hudson Bay and also James Bay				1 inch:ca. 110 miles 14.5 x 16 cm	
93A	A SKETCH OF MAKENZIES TRACK ALONG THE RIVERS UNIJAH, [Peace] AND TACOUTCHE, [Columbia] ACROSS THE ROCKY MOUNTAINS TO THE PACIFIC OCEAN J:W:L:	(1805)	(James Winter Lake)	A1/220, fo.1d	1 inch:ca. 25 miles 37.2 x 37.5 cm	
94A	SKETCH OF A MAP, SHEWING THE EXTENDED TRACK OF TRADE OF THE NORTH WEST COMPANY, FROM LAKE WINIPIQUE, TO THE PACIFIC OCEAN. J:W:L:	(1805)	(James Winter Lake)	A1/220, fo.36d	1 inch:ca. 140 miles 32 x 66.8 cm	
	Attached Map: THE ENTRANCE OF COLUMBIA RIVER FROM THE NORTH PACIFIC OCEAN DISCOVERED BY CAPT. VANCOUVER. JWL.				12.5 x 23.3 cm	

NO.	TITLE OR DESCRIPTION	DATE	CARTOGRAPHER	REFERENCE	SCALE/SIZE	TEXT
95A	Map across Rocky Mountains to Columbia River from Athabasca River to Red Deer River *Drawn by Jean Findley 1806*	1806	Peter Fidler	E3/4, fos.16d–17	1 inch:ca. 10 miles 37.5 x 49.5 cm	65–6
96A	Indian map *Drawn by Cha chay pay way ti May 1806*	1806	Peter Fidler	E3/4, fo.13d	1 inch:ca. 22 miles 37.5 x 24 cm	66,22
97A	*a Sketch of the Road from the Sandy or Northern Indian Lake, to where Mr John Charles built in 1806 – drawn by himself.*	1806	Peter Fidler	E3/3, fo.49	20.3 x 24.1 cm	66
98A	South part of Moose Lake	1806–7	Peter Fidler	B49/a/32b, fo.35	1 inch:ca. 0.4 mile 15.3 x 15.3 cm	
99A	Indian map *Drawn by oo ke mow a thin 1807* from Painted Stone, Hairy Lake, along Wop poo min na kek River, through Gull Lake	1807	Peter Fidler	E3/4, fo.11d	10 x 24 cm	65
100A	147 sketch maps from Frog Portage to Lake Athabasca, and return to Fort Churchill	1807	Peter Fidler	E3/3, fos.2d–41		64
101A	*This Sketch drawn by a Jepewyan at Churchill 17th Augt. 1807. from mouth of Seal river to the Southern Indian Lake – the same road Mr Stayner passed. 1794* (in two sketches)	1807	Peter Fidler	E3/3, fo.49	15.8 x 24.1 cm	65
102A	Map of Trout Lake from Thomas Bunn's journal	1807	Peter Fidler	E3/3, fo.47d	1.3 x 3.8 cm	66
103A	Map accompanying "These distances from Severn to York Factory … determined by Mr Thos. Thomas Senr. …", showing two rivers entering Hudson Bay	1807	Peter Fidler	E3/3, fo.47d	1.3 x 3.8 cm	66
104A	Two sketch maps by Thomas McNab of rivers and lakes in the area of Island Lake, Cross Lake, Red Sucker Lake, and Shamattawa River	1807	Peter Fidler	E3/3, fo.49d	36.8 x 24.1 cm	66
105A	*Sketch of the Road up Jack River, to Jack Lake by Mr Thomas Swain. 1807* (in five sketches)	1807	Peter Fidler	E3/3, fo.50		66
106A	Map to accompany "Journal of a journey … from Cumberland House, by the Frog Portage by the Grey Deers Lake & to the Athapascow thro' Lake Wollaston, & from … Deers River, down the … Churchill River to Fort Prince of Wales."	1808	Peter Fidler	G2/21	1 inch:ca. 7 miles 153 x 192.5 cm	65
107A	Eighteen sketch maps from the mouth of Red River along the east shore of Lake Winnipeg to Playgreen Lake	1808	Peter Fidler	E3/3, fos.58d–64		65

Manuscript Maps in the Hudson's Bay Company Archives

Catalogue A

NO.	TITLE OR DESCRIPTION	DATE	CARTOGRAPHER	REFERENCE	SCALE/SIZE	TEXT
108A	Two sketches of part of the shore of Lake Winnipeg from Mossy Point to about 15 miles beyond Poplar River	(1808)	Peter Fidler	B49/a/32b, fo.24		65
109A	Cedar, Winnipegosis, Dauphin, and Manitoba Lakes with St. Martins Lake connection to Lake Winnipeg *Drawn by Mr John McDonald, late of NW Co. 22nd Jany. 1808*	1808	Peter Fidler	E3/3, fo.48	1 inch:ca. 24 miles 35.5 x 24.1 cm	65
110A	Five sketches of *Red River & its communications Mr John McDonald AMK & Co-1808*	1808	Peter Fidler	E3/4, fo.18	1 inch:25 miles 33 x 24.7 cm	65
111A	Sketch of Hayes River area drawn by *Hugh Sabbeston 1808*	1808	Peter Fidler	E3/4, fo.11d	7.5 x 20.7 cm	66
112A	Map of Poplar River from Lake Winnipeg to Little Grand Rapid, based on George Spence	1808	Peter Fidler	E3/4, fo.18d	12 x 8 cm	66
113A	Route between Osnaburgh House, via Cat Lake to Sandy Lake, based on Andrew Flett	1808	Peter Fidler	E3/4, fo.18d	16 x 10.2 cm	66
114A	Six sketches of *Lake Winnipeg from Dogs Head to the Grand Rapid*	(1808)	Peter Fidler	E3/3, fos.66–66d		65,21
115A	Indian map *Drawn by Thoo ool del 29th April 1809* of tracks from Lake Athabasca, Wollaston, and Reindeer Lake to the sea coast at Churchill and to the north	1809	Peter Fidler	E3/4, fo.16	1 inch:ca. 36 miles 26.3 x 24.7 cm	65
116A	Three sketch maps *All Drawn by Chee hooze.i.az za 21st. April 1809* of rivers and lakes in Reindeer Lake area	1809	Peter Fidler	E3/4, fo.15d		65
117A	*Clouston Dicksons sketch 1809* of Egg Lake and Island Lake toward Methye Portage	1809	Peter Fidler	E3/4, fo.17d	1 inch:ca. 15 miles 15.7 x 24.7 cm	66
118A	Egg Lake, toward Methye Portage *Sketched by Mr Jno Charles 1809*	1809	Peter Fidler	E3/4, fo.17d	7.6 x 14 cm	66
119A	Sketch map of Island Lake and Egg Lake, toward Methye Portage	1809	Peter Fidler	E3/4, fo.14d	10.2 x 9.5 cm	
120A	*These two sketches by Chynk,y,es,cum a Bungee Chief 29th May 1809* of area from North Saskatchewan, Setting River, and across to the Beaver River	1809	Peter Fidler	E3/4, fo.15	1 inch:ca. 9 miles 21 x 24.7 cm	65
121A	*These two sketches by Chynk,y,es,cum a Bungee Chief 29th May 1809* of area from Little Slave Lake, Athabasca River to east edge of Stony Mountain	1809	Peter Fidler	E3/4, fo.15	1 inch:ca. 32 miles 20 x 24.7 cm	65

NO.	TITLE OR DESCRIPTION	DATE	CARTOGRAPHER	REFERENCE	SCALE/SIZE	TEXT
122A	Sketch of area from North Saskatchewan River, at Buckingham House, north to Beaver River and Cold Lake	1809	Peter Fidler	E3/4, fo.14	1 inch:ca. 6 miles 13 x 24.7 cm	
123A	Sketch of North Saskatchewan River from old Edmonton House to Acton House, and the area of the upper Red Deer and upper Battle River	1809	Peter Fidler	E3/4, fo.14	1 inch:ca. 12 miles 16.5 x 24.7 cm	
124A	Thirty-nine sketches to illustrate journey from Cumberland House via the Nelson River to York Fort	1809	Peter Fidler	E3/4, fos.2d,3d–10		65
125A	The plantation at Flamboro House, Nelson River entrance	1809	Peter Fidler	E3/4, fo.10	1 inch:ca. 85 feet 6.3 x 7.6 cm	
126A	*Sketch of Pine River from Moose to the Cross Lake drawn by Ah chap pee Bungee boys son 12th June 1809*	1809	Peter Fidler	E3/4, fo.11	1 inch:ca. 3 miles 28 x 24.7 cm	65
127A	Two sketch maps *Drawn by Tow we kish e quib a Nelson River Indian 24th June 1809* of the Nelson to Churchill rivers, in Limestone River area	1809	Peter Fidler	E3/4, fo.11d	1 inch:ca. 4 miles 30.5 x 24.7 cm	65
128A	Sketch *Drawn by See seep 24th June 1809* of rivers and lakes from Nelson River at Leaf Portage to Steel River	1809	Peter Fidler	E3/4, fo.12	1 inch:ca. 4 miles 24.7 x 37.7 cm	65
129A	Sketch *Drawn by Ay kay mis- a Severn Ind. 24th June 1809* from Hayes River, through Sturgeon Lake, to Severn Fort	1809	Peter Fidler	E3/4, fo.13	1 inch:ca. 30 miles 15.7 x 24.7 cm	65
130A	*Sketch of the Track from Split Lake by Foxes Lake & down to the mouth of Hill river also two other sketches of the Indian Canoe Track from bottom of Kettle Falls, into foxes Lake & up the small river above the Moose Nose, up to Foxes Lake – all Drawn by Mr. Cook who has passed them all in Canoes – 24th June 1809 –*	1809	Peter Fidler	E3/4, fo.12d	37.5 x 24.7 cm	66
131A	*Split & As se an Lakes Drawn by Mr Cook 24th June 1809*	1809	Peter Fidler	E3/4, fo.13	1 inch:ca. 3 miles 22 x 24.7 cm	66
132A	Sketch map of Canoe Building and Vermilion Lake, Chatham House	1809	Peter Fidler	E3/4, fo.14	1 inch:ca. 4 miles 11.5 x 24.7 cm	66
133A	Sketch of Hudson Bay shore from Cape Churchill to Knight's Hill	1809	Peter Fidler	E3/3, fo.41	5.1 x 14 cm	66
134A	*Sketch of the Rocks about Churchill &c.* extending from Cape Churchill to North River	1809	Peter Fidler	E3/3, fo.65	15.2 x 24.1 cm	66

Manuscript Maps in the Hudson's Bay Company Archives

NO.	TITLE OR DESCRIPTION	DATE	CARTOGRAPHER	REFERENCE	SCALE/SIZE	TEXT
135A	*Iskemo sketch* of coast of Hudson Bay from Churchill river north past Chesterfield Inlet, in 2 sections *Drawn by Nay hek til lok an Iskemo 40 years of age 8th July 1809*	1809	Peter Fidler	E3/4, fo.16	1 inch:ca. 20 miles 19 x 24.7 cm	66
136A	Rough pencil sketch of coast of Hudson Bay from Churchill River north past Chesterfield Inlet, in three sections	1809	(Nayhektil lok)	F3/2, fos.108d–109	1 inch:ca. 15 miles 30.5 x 37.5 cm	66
137A	Sketch map of Hudson Bay shore from Nelson River to Churchill River, with the Owl and Broad Rivers, and the Little Churchill River. *This Sketch Drawn by a YF Indian when I passed Owl River 29th July 1809.*	1809	Peter Fidler	E3/3, fo.65d	1 inch:ca. 15 miles 26.7 x 24.1 cm	66
138A	Sketch map of Hudson Bay shore at mouth of Owl River	1809	Peter Fidler	E3/3, fo.65d	2.5 x 6.3 cm	66
139A	Two sketches from York Factory to Severn Factory *by Mr Thos. Thomas Chief at Severn walking along shore at high water mark*	1809	Peter Fidler	E3/4, fo.19d		66
140A	Sketch map of east end of Lake Athabasca along the Cree River to Cree Lake and south to Churchill River *Drawn by Cot.aw.ney.yaz.zah. a young man. Jepewyan Feby 17th. 1810*	1810	Peter Fidler	E3/4, fo.14d	1 inch:ca. 20 miles 37.4 x 24.7 cm	66
141A	Sketch map *by Ageenah 4th June 1810* from Wollaston and Reindeer Lakes, the Cochrane River, and the portage east to the "Seal river Track"	1810	Peter Fidler	E3/4, fo.17d	1 inch:ca. 83 miles 12.5 x 17.5 cm	66
142A	Sketch of upper Clearwater River, Methye Portage and Peter Pond, and Churchill Lakes, probably by Ageenah	1810	Peter Fidler	E3/4, fo.17d	1 inch:ca. 10 miles 25.5 x 14 cm	66
143A	Sketch map *by Ageena* of Arctic coast from the Mackenzie River past Coppermine River including part of Great Bear Lake	1810	Peter Fidler	E3/4, fo.17	1 inch:ca. 175 miles 10 x 16 cm	66
144A	"Nelson River, Lake Winnipeg & 3 Sheets by P. Fidler 1810"	1810	Peter Fidler	G1/28	1 inch:ca. 8 miles 1. 51.5 x 109.7 cm 2. 76 x 53.3 cm 3. 53 x 109.7 cm	65
145A	Sketch map *Drawn by Old Pumbles 10th June 1810* of two routes via Beaver River and Waterhen River from Green Lake to Cold Lake	1810	Peter Fidler	E3/4, fo.11	1 inch:ca. 16 miles 9.5 x 24.7 cm	65

NO.	TITLE OR DESCRIPTION	DATE	CARTOGRAPHER	REFERENCE	SCALE/SIZE	TEXT
146A	"Map of Hayes & Nelson River (unknown) abt. 1811"	(c. 1811)	c.u.	G2/22	1 inch:ca. 8 miles 86 x 77 cm	
147A	"Chart of Lake Winipic (unknown) abt. 1811"	(c. 1811)	c.u.	G1/32	1 inch:ca. 6 miles 48.7 x 96 cm	
148A	Map attached to deed of June 12, 1811 conveying Assiniboia to the Earl of Selkirk.	1811	(Aaron Arrowsmith)	E8/1, fo.6 E6/1, fo.9	1 inch:ca. 37 miles 43.2 x 49.5 cm	67
149A	Entrances to Nelson and Hayes Rivers showing position of buildings and of anchorages for ships, 1811–12	(1811–12)	(William Auld)	B42/b/57 (inside front cover)	1 inch:ca. 0.3 miles 32 x 20 cm	72
150A	Map of south end of James Bay, showing Moose Fort to Albany Fort, to Charlton Island and Eastmain Factory	1812	Henry Hanwell Jr	G1/162	1 inch:ca. 9 miles 29 x 40 cm	72
151A	*STRUTTON SOUND Surveyed by Henry Hanwell. Junr. in the Winter. 1811 & 1812*	1812	Henry Hanwell Jr	G1/162	1 inch:ca. 170 yards 25.2 x 40 cm	72
152A	"Nelson River Chart of" entrance to	1812	(William Hillier)	G1/3	1 inch:1 mile 68.6 x 90.5 cm	72
153A	"Hudsons Bay Chart of Scale of 1 Inch to a Mile abt 1811." Six sheets showing the Nelson River from the mouth to Split Lake	1812	(Joseph Howse and William Auld)	G1/31	1 inch:1 mile 1. 59 x 47.3 cm 2. 47.3 x 59 cm 3. 59 x 47.3 cm 4. 47.3 x 59 cm 5. 47.3 x 59 cm 6. 47.3 x 59 cm	68
154A	"Nelson River Chart of Entrance to"	(1813)	(William Hillier)	G1/163	1 inch:ca. 2 miles 30 x 50.7 cm	72
155A	"Six blank maps" of much of interior, inland from York and Churchill	(post–1814)	c.u.	G1/7, G1/8, G1/9, G1/10, G1/11, G1/12	1 inch:ca. 15 miles 58.5 x 58 cm	
156A	"Ground Plan of Carlton House" and the ground adjacent	(1815)	c.u.	G1/76	1 inch:ca. 150 feet 34.5 x 50 cm	
157A	"Chart of Mr Vincents Journey from Albany to Henley from thence to Capenocoggamy, and so to New Brunswick and Moose Factory Performed in the summer of 1815"	(1815)	(Thomas Vincent)	G1/39	1 inch:ca. 6 miles 75.6 x 54.7 cm	

Manuscript Maps in the Hudson's Bay Company Archives

NO.	TITLE OR DESCRIPTION	DATE	CARTOGRAPHER	REFERENCE	SCALE/SIZE	TEXT
158A	Map of Severn District laid down from Indian information by Js. Swain 1815	1815	James Swain Sr	G1/35	1 inch:ca. 8.5 miles 76.8 x 55 cm	71,26
159A	Map in two sheets: Map of Severn District laid down from Indian information by James Swain. Augst 1815	1815	James Swain Sr	G1/33	1 inch:ca. 7 miles 54.7 x 76 cm	71
	Map of Severn District taken down from Indian Information by James Swain 1815			G1/34	1 inch:ca. 7 miles 54.7 x 76 cm	
160A	"Severn Map in 2 sheets 1815"	(1815)	c.u.	G1/36	1 inch:ca. 10 miles 1. 60.4 x 48 cm 2. 60 x 48 cm	71
161A	Sketch of York District	(1815)	(William Cook)	G1/26	1 inch:ca. 15 miles 48.3 x 70 cm	71
162A	Sketch of Cumberland District	1815	(Alexander Kennedy)	G1/30 Map 1	1 inch:ca. 12 miles 37.5 x 47 cm	71
163A	Map of Cumberland House Lake & Beaver Lake	1815	(Alexander Kennedy)	G1/30 Map 2	1 inch:ca. 2 miles 37.5 x 47 cm	71
164A	A Map of the Journeys of James Sutherland and Richard Perkins in the Years 1784, 1786 and 1790 extracted from their Journals and laid down by Peter Fidler, March 1815	1815	Peter Fidler	G1/37	1 inch:ca. 10 miles 66.5 x 101 cm	67, 279n17
165A	Richard Perkin's track from Cat Lake to Severn Factory in 1789 laid down by Peter Fidler April 1815	1815	Peter Fidler	G1/38	1 inch:ca. 10 miles 101 x 66 cm	57,67
166A	Map of Gloucester House District	1816	c.u.	B78/e/3	1 inch:ca. 12 miles 39.3 x 50 cm	71
167A	Rivers and lakes in Osnaburgh District	1816	c.u.	B155/e/4	1 inch:ca. 25 miles 34.3 x 48.3 cm	71
168A	A Plan of Surveys in the Neighbourhood of Mistassinnie Lake by Jas. Clouston and Jas. Robertson	1816	James Clouston and James Robertson	B133/e/1	1 inch:ca. 11 miles 53.7 x 51.7 cm	69
169A	A Chart of the Survey of the Eastermost Branch of Great Whale River, Eastmain Coast, Hudson's Bay; performed by Geoe. Atkinson Senr. in June 1816	1816	(George Atkinson II)	G1/40	1 inch:8 miles 52.4 x 66 cm	69
170A	Site of Neoskweskau Fort and surrounding area	1816	James Clouston	B143/e/3, fo.1	1 inch:ca. 240 feet 23 x 37 cm	69,24

NO.	TITLE OR DESCRIPTION	DATE	CARTOGRAPHER	REFERENCE	SCALE/SIZE	TEXT
171A	Sketch of Carleton District	1817	(James Bird)	G1/27	1 inch:ca. 8 miles 62 x 50 cm	71
172A	Routes taken by James Tate while travelling "through the principal parts of this District [Long Lake] during the preceding season"	1816–17	James Tate	B117/a/2, after page 55	1 inch:ca. 10 miles 48.7 x 59.6 cm	71
173A	A Map of the rivers and Lakes from Naosquiscaw to Nitchequon By James Clouston	1817	James Clouston	B143/a/19, fo.1	1 inch:ca. 14 miles 34.5 x 48.2 cm	69
174A	A Plan of the Route pursued by the Halfbreeds and other servants of the North West Company on the 19 June 1816 according to the information of Antoine Decharme who drove one of their two Carts on that occasion, and referred to in the affidavit of Peter Fidler of the 4th. of August 1817– signed P. Fidler	1817	Peter Fidler	E8/6, fo.95	1 inch:ca. 6000 feet 25 x 19.3 cm	67
175A	Map attached to the deed of 18 July 1817 conveying land adjoining the Red and Assiniboine rivers from Indian chiefs to Lord Selkirk (three copies)	1817	(Peter Fidler)	E8/1, fo.11 E6/1, facing p.15 E8/2, fo.24	1 inch:ca. 20 miles 30.5 x 20.3 cm	67
176A	Sketch map of Eastmain coast from Eastmain River to Richmond Gulf	(1818)	(George Gladman)	B77/e/2b	1 inch:ca. 15 miles 60 x 45.7 cm	
177A	Sketch of the Lead Mine &c. at Little Whale River	(1818)	(George Gladman)	B77/e/2b	20.5 x 23.5 cm	
178A	Sketch of Beren's River District 1819	1819	Donald Sutherland	B16/e/2, fos.3d,4	1 inch:ca. 15 miles 39.8 x 32 cm	71,27
179A	Map of east coast and interior of Hudson Bay	(1819)	(George Atkinson II)	G1/42	1 inch:ca.4.5 miles 183.5 x 81 cm	70
180A	A plan of a Survey from Nimescaw Lake, in Ruperts River, to Waswanapy Lake. Made in June 1819 by James Clouston	1819	James Clouston	G1/43	1 inch:ca. 11 miles 50.3 x 44 cm	70
181A	A Map of Red River District 1819	1819	Peter Fidler	B22/e/1 fo.1	1 inch:ca. 24 miles 20.2 x 32.7 cm	67,71
182A	A Map of Man,ne,tow,oo,pow Lake 1 sheet Peter Fidler 1819	1819	Peter Fidler	G1/41	1 inch:ca.8.8 miles 74.8 x 53.7 cm	67
183A	Sketch of Lesser Slave Lake District	(1819–20)	(Robert Kennedy)	B115/e/1, fo.1d	Scale varies 30.4 x 18.6 cm	71
184A	A sketch of New Brunswick and Waupissattiga Lakes, with the Route to Capush Cushee Lake the latter being about 120 Miles from New Brunswick Fort, in a southerly direction- Taken April 1820, by, John Corcoran	1820	John Corcoran	G1/44	1 inch:ca.2 miles 42 x 66.8 cm	

Manuscript Maps in the Hudson's Bay Company Archives

NO.	TITLE OR DESCRIPTION	DATE	CARTOGRAPHER	REFERENCE	SCALE/SIZE	TEXT
185A	Map of coastline between Churchill and York by T. Thompson, 9 August 1820	1820	Thomas Thompson	G1/51	1 inch:ca. 20 miles 25.6 x 26.7 cm	72
186A	A Set of Maps, Exhibiting the Rivers, Lakes &c. from York Factory to Cumberland House; as surveyed in the Summer of 1819 by Lieut. J. Franklin R.N. and in the Winter of 1819 & 20 by T. Thompson, Master of the Brig Wear of London. To William Williams Esqr. Governor and Chief, of the Honourable Hudson's Bay Company's Territories, in North America; By his Obedient, Humble Servant, April 7th.. 1820 Thomas Thompson	1820	Thomas Thompson	G1/45 G1/46 G1/47 G1/48 G1/49 G1/50	1 inch:ca. 8 miles 27.1 x 29.8 cm 27.1 x 29.8 cm 27.5 x 31.5 cm 27.5 x 31.5 cm 26.5 x 31.1 cm 26.5 x 31.1 cm	72
187A	A Sketch a la Savage of the Manetoba District 1820	1820	Peter Fidler	B51/e/1, fos.1d–2	1 inch:ca.23 miles 31.7 x 39.5 cm	67,23
188A	A Map of a Journey in the Interior of New Britain from Ruperts House in Latitude 51°.27' North Longitude 79° West to the Mouth of Caniapuscaw River in Lat 57°..56' North and Longitude 67° 40' West. and from thence to Little Whale River in Lat 55°.59' North Longitude 77° West. in the years 1819 and 1820 By James Clouston	(1820–1)	James Clouston	G2/23	1 inch:ca. 22 miles 80.2 x 67 cm	70
189A	Peak View September 18th. 1824 showing relative positions of the sources of rivers from "the top of a very lofty peak situated between the sources of River Malade & Salmon River"	1824	(Alexander Ross)	B202/a/1, fo.49d	1 inch:ca. 25 miles 5.1 x 19.7 cm	87
190A	Liard River from sources to confluence with Mackenzie River	(1824)	(Murdoch McPherson)	B116/a/2, fo.26	1 inch:ca. 27 miles 32 x 40 cm	77
191A	A Chart of the McKenzie's River Department. Comprehending the situation of Athabasca, Slave Lake, & Peace River: – Exhibiting the supposed situation of the Main Ridge Rocky Mountain Compiled chiefly from Indian Authorities –at Fort de Liard 1824	1824	(Murdoch McPherson)	G1/52	1 inch:ca. 42 miles 76 x 55 cm	77,32
192A	Sketch of the route from York factory to Red River by Wm. Kempt 1824	1824	William Kempt	G1/53	1 inch:ca. 18 miles 91.5 x 47.3 cm	83–4
193A	Sketch of the Trout Fall. ...Wm. Kempt 1824	1824	William Kempt	G1/54	1 inch:ca.60 feet 67 x 23.2 cm	83–4
194A	"Severn Sketch of by Work 1824/25"	1824–5	(John Work)	G1/55	1 inch:ca.14 miles 74.7 x 53.3 cm	85–6

NO.	TITLE OR DESCRIPTION	DATE	CARTOGRAPHER	REFERENCE	SCALE/SIZE	TEXT
195A	*A Sketch of Surveys on the Eastmain of Hudson's Bay made at different times between the years 1811 and 1825 by James Clouston*	1825	James Clouston	G1/57	1 inch:ca. 22 miles 113 x 93.6 cm	70,75, 120,25
196A	Map of Missouri, Yellowstone, Bighorn, and Platte rivers (originally pasted to dorse of 198A)	(1825)	c.u.	G1/58	1 inch:ca. 80 miles 38 x 46 cm	
197A	Map of Missouri, Yellowstone, Bighorn and Platte Rivers (drawn on dorse of 198A)	(1825)	c.u.	G1/58	1 inch:ca. 80 miles 38 x 46 cm	
198A	Map with Indian place names, likely of Upper Bighorn River, upper Platte and upper Green River tributaries, at South Pass over the Wind River, *Mars 19, 1925*	1825	c.u.	G1/58	38 x 46 cm	
199A	Map of the Snake River Country	(1825)	(William Kittson)	B202/a/3b	1 inch:20 miles 79.3 x 51.2 cm	88,43
200A	*White Falls Portage. Wm. Kempt, 1825*	1825	William Kempt	G1/56	1 inch:264 feet 76.8 x 67.5 cm	83–4,37
201A	*A Sketch of Thompson's River District 1827. by Archd. McDonald*	1827	Archibald McDonald	B97/a/2, fo.40	1 inch:ca. 20 miles 40.4 x 25.3 cm	93,48
202A	Twenty-one sketch maps of regions inland from York Fort, up the Nelson River to the north and south of the river to Foxes Lake and Steel River, 26 Mar.–5 May 1827	1827	George Taylor Jr	B239/a/138, fos.1d–3,4–6, 7–7d		85
203A	Fifty-five sketch maps of route taken by George Taylor Jr from York Factory to Red River settlement through Hayes River, Holy Lake, Hairy Lake, Lake Winnipeg, 18 Sept.–24 Oct. 1827	1827	George Taylor Jr	B235/a/10, fos.1–25d	1 inch:1 mile	84,38
204A	Rough drafts of 55 sketch maps of route taken by George Taylor Jr from York Factory to Red River settlement, 18 Sept.–24 Oct. 1827	1827	George Taylor Jr.	B235/a/11, fos.1–23d	1 inch:1 mile	84
205A	Sketch of location of oak tree used as latitude marker at west end of Turtle Mountain and of route along north side of mountain to east end.	1827	George Taylor Jr	B235/a/10, fo.29d	1 inch:ca. 10 miles 7.7 x 17.8 cm	83
206A	Sketch of location of oak tree used as latitude marker at west end of Turtle Mountain and of route along north side of mountain to east end, with addition of symbol and name Hairy Hill	1827	George Taylor Jr	B235/a/11, fo.27d	1 inch:ca. 10 miles 7.7 x 17.8 cm	83
207A	Sketch of location of oak tree used as latitude marker on east bank of creek, west end of Turtle Mountain	1827	George Taylor Jr	B235/a/10, fo.29	7.7 x 17.8 cm	83

Manuscript Maps in the Hudson's Bay Company Archives

NO.	TITLE OR DESCRIPTION	DATE	CARTOGRAPHER	REFERENCE	SCALE/SIZE	TEXT
208A	Sketch of location of oak tree used as latitude marker on east bank of creek, west end of Turtle Mountain	1827	George Taylor Jr	B235/a/11, fo.27	7.7 x 17.8 cm	83
209A	York Fort, Hayes River, Lake Winnipeg, to Pembina	(1827)	(George Taylor Jr)	G1/4	1 inch:ca. 15 miles 151.5 x 55.4 cm	84
210A	River communications between La Cloche and Temiscamingue District posts, showing rivers and lakes between Lake Huron and Wanapitei Lake and the Spanish River	(1827)	(John McBean)	D5/2, fo.257	1 inch:ca. 8 miles 22 x 32.8 cm	85,40
211A	Seventy-eight sketch maps of area from Norway House to Jasper House, Rocky Mountains	(1828)	George Taylor Jr	B235/a/11, fos.37d–65d	1 inch:1 mile	84,38
212A	Northwest coast of America from Lat. 52° to Lat. 57° 30′	(1827 to early 1831)	c.u.	G1/158	1 inch:c.15 miles 75.5 x 55 cm	
213A	*Sketch of New Dungeness*	(1828)	(Aemilius Simpson)	G1/159	1 inch:3 miles 49.2 x 73m	91
214A	*Chart of the Route of the Inland Expedition to the Mouth of South River during the Summer of 1828 A.D. Conducted by William Hendry of the Honble. H.B.C. Service*	1828	William Hendry	G1/60	1 inch:ca. 15 miles 68 x 101.7 cm	74–5
215A	*Chart of the Eastmain Coast with route of the Expedition inland to South River undertaken during the Summer of 1828, by W W Hendry*	(1828)	William Hendry	G1/266	1 inch:ca. 15 miles 126 x 97 cm	74–5,29
216A	*P.S. Ogdens Camp Track 1829*	1828–9	Peter Skene Ogden	B202/a/8, fo.1	1 inch:ca. 25 miles 41.4 x 52.3 cm	89
217A	Map illustrating a journal of an expedition up the West Branch of the Liard River, summer 1831	1831	John M. McLeod	B200/a/14, fos.1d–2	1 inch:ca. 25 miles 32.2 x 39.2 cm	78,33
218A	*Columbia River, By Thomas Sinclair in 1831. Cape Disappointment. In Lat 46=17=30. N Longitude 123=53=00.W*	1831	Thomas Sinclair	G1/66	1 inch:ca. 1.6 miles 36.8 x 45.5 cm	91
219A	Sketch map of coastline from Moose Factory to Rupert's House showing route taken by Swanson and party to Hannah Bay House in 1832 and to Rupert's House in the same year	1832	(William Swanson)	B135/a/138, fo.1	1 inch:ca. 15 miles 19.2 x 32 cm	85
220A	Entrance of the Hayes River, York Factory	(1832)	(Benjamin Bell)	G1/118	1 inch:0.5 mile 62.5 × 75.7 cm	
221A	*Plan of Lachine Village*	(ca. 1832)	c.u.	G1/238	1 inch:396 feet 32.7 x 67.5 cm	85,41

NO.	TITLE OR DESCRIPTION	DATE	CARTOGRAPHER	REFERENCE	SCALE/SIZE	TEXT
222A	*Tumgass Harbour By Thomas Sinclair In 1832. Latitude. 55=0.3.N. Longitude. 131=21 W.*	1832	Thomas Sinclair	G1/167	1 inch:ca. 0.7 mile 43 x 28.4 cm	91
223A	*A Sketch of Port George & Cossack Harbour,s the latter in Latitude 53=33. North Longitude. 129=44. West by Thomas Sinclair Jany. 28th. 1832*	1832	Thomas Sinclair	G1/168	1 inch:ca.0.9 mile 23 x 37 cm	91
224A	*Skeategat,s Harbour Queen Charlotte,s Island Latitude 53=22. North Longitude 131=37. West by Thomas Sinclair. July 24th. 1832*	1832	Thomas Sinclair	G1/169	1 inch:ca. 0.85 mile 37 x 22.8 cm	91
225A	*Pearl Harbour. In Latitude. 54=30=00. North. In Longitude. 130=17=00. West. A. Copy By Thomas Sinclair. in Jany. 27th. 1832*	1832	Thomas Sinclair	G1/170	1 inch:ca. 0.45 mile 22.8 x 37 cm	91
226A	*Winter Road from Foxes River to Oxford House 1833 G. Taylor*	1833	George Taylor Jr	G1/61	1 inch:ca. 4.5 miles 74 x 54 cm	84–5,39
227A	Sketch map to illustrate "Journal of Voyage to Simpson's [Bulkley] River by land Summer 1833" by *Simon McGillivray Chief Trader Stewart's Lake 15 July 1833*	1833	Simon McGillivray	D4/126, fo.45	1 inch:ca. 7 miles 20 x 32.4 cm	93–4
228A	"Map of Meshickemac or N.W. River and Whale River received from Ungava 1834"	(1834)	(Erland Erlandson)	G1/62	1 inch:20 miles 66.5 x 56 cm	75, 281n4
229A	"Map.Ungava, of route from Ft. Chimo to Esquimaux Bay & return by E. Erlandson 1834"	(1834)	(Erland Erlandson)	G1/64	1 inch:20 miles 65.5 x 53.5 cm	75, 281n4
230A	Map of a voyage from Fort Chimo, Ungava to Esquimaux Bay, Labrador and back to the fort, by Erland Erlandson, clerk, 6 April–17 July 1834	(1834)	(Erland Erlandson)	G1/236	1 inch:18.5 miles 66.7 x 55 cm	75,30
231A	Pencil sketch of daily route, 6 April to 5 June, from Fort Chimo as far as Meshickemau Lake	(1834)	(Erland Erlandson)	G1/232	1 inch:ca. 25 miles 66.7 x 53.5 cm	75
232A	*Map of the Weenisk Country. Geo.:Barnston June 1834*	1834	George Barnston	G1/63	1 inch:ca. 25 miles 49.8 x 44.6 cm	86
233A	*Indian Chart* showing West Branch of the Liard from Fort Halkett to source, Dease's Branch, Dease Lake, and Frances River	1834	John M. McLeod	B85/a/6, fo.10	1 inch:ca. 20 miles 20 x 32 cm	78
234A	Red River Settlement	(1835)	(George Taylor Jr)	G1/59a,b	1 inch:0.5 mile a 66.8 x 85 cm b 66.8 x 84.5 cm	82

Catalogue A

NO.	TITLE OR DESCRIPTION	DATE	CARTOGRAPHER	REFERENCE	SCALE/SIZE	TEXT
235A	Red River Settlement (signed by the Earl of Selkirk and witnessed by George Simpson, John Dease, and H.W. Palmer, 4 May 1836) (two sheets)	(1835)	(George Taylor Jr)	E6/13, fo.1 fo. 2	1 inch:0.5 mile 62.4 x 84.5 cm 64 x 83.6 cm	82
236A	Plan of Red River Colony Surveyed in 1836, 7 & 8 by George Taylor	1836–8	George Taylor Jr	E6/14	1 inch:0.5 mile 157.5 x 163.5	82–3,36
237A	Map Of the Arctic Coast of AMERICA from Return Reef of Sir John Franklin to Point Barrow of Captn. Beechey Explored by The Honble = Hudson's Bay Company's Northern Discovery Expedition A.D. 1837	1837	(Thomas Simpson and Peter W. Dease)	G1/180	1 inch:8 miles 51.6 x 67 cm	79–80, 34
238A	Map of the Arctic coast of America from Return Reef to Point Barrow	(post–1837)	c.u.	G/179	1 inch:ca. 19 miles 37.2 x 47.7 cm	80
239A	A CHART OF the Entrance OF COLUMBIA RIVER	(ca. 1838)	c.u.	G1/171	1 inch:1 mile 73 x 54.2 cm	91
240A	Copy. – Survey of Charlton as laid down by Kennawap & Cauc-chi-chenis themselves in pencil whilst at Charlton, February and March 1839. In[k]'d over their own marks by C.T.R. Miles at Ruperts House 16 April 1839 … signed Robt. Miles. Copied by henry Connolly	1839	Henry Connolly	G1/65	1 inch:0.75 mile 53.4 x 65.5 cm	86
241A	"Plan of Land Eastward of the Catholic Church [Red River settlement] – March 31st 1839"	1839	c.u.	G1/264	1 inch:ca. 880 feet 32.7 x 19 cm	83
242A	A Map of Labrador shewing The Interior route from Ft. Smith, Esquimaux to Ft. Chimo Ungava Bay	(1840)	(John McLean)	G1/66	1 inch:ca. 30 miles 66.1 x 53.4 cm	75
243A	Map of Kaipokok (Kibokok) Bay, Labrador	(1840)	c.u.	G1/234	1 inch:ca.3 miles 24.2 x 37.5 cm	76
244A	Map showing Lake Melville, Labrador, with distances from North West River House	(1840)	(W.H.A. Davies)	G1/235	1 inch:10 miles 32.5 x 53 cm	76
245A	"A Chart of George's River Mouth"	(1840)	(William Kennedy)	G1/67	1 inch:1 mile 52.7 x 33 cm	75
246A	Two "Arctic Charts Containing Dease and Simpson's Strait, Victoria Land, Boothia Felix etc."	(post April 1840)	c.u.	G1/5, G1/6	1 inch:ca. 45 miles 34 x 50.4 cm	80
247A	Sketch of a Map as pencilled out by Thomas J. Beads at Canniapiscaw Winter 1841	1841	Thomas John Beads and John Spencer	G1/69	1 inch:21 miles 65.5 x 104 cm	76–7,31
248A	A Chart of the Entrance of Columbia River by James A Scarborough June 20th. 1841	1841	James A. Scarborough	G1/172	1 inch:0.6 mile 55.4 x 76 cm	91

NO.	TITLE OR DESCRIPTION	DATE	CARTOGRAPHER	REFERENCE	SCALE/SIZE	TEXT
249A	Endorsed: "Map of Cowlitz & Nasqually" (two copies)	1841	(Adolophus Lee Lewes)	F25/1, fos.12,13	1 inch:4.5 miles 39 x 31.9 cm	91
250A	*Sketch of the Cowelitz Farm* (two copies)	(1841)	c.u.	F25/1, fos.9a, 9b	31.8 x 39 cm	92
251A	*Sketch Simpson's Beaver-preserve Charlton Island as originally laid down by the Indians 1838/39 Red Ink additions thereto by them 1839/40*	(1842)	(Robert Miles)	G1/68	1 inch:ca. 0.75 mile 52.8 x 65.3 cm	86
252A	GROUND PLAN OF PORTION OF VANCOUVERS ISLAND SELECTED FOR NEW ESTABLISHMENT TAKEN BY JAMES DOUGLAS ESQR.	1842	James Douglas and Adolphus Lee Lewes	G2/25	1 inch to 0.25 mile 94.4 x 121.3 cm	95,97,50
253A	Sketch to accompany application for timber rights on Lake Timiskaming at mouth of the Ottawa River	1842	Angus Cameron	B134/c/52, fo.16	1 inch:ca. 2 miles 19.0 x 17.7 cm	83
254A	Map of country between Ungava, Esquimaux Bay, and Gulf of St. Lawrence showing main routes and intervening posts	(ca. 1843)	(John McLean)	G1/237	1 inch:ca. 30 miles 65.2 x 52.6	75
255A	Map of country between Ungava, Esquimaux Bay, and Gulf of St. Lawrence showing main routes and intervening posts	(ca. 1843)	c.u.	G1/233	1 inch:ca. 155 miles 20.2 x 32.8 cm	75
256A	*Island Ministickwattam Nabowisho's preserve as laid down by him Summer 1842 all eligible for Beaver exclusive of other Lakes and places he did not then examine*	(1843)	Nabowisho (with Robert Miles)	G1/70	1 inch:ca. 0.9 mile 53.3 x 65.7 cm	86,166
257A	*Map of Cowelitz Farm*	1843	c.u.	F12/2, fos.10Ad–10B	32 x 39.7 cm	92
258A	*Copy of the Chart of the Cowlitz Farm as to be sown in the Fall 1843 and Spring 1844* (two copies)	1843	c.u.	F25/1,fo.10; F26/1, fo.66d-67	31.9 x 40 cm 32 x 40 cm	92,44
259A	West coast of Hudson Bay from Churchill to Repulse Bay	(1844)	(Robert Harding)	D5/12, fo.1130d	1 inch:ca. 290 miles 23.8 x 19.6 cm	
260A	Rough sketch of Red River settlement on back of an envelope, addressed to Archibald Barclay, Hudson's Bay House, London, to show location and extent of land of the Catholic Mission	(ca. 1844)	c.u.	E8/8,fo.37	1 inch:ca. 1.6 miles 22.2 x 20.8 cm	83
261A	*A Sketch of the course of the West Branch of the Liard River and Pellys' River so far as discovered By Your obedient Humble Servant R. Campbell*	(1844)	Robert Campbell	G1/71	1 inch:ca. 30 miles 53 x 66 cm	78

Manuscript Maps in the Hudson's Bay Company Archives

NO.	TITLE OR DESCRIPTION	DATE	CARTOGRAPHER	REFERENCE	SCALE/SIZE	TEXT
262A	Copy of the Chart of the Cowelitz Farm as cultivated in 1844 and spring 1845 (two copies)	(1844)	c.u.	F12/2,fo.57; F12/2,fo.104	31.7 x 39.5 cm	92
263A	MAP OF OREGON CITY as surveyed in the Spring of the Year 1844	1844	Adolphus Lee Lewes	G1/121	1 inch:240 feet 41.4 x 73 cm	
264A	SKETCH OF THE ENVIRONS OF FORT VANCOUVER, embracing a section of about 16 miles in length, shewing the course of the great Conflagration, by which the Fort was nearly destroyed, on the 27th. day of September 1844. H.N. Peers	(1844)	H.N. Peers	G1/125	1 inch:ca. 4000 feet 39 x 66.5 cm	92
265A	SKETCH of the Environs of FORT VANCOUVER embracing a section of about 16 Miles in length shewing the course of the great conflagration by which the Fort was nearly destroyed on the 27 day of September 1844	(1844 or later)	(H.N. Peers)	G1/124	1 inch:ca. 4000 feet 49.3 x 72.3 cm	92, 283n15
266A	Rough sketch of part of Kootenay Lake showing discharge of Kootenay River and location of ore body	1844	Archibald McDonald	A11/70, fo.86d	1 inch:ca. 4 miles 20.5 x 20.2 cm	
267A	Sketch of Fort Vancouver and Plain Representing the Line of Fire in September 1844	(1844 or 1845)	(Adolphus Lee Lewes)	G1/123	1 inch:900 feet 31.7 x 39.5 cm	92
268A	Sketch map to illustrate report on "Charlton Beaver Preserve by the Indian Tom Pipes who resided there Winter 1844/45 and returned to Ruperts House March 14th 1845"	1845	Joseph Gladman	B186/b/49, fo.27	1 inch:0.75 mile 53.6 x 66.3 cm	86
269A	Plan of Fort Colville Columbia River M Vavasour Liut Royal Engr	(1845)	Mervin Vavasour	G1/193	1 inch:100 feet 59.5 x 47.4 cm	92
270A	Plan of mud Fort at Walla Walla Columbia River M. Vavasour Lieut Royal Engr	(1845)	Mervin Vavasour	G1/194	1 inch:100 feet 44.5 x 36.6 cm	92
271A	Sketch of Fort Vancouver and adjacent Plains Lat 45.° 36'N. Long. 122.° '37'W M Vavasour Lieutn. Royal Engr 1845 Inset: Plan of Fort Vancouver	(1845)	Mervin Vavasour	G1/195	1 inch:0.25 mile 53.8 x 68 cm 1 inch:100 feet 15.7 x 20.7 cm	92
272A	Eye sketch of the Route from the Columbia River to Nisqually on Puget's Sound. H.J. Warre Lt 14 Regt.-M. Vavasour Lt Royal Enge	(1845)	Henry J. Warre and Mervin Vavasour	G1/196	1 inch:5 miles 35.8 x 52 cm	92

NO.	TITLE OR DESCRIPTION	DATE	CARTOGRAPHER	REFERENCE	SCALE/SIZE	TEXT
273ᴬ	Sketch map accompanying "Suggestions for the exploration of a new route of communication by which ... the transport of the supplies and returns to and from the Districts of New Caledonia and Thompson's River might be advantageously carried on in connexion with Fort Langley and the new establishment of Victoria"	1845	Alexander C. Anderson	B5/z/1, fo.2	1 inch:ca. 20 miles 20.5 x 20 cm	94,49
274ᴬ	Endorsed: "Sketch of Cowelitz Farm 1846," copy of a map of the Cowlitz Farm as cultivated in 1845 and spring 1846	(1845)	c.u.	F25/1,fo.11	31.8 x 39.8 cm	92
275ᴬ	*Chart of Charlton Beaver Preserve with the Lodges seen Winters '44/'45 and '45/'46 noted Rupert's House, 23d March 1846*	1846	Joseph Gladman	B186/b/51, fo.25	1 inch:0.75 mile 54.2 x 67.3 cm	86
276ᴬ	"Chart [of] Charlton Beaver Preserve with the Lodges seen Winters 1844/45 and '45/46 noted." Rupert's House, 23rd March 1846 Joseph Gladman	1846	Joseph Gladman	B186/b/52, fo.17	1 inch:0.75 mile 52.8 x 64.8 cm	86
277ᴬ	*Fort Vancouver and Village in 1846 Drawn by R Covington* Inset: Map of village	1846	Richard Covington	G2/24	1 inch:2112 feet 63 x 99 cm 24 x 37 cm	92
278ᴬ	Plan of Fort Vancouver	(c. 1846)	c.u.	G1/126	28 x 37.5 cm	92
279ᴬ	*Eye Sketch of the Plains &c about Nisqually at the Head of Puget's Sound M Vavasour Leiut Royal Engr*	(1846)	Mervin Vavasour	G1/197	1 inch:1 mile 54 x 70.8 cm	92
280ᴬ	*Sketch of Cammusan Harbour, Vancouvers Island. shewing the position of Fort Victoria, from a Drawing of Js.Scarboro Capt H.H.B.C. M Vavasour Leiut Royal Engrs* Inset: Plan of the fort	(1846)	Mervin Vavasour	G1/198	1 inch:600 feet 49.2 x 67 cm 1 inch:100 feet 15.7 x 13.6 cm	92
281ᴬ	*Rough Chart of the Columbia River, from the head of the Navigation to the Pacific Ocean. From a sketch belonging to the H.B.C. to which have been added the soundings opposite Vancouver, and at the mouth of the Willamette, also the alterations in the sands at the mouth of the River*	(1846)	Mervin Vavasour	G1/199	1 inch:1 mile 97 x 65.5 cm	92
282ᴬ	*Plan of Tongue Point, on the South Shore of the Columbia River. M. Vavasour Lieut Royal Engr*	(1846)	Mervin Vavasour	G1/200	1 inch:330 feet 69.7 x 53 cm	92

Manuscript Maps in the Hudson's Bay Company Archives

NO.	TITLE OR DESCRIPTION	DATE	CARTOGRAPHER	REFERENCE	SCALE/SIZE	TEXT
283A	Plan of CAPE DISAPPOINTMENT	(1846)	Mervin Vavasour	G1/201	1 inch:0.1 mile 63.4 x 77.3 cm	92
284A	Plan of Fort Ellice on Beaver Creek near the Assenneboine River M Vavasour Lieut Royal Engr 1846	1846	Mervin Vavasour	G1/190	1 inch:100 feet 53 x 36 cm	92
285A	Plan of Fort Carlton on the Saskatchewan River M Vavasour Lieut Royal Engr 1846	1846	Mervin Vavasour	G1/191	1 inch:100 feet 53.5 x 36 cm	92
286A	Plan of Fort Edmonton on the Saskatchewan River M Vavasour Lieut Royal Engr. 1846	1846	Mervin Vavasour	G/192	1 inch:100 feet 39.5 x 60 cm	92, 45
287A	"No.1 Diagram of Nisqually claims"	(1846)	c.u.	B223/z/5, fo.247	30 x 32.4 cm	93
288A	Sketch of claims registered in the names of Henry N. Peers, John Lee Lewes, Donald Manson, Daniel Harvey, and George Harvey at Fort Vancouver	(1846)	c.u.	B223/z/5, fo.158	8.7 x 19.8 cm	93
289A	Claims at Fort Vancouver. Surveyed by A. Lee Lewes 1845 & 1846	(1846)	Adolphus Lee Lewes	G1/127	1 inch:1584 feet 35.2 x 59 cm	93
290A	Small sketch of claims by some settlers at Fort Victoria	(ca. 1846)	c.u.	B223/z/5, fo.164	18 x 19.7 cm	96
291A	"a rough draft on tracing paper of my discoveries"	1847	John Rae	E15/3, fo.40	1 inch:ca. 22 miles 48.6 x 36.8 cm	80–1
292A	Endorsed: "Rae's discoveries Admiralty Sketch," based on Rae's rough sketch	(1847)	(John Rae)	G1/177	1 inch:ca. 22 miles 35.5 x 45 cm	80–1, 35
293A	Map showing Committee Bay, Rae Isthmus, and Repulse Bay, a copy of Rae's rough sketch	(1847)	c.u.	G1/178	1 inch:ca. 22 miles 37.4 x 38.8 cm	80–1
294A	No.1. Ground Plan of Puget's Sound Co's principal establishment at Nisqually	(1847)	(William F. Tolmie)	F25/1, fo.31	1 inch:0.1 mile 52.2 x 32 cm	93, 46
295A	No.2. Map of Pasture land adjoining Cattleherds' Station at Spanuch, Nisqually	(1847)	(William F. Tolmie)	F25/1, fo.36	1 inch:0.3 mile 32.5 x 52.2 cm	93
296A	No.3. Map of some of the Pasture Land adjoining Shepherd's station at Tlilthlow, Nisqually	(1847)	(William F. Tolmie)	F25/1, fo.33	1 inch:ca. 750 feet 52 x 32.7 cm	93
297A	No.4. Map of some of the Pasture Land adjoining Shepherd's station at "Muck," Douglas-Burn, Nisqually	(1847)	(William F. Tolmie)	F25/1, fo.34	1 inch:ca. 750 feet 52.3 x 32.8 cm	93
298A	No.5. Map of some of the Pasture Land adjoining Shepherd's station at 'Sastuk' Nisqually	(1847)	(William F. Tolmie)	F25/1, fo.35	1 inch:0.3 mile 32.6 x 52 cm	93, 47

NO.	TITLE OR DESCRIPTION	DATE	CARTOGRAPHER	REFERENCE	SCALE/SIZE	TEXT
299A	*Portion of Lake Huron showing La Cloche Island*, Crown Land Department, Montreal	1847	Thomas Bouthillier	D5/20,fo.414	1 inch:2.5 miles 51.7 x 41.4 cm	
300A	*Rough Chart of Red River Settlement shewing the unoccupied Land in the vicinity of Fort Garry March 1848*	1848	(Edward M. Hopkins)	G1/320	1 inch:1440 feet 25 x 40.7 cm	83
301A	*GROUND PLAN of portion of VANCOUVER'S ISLAND selected for NEW ESTABLISHMENT taken by James Douglas Esqr. Drawn by A. Lee Lewes L.S. Traced from a M.S. Sept 1848 R.A.*	1848	R.A.	G2/40	1 inch:0.25 mile 75.7 x 98.5 cm	96, 284n21
302A	*Survey plat of E. Spencer's claim at Fort Vancouver area by P.W. Crawford as directed by Mr. Taylor with accompanying field notes*	1848	P.W. Crawford	B223/z/5, fos.132d,133	1 inch:528 feet 40 x 32.5 cm	93
303A	*Figurative plan exhibiting the pretensions of the parties in action en Bornage for the Seigneury of Mille Vaches Comte du Saguenay District of Quebec Montreal 10th. Oct 1847. 17th. Jany 1848*, a copy	1848	George Barnston	B214/z/1, fo.5	1 inch:15 arpents 41.7 x 26 cm	83
304A	*Sketch of the Country near the mouth of the Quaqualla [Coquihalla] River*	(1849)	c.u.	B113/c/1, fo.1	1 inch:0.3 mile 21.3 x 26.8 cm	
305A	*Chart and Directions for entering the Columbia River – By William Henry McNiell* [McNeill]	(1848–9)	William H. McNeill	G1/157	1 inch:ca. 1.3 miles 53.7 x 65.4 cm	91
306A	*Sketch map showing routes from Great Bear Lake to the Coppermine River*	1849	John Rae	E15/5,fo.33	1 inch:ca. 20 miles 11.5 x 20 cm	81
307A	*Hudson's Bay Company's Reserve at Tadoussac*	1849	George Gladman	D5/26, fos.470d–471	1 inch:ca. 400 feet 26.8 x 41.7 cm	83
308A	*Map of the Victoria District Vancouvers Island 1850 W. Colqn. Grant*	1850	Walter Colquhoun Grant	G1/256	1 inch:600 feet 149 x 152 cm	97
309A	*Map of southern tip of Vancouver Island, showing division of land into squares representing 640 acres, numbered 1–196*	(1850)	c.u.	G1/134	1 inch:1 mile 86.5 x 67.5	
310A	*VICTORIA & PUGETSOUND DISTRICTS Sheet No.1*	(1851)	(Joseph D. Pemberton)	G1/131	6 inches:1 mile 74 x 92.5 cm	100
311A	*VICTORIA & PUGETSOUND DISTRICTS Sheet No 2*	(1851)	(Joseph D. Pemberton)	G1/132	6 inches:1 mile 75.5 x 98 cm	100
312A	*MAP OF VICTORIA & PUGETSOUND DISTRICTS Sheet No. 3*	(1851)	(Joseph D. Pemberton)	G1/133	6 inches:1 mile 105 x 97.5 cm	100

Manuscript Maps in the Hudson's Bay Company Archives

Catalogue A

NO.	TITLE OR DESCRIPTION	DATE	CARTOGRAPHER	REFERENCE	SCALE/SIZE	TEXT
313A	VICTORIA DISTRICT.LOT No.1. Section No. (VI)	1851	(Joseph D. Pemberton)	H1/1,fo.4	6 inches:1 mile 25 x 7.7 cm	100
314A	*Victoria District. Section I. Lot No. 2*	1851 or 1852	(Joseph D. Pemberton)	H1/1,fo.6	6 inches:1 mile 32 x 22.7 cm	100,51
315A	*License of Occupation of Tadoussac dated 2.December 1847*	(1851)	(George Gladman)	D5/31, fo.196	1 inch:ca.330 feet 22.6 x 20.7 cm	
316A	*Sketch of HBC Reserve at Tadoussac showing area fenced in*	(1851)	George Gladman	D5/31, fo.452	1 inch:ca.880 feet 20 x 24.8 cm	
317A	*Map of Lands, claimed by the Puget Sound Company at Nisqually – NISQUALLY PLAINS – Edward Huggins del:*	(1851)	Edward Huggins	F25/1, fo.32	1 inch:0.87 mile 50.5 x 62.6 cm	108
318A	*PLAN OF RESERVE … Cadboro Bay*	(1852)	(Joseph D. Pemberton)	G1/258l	6 inches:1 mile 38 x 50.5 cm	100
319A	Map showing Haro Strait and Cordova Bay	(1852)	(Joseph D. Pemberton)	G1/130	6 inches:1 mile 74 x 98.8 cm	100
320A	*THE NEIGHBOURHOOD OF VICTORIA drawn with reference to Levels only, which are slightly exaggerated*	(1852)	(Joseph D. Pemberton)	G1/258a	6 inches:1 mile 50 x 60.6 cm	101
321A	*WATER SUPPLIES AT ESQUIMALT, SAANICH INLET & THE COWITCHIN COUNTRY ROUGHLY SKETCHED*	(1852)	(Joseph D. Pemberton)	G1/258o	1 inch:1 mile 73 x 54 cm	101
322A	*MAP OF THE COUNTRY AROUND ESQUIMALT HARBOUR*	(1852)	(Joseph D. Pemberton)	G1/258b	1 inch:0.5 mile 86.8 x 67.2 cm	101
323A	*A PLAN OF THE TOWN OF VICTORIA SHEWING PROPOSED IMPROVEMENTS*	(1852)	(Joseph D. Pemberton)	G2/38	1 inch:132 feet 90.4 x 156.5 cm	101,52
324A	Saaquash coal bores	1852	Boyd Gilmour	A11/73,fo.31d	8.5 x 20 cm	103
325A	Fort Rupert coal bore	1852	Boyd Gilmour	A11/73	1.3 x 3 cm	103
326A	*An Explanation and skitch of The Mettles from Quatsey to McNeils Harbour. Shewing the different places where I have made trials of them. … surveyed by B Gilmour 1852*	1852	Boyd Gilmour	G1/138	1 inch:0.5 mile 66 x 158 cm	103,53
327A	*MAP OF THE COUNTRY ROUND NANAIMO HARBOUR*	(1852)	(Joseph D. Pemberton)	G1/258c	4 inches:1 mile 100.7 x 142 cm	101
328A	Chart of Nanaimo Harbour and the neighbouring coast.	(1852)	(Joseph D. Pemberton)	G1/140	4 inches:1 mile 43 x 63.3 cm	102

NO.	TITLE OR DESCRIPTION	DATE	CARTOGRAPHER	REFERENCE	SCALE/SIZE	TEXT
329A	*Sketch of Wentuhuysen Inlet East coast of Vancouver's Island Lat. 49°10' Long. 123°.37'W*	(1852)	(Joseph D. Pemberton)	G1/258e	1 inch:0.4 mile 24.5 x 38 cm	102
330A	Victoria District, *Lot.3. Section 11*	(1852)	(Joseph D. Pemberton)	H1/1,fo.9	6 inches:1 mile 26.7 x 20.7 cm	99
331A	*VICTORIA DISTRICT LOT No.3. Section No.11*	(1852)	(Joseph D. Pemberton)	H1/1,fo.12	6 inches:1 mile 30.3 x 23.6 cm	99
332A	Victoria District, *Lot No.4 Section III.*	(1852)	(Joseph D. Pemberton)	H1/1,fo.14	6 inches:1 mile 20.5 x 23.8 cm	99
333A	*VICTORIA DISTRICT. Lot No. 2(38). Sections nos.1,1a & 1b*	(1852)	(Joseph D. Pemberton)	H1/1, fo.73	4 inches:1 mile 21.2 x 15.3 cm	99
334A	*Victoria District. LOT no.4 Section no. III*	(1852)	(Joseph D. Pemberton)	H1/1,fo.17	6 inches:1 mile 24.5 x 27 cm	99
335A	Esquimalt District, *LOT.5. Section IX*	(1852)	(Joseph D. Pemberton)	H1/1,fo.19	6 inches:1 mile 26 x 22.7 cm	99
336A	*Esquimalt District. LOT nr.5 Section IX*	(1852)	(Joseph D. Pemberton)	H1/1,fo.22	6 inches:1 mile 28 x 27 cm	99
337A	Victoria District *LOT-No.6. Sections.-no. VII & VIIa*	(1852)	(Joseph D. Pemberton)	H1/1,fo.24	6 inches:1 mile 22.9 x 24 cm	99
338A	*Victoria District. LOT No 7 Section V*	(1852)	(Joseph D. Pemberton)	H1/1,fo.27	6 inches:1 mile 24.7 x 19 cm	99
339A	[Esquimalt District] *ESQUIMALT HARBOUR LOT No. 9 section No VIII*	(1852)	(Joseph D. Pemberton)	H1/1,fo.29	4 inches:1 mile 18.6 x 14.5 cm	99
340A	*Esquimalt District. LOT No. 9 Section No. VIII (two copies)*	(1852)	(Joseph D. Pemberton)	H1/1,fo.31 H1/1,fo.33	ca. 25.4 x 20.3 cm	99
341A	*Esquimalt District LOT 10. Sectn. VII*	(1852)	(Joseph D. Pemberton)	H1/1,fo.37	6 inches:1 mile 19.6 x 16 cm	99
342A	*Esquimalt District. LOT No. 11. Section No. VI*	(1852)	(Joseph D. Pemberton)	H1/1,fo.39	6 inches:1 mile 23.2 x 19.5 cm	99
343A	*Victoria District LOT No. 14 Section No. XIV*	(1852)	(Joseph D. Pemberton)	H1/1,fo.41	4 inches:1 mile 19.8 x 13 cm	99
344A	*Victoria District LOT No 15. Section No. X*	(1852)	(Joseph D. Pemberton)	H1/1,fo.43	4 inches:1 mile 20.8 x 16 cm	99

Manuscript Maps in the Hudson's Bay Company Archives

NO.	TITLE OR DESCRIPTION	DATE	CARTOGRAPHER	REFERENCE	SCALE/SIZE	TEXT
345A	*Victoria District. LOT No. 16. Section no. XI*	(1852)	(Joseph D. Pemberton)	H1/1,fo.45	4 inches:1 mile 12.3 x 20.5 cm	99
346A	*Victoria District LOT 17 Secn. XII*	(1852)	(Joseph D. Pemberton)	H1/1,fo.47	4 inches:1 mile 21 x 16.7 cm	
347A	*Victoria District. LOT No. 18. Section No. XVII.*	(1852)	(Joseph D. Pemberton)	H1/1,fo.49	4 inches:1 mile 20.5 x 13.8 cm	
348A	*Victoria District LOT No. 20. Section No. XV*	(1852)	(Joseph D. Pemberton)	H1/1,fo.51	4 inches:1 mile 18.5 x 16.2 cm	
349A	*Esquimalt District LOT 21. Sectn III.*	(1852)	(Joseph D. Pemberton)	H1/1,fo.53	4 inches:1 mile 21.7 x 17.8 cm	
350A	*Victoria District LOT 22. Section no IV*	(1852)	(Joseph D. Pemberton)	H1/1,fo.55	4 inches:1 mile 18.3 x 24.5 cm	
351A	*Sketch of the H.B. Companys Claim at Okinagan O.S. Surveyed by A.C.A. Esqe. 1852*	1852	Alexander C. Anderson	G1/120	1 inch:1 mile 32.3 x 20.3 cm	108
352A	*PLAN of the HUDSON'S BAY COMPANY'S CLAIM at Vancouver Surveyed by A. Lee Lewes 1852*	1852	Adolphus Lee Lewes	G1/258j	1 inch:0.6 mile 50 x 101 cm	108
353A	Outline of the southeast coast of Vancouver Island between Victoria and Valdez Inlet	(1853)	(Joseph D. Pemberton)	A11/74,fo.96	1 inch:ca. 17 miles 25.3 x 19 cm	102
354A	Victoria and Puget Sound (Esquimalt) Districts – copy of 1851 original, on a single sheet, reduced scale	(1853)	(Joseph D. Pemberton)	G1/181	4 inches:1 mile 95 x 138.5 cm	102
355A	*PENINSULA OCCUPIED BY THE PUGET SOUND COMPANY*	(1853)	(Joseph D. Pemberton)	G1/258m	4 inches:1 mile 50.6 x 66.4 cm	102–3
356A	*DISTRICT MAP OF METCHOSIN*	(1853)	(Joseph D. Pemberton)	G1/258f	4 inches:1 mile 74.5 x 114.5 cm	102
357A	Plan of Sooke District, Vancouver Island	(1853)	(Joseph D. Pemberton)	G1/258k	2 inches:1 mile 100 x 71.5 cm	102
358A	*NANAIMO HARBOUR AND PART OF THE SURROUNDING COUNTRY*	(1853)	(Joseph D. Pemberton)	G1/141	4 inches:1 mile 98.5 x 148 cm	102
359A	Map of Saanich Peninsula, and Finlayson Arm, Vancouver Island	(1853)	(Joseph D. Pemberton)	G1/258d	2 inches:1 mile 95.7 x 73.7 cm	102
360A	Map of southern tip of Vancouver Island from Sooke Bay to Cordova Bay	(1853)	(Joseph D. Pemberton)	G1/137	1 inch:1 mile 35.4 x 65 cm	102

NO.	TITLE OR DESCRIPTION	DATE	CARTOGRAPHER	REFERENCE	SCALE/SIZE	TEXT
361A	*Metchosin District LOT 23 Secn. 1*	(1853)	(Joseph D. Pemberton)	H1/1, fo.57	4 inches:1 mile 24 x 17.7 cm	102
362A	*Victoria District Lot 24 Section XVIII*	(1853)	(Joseph D. Pemberton)	A37/42, fos.190B–191	4 inches:1 mile 37.4 x 27.2 cm	102
363A	*Victoria District. Lot 24 Sect XVIII* (two copies)	(1853)	(Joseph D. Pemberton)	H1/1, fo.59 fo.61	4 inches:1 mile 37.2 x 25.6 cm 43.6 x 35.6 cm	102
364A	OUTLINE OF WORKS & IMPROVEMENTS EFFECTED AT NANAIMO. *May 1853* (two copies)	1853	(Joseph W. McKay)	G1/258g	1 inch:ca. 1000 feet 57.4 x 48 cm	103
365A	*No. 2, Shoreline of Bellingham Bay in the vicinity of the coal measures, and the general conformation of the ground by horizontal curves at intervals of 60 feet*	(1853)	(W.P. Trowbridge)	A10/35, fo.197	55 x 40.3 cm	103
366A	Yukon District showing positions of various Indian tribes in 1853	1853	William L. Hardisty	D5/38, fo.77	1 inch:ca. 230 miles 15.5 x 20 cm	106–7, 57
367A	Plan of new Fort Simpson, Mackenzie River	1853	James Anderson (A)*	D5/36, fo.462	1 inch:ca. 52 feet 15.5 x 19.8 cm	
368A	*Copy Of SECTION OF MAP OF OREGON AND UPPER CALIFORNIA From the Surveys of JOHN CHARLES FREMONT And other Authorities DRAWN BY CHARLES PREUSS Under the Order of the SENATE OF THE UNITED STATES Washington City 1848 Exhibiting the Boundary line between the possessions of Great Britain and the United States as therein laid down* Inset map: *A PART OF THE ARRO ARCHIPELAGO ENLARGED*	(1853)	c.u.	G1/155	1 inch:ca. 48 miles 44.8 x 39 cm 1 inch:ca. 11 miles 10 x 21 cm	108
369A	Map of southern tip of Vancouver Island and Oregon Territory showing boundary line between possessions of Great Britain and United States: "Traced from a Map drawn under the order of the Senate of the United States at Washington 1848 (J. Arrowsmith)"	(1853)	c.u.	G1/156	1 inch:ca. 48 miles 13.2 x 14.5 cm	108
370A	*PLAN OF COLVILE TOWN SHEWING IMPROVEMENTS MADE PREVIOUS TO MAY 8, 1854*	(1854)	(Joseph D. Pemberton)	G1/258h	1 inch:ca. 200 feet 97.4 x 73.4 cm	103
371A	*PLAN OF COLVILE TOWN SHEWING IMPROVEMENTS MADE PREVIOUS TO MAY 8 1854*	(1854)	(Joseph D. Pemberton)	G1/258i	1 inch:ca.200 feet 95 x 70 cm	103

* The HBC used (A) and (B) to distinguish between two James Andersons. See also 412A, 467A, 479A.

Manuscript Maps in the Hudson's Bay Company Archives

Catalogue A

NO.	TITLE OR DESCRIPTION	DATE	CARTOGRAPHER	REFERENCE	SCALE/SIZE	TEXT
372A	"A Rough Outline shewing lands at Nanaimo resurveyed and allotted to the H.B. Co. Previous to May 10th.54"	(1854)	(Joseph D. Pemberton)	A11/75,fo.232	10 x 20 cm	103
373A	PLAN OF SOOKE DISTRICT, VANCOUVER ISLAND	(1854)	(Joseph D. Pemberton)	G1/136	2 inches:1 mile 58.2 x 84 cm	103
374A	PLAN OF VICTORIA AND ESQUIMALT DISTRICTS – VANCOUVER ISLAND	(1854)	(Joseph D. Pemberton)	G1/253	2 inches:1 mile 65 x 99.5 cm	103
375A	Map of Esquimalt District showing Esquimalt Farm, Craigflower Farm, Constance Cove Farm, and Viewfield Farm	(1854)	(Joseph D. Pemberton)	F25/1,fo.16	4 inches:1 mile 76.4 x 67.8 cm	103
376A	CRAIGFLOWER FARM	(1854)	(Joseph D. Pemberton)	F25.1,fo.,14	4 inches:1 mile 30.3 x 24.3 cm	103
377A	Victoria DISTRICT, Lot 24 Section XVIII	(1854)	Joseph D. Pemberton	A11/75, fo.302	4 inches:1 mile 24.5 x 33.9 cm	103
378A	Esquimalt District LOT 26 Section I	(1854)	(Joseph D. Pemberton)	H1/1,fo.69	4 inches:1 mile 25 x 32.6 cm	105,56
379A	Esquimalt District Lot 27 Section II	1854	(Joseph D. Pemberton)	A37/42, fo.189	4 inches:1 mile 25 x 35 cm	103
380A	Esquimalt District LOT 28 Section X	(1854)	(Joseph D. Pemberton)	H1/1,fo.71	4 inches:1 mile 23.6 x 34 cm	103
381A	Esquimalt District LOT 29 Section XI	(1854)	(Joseph D. Pemberton)	H1/1,fo.75	4 inches:1 mile 22.4 x 30.5 cm	103
382A	Victoria District LOT 30 Section XXI	(1854)	(Joseph D. Pemberton)	H1/1,fo.77	4 inches:1 mile 23.4 x 26.5 cm	103
383A	Victoria District Lot 30 Sect. XXI (two copies)	(1854)	(Joseph D. Pemberton)	F16.2, fo.216 fo.218	4 inches:1 mile 18.5 x 12.5 cm 18.8 x 12.2 cm	103
384A	Victoria District LOT 33 Section XXIX	(1854)	(Joseph D. Pemberton)	H1/1,fo.85	4 inches:1 mile 34.8 x 33.5 cm	103
385A	Sketch of the proposed coal workings upon Newcastle Island, Nanaimo by Boyd Gilmour. (two copies)	(1854)	Boyd Gilmour	A11/75, fos.417–18, 423–4	21.3 x 27.2 cm	104

NO.	TITLE OR DESCRIPTION	DATE	CARTOGRAPHER	REFERENCE	SCALE/SIZE	TEXT
386A	Sketch of part of The Reserve of Tadoussac Shewing the Reserve for R. Catholic Church & Burying Ground.	1854	A. Morin	B214/z/1, fo.10	35.5 x 44.5 cm	
387A	Tracing of Map of the southeast corner of Vancouver's Island from Soke to Cowitchin, including the Saanich country and Inlet	1855	Fr.Augt.	G2/33	2 inches:1 mile 174 x 135 cm	104
388A	Esquimalt District LOT 9 Section VIII	(1855)	(Joseph D. Pemberton)	H1/1,fo.35	4 inches:1 mile 25.5 x 17.4 cm	103
389A	Victoria District. LOT 25. Section. XIX	(1855)	(Joseph D. Pemberton)	H1/1,fo.66	4 inches:1 mile 12.8 x 16 cm	103
390A	Victoria District LOT 25. Section no. XIX	(1855)	(Joseph D. Pemberton)	H1/1,fo.67	4 inches:1 mile 17.7 x 15.5 cm	103
391A	Victoria District LOT. 25. Section XIX	(1855)	(Joseph D. Pemberton)	H1/1,fo.63	4 inches:1 mile 16.7 x 19.8 cm	103
392A	Victoria District LOT No 28 Section Nos. 1, 1a, 1b.	(1855)	Joseph D. Pemberton	H1/1,fo.73	4 inches:1 mile 20.3 x 15.3 cm	103
393A	Victoria District LOT 31. Section XXXI	(1855)	(Joseph D. Pemberton)	H1/1,fo.79	4 inches:1 mile 37.7 x 25 cm	103
394A	Victoria District LOT 32 Section XXXII	(1855)	(Joseph D. Pemberton)	H1/1,fo.81	4 inches:1 mile 32.8 x 24.5 cm	103
395A	Victoria District LOT 33 Section XXIX	(1855)	(Joseph D. Pemberton)	H1/1,fo.83	4 inches:1 mile 36 x 26.5 cm	103
396A	Victoria District LOT 34 Section XXX	(1855)	(Joseph D. Pemberton)	H1/1,fo.87	4 inches:1 mile 37.6 x 28 cm	103
397A	Victoria District LOT 35 Section XXII	(1855)	(Joseph D. Pemberton)	H1/1,fo.89	4 inches:1 mile 22.5 x 17 cm	103
398A	Victoria District LOT 36 Section XXIII	(1855)	(Joseph D. Pemberton)	H1/1,fo.91	4 inches:1 mile 23 x 18.6 cm	103
399A	Victoria District LOT 37 Section XXIV	(1855)	(Joseph D. Pemberton)	H1/1,fo.95	4 inches:1 mile 20.4 x 17 cm	103
400A	Victoria District LOT 45. Section IX	(1855)	(Joseph D. Pemberton)	H1/2,fo.16	4 inches:1 mile 21.8 x 18 cm	103

NO.	TITLE OR DESCRIPTION	DATE	CARTOGRAPHER	REFERENCE	SCALE/SIZE	TEXT
401^A	*Esquimalt District Lot 46, Section* XII	(1855)	(Joseph D. Pemberton)	H1/2,fo.18	4 inches:1 mile 18.2 x 14.8 cm	103
402^A	*Esquimalt District LOT 47 Section* XIII	(1855)	(Joseph D. Pemberton)	H1/2,fo.20	4 inches:1 mile 16.5 x 14 cm	103
403^A	*Victoria District LOT 48 Section* XXXIII	(1855)	(Joseph D. Pemberton)	H1/2,fo.22	4 inches:1 mile 31 x 24 cm	103
404^A	*ESQUIMALT DISTRICT LOT 50. Sec.* IV.	(1855)	(Joseph D. Pemberton)	H1/2,fo.24	2 inches:1 mile 13.8 x 13 cm	103
405^A	*ESQUIMALT DISTRICT LOT 51. Section* XIV.	(1855)	(Joseph D. Pemberton)	H1/2,fo.26	4 inches:1 mile 23 x 18.4 cm	103
406^A	*Tracing from the Plan of Work done at the Nanaimo Colliery during the Quarter Ending March 31/55*	1855	George Robinson	A11/75,fo.595	1 inch:132 feet 29.8 x 50 cm	104
407^A	*Plan of the Works at Nanaimo and Pemberton's Encampment – July 16/55.*	1855	George Robinson	A11/75,fo.671	1 inch:132 feet 21.5 x 43 cm	104
408^A	*Plan of the Coal Works at Newcastle Island. July 16/55. Geo^e Robinson Mine Agent*	1855	George Robinson	A11/75,fo.672	21.8 x 24.7 cm	104
409^A	*Copy of the Mining Surveys at the Small Shaft and at Pembertons Encampment, Nanaimo-Colliery – V.I. Sepr. 29/55 George Robinson Mine Agent*	1855	George Robinson	A11/75,fo.809	1 inch:132 feet 24.5 x 46 cm	104
410^A	*Tracings of the workings at the Small Shaft, and at Pembertons Encampt-Nanaimo.... Geo. E. Robinson Novr.20/55*	1855	George Robinson	A11/75,fo.876	1 inch:132 feet 25.3 x 46 cm	104
411^A	*Township of Egan on rivers Gatineau and Desert, showing lands required by* HBC	1855	Archibald McNaughton	D5/40,fo.129	2.5 inches:1 mile 12.2 x 18.5 cm	109
412^A	*Rough Sketch of the course of the Beghulatesse or Inconnue River and adjacent Country*	1855	James Anderson (A)	B200/b/31, p. 68	1 inch:ca. 58 miles 19.5 x 31.5 cm	107,58
413^A	*Tracing of map of Vancouver Island from Nanaimo and Qualicum across to Barclay Sound and Port San Juan*	(1856)	(Joseph D. Pemberton)	A11/76, fo.403B	1 inch:3.5 miles 82 x 65.5 cm	104
414^A	*Esquimalt District LOT no. 53. Section no.* XVIII	1856	(Joseph D. Pemberton)	H1/2,fo.28	4 inches:1 mile 16.7 x 18 cm	105
415^A	*Esquimalt District LOT no. 53. Section no.* XVIII	1856	(Joseph D. Pemberton)	H1/2,fo.30	4 inches:1 mile 16.3 x 18 cm	105

NO.	TITLE OR DESCRIPTION	DATE	CARTOGRAPHER	REFERENCE	SCALE/SIZE	TEXT
416A	Esquimalt district. LOT no. 59. Sections nos. XXI, XXII.	1856	(Joseph D. Pemberton)	H1/2,fo.39	4 inches:1 mile 9.7 x 7.8 cm	105
417A	Survey of the Hone-Hudson's Bay Comps Coal Mines, at Park-Head, Nanaimo. Vancouvers Island Decr 31st–1856. Geo E Robinson. Mine Agent	1856	(George Robinson)	A11/76,fo.451	1 inch:132 feet 38.4 x 50.5 cm	104
418A	Coal formation at Nanaimo and Newcastle Island … March 3rd 1856. Geo E Robinson Mine Agent (two copies)	1856	George Robinson	A11/76a, fos.85–85d, 111–111d	1 inch:¼ mile	104
419A	"a tracing by Mr. Ross of part of a Map, Sent by Sir J. Richardson to Mr Murray" showing part of Alaska and the Yukon	(1856)	(Bernard Ross)	D7/1,fo.238b	1 inch:220 miles 10.8 x 20 cm	107
420A	Copy of map alluded to in preceding letter–I can make nothing of it JA (Letter refers to Indian tale of finding encampment, thought to be Franklin's but probably James Anderson's (A), on the shores of the Thlewycho desse)	1856	Lawrence Clarke	B200/b/32, p.158	15 x 19.5 cm	
421A	Plan of [mining] Location at Current River Thunder Bay – Lake Superior	(1856)	John Mackenzie	B134/b/13, fo.294	Scale varies 16.3 x 21.7 cm	
422A	Plan of Fort Rupert Vancouver's Island	1857	(Joseph D. Pemberton)	A11/76,fo.669	1 inch:50 feet 37.4 x 48 cm	104,113
423A	Plan of Fort Rupert, Vancouver's Island	1857	(Joseph D. Pemberton)	A11/76,fo.672	1 inch:50 feet 30.6 x 46.3 cm	104
424A	MAP.OF.HON'BLE. HUDSON'S. BAY. CO'S. BUILDINGS. AT. VICTORIA	(1857)	(Joseph D. Pemberton)	A11/76,fo.694	4 inches:1 mile 28.4 x 39.4 cm	104
425A	MAP. OF. HON'BLE. HUDSONS BAY. CO'S. BUILDING'S. AT. VICTORIA	(1857)	(Joseph D. Pemberton)	A11/76,fo.700	4 inches:1 mile 28.5 x 41.6 cm	104
426A	Map showing part of Vancouver Island, between Straits of Juan de Fuca and Cowichin Harbour	(1857)	(Joseph D. Pemberton)	G1/135	1 inch:3.5 miles 41 x 64.5 cm	105
427A	Sketch of Journey of Exploration through the Cowetchin Valley Vancouver's Island, made by Mr. J.D. Pemberton Colonial Surveyor, October 1857	1857	Joseph D. Pemberton	G1/139	43.5 x 67.5 cm	105,55
428A	Victoria District. LOT no. 55 Section no. XLVI	(1857)	(Joseph D. Pemberton)	H1/2,fo.36	4 inches:1 mile 15.5 x 17.8 cm	105
429A	Victoria District Section nos 1c.1d. LOT no. 60.	(1857)	(Joseph D. Pemberton)	H1/2,fo.42	4 inches:1 mile 21 x 15.5 cm	105

Manuscript Maps in the Hudson's Bay Company Archives

NO.	TITLE OR DESCRIPTION	DATE	CARTOGRAPHER	REFERENCE	SCALE/SIZE	TEXT
430A	*Lake District Section II LOT no. 62.*	(1857)	(Joseph D. Pemberton)	H1/2,fo.44	4 inches:1 mile 14.8 x 9.8 cm	105
431A	*Lake District. Sect III LOT no.63.*	1857	(Joseph D. Pemberton)	H1/2,fo.46	4 inches:1 mile 13.4 x 9.5 cm	105
432A	*Lake District Section IV LOT no.64*	(1857)	(Joseph D. Pemberton)	H1/2,fo.48	4 inches:1 mile 14.7 x 9 cm	105
433A	*Lake District. Section VI LOT no.65*	(1857)	(Joseph D. Pemberton)	H1/2,fo.50	4 inches:1 mile 13.4 x 14.4 cm	105
434A	*Esquimalt District LOT no. 67. Section no. XX*	(1857)	(Joseph D. Pemberton)	H1/2,fo.55	4 inches:1 mile 7.5 x 7.5 cm	105
435A	*Survey of the Hon^e Hudson's Bay Compys Coal Mines at Park-Head, Nanaimo, Vancouvers Island May 23rd 1857. George Robinson. Mine Agent.*	1857	George Robinson	A11/76,fo.614	1 inch:132 feet 38 x 50.5 cm	104
436A	*Survey of the Hon^e Hudson's Bay Compy's Coal Mines at Park-Head, Nanaimo, Vancouver's Island. May 23rd 1857. George Robinson. Mine Agent.*	1857	George Robinson	A11/76,fo.641	1 inch:132 feet 37.1 x 50 cm	104
437A	*"a small diagram of the Lands as surveyed by Mr McArthur in behalf of Mr Rooney ..."* at River Desert	(1857)	(Archibald McNaughton)	B134/c/74, fo.204	5 inches:1 mile 13.6 x 16 cm	109,60
438A	*Part of Township of Egan*: Lots 1 and 2, Range A, showing improvements by Hudson's Bay Company and Mr Aubert	1857	E.P. Tache	D5/44,fo.25	10 inches:ca.1 mile 17.7 x 22.8 cm	109
439A	*Victoria District LOT 37 Sections XIII–XV. A–XXIV*	(1858)	(Joseph D. Pemberton)	H1/1,fo.93	4 inches:1 mile 18.4 x 12.3 cm	105
440A	*Victoria District LOT 39. Section XXVI*	(1858)	(Joseph D. Pemberton)	H1/2,fo.2	4 inches:1 mile 17.3 x 21.7 cm	105
441A	*Victoria District LOT 40 Section XXVII*	(1858)	(Joseph D. Pemberton)	H1/2,fo.4	4 inches:1 mile 22.3 x 18.6 cm	105
442A	*Victoria District Lot 44 Section LXXI*	(1858)	(Joseph D. Pemberton)	H1/2,fo.14	4 inches:1 mile 18.7 x 12.2 cm	105
443A	*Esquimalt District LOT no. 54 Section no. XV*	(1858)	(Joseph D. Pemberton)	H1/2,fo.34	8 inches:1 mile 25 x 26.4 cm	105

NO.	TITLE OR DESCRIPTION	DATE	CARTOGRAPHER	REFERENCE	SCALE/SIZE	TEXT
444^A	*Esquimalt District LOT no.66 Section no. V*	(1858)	(Joseph D. Pemberton)	H1/2,fo.53	4 inches:1 mile 15 x 21 cm	105
445^A	*Esquimalt District LOT no.66 Section no. V*	(1858)	(Joseph D. Pemberton)	H1/2,fo.54	4 inches:1 mile 15 x 21.5 cm	105
446^A	*Esquimalt District LOT no.68 Section no. XIX*	(1858)	(Joseph D. Pemberton)	H1/2,fo.57	4 inches:1 mile 15.3 x 8.7 cm	105
447^A	*Victoria District LOT 69 Section LI*	(1858)	(Joseph D. Pemberton)	H1/2,fo.59	4 inches:1 mile 8.8 x 12.4 cm	105
448^A	*Victoria District LOT 72. Section LXXII*	(1858)	(Joseph D. Pemberton)	H1/2,fo.61	4 inches:1 mile 18.6 x 12 cm	105
449^A	*Victoria District LOT 79. Section LXI*	(1858)	(Joseph D. Pemberton)	H1/2,fo.63	4 inches:1 mile 18.6 x 12 cm	105
450^A	*Victoria District LOT 81 Sections LXII. LXIII*	(1858)	(Joseph D. Pemberton)	H1/2,fo.65	4 inches:1 mile 18.7 x 15.5 cm	105
451^A	*Lake District LOT 85 Sect. XIV.*	(1858)	(Joseph D. Pemberton)	H1/2,fo.67	4 inches:1 mile 19 x 12.2 cm	105
452^A	*Lake District Lot 86 Section XV*	1858	(Joseph D. Pemberton)	H1/2,fo.69	4 inches:1 mile 18.6 x 12.3	105
453^A	*Lake District Lot 87 Section XVI*	1858	(Joseph D. Pemberton)	H1/2,fo.71	4 inches:1 mile 18.5 x 12.3	105
454^A	*Lake District Lot 88 Section XVII.*	1858	(Joseph D. Pemberton)	H1/2,fo.73	4 inches:1 mile 18.6 x 12.3 cm	105
455^A	*Victoria District Lot 95 Sections XXXVI, XLII and XLIII*	1858	(Joseph D. Pemberton)	H1/2,fo.75	4 inches:1 mile 18.6 x 18.1 cm	105
456^A	*Victoria District Lot 99 Section LXV*	(1858)	(Joseph D. Pemberton)	H1/2,fo.77	4 inches: 1 mile 18.6 x 12.3 cm	105
457^A	*Esquimalt District Lot 105 Section XVI*	(1858)	(Joseph D. Pemberton)	H1/2,fo.79	4 inches:1 mile 18.4 x 12 cm	105
458^A	*Metchosin District Lot no.110 Section no II*	(1858)	(Joseph D. Pemberton)	H1/2,fo.81	4 inches:1 mile 18.5 x 12.4	105
459^A	*Esquimalt District Lot no111 Section no XXIX*	(1858)	(Joseph D. Pemberton)	H1/2,fo.83	4 inches:1 mile 18.4 x 12.4 cm	105

Manuscript Maps in the Hudson's Bay Company Archives

NO.	TITLE OR DESCRIPTION	DATE	CARTOGRAPHER	REFERENCE	SCALE/SIZE	TEXT
460A	*Esquimalt District Lot 112 Section xxx*	(1858)	(Joseph D. Pemberton)	H1/2,fo.85	4 inches:1 mile 18.5 x 12.1 cm	105
461A	*Esquimalt District Lot no.117 Section no LXVI*	(1858)	(Joseph D. Pemberton)	H1/2,fo.87	4 inches:1 mile 16.8 x 12 cm	105
462A	*VANCOUVER'S ISLAND. PUGET SOUND COMPANY'S FARMS roughly sketched*	(1858)	(Alexander G. Dallas)	F25/1,fo.39	1 inch:1 mile 28 x 34.2 cm	105
463A	*Survey of the Hone Hudson's Bay Compys Coal mine at the No.3 Shaft, Nanaimo. Vancouvers Island. Jany 23rd 1858 George Robinson Mine Agent.* (two copies)	1858	George Robinson	A11/76,fo.,978; A11/76,fo.982	1 inch:132 feet 25 x 38.5 cm	104,54
464A	Plan of town lots at Fort Hope	1858	(Alexander G. Dallas)	F12/4,fo.544	22.2 x 14.5 cm	107
465A	Plan of town lots at Fort Yale	1858	(Alexander G. Dallas)	F12/4,fo.545	12 x 22.7 cm	107, 59
466A	*A rough plan of the english narrows in the Messier channel on the West coast of Patagonia between about 49° to 50° South*	(1858)	J.F. Trivett	C7/79,fo.21	1 inch:ca.360 feet 32.4 x 40.5 cm	109
467A	*CHART OF A PORTION OF THE EASTMAIN COAST 1858, north of Great Whale River to Mosquito Bay, from observations and notes of James Anderson (B)*	1858	Alexander McDonald	G2/49	1 inch:ca. 9.5 miles 33 x 123.4 cm	109
468A	*CHART of Entrance to LITTLE WHALE RIVER*	(1858)	Alexander McDonald	G1/284	1 inch:600 feet 32 x 60.2 cm	109
469A	Plan of part of town of Victoria, Bastion St., Langley St. area, showing town lots and outlines of some buildings	(1859)	(Joseph D. Pemberton)	H1/9,fo.17	1 inch:33 feet 29.5 x 33.5 cm	105
470A	Plan of two sections in the North Saanich District, alloted to John Irving, i.e., Sect. III E 13 N, IV E 13 N	(1859)	(Joseph D. Pemberton)	A11/77,fo.164	2 inches:1 mile 14 x 14.7 cm	105
471A	*FORT VANCOUVER and U.S. MILITARY POST with Town Environs &c 1859*	1859	Richard Covington	G1/128	2.5 inches:1 mile 61.2 x 87.4 cm	108
472A	*An outline Chart shewing the Track of the Honble. Hudsons Bay Company's Steam Vessel "Labouchere" from the Atlantic to the Pacific Oceans through the Straits of Magelhaens, and the inner Passages of Sarmiento, and Smyth's Channels in November and December.1858 J.F. Trivett*	(1859)	J.F. Trivett	G1/173	1 inch:ca. 24 miles 66 x 53 cm	109
473A	*An outline Chart shewing the tracks of the Honble. Hudsons Bay Company's Ship "Princess Royal" from Vancouvers Island to London*	(1859)	J.F. Trivett	G1/175	1 inch:ca. 425 miles at Equator. 53 x 65 cm	109

NO.	TITLE OR DESCRIPTION	DATE	CARTOGRAPHER	REFERENCE	SCALE/SIZE	TEXT
474A	Chart showing the route of the HBC steamship *Labouchere* from London to Victoria, 8 September 1858 to 28 January 1859, via the Strait of Magellan and Inner Passage, Chile; and of the *Princess Royal* from London to Victoria, 22 September 1858 to 31 January 1859, around Cape Horn	(1859)	J.F. Trivett	G1/174	1 inch:ca. 425 miles at Equator 53 x 65.7 cm	109
475A	"Chart of country from Saint Cloud to Georgetown – 1859"	1859	c.u.	G1/239	1 inch:12 miles 33.5 x 36.7 cm	
476A	PLAN OF BUILDINGS ON THE PUGET SOUND COMPANY'S FARM AT ESUIMALT	1860	c.u.	F25/1,fo.22	1 inch:116 feet 37.5 x 42 cm	112
477A	Ground plan of government buildings on HBC land, on Langley, Government, and Bastion Streets, Victoria	1860	c.u.	A11/77,fo.524	1 inch:33 feet 31.7 x 35.2 cm	105,110
478A	NANAIMO DISTRICT	(1860)	c.u.	G1/258	55.5 x 31.6 cm	113
479A	Mingan River mouth, Mingan Island, Mingan Post on St. Lawrence River, and the proposed HBC claim	(1860)	(James Anderson (A))	D5/52, fo.577d	1 inch:1800 feet 20.8 x 32.8 cm	114–15, 63
480A	*Outline Chart shewing the Outward and Homeward tracks of the Honble. Hudsons Bay Company's Ship Princess Royal. between England and Vancouver Island 1859–60 JFT*	(1860)	J.F. Trivett	G1/176	1 inch:ca. 425 miles at Equator 66 x 53.2 cm	109
481A	Plan of warehouse property and adjoining properties, Victoria, waterfront, Wharf and Bastion streets area	1861	Alexander G. Dallas	A11/78,fo.193	1 inch:ca. 35 feet 23.7 x 19.7 cm	
482A	Plan of Government Reserve and surrounding property, south of James Bay, Victoria, sold by the HBC, 1859–1861	(1861)	c.u.	A11/80,fo.627	1 inch:ca. 220 feet 34 x 38 cm	110
483A	Plan of part of Victoria, showing government buildings, Governor Douglas's residence, and other properties. "F.W. Green.C.E. Victoria 1861"	1861	F.W. Green	G1/145	1 inch:264 feet 28.5 x 32 cm	111
484A	Plan of Maple Point School Reserve and Grant to Thomas Russell on Portage Inlet, Vancouver Island	(1861)	(Land Office, Victoria)	F16/2,fo.261	1 inch:164 feet 33.5 x 23 cm	
485A	PLAN OF NANAIMO	ca. 1861	c.u.	G1/142	1 inch:264 feet 62.6 x 96 cm	113, 285n15
	Inset map: PLAN OF HUDSON'S BAY COY LAND. NANAIMO				1 inch:1 mile 21.5 x 32 cm	

Manuscript Maps in the Hudson's Bay Company Archives

NO.	TITLE OR DESCRIPTION	DATE	CARTOGRAPHER	REFERENCE	SCALE/SIZE	TEXT
486A	Tracing of land owned by HBC, Nanaimo, attached to draft agreement for sale of estate to James Nicol, in connection with proposed Vancouver Coal Mining and Land Company	(1861)	c.u.	F33/1,fo.1	2 inches:1 mile 35.2 x 26.7 cm	113
487A	*Plan of Victoria District Lot 24 Sec.18. No.2.*	(Post 1861)	(Land Office, Victoria)	G1/146	1 inch:528 feet 59.4 x 73.5 cm	
488A	Sketch plan of part of central Nanaimo	(1861–2)	c.u.	A11/78,fo.2	24.2 x 35.4 cm	113
489A	Plan of lots around Indian Reserve, Nanaimo	(1861–3)	c.u.	A11/78,fo.1	25.6 x 25.5 cm	113
490A	*No 1. Harbour Master's office, Lot 15, Block 70, on the map of Victoria Town, and coloured red in the accompanying tracing. Victoria, Vancouver Island, 29 July, 1862*	1862	c.u.	A11/78,fo.481	1 inch:ca. 230 feet 22.8 x 32.8 cm	111
491A	*No 2. Police barrack and prison lot, as described and coloured in the accompanying tracing. Victoria, Vancouver Island, 29 July, 1862*	1862	c.u.	A11/78,fo.484	1 inch:ca. 65 feet 25.2 x 27 cm	111
492A	*No. 3. Post Office lots, Lots 1603, 1605, and 1607 in Block one, on the map of Victoria Town, and coloured in the accompanying tracing. Victoria, Vancouver Island, 29 July, 1862*	1862	c.u.	A11/78,fo.487	1 inch:ca. 35 feet 19.3 x 30.4 cm	111
493A	*No. 4. Public park and school reserve, Victoria, Vancouver Island, 29 July, 1862*	1862	c.u.	A11/78,fo.490	1 inch:1320 feet 21.5 x 21.8 cm	111
494A	*No. 6. Government offices, Victoria, Vancouver Island, 29 July, 1862*	1862	c.u.	A11/78,fo.493	25.2 x 24 cm	111
495A	Plan of land owned by HBC, Nanaimo, attached to indenture between the company and the Vancouver Coal Mining and Land Company, on sale of estate (two copies)	(1862)	c.u.	F33/1,fo.31A; F33/1,fo.36	2 inches:1 mile 41 x 32.5 cm	113
496A	Sketch plan of Blocks 75 and 76, Victoria townsite, highlighting Lot 1, Block 76	1862	c.u.	B226/z/1,fo.79	6 x 7.5 cm	
497A	Sketch plan of Blocks 75 and 76, Victoria Townsite, highlighting Lot 17, Block 76	1862	c.u.	B226/z/1,fo.80	6 x 7.5 cm	
498A	Sketch plan of Blocks 75 and 76, Victoria Townsite, highlighting Lot 18, Block 76	1862	c.u.	B226/z/1,fo.81	6 x 7.5 cm	

NO.	TITLE OR DESCRIPTION	DATE	CARTOGRAPHER	REFERENCE	SCALE/SIZE	TEXT
499A	Sketch plan of Blocks 75 and 76, Victoria Townsite, highlighting Lot 7, Block 75	1862	c.u.	B226/z/1,fo.82	6 x 7.5 cm	
500A	Sketch plan of Blocks 75 and 76, Victoria Townsite, highlighting Lot 8, Block 75	1862	c.u.	B226/z/1,fo.83	6 x 7.5 cm	
501A	Plan of lot sold to Capt. W. H. Franklyn, in Nanaimo, adjoining the Weslyan parsonage lot, by the HBC, 3 March 1862	(1862)	c.u.	H1/9,fo.21	1 inch:132 feet 25 x 20 cm	113
502A	Plan of "Lot of Land sold to William Hales Franklyn by the HBCy. Nanaimo"	(1862)	c.u.	H1/9,fo.23	1 inch:132 feet 26.5 x 20.5 cm	113
503A	Beckley Farm, Victoria, showing land unsold in 1861. "Red river Feby..63. "A.G.D."	1863	Alexander G. Dallas	A12/43,fo.86	15.3 x 24.8 cm	111, 285n3
504A	Copy of *Map of the Subdivisions of Beckley Farm. and the Lands south and west of James Bay Victoria, Vancouver Island. 7th.January 1863*	1863	(F.W. Green)	A11/80,fo.429	1 inch:ca. 200 feet 73 x 100.5 cm	111
505A	*TOWN OF VICTORIA V.I. SUBDIVISION OF SECTION NO. 1. 1863*	1863	H.O. Tiedemann	G2/35	1 inch:132 feet 205.7 x 218 cm	112
506A	*TOWN OF VICTORIA V.I. SUBDIVISION OF SECTION NO 1. 1863*	1863	(H.O. Tiedemann)	G1/252	1 inch:528 feet 59 x 61.2 cm	112
507A	*Map or Tracing of Land at Nanaimo Vancouver Island Victoria V.I. 3 February 1863*	1863	c.u.	A11/79,fo.38	1 inch:1 mile 31.3 x 26 cm	113
508A	Tracing of land lots at Nanaimo with enclosure drafting reconveyance of land and hereditaments in the District of Nanaimo	(1863)	c.u.	F33/1,fo.63	1 inch:200 feet 38 x 50.5 cm	113
509A	Plan of part of Nanaimo, Vancouver Island, showing *Property released by Deed from H.B. Co. dated 28th, August 1863*	(1863)	c.u.	G1/154	1 inch:ca.380 feet 35.5 x 65.5 cm	113
510A	*PLAN OF ONE HUNDRED ACRES OF LAND Preempted by THE HONLE. HUDSON'S BAY CO. at Fort Rupert, V.I. P. N. Compton Fort Rupert V.I. Octr. 1863.*	1863	P.N. Compton	G1/231	1 inch:132 feet 53 x 66.2 cm	113,61
511A	*MAP OF MONTANA WASHINGTON IDAHO AND OREGON SHEWING THE HUDSON'S BAY AND PUGET'S SOUND AGRICULTURAL COMPANIES CLAIMS.*	(1863–64)	c.u.	G2/34	1 inch:16 miles 200.7 x 261.6 cm	114
512A	*MAP OF THE DISTRICT OF VICTORIA &c. IN VANCOUVER ID. No. 1*	(1864)	c.u.	A11/80,fo.423	2 inches:1 mile 57 x 54 cm	112

Manuscript Maps in the Hudson's Bay Company Archives

NO.	TITLE OR DESCRIPTION	DATE	CARTOGRAPHER	REFERENCE	SCALE/SIZE	TEXT
513^A	Tracing of District School Reserve showing different survey lines related to land dispute between HBC and Colonial Government	1864	c.u.	A11/79,fo.491	1 inch:120 feet 36.3 x 38.3 cm	112
514^A	*Victoria District LOT 42 Section XVIII A* (two copies)	(1864)	(Land Office, Victoria)	H1/2,fo.8 H1/2,fo.6	4 inches:1 mile 17.2 x 12.4 cm	112
515^A	*Victoria District LOT 43 Section XIX A*	(1864)	(LAND OFFICE, VICTORIA)	H1/2,fo.10	4 inches:1 mile 21.1 x 16.4 cm	112
516^A	*Victoria District LOT 44 Section XX*	(1864)	(Land Office, Victoria)	H1/2,fo.12	4 inches:1 mile 21.4 x 18.5 cm	112
517^A	*Victoria District LOT NO.57. Section no. XLVIII.*	(1864)	(Land Office, Victoria)	H1/2,fo.38	4 inches:1 mile 18.2 x 14.5 cm	112
518^A	*Victoria District Lot no.118 Section no. LXXXIII*	(1864)	(Land Office, Victoria)	H1/2,fo.89	4 inches:1 mile 15.3 x 10.1 cm	112
519^A	*Victoria District Lot no.119 Section no. LXXX*	(1864)	(Land Office, Victoria)	H1/2,fo.91	4 inches:1 mile 18.4 x 13 cm	112
520^A	*Victoria District Lot no.120 Section no. XX*	(1864)	(Land Office, Victoria)	H1/2,fo.93	4 inches:1 mile 16.8 x 11.1 cm	112
521^A	*Sooke District Lot no. 121. Section no. IV.*	(1864)	(Land Office, Victoria)	H1/2,fo.95	4 inches:1 mile 18.4 x 15 cm	112
522^A	*Victoria District Lot no.122 Section no XXXIV*	(1864)	(Land Office, Victoria)	H1/2,fo.97	4 inches:1 mile 15.4 x 11.1 cm	112
523^A	*Esquimalt District Lot no 123. Sections XCI. XCII. XCIII C.*	(1864)	(Land Office, Victoria)	H1/2,fo.99	4 inches:1 mile 25.8 x 27.2 cm	112
524^A	*Esquimalt Dist. Lot No 124. Section No. XXVII*	(1864)	(Land Office, Victoria)	H1/2,fo.101	4 inches:1 mile 18.3 x 16.4 cm	112
525^A	*Victoria Dist. Lot No. 125. Section No. LXXVII*	(1864)	(Land Office, Victoria)	H1/2,fo.103	4 inches:1 mile 16.3 x 14.3 cm	112
526^A	*Esquimalt Dist. Lot No. 126. Section No. XXXII*	(1864)	(Land Office, Victoria)	H1/2,fo.105	4 inches:1 mile 14.3 x 14 cm	112
527^A	*Esquimalt Dist. Lot No. 127 Section No XXXI.*	(1864)	(Land Office, Victoria)	H1/2,fo.107	4 inches:1 mile 18.5 x 16 cm	112
528^A	*Victoria District Lot nr. 128 Section Nos. LXX. LXXXVII. LXXXVIII*	(1864)	(Land Office, Victoria)	H1/2,fo.109	4 inches:1 mile 25.7 x 24.5 cm	112

NO.	TITLE OR DESCRIPTION	DATE	CARTOGRAPHER	REFERENCE	SCALE/SIZE	TEXT
529A	*Victoria District Lot no. 128 Section nos. LXX, LXXXVII, LXXXVIII*	(1864)	(Land Office, Victoria)	H1/2,fo.111	4 inches:1 mile 24.4 x 25.5 cm	112
530A	*Esquimalt Dist. Lot No 129. Section No XCVI*	(1864)	(Land Office, Victoria)	H1/2,fo.113	4 inches:1 mile 18.3 x 16.4 cm	112
531A	*Lake Dist. Lot No. 130 Section No. 1.*	(1864)	(Land Office, Victoria)	H1/2,fo.116	4 inches:1 mile 15.9 x 18.6 cm	112
532A	*Lake Dist. Lot No. 130 Section No. 1.*	(1864)	(Land Office, Victoria)	H1/2,fo.117	4 inches:1 mile 16.6 x 18.5 cm	112
533A	Esquimalt Dist Lot No. 131 Section No XXVIII.	(1864)	(Land Office, Victoria)	H1/2,fo.119	4 inches:1 mile 17.4 x 13.3 cm	112
534A	"Plan of Claim at Fort Shepherd B.C. August 26th. 1864"	(1864)	(W.F. Tolmie)	G1/318	1 inch:ca.570 feet 19.6 x 31.5 cm	114,62
535A	Rough sketch to show position of lot purchased by John Schultz, Red River Settlement	(1864)	(William McTavish)	E8/8,fo.107Bd	1 inch:ca. 420 feet 19.8 x 23.3 cm	115
536A	*WORKING PLAN of the proposed TELEGRAPH LINE from FORT GARRY to the ROCKY MOUNTAINS. 1864.*	1864	(A.W. Schwieger)	G1/327	1 inch:ca. 24 miles 71 x 122.4 cm	116,64
537A	*ESQUIMALT Suburban LOTS H. O. Tiedemann C.E.*, survey plan of Viewfield Farm	(1865)	H.O. Tiedemann	F25/1,fo.27	1 inch:ca. 240 feet 69.7 x 94 cm	112
538A	*Victoria District, Lot 24 Section XVIII, Tracing No. 1*	(1865)	(H.O. Tiedemann)	A11/80,fo.165	4 inches:1 mile 48.5 x 47 cm	111
539A	*Victoria District, Lot 24, Section XVIII, Tracing No. 2*	(1865)	(H.O. Tiedemann)	A11/80,fo.428	4 inches:1 mile 38.8 x 44 cm	111
540A	*Victoria District. Lot No. 1, Section No. VI. Tracing No. 2*	(1865)	(H.O. Tiedemann)	A11/80,fo.168;	6 inches:1 mile 30.6 x 23.5 cm	111
541A	*Victoria District. Lot No. 1, Section No. VI (duplicate)*	1865	(H.O. Tiedemann)	A11/80; fo.424	6 inches:1 mile 29.5 x 26.5 cm	111, 285n6
542A	Indian Reserve, Victoria: "Copy of an original plan in the Land Office at Victoria V.I and there Known as the Town Map of 1855." Tracing No. 3	(1865)	(H.O. Tiedemann)	A11/80,fo.167	1 inch:132 feet 25 x 69 cm	111
543A	Indian Reserve, Victoria: "Copy of an Original plan in the Land Office at Victoria, V.I. And there Known as the Town Map (official) of 1858" (duplicate)	(1865)	(H.O. Tiedemann)	A11/80,fo.425	1 inch:132 feet 26 x 67.7 cm	111

Manuscript Maps in the Hudson's Bay Company Archives

NO.	TITLE OR DESCRIPTION	DATE	CARTOGRAPHER	REFERENCE	SCALE/SIZE	TEXT
544A	Plan of property of Governor Douglas and Government Reserve, Victoria: "Copy of an Original plan in the Land Office at Victoria, V.I. And there Known as the Town Map (official) of 1858." *Tracing No. 4*	(1865)	(H.O. Tiedemann)	A11/80,fo.166	1 inch:264 feet 40.7 x 26.3 cm	111
545A	Plan of property of Governor Douglas and Government Reserve, Victoria: "Copy of an Original Plan in the Land Office at Victoria, V.I. And there Known as the Town Map (official) of 1858." (duplicate)	(1865)	(H.O. Tiedemann)	A11/80,fo.426	1 inch:264 feet 23.2 x 39.8 cm	
546A	Large scale tracing showing changes in Government Reserve and Governor Douglas's property between 1855 and 1862. *Tracing No. 5*	(1865)	(H.O. Tiedemann)	A11/80,fo.169a	1 inch:132 feet 53.3 x 63 cm	111
547A	Large scale tracing showing changes in Government Reserve and Governor Douglas's property between 1855 and 1862 (duplicate)	(1865)	(H.O. Tiedemann)	A11/80,fo.427	1 inch:132 feet 52 x 71 cm	111
548A	Plan showing boundaries of Reserve Lot VI and Lot z, on James Bay, Victoria	1865	(Alexander G. Dallas)	A13/14,fo.294	1 inch:ca. 120 feet 12.6 x 20.4 cm	111
549A	Map of the subdivisions of the Beckley Farm, showing (in pink) HBC property; lots sold to individuals or surrendered formally to the Crown	(1865)	c.u.	A12/43,fo.229	1 inch:ca. 530 feet 38.4 x 37 cm	111
550A	Survey of Telegraph Creek and upper part of Shuswap River near Tete Jaune Cache, May 1865	(1865)	(Joseph W. McKay)	E15/13,fo.42	1 inch:8 miles 30.7 x 29.6 cm	116
551A	*MAP OF THE COUNTRY BETWEEN RICHFIELD AND TETE JAUNE'S CACHE Explored by J. W. McKay, For Overland Telegraph Line April May and June 1865*	(1865)	(Joseph W. McKay)	E15/13,fo.41	1 inch:8 miles 30.6 x 53.8 cm	116,65
552A	*SKETCH OF KAMLOOPS, AND SHUSWAP LAKES WITH PART THOMPSONS RIVER By W. A. Mouat* Endorsed "WHN"	(1865)	William H. Newton	A11/80,fo.214	1 inch:10 miles 90 x 60 cm	116
553A	*SECTION XVIII VICTORIA DISTRICT*	(1866)	H.O. Tiedemann	A11/80,fo.430	1 inch:396 feet 74 x 82.5 cm	111–12
554A	*VICTORIA DISTRICT LOT 32 Section XXXII*	(1866)	c.u.	A11/80,fo.649	4 inches:1 mile 37.7 x 33.2 cm	
555A	*VICTORIA DISTRICT LOT 31 Section XXXI*	(1866)	c.u.	A11/80,fo.650	4 inches:1 mile 37.8 x 33.2 cm	

NO.	TITLE OR DESCRIPTION	DATE	CARTOGRAPHER	REFERENCE	SCALE/SIZE	TEXT
556A	VICTORIA DISTRICT LOT 24 Section No.XVIII	(1866)	c.u.	A11/80,fo.651	4 inches:1 mile 36.2 x 32.5	
557A	Tracing shewing property taken possession of at Fort Vancouver – W. T. by the U.S. Military Authorities March 1860	(1866)	c.u.	B223/z/5,fo.49	39.4 x 24.9 cm	114
558A	PLAN accompanying Report of Exploration from Saguenay River to Lake Pipmouagan, & down Bersimis River to St. Lawrence	(1866)	c.u.	G1/326	1 inch:6 miles 54 x 93 cm	
559A	Rough plan showing location of HBC lots at Georgetown, Minnesota	(1866)	(W. Mactavish)	A12/44,fo.132	1 inch:0.5 miles 25.3 x 20.8 cm	115
560A	Savana's Ferry to Seymour The line of navigation of the steamer "Marten" shewing all the information in the possession of the Lands and Works Department, New Westminster, B. C. Joseph W. Trutch Chief Commissioner of Lands & Works July 19th. 1866 Drawn by J. B. Launders	1866	J.B. Launders	G1/324	1 inch:5 miles 48 x 80 cm	116,66
561A	Map of part of Nanaimo, showing land around Commercial Harbour to be reconveyed by HBC to the Vancouver Mining & Land Company.	(1867)	c.u.	F33/1,fo.122	1 inch:ca. 200 feet 39.5 x 83.5 cm	113
562A	Map of part of Nanaimo, showing land around Commercial Harbour, to be reconveyed by HBC to the Vancouver Mining & Land Company	(1867)	(c.u.)	F33/1,fo.116	1 inch:ca. 200 feet 40.5 x 83 cm	113
563A	PLAN OF LANGLEY FARM	(1867)	(William H. Newton)	A11/82,fo.186B	40.6 x 30.5 cm	114
564A	Plan Of Langley Farm	(1867)	(William H. Newton)	B226/b/39	34 x 33.2 cm	114
565A	Notarially attested copy (F) of plan of 12 lots purchased by John McLoughlin at Champoeg, Willamette River, Oregon, attached to letter of July 15, 1848	(1867)	(Montague W.T. Drake)	F16/2,fo.304	1 inch:ca.80 feet 10.5 x 20.2 cm	114
566A	Notarially attested copy (G) of plan of lots purchased by John McLoughlin at Champoeg, Williamette River, Oregon, resurveyed, and plan attached to letter of July 15, 1848	(1867)	(Montague W.T. Drake)	F16/2,fo.306	1 inch:ca. 88 feet 30 x 20.2 cm	114
567A	ROUGH SKETCH OF THE EAGLE CREEK VALLEY Explored by James Bissett Month. July 1867	1867	William H. Newton	G1/322	1 inch:4 miles 27.7 x 42.2 cm	116–17
568A	ROUGH SKETCH OF THE NORTH BRANCH OF THOMPSON RIVER, AND OF THE CLEARWATER RIVER TO QUESNELLE LAKE Explored by James Bissett, September 1867	1867	William H. Newton	G1/323	1 inch:8 miles 54.5 x 50.7 cm	117

Manuscript Maps in the Hudson's Bay Company Archives

Catalogue A

NO.	TITLE OR DESCRIPTION	DATE	CARTOGRAPHER	REFERENCE	SCALE/SIZE	TEXT
569A	*ROUGH SKETCH OF THE COUNTRY BETWEEN GRENVILLE CANAL AND LAC DES FRANCAIS Explored by William Manson. 1867*	1867	William H. Newton	G1/321	1 inch:528 feet 46.5 x 76.4 cm	117
570A	Land on Esquimalt Harbour, selected by HBC for erection of wharf and warehouse, originally shown on Tiedemann's 1866 map of the PSAC farms, Craigflower, Constance Cove, and Viewfield. "H.O. Tiedemann C.E."	1868	H.O. Tiedemann	F25/1,fo.3;	1 inch:99 feet 77.5 x 100.3 cm	112
571A	Land on Esquimalt Harbour, selected by HBC for erection of wharf and warehouse, originally shown on Tiedemann's 1866 map of the PSAC farms, Craigflower, Constance Cove, and Viewfield	(1868)	H.O. Tiedemann	F25/1,fo.25;	1 inch:99 feet 74.7 x 87 cm	
572A	*Plan Shewing The Hon. Hudsons Bay Cos. Warehouse Barkerville. B.C. 13th. October 1868*	1868	W.H. Newton	A10/77,fo.141		112
573A	*Ground plan of the Hudsons Bay Company's Property Quesnelle B.C.*	1869	c.u.	A11/84,fo.540	1 inch:20 feet 25.4 x 58 cm	114
574A	*PLAN OF THE FORT [Victoria] AND THE ADJACENT PROPERTY. H.B.CO.*	1870	c.u.	G1/251	1 inch:40 feet 50.5 x 77.4 cm	114
575A	Plan of property, on Spring Road, Victoria, in the vicinity of the springs	not known	c.u.	H1/9,fo.24	19.8 x 21.6 cm	
576A	Endorsed: "Rough Sketch of Fishery for Porpoises, and Salmon Weir"	(post 1867)	c.u.	G1/265	32.2 x 20.5 cm	
577A	Chart of Bering Sea/Kamchatka	(post 1867)	c.u.	G1/287	1 inch:ca. 85 miles 63.5 x 78.3 cm	
578A	Plan showing an area of 105.9 acres bounded by Esquimalt Harbour on east, Section 88 on north, Section 96 on south, Section 1 on west	not known	c.u.	F25/1,fo.26	1 inch:200 feet 44.2 x 46.1 cm	
579A	Unfinished sketch of river and lake area, with some terrain forms	not known	c.u.	G1/258t	39.5 x 72 cm	
580A	*Victoria District Lot 30 Section XXI*	not known	(Joseph D. Pemberton)	F16/2,fo.216	4 inches:1 mile 17.8 x 12.7 cm	
581A	*Victoria District Lot 30 Section XXI*	not known	(Joseph D. Pemberton)	F16/2,fo.218	4 inches:1 mile 17.8 x 12.7 cm	

CARTOGRAPHER	SOURCE	REFERENCE	HBCA REF.	TEXT
(J.D. Pemberton) Tiedemann	Victoria, Surveys and Land Records	37T2 Large Tray, East Coast (Vancouver Island)		105
(J.D. Pemberton) Tiedemann	Victoria, Surveys and Land Records	Formerly 28T2 Victoria Town		
Homfray	Victoria, Surveys and Land Records	22T1 Victoria Town		
	British Columbia Archives	8000 V14		114
	Victoria, Surveys and Land Records	12T2		
Mactavish	British Columbia Archives	S616.9(88) M175ma 1863		111, 285n4
McDonald	Victoria, Surveys and Land Records	11T3 Miscellaneous		

Catalogue B

Hudson's Bay Company Manuscript Maps (1670–1870) in Archives Other than the Hudson's Bay Company Archives*

Maps in this catalogue are referred to in the text by numbers with a superscript B. The text column in the catalogue records plate numbers (in italic) and the text page(s) on which a map is discussed. The reference column gives the manuscript catalogue number used by the archive where the map is located or the page reference when a publication is the only known source of a map. The HBCA reference number identifies the journals, letters, or other documents in the HBCA where the map is mentioned. The printing conventions used in this catalogue are:

1. A title printed in italic reproduces as nearly as possible the inscription on a map. Example: 4B *A Plan of the Coppermine River by Sam Hearne July 1771*.
2. A title printed in roman and enclosed in quotation marks is derived from letters, journals, or other documents in the HBCA. Example: 2B "Map of Hudsons Bay."
3. A title in roman without quotation marks is derived from the catalogue of the archive in which a map is located. Example: 1B West side of James Bay.
4. When the date of the drafting of a map or the cartographer's name have been postulated on the basis of firm evidence but are not certainly known, the name and date are given in parentheses. The designation c.u. for "cartographer unknown" is used when the evidence does not permit even an informed guess regarding the identity of the cartographer.

* Two of the maps in this catalogue, 10B and 11B, are available now, so far as is known, only in their printed form.

NO.	TITLE OR DESCRIPTION	DATE	CARTOGRAPHER	SOURCE	REFERENCE	HBCA REF.	TEXT
1B	West side of James Bay	(1678)	Thomas Moore	British Museum	Add. MSS vol. 5027A, fo.64		25–6,1, 276n5
2B	"Map of Hudsons Bay"	1685	John Thornton	British Museum	Add. MSS vol. 5414, fo.20	A1/8,fo.20	26
3B	Richmond Fort and parts Adjacent as far south as Great Whale River	(1750)	William Coats	Hydrographic Department Admiralty, Great Britain	*See also*, G. Williams, *The Beaver*, Winter 1963, p. 8.		34–5
4B	*A Plan of the Coppermine River by Sam Hearne July 1771*	(1772)	Samuel Hearne	James Ford Bell Collection, University of Minnesota		A6/11,fo.172; A5/1/.fo.151	44,10
5B	*Chart of part of Hudsons Bay and Rivers and Lakes falling into it by Philip Turnor*	(1782–3)	Philip Turnor	British Museum	P 7496		51, 278n4

NO.	TITLE OR DESCRIPTION	DATE	CARTOGRAPHER	SOURCE	REFERENCE	HBCA REF.	TEXT
6B	Hudson Bay Company Settlements	(1789–90)	(Edward Jarvis)	Clements Library, University of Michigan			
7B	Map of Columbia River Basin (with additions in 1825 and 1849)	1821	Alexander Ross	British Museum	Add. MSS 31,358B		87
8B	British Columbia Interior – Fraser, Thompson River Area	1835	Samuel Black	British Columbia Archives	P615pBC, R 888t 1861	A10/17,fo.79; A1/63,fo.98; A5/14, p. 187	
9B	Winter Discoveries on Great Bear Lake and Routes through the Barren Lands to the Coppermine River AD 1837–39	1839	Thomas Simpson	Royal Geograpical Society, London	Canada S8		80
10B	Chart of Simpson's journey (during which he was killed) from Red River towards St. Peters, Mississippi River	1840	Thomas Simpson	Narrative of the Discoveries on the North Coast of America by Thomas Simpson	p. XVI		80
11B	Sketch map of Peel River	1845	Alexander K. Isbester	Journal Royal Geographical Society	Vol. 15, 1845 p. 332		79
12B	An approximate sketch of the route upon an enlarged scale (showing Black-eye's trail)	1846	Alexander C. Anderson	British Columbia Archives	P 616.5gmbh A546a 1846		94, 284n20
13B	Original sketch of exploration between 1846 and 1849 (Kamloops to Fort Langley, including Harrison-Lilooet route)	(ca. 1849)	Alexander C. Anderson	British Columbia Archives	8000 L10		94–5
14B	VICTORIA DISTRICT & PART OF ESQUIMALT	(1851)	(Joseph D. Pemberton)	Victoria, Surveys and Lands Records	5 Locker L	A11/73, fos.61,73 B226/b/8, fo.18	100
15B	A PLAN OF THE TOWN OF VICTORIA SHEWING PROPOSED IMPROVEMENTS	1852	(Joseph D. Pemberton)	Victoria, Surveys and Land Records	7 Locker 9		101
16B	Map of the LAND claimed by THE PUGET SOUND AGRICULTURAL ASSOCIATION in PIERCE COUNTY. w.t.	(1852)	(Edward Huggins)	National Archives Washington, D.C.	RG 76,Series 71, Map 12		108
17B	Approximate Survey of Land claimed by the Puget Sound Agricultural Association on the north bank of the Cowlitz River	(1852)	(c.u.)	National Archives Washington, D.C.			
18B	Unfinished map of Cordova Bay to Sooke, Vancouver Island	(1853)	(Joseph D. Pemberton)	Victoria, Surveys and Land Records			
19B	Victoria Town Incomplete	(1853)	(Joseph D. Pemberton)	Victoria, Surveys and Land Records			
20B	Esquimalt District, Lot 26, Sect.I	1854	(Joseph D. Pemberton)	British Columbia Archives			
21B	Esquimalt District, Lot 27, Sect.II	1854	(Joseph D. Pemberton)	British Columbia Archives			
22B	Esquimalt District, Lot 28, Sect.X	1854	(Joseph D. Pemberton)	British Columbia Archives			
23B	Esquimalt District, Lot 29, Sect.XI	1854	(Joseph D. Pemberton)	British Columbia Archives			
24B	Victoria District, Lot 30, Sect.XI	(1854)	(Joseph D. Pemberton)	British Columbia Archives			
25B	"Old Map Victoria Harbour" Town Plan of Fort Victoria	(1855)	(Joseph D. Pemberton)	Victoria, Surveys and Land Records			
26B	Plan of PSAC Land Claim at Nisqually, Washington Territory	(1855)	c.u.	National Archives Washington, D.C.			
27B	"tracing of your [K. McKenzie's] proposed Purchase at Lake Hill"	1856	(Joseph D. Pemberton)	British Columbia Archives			
28B	A Rough Tracing of part of British Columbia from the Columbia River to Fraser River etc. from one in possession of the Hudson Bay Co.	(1858)	HBC Employee	Victoria, Surveys and Land Records			
29B	Reconnaissance Sketch of the Fraser River, between Fort Hope and Fort Yale; taken on the 13th. & 14th. Septr. 1858	1858	(James Yale)	Victoria, Surveys and Land Records			

Catalogue C

Hudson's Bay Company Manuscript Maps (1670–1870) Not Located*

Maps in this catalogue are referred to in the text by numbers with a superscript c. The text column in the catalogue records the page(s) on which a map is discussed. The reference column identifies the HBC journals, letters, or other documents in which a map is mentioned. The printing conventions used in this catalogue are:

1. A title taken from HBC journals, letters, or other documents, is printed in roman and enclosed in quotation marks. Example: 1^C "Map of Hudsons Bay according to Capt. James's Description."

2. A title printed in roman without quotation marks has been assigned by the author, based on information in the HBCA archives. Example: 8^C Map of Moose River for 24 miles upwards from Moose Fort.

3. Where appropriate, the author has assigned titles or descriptions that include quotations from HBC archival sources. Example: 6^C Copy of "Mapp of Hudsons Bay," to be sent to "Lord Marlebrough" in Holland.

4. When the date of the drafting of a map or the cartographer's name have been postulated on the basis of firm evidence but are not certainly known, the name and date are given in parentheses. The designation c.u. for "cartographer unknown" is used when the evidence does not permit even an informed guess regarding the identity of the cartographer.

* These maps are perhaps no longer extant. See chapter 1 and appendices 2 and 4.

NO.	TITLE OR DESCRIPTION	DATE	CARTOGRAPHER	REFERENCE	TEXT
1^C	"Map of Hudsons Bay according to Capt. James's Description"	1669	Norwood	A14/1,fo.108	25
2^C	Three "Mapps of the bottome of The Bay"	(1680)	(John Thornton)	A1/3,fos.30,30d	26
3^C	"new Map of Port Nelson"	1686	(John Thornton)	A1/84,fo.27	26
4^C	Two "mapps of Hudsons Bay"	1699	John Thornton	A1/22,fo.19	26
5^C	Two "Mapps of hudsons Bay"	1701	John Thornton	A1/23,fo.35d	26,28, 276n13
6^C	Copy of "Mapp of Hudsons Bay," to be sent to "Lord Marlebrough" in Holland	1701	(John Thornton)	A1/23,fo.35d	26,28, 276n13
7^C	"Mapp of Hudsons Bay"	1702	John Thornton	A1/24,fo.9	26,28, 276n13
8^C	Map of Moose River for 24 miles upwards from Moose Fort	(1740)	(George Howy)	B135/a/10,fo.11d	35–6
9^C	"a draught of ye 2 rivers," Nelson and Hayes	1745	Joseph Robson	A11/114,fo.121	36

Catalogue C

NO.	TITLE OR DESCRIPTION	DATE	CARTOGRAPHER	REFERENCE	TEXT
10C	A map of Cape Diggs to Little Whale River	1749	(William Coats)	A1/38,pp.285,331	34–5
11C	Map of Eastmain coast and Richmond Gulf area.	(1750)	John Yarrow	A6/7,fos.154d,169.	35
12C	Three copies of J. Robson's map (9C) of the Nelson-Hayes River Mouth	1750	c.u.	A6/8,46d	36
13C	Two maps showing measurements up the Nelson and Hayes Rivers from York Fort, ca. 50 miles	1750–51	James Isham	A11/114,fos.148d,155; A6/8,fo.96d; B239/a/34, fos.14d,16,22d,23	36
14C	Map of Nelson River inland some distance from York Fort	1754	James Isham	A6/9,fo.10;	36–7
15C	Map of the lower twelve miles of the Severn River, from a journey by Richard Ford and Christopher Atkinson	1754	Richard Ford	B239/a/37,fo.31; A11/114, fo.162d; A6/9,fos.9d,33	36
16C	Map of Henday's journey inland from York Fort to the North Saskatchewan River region	1755	Anthony Henday	A11/114,fos.190,197; A6/9,fo.33d	38
17C	Copy of Henday's map of a journey inland from York Fort to the North Saskatchewan River region	1775	James Isham	A11/114,fo.197	38
18C	Sketch of Smith and Waggoner's journey from York Fort to the Saskatchewan grasslands, west of the upper Assiniboine River	1757	(James Isham)	A11/115,fo.10d	38
19C	Draft of Chesterfield Inlet called by the Indians Kis-catch-ewen	1761	William Christopher	B42/a/55,fo.48	41
20C	"Draught of Rankin's Inlet"	1764	Magnus Johnston	A11/4,fos.1d,3,7; A5/1/fo.68	41–2
21C	Chart of the west coast of Hudson Bay, north of Churchill	1765	Magnus Johnston	A5/1,fo.75; A6/10,fo.118d	42
22C	Draft on deerskin of the country to the northward of Churchill River by the Northern Indians, Meatonabee and Idotly'azee	1767	Meatonabee and Idotly'azee	B42/a/64,fo.63; A11/14,fo.78d	42–3
23C	"A Map of the Rout from Severn to York Fort by the rivers W. Falconer." Later MS note "Missing 1819"	1767	William Falconer	A64/45, #11	40
24C	"A Map of W. Tomisons rout to Lake Winnipeg from Severn W. Falconer"	1768 or 1770	William Falconer	A64/45,#41,chest 1, p.129	38,40
25C	Map of a journey across the Barren Grounds from Churchill Fort	1770	Samuel Hearne	B42/a/80,fos.21,22; A5/1, fo.143; A6/11,fo.152	43–4

NO.	TITLE OR DESCRIPTION	DATE	CARTOGRAPHER	REFERENCE	TEXT
26C	"Plan of Tracts between Basquiau and York Fort," a copy of 21A	1772 or 1773	c.u.	A1/44,fos.79,79d	40
27C	"The Interior of Hudsons Bay." Later MS note: "Missing 1819"	1774	William Falconer	A5/1,fo.169; A64/45, #11	48
28C	Map of the Road to Abitibi Pedlar settlement from Moose Fort	1774	Coochenau	B135/a/55,fo.7	
29C	Map of the journey of Edward Jarvis from Albany Fort to Moose Fort via the Chepy Sepy	1775	Edward Jarvis	A11/4,fo.25; B3/b/13,fo.31	45
30C	Map of the mouth of the Albany River	1775	John Hodgson	A11/4,fo.15d; B3/a/68,fos.28d, 29,29d	45–6
31C	Map of the Albany River from Albany Fort to Henley House	1775	John Hodgson	B3/a/70,fo.7; B3/b/13,fo.23; B86/a/27,fo.1d; B86/b/1,fo.4d	46
32C	Draft of Albany River for three miles above old Henley House	1775	John Hodgson	B86/a/27,fo.3; B86/b/1,fo.3d	46
33C	Two copies of Samuel Hearne's map of principal lakes and rivers leading from York Fort to Basquiaw	1776	Henry Hanwell Sr	A1/45,fo.12d; A64/45,#39,Chest 1,p.129. A64/52,#9,Chest 1,p.177	45
34C	"A Map of James's Bay Mr. Hanwell"	(1776)	Henry Hanwell Sr	A64/45,#38,Chest 1,p.129	45
35C	Plan of the route taken by Edward Jarvis from Albany Fort to Michipicoten	1776	Edward Jarvis	A5/2,fo.25; A11/4 fos.28,37; B3/b/14,fo.15	46
36C	Copy of a plan of the route taken by Edward Jarvis from Albany Fort to Michipicoten	1776	John Hodgson	B3/b/14,fo.11d; A11/4,fo.28	46–7
37C	Draft of the Albany River up past Henley House	1776	John Hodgson	B86/b/1,fo.12d; B3/a/70,fo.34	46
38C	Map of Nodaway (Nottaway) River	1776	John Hodgson	B3/b/13,fo.55; B3/a/70,fos. 34d,36d; A11/4,fo.33	46
39C	Two maps prepared in London of the area from James Bay to Lake Winnipeg, Lake Superior, Rainy Lake, Lake of the Woods, and the Albany and Moose area	1776	c.u.	A5/2,fos.13,14,16; B3/b/14,fo.2	48
40C	"rough Sketch of the track up to Basquiau by Chuck,e,ta,nau River & Winnipeg Lake from the information of some Assinipoet [Assiniboine] Indians," and from John Cole, 1772	1776 (1772)	Thomas Hutchins	A11/3,fos.40,40d	

Manuscript Maps Not Located

NO.	TITLE OR DESCRIPTION	DATE	CARTOGRAPHER	REFERENCE	TEXT
41C	Map of Missinaibi River from Moose Fort to Wappiscogamy, in two sections, (from Moose Fort to Souweska Sepy, and from Souweska Sepy to Wappiscogamy)	1777	George Donald	B135/b/5,fo.13d; A11/44,fo.64; A5/2,fo.41	47
42C	Map of Missinaibi River from Wappiscogamy to Missinaibi Lake	1777	George Donald	A11/44,fos.44,64; B135/a/58, fos.36,39; B135/b/5,fos.36d,39,41d	47–8
43C	Map of route from Missinaibi Lake to Lake Superior, with portages, for use of Indian guides	1777	Pequatisahaw	B135/b/5,fo.39d	47–8
44C	Map of Moose area, including Lake Kenogamissi	1777	Tuck, go, my	A11/44,fo.57d	
45C	Plan of the Albany River from Albany Fort to Lake Upashewa, to establish Gloucester House	1777	(John Hodgson)	B86/b/2,fo.26; A11/4,fo.55	46–7
46C	Map of Sutherland's Track from Albany Fort to Lake Winnipeg	1778	George Sutherland	A11/4, fos.73,77d,87	47
47C	"Draught or Chart prepared from the best Evidence and information collected from the Journals," probably a chart of the coast of Hudson and James bays and of the rivers flowing into them, sent from London to Albany	1778	c.u.	A6/12,fo.100; A11/4,fo.87	48
48C	Chart of "the Coast, Shoals and Soundings from Cape Churchill to Cape Tatnam"	1783	John Marley	A5/2,fo.86d	49, 278n1, 280n25
49C	"Tract from London to the Settlements in the Bay"	1783	John Marley	A5/2,fo.86d	49, 278n1
50C	Map of the route from Gloucester House to Pascocoggan Lake and from Gloucester House to Lake St. Ann (Lake Nipigon)	1784	James Sutherland	A5/2,fo.123	55
51C	"Map of the Road to St. Ann's Lake (Nipigon) from Pascocoggan Lake"	1784	Young Canadian trader	B78/a/11,fo.7d	279n16
52C	"A Chart of Hudsons Bay with the Rivers & Lakes by P. Turnor 1789"	1789	Philip Turnor	A64/45,#55, Chest 3,p.135	
53C	Richard Perkins track from Henley to Lake St. Ann's (Lake Nipigon)	(1790)	(John Hodgson)	B86/a/44,fos.21d,56d; B3/b/27, fo.55	56
54C	Map of the route between Manchester House and South Branch House, Saskatchewan River area	1790	Thomas Stayner	B42/b/44,fo.30; A11/117,fo.49	52

NO.	TITLE OR DESCRIPTION	DATE	CARTOGRAPHER	REFERENCE	TEXT
55C	Map of the journey from Albany Factory to Lake St. Ann's (Lake Nipigon)	1791	John McKay and John Hodgson	A11/5,fo.189	56–7
56C	Map of the journey from Albany Factory towards Lake Wepiscauacaw (Mepiskauacaw)	1791	John Hodgson John Knowles	A11/5, fos.174d,189	56–7
57C	Map of area from Lake Nipigon, Rainy Lake, and Lake of the Woods to Portage de l'Isle region and to Osnaburgh House	1791	Donald McKay	A11/5,fo.175d	57–8, 120
58C	"Draught of the Lakes and Rivers of the South track from Cumberland House to York Factory"	1791	David Thompson	A11/117, fos.54,109d; A5/3, fo.64d	59
59C	"The Track from Cumberland House to Isle la Crosse with the Magnetic Bearings on 14 sheets"	1792	Philip Turnor	A64/45,#42,Chest 1, p.130	54
60C	"A Map of Hudson's Bay and Straits with the Chesterfield Corbets Inlet or Bay"	1792	Charles Duncan	A1/47,fo.7; A64/45,#44,Chest 1, p.130	49, 278n1, 280n25
61C	Rough chart of Seal River	1792	Charles Duncan	B239/b/54,fo.5	49, 278n1
62C	"Chart of the North West and the Interior Country"	(1792)	(Donald McKay)	B3/a/93b,fos.10,12	57–8
63C	"A Map of the Interior from Moose Fort Containing Some new discoveries," the journeys to Sowowominicaw, and to Kinogamesee	1793	John Mannall	A64/45,#52,Chest 1,p.131	55
64C	A map of the journey from Micabanish (New Brunswick House) to River Pique (Pic River).	1793	Philip Good	B145/a/6,fos,51,51d; B135/b/23, fo.55	55
65C	"A Map of his Journey to the Stony Mountains & with the River Saskatchewan"	1795	Peter Fidler	A11/52, fos.1–2d; A64/52,#53, Upright Case,p.182	61–2
66C	"Chart of Chesterfield Inlet"	1796	George Taylor Sr	B239/b/57,fo.4	
67C	Map of the Beaver and Athapescow (Athabasca) rivers and from there to Edmonton and Buckingham Houses	1800	Peter Fidler	E3/2,fo.60	62
68C	Draft from York Fort to Edmonton House for Mr. Ballenden	1796–97	Peter Fidler	B39/a/2,fos.22,23d	280n5
69C	Draft from Buckingham House to the Rocky Mountains for Mr. Lean	1802	Peter Fidler	B39/a/2,fos.22,23d	

Manuscript Maps Not Located

NO.	TITLE OR DESCRIPTION	DATE	CARTOGRAPHER	REFERENCE	TEXT
70C	"A Map from Peter Fidler 1802"; a composite "Map of … Journey from Buckingham House to the Rocky Mountains … in six sheets: … [with] an extra sheet and half annexed … shewing the rivers and other remarkable places to the Mis sis sury river, which is taken solely from Indian information"	1802	Peter Fidler	A5/4,fo.103d; A11/52,fos.1–2d; B39/a/2,fos.22,23d; A64/52,#53	63–4
71C	"Sketch of the Road from the Sandy or Northern Indian Lake, to where Mr John Charles built in 1806 – drawn by himself"	1806	John Charles	E3/3,fo.49	
72C	Two associated sketches of rivers and lakes in the area of Island Lake, Cross Lake, Red Sucker Lake, and Shamattawa River	1807	Thomas McNab	E3/3,fo.49d	
73C	Twelve sheet "Map of the Communication between … Frog Portage and the Athapescow Lake (East End). Thro' the Grey Deers Lake and Lake Wollaston – also the Track from the lower end of Deers River, down thro' the Missinnippe or Churchill River to that factory"	1808	Peter Fidler	E3/5,fo.1d	64–5
74C	Sketch of Hayes River area	1808	Hugh Sabeston	E3/4,fo.11d	
75C	Map of Poplar River from Lake Winnipeg to Little Grand Rapid	1808	George Spence	E3/4,fo.18d	
76C	Route between Osnaburgh House, via Cat Lake to Sandy Lake	1808	Andrew Flett	E3/4,fo.18d	
77C	"Chart of Hudson's Bay"	1808	George Roberts	A1/49,fo.58	72
78C	"Map of the South Branch of Saskatchewan by Jos Howes 1809"	1809	Joseph Howse	A64/52,#73,Portfolio B,p.185	68
79C	Hudson and James bays coast between Albany and Severn Factories "with the different Inland Communications," based on Thomas Bunn, George Taylor Sr, and William Tomison	1812	Peter Fidler	A10/1,fo.107d	66
80C	Hudson Bay coast between Severn and York Factories, with some of the rivers communicating with the interior, based on Fidler, James Swain, Thomas Swain, Thomas Thomas, George Taylor Sr, and several Indians	1812	Peter Fidler	A10/1,fo.107d	66
81C	The Rocky Mountains from Acton House to Howse's House and Great Fall, based on James Whitway and David Thompson	1812	Peter Fidler	A10/1,fo.107d	66

NO.	TITLE OR DESCRIPTION	DATE	CARTOGRAPHER	REFERENCE	TEXT
82C	"Rocky Mountains Sketch of, shewing the connection of the Athapescow, Saskatchewan and Missouri Rivers"	1812	Joseph Howse	A64/52,#58	68
83C	"Map of Ruperts and slude Rivers together with various lakes"	1814	James Clouston	B143/e/1,fol.1; B59/e/1,fo.2; A6/18,fo.271	69,71
84C	"Columbia River Sketch of by Josh Howes 1815"	1815	Joseph Howse	A64/52,#82, Portfolio B,p.185	68
85C	"Traced Copies of all those parts of the Company's Territories where it is desirable to gain further Information"	1815	Aaron Arrowsmith	A1/51,fo.22d	71
86C	"A Rough Sketch of the York Department"	1815	William Cook	B239/e/1,fo.4d	
87C	Rough sketch of Great Whale River made on a journey, 1816	(1816)	(George Atkinson III)	B135/b/36,fos.6d–7,9d	69
88C	"plain Mercator Charts divided into 32 Compartments"	1818	c.u.	A6/19,fos.63d,64	71
89C	"draft ... received from an Indian of the track to Big Lake House"	1818	Indian	B19/a/1,fo.4	
90C	Map of journey from Red River Colony to Martin's Falls	1818	Peter Fidler	B22/a/21,fo.5	67
91C	Draft of area in Mistassini Lake country	(1819)	Robert McCulloch	B143/a/20,fo.6	
92C	Chart of New Caledonia by Tete Jaune, an Iroquois Indian	1819	Tete Jaune	B190/a/2,fo.24d	
93C	"Lesser Slave Lake & adjoining parts taken from the information of the Natives 1820 & 21 – a rough sketch by Robt Kennedy"	(1820–21)	Robert Kennedy	A64/52,Portfolio A,p.183	
94C	Map drawn by Indian guides during James Clouston's 1820–21 journey; information from it was "laid down in pencil" on Clouston's map, i.e., on 188A	1820–21	Indian guides	G2/23	70
95C	Plan of a village on the Assiniboine River	1823	William Kempt	E.H. Oliver, *The Canadian North-West*, pp.45, 234 PAM, Bulger Corres. III M151, pp.156–61	82
96C	Plan of the Red River settlement, related to the index to the plan	1823–24	William Kempt	E6/11	82
97C	Copy of the "Sketch of the Trout Fall ... Wm. Kempt 1824"	1824	William Kempt	G1/54	83–4, 282n37
98C	"Chart of Main Ridge Rocky Mountains"	1824–25	c.u.	A64/52,#101,Portfolio B,p.186	87

Manuscript Maps Not Located

Catalogue C

NO.	TITLE OR DESCRIPTION	DATE	CARTOGRAPHER	REFERENCE	TEXT
99C	"Map of Country lying West of Rocky Mountains 5 sheets by Archibald Macdonald 1824/25"	1824–25	Archibald McDonald	A64/52,#102,Portfolio B,p.186; D4/88,fo.41; A12/1,fo.144d	87–8
100C	"Chart of NW Coast of America from Columbia River to Frasers River by Jn Work 1824/25"	1824–25	John Work	A64/52,#100,Portfolio B,p.186; A12/1,fo.144	90
101C	"Snake Country – Chart of – describing the route of Donald McKenzie in 1819/20 also of Finan McDonald in 1823 & of Alexr Ross – 1824 – in 3 sheets by Alexr Ross – 1825"	1825	Alexander Ross	A64/52,#98,Portfolio B,p.186	87
102C	Map of area between the Columbia and Fraser Rivers	1825	Francis N. Annance	A64/52,#104,Portfolio B,p.186; D4/7,fo.94	90
103C	"Map of N.W. Coast of America"	1825	Thomas McKay	A64/52,#105,Portfolio B,p.186	90
104C	Two maps of the Snake River area	1825	Archibald McDonald	D4/6,fo.50; D4/119,fo.46	88
105C	"Survey of Columbia River by Henry Hanwell 1825"	1825	Henry Hanwell Jr	A64/52,#106,Portfolio B,p.186	90
106C	"Survey of Observatory Inlet by Henry Hanwell 1825"	1825	Henry Hanwell Jr	A64/52,#107,Portfolio B,p.186	90
107C	"Survey of Gulf of Georgia by Henry Hanwell 1825"	1825	Henry Hanwell Jr	A64/52,#108,Portfolio B,p.186	90
108C	Copy of "White Fall Portage. Wm. Kempt, 1825"	1825	William Kempt	G1/56	84
109C	Map of Second Snake Country Expedition, Peter Skene Ogden 1825–26 (2 copies)	1826	Peter Skene Ogden	B223/b/2,fos.21d,29	89
110C	"Map of the Snake Country 1826"	1826	Peter Skene Ogden	A64/52,#109,Portfolio B; B202/e/2,fos.1,1d,2	89
111C	Sketch of the Thompson River, to the forks of the Fraser River and country south along the Similkameen River to the Okanogan	1826	Archibald McDonald	B97/a/2,fo.29	93
112C	Oregon Coast, Columbia River south to the Umpquah River	1826	Alexander Roderick McLeod	B223/b/2,fo.23	89
113C	"Map of Mr Ogdens Track 1826–27"	1826–27	Peter Skene Ogden	A64/52,#113,Portfolio B, p.186	89
114C	Chart of the Columbia River from Cape Disappointment to Fort Vancouver	1827	Aemilius Simpson	B223/b/3,fos.13–14d	90
115C	Survey of Fort Vancouver and farm	1827	(Aemilius Simpson)	A64/52,#111,Portfolio B, p.186	90

Manuscript Maps Not Located

NO.	TITLE OR DESCRIPTION	DATE	CARTOGRAPHER	REFERENCE	TEXT
116C	Map of trip inland from York Fort from separate sketches, showing location of beaver houses and of fur trading posts observed en route.	1827	George Taylor Jr	B239/a/138,fos.3d,6d,8	85
117C	Sketch maps of area inland from York Fort, from which a composite map was made	1827	George Taylor Jr	B239/a/138,fos. 1d–3,4–6,7–7d	85
118C	Sketch of Sturgeon River and a lake	1827	Sturgeon River Indian	B239/a/138,fo.8	85
119C	Map of the country in the vicinity of Pike Lake	1827	James Robertson	B186/b/13,fos.55,56,62	
120C	"Chart of north end of Vancouver Island Port Bull"	(1828)	Aemilius Simpson	A64/52,#120,Portfolio B,p.187	91
121C	"Copy of Port George & Cossack Harbour by M. Pierce June 1825"	1828	Aemilius Simpson	A64/52,#125,Portfolio B,p.187	91
122C	"Tumgaise Harbour Aug. 1828"	1828	Aemilius Simpson	A64/52,#127,Portfolio B,p.187	91
123C	"Copy of Skidgetts Harbour by Capt. Pierce"	1828	Aemilius Simpson	A64/52,#129,Portfolio B,p.187	91
124C	"Chart of Fraser River"	1829	Aemilius Simpson	A64/52,#133,Portfolio B,p.187	90
125C	Chart of the entrance to the Nass River, British Columbia	1829	Aemilius Simpson	B223/c/1,fos.25,26,26d; D4/16, fo.22d; A12/1,fo.378d	91
126C	Map of the rivers and lakes of the Lake Superior District.	1830	Duncan Haggart	B129/b/4,fo.36; B129/b/5,fo.2d	
127C	Survey of winter road inland from Oxford House	1830	George Taylor Jr	B156/a/12,fos.64,64d	84
128C	Sketch of the Colvile District	1830	John Work	A64/52,#134,Portfolio B,p.187	
129C	Chart of the Babine Country	1833	Terrewill, an Indian	D4/126,fo.36	
130C	Rough sketch of Esquimaux Bay and Kaipokok (Kibokok) Bay, Labrador	1838	W.H.A. Davies	B153/b/1,p.20	76
131C	Rough sketch of a route from North West River, Labrador to Masquaro (Musquaro), Gulf of St. Lawrence	1838	George Duberger	B153/a/3,fo.48; B153/b/1,p.21	76
132C	Draft of a route from Lake Mistassini leading to "Muschowiaugan River" in Quebec	1838	Petaibish (Peataibish)	B215/a/3,fo.18	
133C	Sketch of the trip taken by Beads, June 14,1837–December 7,1837, from Nitchequon post through Moshewanaugan to the shore of the St. Lawrence at Tadoussac and return to Nitchequon (3 copies)	1838	Thomas John Beads	B186/b/35, pp.58,59	76

NO.	TITLE OR DESCRIPTION	DATE	CARTOGRAPHER	REFERENCE	TEXT
134C	General map of region from Hudson Bay on the west, Ungava on the north, and St. Lawrence River on the south, showing some of the main rivers and lakes (3 copies)	1839	Thomas John Beads and John Spencer	B186/b/38,pp.39,40,41	76
135C	Survey of Charlton Island and two small adjacent islands prepared in pencil by Cauc-chi-chenis and Kennewap in February-March-April, 1839, and inked over on April 16,1839 by Robert Miles	1839	Cauc-chi-chenis, Kennewap, and Robert Miles	B186/a/58,p.33	86
136C	Sketches of two routes between North West River House, Melville Lake, Labrador, and Kaipokok (Kibokok), Labrador	1840	W.H.A. Davies	B153/a/3,fos.6,8; B153/b/1, pp.41–2	76
137C	"Sketch of the Prairie Land about Nasqually made under the direction of James Douglas from a tour of inspection. Accompanied by Capt. Wm H McNeil"	1841	(Adolphus Lee Lewes)	F12/4,fo.30	91
138C	Sketch of the country from the Rocky Mountains in the Bow River area to Fort Colvile, via Flatheads, Coeur d'Alene, Pend Oreille, sent to George Simpson	1841	Archibald McDonald	D5/6,fos.128,128d	
139C	Chart of Comoosan Harbour	(1842)	(James Scarborough)	G1/198	96
140C	Chart of the Beaver Preserve of Charlton Island	1844	John Clouston	B186/b/49,fo.5d	
141C	Plan of the Cowlitz Farm showing sowing plan 1844	1844	c.u.	F12/2,fo.40	92
142C	Plan of proposed limits of HBC land claims at Sault Ste. Marie, required for their use there	1846	c.u.	D4/35,fos.59,59d	83
143C	Plan of Oregon City #8	1846	Mervin Vavasour	B225/z/4,p.232d	92
144C	The entrance to the Columbia River #15	1846	Mervin Vavasour	B225/z/4,p.232d	92
145C	Plan of Fort Pitt #3	1846	Mervin Vavasour	B225/z/4,p.232d	92
146C	Map of route of brigade trail explored by Alexander Anderson from Fort Kamloops to Fort Langley via Harrison Lake, and the Coquihalla to the Similkameen River, sent to Governor Simpson	1846	Alexander C. Anderson	D5/19,fo.288	94
147C	John Bell's route from Peel River to junction of Porcupine and Yukon Rivers	Between 1846 and 1849	John Bell	D5/22,fo.162d	79
148C	Chart of Akkawmiskee Island	1847	Thomas Corcoran	D5/20,fo.15d	

Manuscript Maps Not Located

NO.	TITLE OR DESCRIPTION	DATE	CARTOGRAPHER	REFERENCE	TEXT
149C	Chart drawn by Eskimo for John Rae, at Gibson's Cove, head of Repulse Bay, Hudson Bay, of the isthmus of Melville Peninsula, across to Committee Bay of the Arctic Ocean	1847	An Eskimo	E15/3,fo.46	80–1
150C	"ground plan of the Country in the vicinity of Nisqually"	1847	c.u.	A11/72,fo.29d	
151C	Sketch of travels of summer 1847, indicating Anderson's proposed brigade trail	1847	Alexander C. Anderson	D5/21,fo.294d	95
152C	"chart of Vancouver's Island … which shows the situation of the Coal beds, indicated by a line extending from McNeills to Beaver harbour"	1848	James Sangster	A11/72,fo.44; B223/b/38, fos.64d,65	
153C	Two sketches of Peers's route from Kamloops to Fort Hope, being a revision of A.C. Anderson's 1846 Coquihalla – Tulameen – Nicola Lake route	1848	Henry N. Peers	B223/b/38,fo.65; B113/c/1,fo.8	95
154C	Sketch of the mountainous part of the brigade trail from Fort Hope to the Horse Guard on the Tulameen River, British Columbia	1849	c.u.	A11/72,fo.174	95
155C	Diagram of the Fur Trade Reserve at Fort Victoria	1850	Walter Colquhoun Grant	A11/72,fo.322	98
156C	"Copy on tracing paper of the survey already made by me, with a small portion of the Southern part of the Coast etc. roughly filled in"	1850	Walter Colquhoun Grant	A11/62,fo.596	98
157C	"ground plan of the south coast of Vancouver's Island"	1850	Joseph W. McKay	A11/72,fo.205	
158C	Map of Vancouver Island (likely S.E. area), on large scale, divided into sections, with numbers differently arranged from 309A for purpose of land registration	1850	c.u.	A6/28,fo.163	
159C	Rough tracing of the Wollaston Peninsula Coast, Victoria Land, explored by Rae	1851	John Rae	E15/8,fo.60d	81
160C	Rough chart of the coast of Victoria Land from Pt. Back, south coast, to Pt. Pelly, east coast, explored by Rae	1851	John Rae	E15/8,fo.66	81
161C	"Three sketches of portions of Victoria District … marked out as the Fur trade Reserve"	1851	Joseph D. Pemberton	A11/73,fos.66,214,437,487,487d	100
162C	Rough plan of the country around Charles Wren's deemed trespass on PSAC lands at Nisqually	1851	William F. Tolmie	B151/b/1,fo.44d	107

NO.	TITLE OR DESCRIPTION	DATE	CARTOGRAPHER	REFERENCE	TEXT
163C	Sketch of Mitchell Harbour, Englefield Bay, west coast of Queen Charlotte Islands	1851	George M. Nutt	A11/73,fos.67d,181d,222	106
164C	Sketch of "Skiddigate's passage," Queen Charlotte Islands	1851	Charles Edward Stuart	A11/73,fo.181d	106
165C	Map of lands claimed by the PSAC at Nisqually, Nisqually Plains	1851	Edward Huggins	F12/2,fo.421d	108
166C	Rough chart of the coast of Victoria Land explored by Rae in spring and summer 1851	1852	John Rae	E15/8,fos.77d-78	81
167C	Copy of "Victoria and Puget Sound Districts Sheet No.1" (i.e., 310A)	1852	Joseph D. Pemberton	A11/73,fo.487d	100
168C	Sketch map of S.E. Vancouver Island sent to A. Colvile; a copy also sent to Governor Douglas	1852	Joseph D. Pemberton	A6/120,fo.64	101
169C	Chart of PSAC reserve at Esquimalt	1852	c.u.	F12/2,fo.431	
170C	Map of Coast of Vancouver Island from Victoria to Nanaimo	1852	Joseph D. Pemberton	A11/73,fos.574,622,622d,637	102
171C	Unfinished chart of the Koskimo Inlet, Portage and coast of Vancouver Island as far as the Nimpkish River	1852	Hamilton Moffatt	Letter of July 9, 1859, to J. D. Pemberton	103
172C	Corrected outline of map of coast of Vancouver Island from 'Cowitchin' Head to Nanaimo	1852	Joseph D. Pemberton	A11/73,fos.658,660,664d	102
173C	Map of the land claimed by the PSAC in Pierce County, Washington Territory	1852	Edward Huggins	B226/b/4,fos.8,8d,76d	108
174C	Approximate survey of land claimed by the PSAC on the north bank of the Cowlitz River	1852	c.u.	B223/b/40,fo.24; B151/b/2, fo.26	108
175C	Rough chart of the Peel and Bell River area, west of the Mackenzie River	1852	Alexander H. Murray	B200/b/26,pp.11,30	106
176C	Map of Village Bay, Esquimalt Harbour, and best location for erection of boat slip	(1853)	Joseph D. Pemberton	A11/74,fo.69	103
177C	Tracing of "Nanaimo Harbour and Part of the Surrounding Country" (358A), with some additions in the Chase River area	1853	Joseph D. Pemberton	A11/74,fos.418,418d	102
178C	Sketch map of Bellingham Bay and area, and certain of the San Juan Islands	1853	Joseph D. Pemberton	A11/74, fos.416,416d,417	103

NO.	TITLE OR DESCRIPTION	DATE	CARTOGRAPHER	REFERENCE	TEXT
179C	"Map of the S. E. corner of Vancouver's Island from Soke to Cowitchin including the Saanich country and Inlet"	1853	Joseph D. Pemberton	A11/74,fos.455,474,474d, 475,475d	102
180C	Chart of the ship channel from Victoria to Nanaimo, by the Canal de Harro, prepared for Capt. R. Welbank, HBC Committee	1853	Joseph D. Pemberton	A1/68,p.180; A11/74, fos.398, 401,401d	102
181C	Sketch of Newcastle Island, Nanaimo	1853	Boyd Gilmour	B226/b/7,fo.89d	103
182C	Chart of Bellingham Bay; the coal discovered, and its position and approaches	1853	Joseph W. McKay	B226/b/7,fos.104,106d	103
183C	"Improved map of Queen Charlotte Island, containing all the latest discoveries"	1853	(Charles E. Stuart)	B226/b/11,fos.7,12; A11/74, fos.389d,436	106
184C	Chart of Dr. Rae's discoveries during the Arctic expedition from Repulse Bay to the west coast of Boothia Peninsula, as far north as Cape Porter during the summer of 1853	1854	John Rae	E15/9,fo.94d; A1/69,p.165	82
185C	Map of the "South Eastern Coast of Vancouver Island," the coast and the islands of the Strait of Georgia	1854	Joseph D. Pemberton	A11/75,fo.12; A1/69,p.81	103
186C	Survey plan of HBC claims at Sault Ste. Marie	1854	Alexander McDonald	D4/49,fo.50	109
187C	Plan showing proposed alterations to boundaries of HBC claims at Sault Ste. Marie	1854	Alexander McDonald	D4/49,fo.49d	109
188C	Sketch of HBC claims at Batchewana	1854	Alexander Mcdonald	D4/55,fo.132	
189C	"Copy of map of Puget's Sound Agricultural Company's Land claim at Nisqually"	1855	c.u.	B151/b/3,fo.43d	108
190C	Map from Mountain Portage to Sana Hill Bay, Back River, 2–8 July 1855	1855	James Anderson and James Stewart	S. Mickle, *The Hudsons' Bay Company Expedition in Search of Sir John Franklin*, p.45	82
191C	"Sketch of Mr. Horne's route from Qua la-Kum to the west coast of Vancouver's Island"	1856	(Charles E. Stuart)	B226/b/12,fos.122,125d	104
192C	Map of Vancouver Island from Nanaimo and Qualicum across to Barclay Sound and Port San Juan	1856	Joseph D. Pemberton	A11/76,fos.399,403A	104
193C	Sketch of the coal works and coal seams in relation to the discovery of the Park-Head mine outcrop	1856	George Robinson	A11/76a,fos.213,216; B226/b/12,fo.87d	104

Manuscript Maps Not Located

NO.	TITLE OR DESCRIPTION	DATE	CARTOGRAPHER	REFERENCE	TEXT
194C	Plans of HBC claims on Lake Huron and Lake Superior: Mississanque, Nipigon, La Cloche, Michipicoten, Pic, Fort William, Augeuwang River	1856	Alexander McDonald	D4/51,fo.117d; D4/52,fos.6,10d	108–9, 285n10
195C	"sketch of Fraser's River from Fort Hope to the Forks of Thompson's River"	1858	c.u.	A11/76b,fo.961d; B226/b/15, fo.38d	108
196C	Plan of the town of Yale	1859	Ovide Allard	B226/b/19,p.11; B226/b/17, p.128	107
197C	Ground plan of Block 17, town of Yale, showing positions of Hudson's Bay Company's store and other buildings, and line of fence	1859	Ovide Allard	B226/b/19,p.18	107
198C	Sketch of Fort Hope, British Columbia, with enclosure behind, with adjoining and encroaching town lots	1859	Joseph D. Pemberton and John Ogilvie	B226/b/19,p17	107
199C	Rough sketch to illustrate the position of Fort Berens (Lillooet) with reference to other HBC posts, and to the new route into the interior	1859	(Alexander G. Dallas)	A11/77/fol.121	108
200C	"Copy of plan of Yale, shewing Company's claim as laid down by Governor, (Douglas) compared with present holding"	1860	c.u.	A11/77,fo.756	114
201C	Small tracing of lot sold to Mr. Lowenberg, south of James Bay, Victoria, of which he had been dispossessed	1861	c.u.	B226/b/20,p.213; A11/78,fo.131	111
202C	Map of San Juan Island and improvements made by the HBC there	1861	C.J. Griffin	B226/b/22,p.197	
203C	"No.5 Church Reserve," parsonage lot and cemetery lot, Victoria, Vancouver Island	1862	c.u.	A11/78,fo.519	111
204C	Map of land surveyed to the south and west of James Bay, Victoria, on order of Alexander Dallas, before March 24, 1861	1862	c.u.	A11/78,fo.553; B226/b/20,p.313	111
205C	Diagram of Hope, B.C., based on official map of town, indicating two blocks for the HBC, and the 5-acre suburban lot set aside for the HBC on public lands	1862	Dugald Mactavish	B226/b/22,p.388	113–4
206C	Map of the subdivisions of Beckley Farm and the lands south and west of James Bay, Victoria, Vancouver Island	1863	(F.W. Green)	A11/79,fo.17; B226/b/22,p.423	111

NO.	TITLE OR DESCRIPTION	DATE	CARTOGRAPHER	REFERENCE	TEXT
207C	Copy of map of the subdivisions of Beckley Farm to the south and west of James Bay. Likely Copy D	1863	(F.W. Green)	A11/79,fo.63	111
208C	Copy of map of the subdivisions of Beckley Farm south and west of James Bay sent to A. G. Dallas, March 16, 1863. Likely Copy E	1863	(F.W. Green)	B226/b/20,p.345	111
209C	A plan of Langley, indicating the location of HBC holdings	1863	c.u.	A11/79,fo.246d	114
210C	Rough map of the Rae-Schwieger telegraph expedition route from Fort Garry to Fort Edmonton, summer 1864, with proposed alternate routes for the telegraph line indicated.	1864	John Rae	E15/13,fos.20d,21,21d,22d; A1/77,p.7	115
211C	Rough map of the Rae-Schwieger telegraph expedition route from Fort Edmonton to Tete Jaune Cache	1864	John Rae	E15/13,fo.27,27d,28; A1/77,p.5	115
212C	Rough sketch of telegraph routes and distance from Red River Colony to Vancouver Island	1864	John Rae	E15/13,fos.35,43; A1/76,p.209	116
213C	Esquimalt suburban lots: survey plan of Viewfield Farm	1865	H.O. Tiedemann	F11/5,fo.281; F11/2,p.219	112
214C	Map of PSAC farms of Craigflower, Constance Cove and Viewfield, Esquimalt District, Vancouver Island	1866	H.O. Tiedemann	F11/5,fos.330d,334d,335,335d	112
215C	Copy 2, Section XVIII, Victoria District, showing HBC land and land reconveyed to Crown, sent to colonial secretary's office, Victoria	1866	H.O. Tiedemann	A11/80,fo.430	111, 285n6
216C	Ground plan, with improvements marked thereon, of land selected by HBC at Esquimalt	1868	H.O. Tiedemann	A11/83,fo.679d; F11/5,fo.438	112
217C	Copy of ground plan, with improvements marked thereon, of land selected by HBC at Esquimalt	1868	c.u.	F11/5,fos.438,444d	112
218C	Ground plan of the HBC property, Barkerville B.C.	1869	c.u.	A11/84,fo.75d,76d	
219C	Map of the subdivision of Uplands Farm, Victoria District, Vancouver Island	1870	c.u.	A11/85,fo.418	
220C	Rough sketch of the country from the mouth of the Stikine River to Dease Lake	1870	William Manson	A11/85,fos.441–443d	117

Manuscript Maps Not Located

PART FIVE

*Appendices, Glossary,
Notes, Bibliography,
and Index*

Appendices

APPENDIX ONE

The Hudson's Bay Company

The Hudson's Bay Company, now 320 years old, is the longest continuously operating commercial enterprise in existence. It was one of a number of joint-stock trading companies that were established in Elizabethan and Stuart times, in a period of colonial expansion with trade carried on over long distances. Although chartered by the crown, these companies were privately financed. The initial eighteen incorporators, or "Adventurers" of the Hudson's Bay Company opened their first stock book in 1667, the name of the Duke of York (later King James II) being the first entry, and Prince Rupert's the second. The Adventurers applied to Rupert's cousin, King Charles II, for a charter, which was granted in 1670. The members became the absolute proprietors of an area, named Rupert's Land, the extent of which, although unknown at the time, was the vast region within the watersheds of Hudson and James bays and of Hudson Strait. Prior to receiving their charter, the stockholders had supported a trading expedition in 1668–9, which, after wintering at the mouth of the newly named Rupert River in James Bay, had returned with a cargo of furs whose value was sufficient enough to encourage them to apply for the royal prerogative to this area. They received among many rights the monopoly of trade in furs, fish, minerals, and other products. The company had the power to make laws for its own employees and posts, but also for all persons in this charter territory. It could administer justice and employ armed force if needed.

The proprietors of the Hudson's Bay Company held a "General Court" annually, at which an executive, called the "Committee," was elected, and general business transacted. The committee consisted of a governor, a deputy governor, and a further seven members. A secretary was also hired. This executive met monthly, or at times weekly, at one of the places of business of a committee member, for example, at the Tower of London and at Prince Rupert's home. Later, the members convened in hired premises, usually a coffee house, at which they also held fur auctions. As the company became more firmly established, it organized its own office, warehouse, and fur auction rooms in the City of London, moving within the City to several different locations over the centuries.

The committee directed all the North American operations of the company: the purchase of goods for the trade (outfits), supplies for the fur posts, the fur sales, the hiring of employees, their contracts, their financing, the purchase and hiring of ships, the preparation for the annual voyages, and the checking-in of the ships upon their return. It prepared all official instructions to captains, all official correspondence with officers, all personal correspondence with officers and servants, and read the contents of packets brought back to London, which consisted of post journals, annual reports, special reports, inventories of trade returns, contracts, maps, and personal letters.

One of the committee's main tasks was overseeing the fitting out of the ships in the spring. About the end of May, some members would go to Gravesend on the Thames estuary, see the outfits and goods on board, place the instructions and packets in the captains' hands, pay wages, and watch the ships weigh anchor for North America. The vessels would then sail along the east coast of Britain, passing by the Orkneys. For many years an agent was maintained in the islands as they were a major source of labour for the company. In the autumn as the ships reached port again at Gravesend, some of the committee members would return to the docks, receive the packets, search the lockers to prevent private trade in furs, and direct the unloading of the cargo, which would be transferred to the warehouse in the City and placed in the custody of the secretary. At the ensuing committee meetings the packets would be opened and all the post journals, reports, correspondence, and inventories of trade returns read; those present would view the maps and other items of interest arriving from various posts. Often, the captains and other senior officers home for furlough, or for retirement, would be interviewed.

In the early years a governor was appointed to act on the company's behalf for a term of years at each of the main posts (factories). These were located at the mouths of major rivers entering Hudson and James bays, viz., Albany, Churchill, Moose, and York factories. Several other adjunct seaside posts were also built at river mouths. Every post developed its own trade area inland (usually based on river basins), although the lesser posts, such as Severn House and Eastmain House, operated under the aegis of their factories. As trading posts were opened inland, each had its own trade tributary region. Eventually the term 'district' was applied to a grouping of these post areas, which together were subservient to a bay factory. In 1810 the company divided Rupert's Land into two departments, northern and southern, each comprised of districts. These departments were administered by governors, one centred at York Factory and the other at Moose Factory. The governors met annually with their district managers.

The stages of advancement for trading personnel were from apprentice clerk, to clerk, to chief trader and finally to chief factor. It was the chief traders and factors who were appointed as post masters and district managers, and these were the personnel who met with the governors annually. Some persons were hired as writers, whose main tasks were to act as secretaries to the post masters by keeping daily post journals (or at least by copying them from their superiors), by writing and copying reports, official correspondence, and letters between post masters. They also kept accounts. Other men were hired as doctors (surgeons), but they also entered the normal trade echelon during their careers. Some servants were hired as tradesmen, such as blacksmiths or carpenters, or as general labourers. These latter carried out much of the physical labour around the posts, went out to cut and haul timber and firewood, to hunt, fish, and provide muscle power in the canoes and, later, in the wooden York boats. Men were hired also to sail the various factory sloops on the bay.

Although the governors and the chief officers had considerable local power of decision, the London committee kept a tight reign over the details of the operation. Only the traders were permitted to have direct contact with the natives while trading. After the end of the seventeenth century, the exchange rate of goods for furs was left to the chief factors in Rupert's Land, but before that the committee and general court determined the standard of trade. This expressed the price of European trade goods in terms of a quantity of "made beaver," this being a prime, whole beaver pelt. The quantities to be traded for various articles were not always adhered to, as chief factors were left some latitude.

When the company expanded beyond the boundary of Rupert's Land into the northern plains, and over into the Athabasca and Mackenzie regions, it received the right of monopoly trade in these areas from the crown, although this was not honoured by Canadian traders. And when the Hudson's Bay Company joined together with the North West Company in 1821, it gained a larger trade territory stretching to the Pacific Ocean and south in the Columbia River basin. Reorganisation followed, and an assessment of all employees was made. Some from each company were offered positions at various levels, and others lost their positions at the end of their contracts. A deed poll was drawn up, which listed the employment levels and outlined duties, rights, and privileges. Although chief traders and chief factors had been given shares in the Hudson's Bay Company's profits some years before this fusion, afterwards officers became partners, holding in common, forty percent of the shares in the company.

The northern and southern departments of Rupert's Land were maintained after 1821, but some district boundaries were altered. At a later date, the districts west of the Rocky Mountains were united into a Western Department, and eventually Fort Victoria became its chief entrepot. Also, Montreal Department was established, made up of posts operating in Upper and Lower Canada, and including the former area of the King's Posts along the north shore of the St. Lawrence River. George Simpson, who had become governor of the Northern Department, became governor of both departments of Rupert's Land in 1826. Very quickly, in 1829, he was appointed governor of Rupert's Land, in effect the chief person in charge of company operations in North America. Simpson purchased a home at Lachine and set up his headquarters there, along with warehousing at that same location. In these new circumstances, basic operating decisions were made in North America, while the executive and general courts became involved only in the more general direction of affairs. However, local decision could be overridden by London.

Formerly, each year the chief officers of departments and districts had come together for council meetings to deal with each area and sub-area. Under Governor Simpson, they convened as a whole. This council passed regulations governing the trade, arranged the logistics for each region, deployed the officers and servants, set up the roster for furloughs, advised on promotions, applied discipline as needed, and worked out contracts. The council members set the standard for the trade, a most important undertaking. Norway House at the exit of the Nelson River from Lake Winnipeg was the depot of supplies for the interior, and councils sometimes met there. On other occasions, meetings were at Fort Garry on the Red River, or at York Fort. Other significant logistical points were Carlton House and Edmonton House. Carlton House, near the Saskatchewan Forks, had the task of obtaining pemmican (buffalo meat) from the plains to the south, and supplying this to the northern brigades passing to Ile à la Crosse on the upper Churchill River system. Edmonton House was the departure point for the Western Department and for the north, via the Athabasca River. However, after the establishment of Fort George on the lower Columbia River, and later Fort Vancouver, also on the Columbia River, as well as Fort Victoria on Vancouver Island, ships came directly around Cape Horn; and some personnel, packets, and goods were sent across the Isthmus of Panama, to go north to San Francisco and the Northwest. In the east, fur brigades set out from Lachine for the upper Great Lakes, via the old Ottawa-Nipissing voyageur route to Georgian Bay. Also, the port of New York was used increasingly as a route of entry and departure for officers and for correspondence from London to Lachine.

In 1869, in response to changing political and economic circumstances, the Hudsons' Bay Company arranged terms for the surrender to Her Majesty the Queen in 1870 of all rights of government and of the monopoly of trade in Rupert's Land, and of trading rights in Canada, the colony of British Columbia, and in other parts of British North America. Later the company moved its headquarters and archives to Canada, closed its fur auction facilities in London, and sold its stores in former fur trading settlements in the north. Essentially, it is now a major department store enterprise in North America.

APPENDIX TWO

General Characteristics of the Cartographic Records

The cartographic records of the Hudson's Bay Company consist of manuscript and printed maps, charts, plans, and atlases, and date from 1669 to the present time. These records may be divided into two groups:

1 Documents prepared by company employees, documents that the company, requiring a map for a specific purpose, contracted for with independent cartographers, or documents received by the company in the course of correspondence with other agencies, usually governmental. The materials in this first group are all manuscript records, with one exception, a map engraved (and possibly drawn) by R.W. Seale in 1748 (8^A).
2 Documents purchased by (but not previously contracted for), or that were gifts to, the company from commercial publishers or from government agencies. Except for the manuscript maps sent to the company by the surveyor-general's office of the Colony of Vancouver's Island and later of the Colony of British Columbia, the materials in this second group are printed. Manuscript notations have been added to some of the printed items, apparently by some member of the Arrowsmith firm.

As far as their present existence is concerned, the cartographic documents in both these groups fit into one of three categories:

a Those that are extant and presently located in the HBCA.
b Those that are not currently in the HBCA but may be found in other archives.
c Those that are not extant or whose whereabouts are unknown. The exact number in this category can never be ascertained, but documentary analysis has revealed that at least 220 of the cartographic documents once prepared for the company have disappeared. Most of these were in manuscript form.

APPENDIX THREE

General Inventory of the Hudson's Bay Company Map Archives

Over the centuries the Hudson's Bay Company amassed what has become the largest private collection of maps in Canada. This collection, since 1974 located in the Provincial Archives of Manitoba at Winnipeg, is also the principal source of fur-trade map documents in North America. The following inventory provides for the first time a general resume of the full extent and range of the cartographic documents in the archives:

	Sub-Total	Total	
Atlases		36	
Maps, charts, plans			+ segmental
1 Catalogued			sketches
a pre-1870 manuscript maps, charts	581		
			373 (Fidler)
			184 (Taylor)
b pre-1870 manuscript plans	133	714	
c post-1870 manuscript maps, charts, and plans	198	912	
d printed maps, charts, and plans (1669–1975)	1063	1975	
2 Uncatalogued maps, charts, plans (manuscript post-1870*; printed 1669–1975†)	2602	4577	+ 557

If the 256 cartographic documents for which there is documentary evidence but that are now in other collections or have disappeared are added to those listed above, then 4833 items plus 557 segmental sketches or 5390 documents in all were prepared for, purchased by, or given to the Hudson's Bay Company over the three centuries.

* Among the items not yet fully catalogued are approximately 2100 manuscript lot survey plats of the Lands Department of the company, drawn after 1870.
† Among the more notable examples of printed maps and charts in the archives are eighty-six separate sheets produced by the Arrowsmith firm.

APPENDIX FOUR

The Catalogues of the Hudson's Bay Company Cartographic Collection

Three distinct catalogues of the collection exist. Only the most recent catalogue, begun in 1924 and presently in use, is a true one in the sense that it actually describes most of the cartographic materials in the archives. The two earlier catalogues are finding aids only, being lists or inventories, with little descriptive data provided.

The first of these, a list titled "Maps, Charts and Drawings," is one part of an "Inventory of Books, etc." (A64/65,fos.128–136). At the time of the inventory, there were three chests in Hudson's Bay House holding fifty-six numbered items, constituting in all, forty-seven maps and charts and thirty-eight plans. Although no actual date was inscribed in the volume, I believe it was done in the year 1796. The reasoning for this is as follows. The most direct evidence is an item in the minutes of the committee meeting of 2 November 1796, chaired by Governor Samuel Wegg (A1/47,fo.80). The secretary was ordered to have twenty guineas paid to

John Brome, the sub-accountant of the company, for the work which he had undertaken, above and beyond his normal duties, in classifying and arranging the company's records and making an inventory and index of the collection (A1/47,fo.45,46). It would appear that Brome must have completed the task just prior to the time of the committee meeting. Moreover, the latest dates for maps entered in the list were 1794 and 1795. Finally, with few exceptions, the latest date Brome gave for items such as post journals, accounts, contracts, indents, and portledge books is 1793. Items such as the Moose and York Factory Account Books, entered in the record after 1793 and 1794 by a different hand, using different inks, are obviously additions to these pages. That some of the maps on this list were no longer in the collection is clear from the notation "Missing 1819," added in red ink. Other maps, known to have existed because they are mentioned in journals and other documents that had been drafted prior to 1794 do not appear on the inventory and must, therefore, have disappeared previously.

The second extant list of the cartographic collection may be dated 1847 or slightly later. This list is entered in a volume, called in the recent catalogue, a "Catalogue of Library, etc.," although no title is inscribed on this document itself (A64/52). The cartographic list which is part of this "Catalogue" is entitled "Maps, Charts and Drawings" (A64/52,fos.177–188). Dating this finding aid is a more complicated task than that posed by the previous list. The volume is formed of sheets bearing four watermark dates, 1815, 1825, 1829 and 1847. Items could not have been entered in the inventory until 1847 or later, because the volume of bound sheets could not have been delivered by the stationer until that date. The latest map date included is 1840. The list repeats most of the maps, charts, and plans given in the 1796 volume but also includes both manuscript and printed document titles added since that date, up to 1840. Approximately one hundred forty-three maps and charts and forty-six plans are itemized. The cataloguer noted that these were housed in the original three cases with the more recent items being held in two Portfolios, A and B. The order of items in the earlier group had been slightly altered; those entered in the portfolios are in roughly chronological order.

The preparation of this inventory was likely not as straightforward as the previous discussion might suggest. In 1819 there appears to have been a search made through the map chests, and those documents missing at that date noted in red ink in the 1796 inventory. Moreover, there was likely an intermediate listing made by 1822 at the latest, for an examination of the 1847 volume shows that most lists extend in the same hand and ink to 1819, 1820, or 1822. Notations beyond this time are in other hands and inks. Presumedly, either the company secretary, or some other designated person, had taken on the task of transferring the details from the inferred 1822 inventory to the new journal purchased in 1847. Further entries were made in this journal at different times by different persons, in some instances as late as the 1860s. It would appear, however, that this map indexing was not a regular office procedure, and the inventory rapidly became outdated. Maps accumulated uncatalogued in Hudson's Bay House until the establishment of the Hudson's Bay Company Archives. After 1924 an archival staff was hired, and the cataloguing system presently in force was formulated.

APPENDIX FIVE

Procedure for Obtaining Latitude on Land by Measuring the Double Meridian Altitude of the Sun

The usual procedure for obtaining latitude required the use of a reflecting plane parallel to the natural horizon. An extensive body of water with a distant shoreline horizon could be used, provided that the surface was calm and the shoreline was not so complicated with islands that a clear horizon could not be observed. When using a natural horizon, the surveyor employed his sextant as though he were observing on sea, not on land. That is, the sun's image was brought down to be in tangent contact with the water horizon. When a suitable natural horizon was unavailable, surveyors could use the artificial horizon, a flat pan, which was filled with quicksilver – or with water, treacle, or similar fluid – when quicksilver was lacking. Since the liquid in the artificial horizon had to be kept very still, it was protected from wind and foreign objects with a roof-shaped glass cover, called parallel glasses. Latitude was less often calculated by observations taken at night, by measuring the height of the moon, pole star, or some other star, data for which was recorded in the *Nautical Almanac*.

The example given here of the procedure for calculating latitude using the double meridian altitude of the sun is taken from the calculations made by George Taylor Jr at Fort Garry, Red River, at noon on 24 October 1827 (B235/a/11,fo.22d). To obtain the double meridian altitude of the sun, Taylor sighted through the telescope of the sextant at the sun's reflected image in the quicksilver in the artificial horizon. The upper limb of this image was brought into visual contact with the lower limb of the sun's direct image, reflected simultaneously by the sextant's mirrors. The two images appeared to touch and be tangent to each other. The index arm, which moved the telescope, was clamped, and the angle read from the vernier scale. This angle, known as the double meridian altitude of the sun, is that between the sun and its reflected image in the artificial horizon. That is, the angle is double the angle known as the altitude of the sun, which is between the sun and the true horizon. As the figures below demonstrate, after corrections, Taylor calculated the latitude to be 49°52'30.5" North:

Double meridian altitude of the sun, taken at noon	57°	28'	30"
Correction for Index Error[a]			+7"
	57°	28'	37"
Observed altitude of the sun's upper limb (divide by two)[b]	28°	44'	18.5"
Apply the sun's semi-diameter[c]		−16'	7.5"
APPARENT ALTITUDE OF SUN'S CENTRE	28°	28'	11"
Correct for refraction in altitude[d]		−1'	45"
	28°	26'	26"
Correct for parallax in altitude			+8"
TRUE CENTRAL ALTITUDE OF SUN	28°	26'	34"
Sun's zenithal distance or co-altitude[e]	61°	33'	26"
Sun's reduced declination[f]	11°	40'	55.5" South
Therefore, LATITUDE was:	49°	52'	30.5" North

[a] This was the amount of instrument error Taylor had calculated previous to the observation, when he was testing the accuracy of the instrument by positioning the various parts of the sextant exactly in their correct horizontal and vertical planes before reading. Since the index reading was incorrect by seven seconds, this amount had to be added to the double altitude angle.

[b] This figure was obtained by halving the double altitude angle, thus giving the observed altitude of the upper limb of the sun.

[c] Taylor obtained the apparent altitude of the sun's centre by subtracting the angle of the sun's semi-diameter (the apparent radius of a generally spherical body). The degree of angle was obtained from the *Nautical Almanac* for the day of observation.

[d] Two errors in observation had to be eliminated. The first was refraction or the bending of the light rays, when they passed through the dense atmosphere to the observer. This Taylor corrected by obtaining the temperature of the air at the time of observation, a figure which is a reflection of air density at that time. Then, by using refraction tables in the *Requisite Tables*, he could obtain the figure needed at that particular air temperature. The degree of the angle is then subtracted. The second observational error was that caused by parallax, the difference in angle between the altitude of the moon or sun, as if viewed from the centre of the earth, and the angle viewed from the observer's position on the surface of the earth. The parallax is at a maximum when the moon is on the observer's horizon and minimum when it is directly overhead. The parallax correction angle was also obtained from the tables, and eight seconds added. The figure resulting from the application of these two corrections was the true meridian altitude of the centre of the sun.

[e] Taylor then calculated latitude in two moves, the first being to subtract the true meridian altitude angle from ninety degrees. This angle (61°33'26") was the sun's zenithal distance or co-altitude.

[f] The second and final calculation was to obtain the sun's declination from the *Nautical Almanac* and subtract this angle from the co-altitude angle. The result was the calculated latitude of the observer at Fort Garry on 24 October 1827.

Usually, further observations would have been made at the same locale, and the mean of the results used. In this instance, Taylor took a further set of noon sun observations two days later and obtained a latitude of 49°52'44" North (B235/a/11,fo.22d). On his way to Turtle Mountain, Taylor and his party stopped at White Horse Plain, and on 29 October, he used a more complicated method of observation to determine the position there. This process involved observing the sun's double altitude at three successive times, two or three minutes apart, just before 10:00 AM, and at five successive times around 11:30 AM. Using the means of these sets of figures, he then calculated the true noontime co-altitude of the sun, that is, the angular distance of the sun from the zenith, measured by the arc of a vertical circle intercepted between the sun and the zenith. It is the complement of the sun's altitude. He then subtracted the sun's noontime declination (the angular distance of the sun north or south of the celestial equator) and obtained the latitude of 49°53'49.3" North.

Meanwhile, at apparent noon on that day, he had also observed the double meridian altitude of the sun, and after processing these figures reached a latitude figure of 49°53'49.5" North. In other words, his mean figure would be 49.4" for White Horse Plain, making this Metis settlement about 1'12" more northerly than the latitude of Fort Garry (B235/a/11,fo.23d). On the following day, 30 October, at a point about five miles out on the plains NWbW from the settlement, he used the clear night sky to sight on the upper limb of the moon, and from this double altitude figure worked out their position to be 49°55'4" North (B235/a/11,fo.25). Thus, in these five days, at three different locations, Taylor had worked out five latitudinal readings, using three different methods of observation. He did not attempt to obtain the longitude coordinates at any of these places.

APPENDIX SIX

Procedure for Obtaining Longitude by Using the Method of Lunar Distances

Obtaining longitude using the method of lunar distances involved observing the motion of the moon among the stars, or of the moon in relation to the sun, as if the moon were the hand of a clock. Longitude was obtained from the calculation of the observed angular distance between the moon and a particular nearby star, or the moon and the sun, this distance being known as lunar distance. Since

Appendices

the bodies, moon and sun or star are apparently moving, they had to be observed simultaneously, or nearly so. The *Nautical Almanac* gives the lunar distances to the sun, larger planets, and brighter stars to the nearest second of arc, for every three hours of apparent time through the year. For any particular time the observer can note which star (as long as its data are given in the *Almanac*) is closely ahead (leading) or closely behind (following) the moon, so that the angle between the objects can be read most easily.

The crucial aspect of longitude determination is the accurate measurement of time, since the observer must find the difference in time between the apparent local time of the observer and the apparent Greenwich Mean Time. This time difference can be translated into degrees, since the sun transits across a meridian at intervals of four minutes for each degree of longitude, or fifteen degrees per hour. The best results come from taking a number of lunar distances, both on leading and following stars. The full instructions for a complete set of 'lunars' calls for five readings in each direction, but ten such readings were seldom done. Observers preferred a clear night, though this was by no means always possible, especially in summer. The clear, bright nights of winter provided better circumstances, if one discounted the very low temperatures at which the recorders must work.

Peter Fidler used the lunar distance methods most often, although on a few occasions he took observations of the eclipses of Jupiter and its satellites, that is, he recorded when Jupiter's shadow began to fall on one of its satellites (immersion), or left the satellite (emersion). In using lunar distances, Fidler observed star, sun, or planet distances. For example, from 30 November 1805 to 26 April 1806, at Nottingham House on Great Slave Lake, he calculated thirty-six longitude observations, sixteen of which were lunar distances from several stars, eleven from several planets, and the remaining nine from the sun.

The example below of the procedure for using the method of lunar distances is based on a set of Fidler's observations taken during the early morning of 30 November 1805, at Nottingham House (E3/2,fo.107d). Although he did not follow the complete textbook instructions, he took all the necessary steps. Nor did he record every computation he made to reach the final longitude figure. Fidler proceeded as follows:

a He determined the rates at which his watches were gaining or losing.
b He determined the index error of his sextant.
c He observed the air temperature ($-20°F$).
d He determined local time by taking three double altitudes of two stars, Lyra and Capella, as close as possible to due east and due west of the moon, with times of observation. He used a different watch for the index error of his sextant, to time each set of altitudes, correcting for the rate of each watch, and thereafter correcting each set of altitudes.
e He then observed the lunar distances between the moon and the star Aquila, eight times, with the times recorded, and similarly, between the moon and the star Aldebaran, these being two stars due east and west of the moon. The means of these lunars and times were obtained. The same watch was used for all these observations.
f He also recorded the temperature during this operation ($-19°F$).
g At this point, altitudes were computed for each of the sets of observations of the two stars, by applying the corrections for the semi-diameter of the moon, for refraction, and for parallax.
h From the *Nautical Almanac*, Greenwich Mean Time of the observations were determined for each of the two sets.
i The time difference between Greenwich and local time was obtained for each set.
j This time difference was translated into degrees, which was the longitude of the place for each set.

These two longitudes were used, along with others calculated at Nottingham House by Fidler, to ascertain its mean longitude co-ordinate.

APPENDIX SEVEN

Persons Who Prepared Hudson's Bay Company Maps or Charts, 1670–1870

NAME*	DATE OF MAP(S)
Allard, Ovide	1859
Anderson, Alexander Caulfield	1845, 1846, 1847, ca. 1849, 1852, 1866
Anderson, James (A)	1853, 1855, 1860
Angt [or Augt], Fr.	1855
Annance, Francis	1825
Arrowsmith, Aaron	1811, 1815
Atkinson, George II	1816, 1819
Atkinson, George III	1816
Auld, William	1811–12, 1812
Barnston, George	1834, 1848
Beads, Thomas John	1838, 1839, 1841
Bell, Benjamin	1829, 1832
Bell, John	1846–8
Best, John	1792
Bird, James	1817
Black, Samuel	1835
Bouthillier, T.	1847
Cameron, Angus	1842
Campbell, Robert	1844
Christopher, William	1761, 1762
Clarke, Lawrence	1856
Clouston, James	1814, 1816, 1817, 1819, 1820–1, 1825
Clouston, John	1844
Coats, William	1749, 1750, 1751

* See the index to locate discussion of persons whose names are in roman. Names in italics are not mentioned in the text.

Colen, J.	1786–7	Henday, Anthony	1755	McDonald, Alexander R.	1854, 1856
Compton, P.N.	1863	Hendry, William	1828	*McDonald, Angus*	1864
Connolly, H.	1839	Hillier, William	1812, 1813	McDonald, Archibald	1824–5, 1825, 1826, 1827, 1841, 1844
Cook, William	1815	Hodgson, John	1775, 1776, 1777, 1789, 1790, 1791	McGillivray, Simon	1833
Corcoran, John	1820			McKay, Donald	1791, 1792
Corcoran, Thomas	1847	Homfray, Robert	1859	McKay, John	1791
Corrigal, Jacob	1791	*Hopkins, Edward M.*	1848	McKay, Joseph William	1850, 1853, 1865
Covington, Richard	1846, 1859	Howse, Joseph	1809, 1812, 1815	McKay, Thomas	1825
Crawford P.W.	1848	Howy, George	1740–1	McLean, John	1840, ca. 1843
		Hudson, James	1788	McLeod, Alexander Roderick	1826
Dallas, Alexander Grant	1858, 1859, 1861, 1863, 1865	Huggins, Edward	1851, 1852	McLeod, John M.	1831, 1834, 1840, 1843
		Hutchins, Thomas	1776	McNaughton, Archibald	1855, 1857
Davies, W.H.A.	1838, 1840			McNeill, William Henry	1848, 1849
Dease, Peter W.	1837	Isbester, Alexander	1845	McPherson, Murdoch	1824
Donald, George	1777, 1786	Isham, James	1750–1, 1754, 1755, 1757	Miles, Robert	1839, 1842, 1843
Douglas, James	1842			Mitchell, Thomas	1744
Drake, Montague W.T.	1867	Jarvis, Edward	1775, 1776, 1789–90, 1791	Moffatt, Hamilton	1852
Duberger, George	1838			Moore, Thomas	1678
Duncan, Charles	1792	Johnston, Magnus	1764, 1765	Morin, A.	1854
		Kempt, William	1823, 1824, 1825	Mouat, W.A.	1865
Erlandson, Erland	1834	*Kennedy, Alexander*	1815	Murray, Alexander Hunter	1852
		Kennedy, Robert	1819–20, 1820–1	Mynd, James	1750
Falconer, William	1767, 1768 or 1770, 1774	*Kennedy, William*	1840		
Fidler, Peter	1791–1821	Kittson, William	1825	Newton, William	1865, 1867, 1868
Ford, Richard	1754	Knight, James	post-1719	Norton, Moses	ca. 1760, 1762, 1767
		Knowles, John	1791	Norwood	1669
Gilmour, Boyd	1852, 1853, 1854			Nutt, George M.	1851
Gladman, George	1818, 1849, 1851	*Lake, James Winter*	1805		
Gladman, Joseph	1845, 1846	Launders, J.B.	1866	Ogden, Peter Skene	1826, 1827, 1828–9
Good, Philip	1793	Lewes, Adolphus Lee	1841, 1842, 1844, 1845, 1846, 1848, 1852		
Graham, Andrew	1772, 1774			Peers, Henry N.	1844, 1848
Grant, Walter Colquhoun	1850			Pemberton, Joseph Despard	1851–1860s
Green, F.W.	1861, 1863	*Mackenzie, John*	1856	*Potts, John*	1759
Griffin, C.J.	1861	*MacTavish, Dugald*	1862, 1863		
		Mactavish, William	1864, 1866	Rae, John	1847–64
Haggart, Duncan	1830	Mannall, John	1793	Roberts, George	1808
Hanwell, Henry Jr	1812, 1825	Manson, William A.	1870	*Robertson, James*	1816, 1827
Hanwell, Henry Sr	1776, 1803, 1805	Marley, John	1781, 1783	Robinson, George	1855, 1856, 1857, 1858
Harding, Robert	1844	McBean, John	1827	Robson, Joseph	1745–6, 1750
Hardisty, William	1853	*McCulloch, Roderick*	1820	Ross, Alexander	1821, 1824, 1825
Hearne, Samuel	1770, 1772, 1774(?), 1775, 1776	McDonald, Alexander	1854, 1856, 1858	Ross, Bernard	1856

Appendices

Appendices

Russell, James	post-1802	
Sangster, James	1848	
Scarborough, James	1841, 1842	
Schwieger, A.W.	1864	
Seale, R.W.	1748	
Simpson, Aemilius	1827, 1828, 1829	
Simpson, Thomas	1837, 1839, 1840	
Sinclair, Thomas	1831, 1832	
Sinclair, William	1795–6	
Spencer, John	1839, 1841	
Stayner, Thomas	1791, 1801	
Stuart, Charles Edward	1851, 1853, 1856	
Sutherland, Donald	1819	
Sutherland, George	1778	
Sutherland, James	1784, 1786	
Swain, James	1815	
Swain, Thomas	1803	
Swanson, William	1832	
Tache, F.P.	1857	
Tate, James	1817	
Taylor, George Sr	1791, 1794	
Taylor, George Jr	1827–38	
Thompson, David	1791, 1794	
Thompson, Thomas	1820	
Thornton, John	1680–1701	
Thornton, Samuel	1709	
Tiedemann, Herman O.	1863–8	
Tolmie, William F.	1847, 1851, 1864	
Trivett, J.F.	1858, 1859, 1860	
Trowbridge, W.P.	1853	
Turnor, Philip	1778–94	
Vavasour, Mervin	1845, 1846	
Vincent, Thomas	1815	
Warre, H.J.	1846	
Work, John	1824, 1825, 1830	
Yale, James	1858	
Yarrow, John	1750	

APPENDIX EIGHT

Hudson's Bay Company Employees or Other Non-Native Persons Who Provided Sketches or Descriptions for Maps Prepared by Peter Fidler

NAME*	DATE OF MAP(S)
Bunn, Thomas	1807
Charles, John	1806, 1809
Cook, William	1809
Dickson, Clouston	1809
Findley, Jean (Jaco Finlay)	1806
Flett, Andrew	1808
Flew	1798
McDonald, John	1808
McNab, Thomas	1807
Sabbeston, Hugh	1808
Spence, George	1808
Stayner, Thomas	1801
Swain, Thomas	1807, 1809
Thomas, Thomas	1807

APPENDIX NINE

Native Persons Who Provided Sketches or Descriptions for Maps Prepared by Peter Fidler and Philip Turnor

NAME*	DATE OF MAP(S)
Ac ko mok ki	1801, 1902
Ageena (Ageenah)	1810
Ah chap pee	1809
Ak ko wee ak	1802
Ak kay mis	1809
Cha chay pay way ti	1806
Chee hooz i az za	1809
Chynk, y, es, cum	1809
Cot aw ney ya zah	1810
Jepewyan (Chipewyan) Indian	1807
Ki oo cus	1802
Nayhektil lok	1809
Old Pumbles	1810
Oo ke mow a thin	1807
See seep	1809
Shew-ditha-da	1791
Thoo ool del	1809
Tow we kish a quib	1809
York Fort Indian	1809

APPENDIX TEN

Native Persons Who Drafted Maps

NAME*	DATE OF MAP(S)
Cauc-chi-chenis	1839
Coochenau	1774
Eskimo at Repulse Bay	1847
Idotly'azee	1767
Indians of Eastmain	1818, 1820–1
Kennewap	1839
Meatonabee	1767
Nabowisho	1843
Nayhektil lok	1809
Pequatisahaw	1777
Petaibish (Peataibish)	1838
Sturgeon River Indian	1827
Terrewill	1833
Tete Jaune	1819
Tuck, go, my	1777

* See the index to locate discussion of persons whose names are in roman. Names in italics are not mentioned in the text.

Glossary

ALTITUDE OF THE SUN. Height of the sun, expressed by its angular distance in a vertical plane above the horizon.

ARTIFICIAL HORIZON. Metal case containing a metal pan, into which mercury (quicksilver) was poured.

ATMOSPHERIC REFRACTION. The deflection of light entering the earth's atmosphere, causing apparent positional elevation of celestial bodies.

BARREN GROUNDS. *See* TUNDRA.

BEADS ON A STRING MAP STYLE. Complicated shorelines of lakes are generalized and rounded, and lakes are connected by simplified and often straight rivers.

BOAT COMPASS. Compass with magnetic needle attached to a card with thirty-two graduated compass points. The needle is free to rotate on a horizontal plane. Next to the card is a lubber's line, lined up with the boat's bow, to show which direction boat is pointing. The pivot, card, and lubber's line are supported by a free-swinging mechanism (gimbals), which allows the card to stay horizontal, even though the boat pitches and tosses.

BOTTOM OF THE BAY. Regional term applied by HBC men to the southern end of James Bay and the several trading posts there.

BRIGADE. Annual movement of company servants, usually by horses, with furs and other items received in trade, from the British Columbia interior to supply centres at the Pacific coast and their return with supplies and new trade goods.

BUFFALO COUNTRY. Regional term applied by fur traders to the grasslands and southern park belt of the interior plains, essentially south of the North Saskatchewan and Saskatchewan rivers, the home of the large buffalo herds.

CABIN. Small, primitive log hut built as a temporary trading post, or as a preliminary shelter before the construction of a larger fur trading post.

CADASTRAL (CADASTRE). Register, survey or map of the ownership of land for tax purposes. The scale is large enough to show property boundaries accurately and usually has measurements indicated.

CARTOGRAPHER. Person responsible for the conception, design, and production of a map. In the case of a manuscript map, the cartographer may be responsible for both the compilation of the data and the drawing of the map. In the case of a printed map, there may be another person as a draftsperson, engraver, lithographer, or printer of the map. In many instances cartographers may be directly involved in all or most of the aspects of production and even the sale of the map in their own shops.

CARTOGRAPHY. Science, technology, and art of making maps, and their study as scientific and historical documents and works of art. It is concerned with the compilation, design, and drafting required to produce a new or revised map, with stages in the reproduction of a map, and with map use.

CARTOUCHE. Decorated frame or panel on a map, enclosing the title, legend, scale, or descriptive text.

CARTRIDGE PAPER. A strong type of paper used for making gun cartridges and also for rough drawing and sketching.

CHART (MARINE CHART, NAUTICAL CHART, NAVIGATION CHART). Map for navigation use, which depicts a part of a sea or lake, indicating the coastal outline, position of rocks, sandbanks and other hazards, channels, anchorages, and water depths.

CHRONOMETER. Instrument to keep accurate time in varying temperatures. It has a better escapement than other time pieces and a compensation balance, so that it remains horizontal whatever the position of the ship. It is used in determining longitude and for other exact observation.

CO-ALTITUDE (OF THE SUN). Angular distance of the celestial object (the sun) from the zenith, measured by the arc of a vertical circle intercepted between the sun and the zenith. It is the complement of the sun's altitude.

COASTING (COASTERS). To sail along a coast, or to trade between coastal ports. A coaster is a master or pilot of a coasting vessel, or the vessel employed in coastal sailing.

COLUMBIA DISTRICT. Hudson's Bay Company district based on the Columbia River and its tributaries, but extending north into Thompson River area, Okanagan area, and south into Snake country.

COMPASS ROSE (SEE ALSO COMPASS CARD). Circle graduated to degrees or quarter points and printed on a chart for reference, usually showing both magnetic and true directions.

COMPASS VARIATION. Angle in degrees between magnetic north and true (geographic) north. It is measured east or west of the true meridian and may increase or decrease annually. Variation differs from place to place.

COMPILE (MAP COMPILATION). To gather maps and other information useful for the design and preparation of a map, including the lay-out of the map base on a graticule, the choice of appropriate map symbols, and the design of the map elements.

CORDILLERA. Series of more or less parallel ranges of mountains with intervening uplands and basins, extending from the Rocky Mountains to the Pacific coast.

COUREUR-DE-BOIS. See WOODRUNNER.

DECLINATION (OF THE SUN). Angular distance of a celestial object (the sun) north or south of the celestial equator.

DORSE. Back of a book page or document.

DRAFT. See MAP.

DRAPERS COMPANY. One of the Worshipful Companies of England. Many drapers had become involved with merchant companies engaged in trade and discovery, and a number of them had become chart makers who trained apprentices. See Campbell, T., "The Drapers' Company Chartmakers."

ENANTIOMORPH. Mirror image.

FACTOR. Title applied by the HBC to the head of a factory.

FACTORY. Term applied by the HBC to the major trading posts that were the centres for large inland trade areas. These were located at the mouths of rivers in Hudson and James bays. Whereas York, at the Nelson-Hayes mouth, was designated a factory, Severn, at the mouth of the Severn River and subordinate to York, was named Severn House.

FAIR COPY. Final draft of a map, indicating the design desired, showing colour scheme, symbols, print styles, and placement of the map elements to be followed by a draftsperson preparing the artwork for the printing of the map. If the map is not to be printed, the fair copy is the final manuscript map.

FATHOM. Unit of length used for measuring the depth of water. While originally based on the distance between the fingertips of a man's outstretched arms, it is generally taken as equal to six feet.

FORT. Trading post of the HBC built in the form of a quadrangle with wooden bastions at each corner that are joined by pallisades of upright logs. Inside this stockade were living accommodations, storehouses, and workshops. The bastions were sometimes used as pelt storehouses.

GRATICULE (PROJECTION GRID). System of latitude and longitude lines by which the curved surface of the sphere is represented as a plane surface.

GREAT PLAINS. Areas of relatively flat country, sometimes rolling, with no prominent elevations, except in the high plains, that is, the western area between three and six thousand feet a.s.l. The plains receive little rainfall and are almost treeless.

HOME INDIANS. Term applied to Indians, mainly the Cree, living in the general region of factories, such as York, Albany, Moose, and other posts on the main rivers. Some gave up hunting as their main livelihood, and undertook tasks such as carrying packets between posts, hunting game for posts, and serving on boat brigades.

HOUSE (fur trade post). Term applied by HBC to its interior fur trading posts.

HYDROGRAPHY (HYDROGRAPHIC, HYDROGRAPHER). Science which studies and describes the surface waters of the earth in terms of their physical features, tides, currents, chemical content of the water, and the form of the sides and bottoms of the bodies of water.

INDENT. Official requisition or order for goods supplied.

INDENTURE MAP. Map or survey plat of property, a formal document that accompanies a deed and is usually executed in two or more copies. See also PLAT.

INLAND WINTERING SYSTEM. HBC fur trading system whereby traders were sent from the Hudson Bay factories to accompany Indian groups returning inland to trap over the winter. They were to encourage these natives to bring their furs back to the factories the following spring and summer.

JAPANNING. The process of varnishing with japan, a quick drying varnish or enamel that is hardened by baking and yields a brilliant (usually black) coating.

LORDS COMMISSIONERS OF TRADE AND PLANTATIONS. Members (who may or may not be peers in the House of Lords) of a board or commission appointed to perform the duties of the high office of state regarding trade and the economies of the colonies.

LUNAR DISTANCES. Distance of the moon from the sun, a planet, or a fixed star, used in calculating longitude.

MAGNETIC BEARING. Line of constant compass direction, usually measured clockwise in degrees from 0 to 360 degrees.

MAGNETIC NEEDLE (COMPASS). Needle, magnetized by a magnet or lodestone, thrust horizontally through a crossbar of wood so that it floats in a bowl of water.

MANITOBA LAKES. Term used in this study to refer to the main prairie lake system: lakes Winnipeg, Manitoba, Winnipegosis, and associated smaller lakes.

MAP. Representation of all or part of the earth's surface, showing physical and/or human features, each point corresponding to a geographic position according to a definite scale and projection.

MAP (MANUSCRIPT). Hand-drawn map that is a unique product. Mechanical instruments, such as a drafting arm and board, stippler, or lettering set, may be used.

MAP (REFERENCE TYPE). Type of map primarily prepared to be used as reference source, for place names, settlement location and size, road, rail, and other transport elements, and location and relationship of physical features, such as rivers, lakes, and terrain features.

MATHEMATICAL TABLES. Tables of distances, vertical angles, base angles, etc., used in observation and calculation of astronomical positions.

MEASURING WHEEL (PERAMBULATOR, ODOMETER). Wheel used by surveyors to measure the distance traversed.

MERCATOR CHART. Chart crafted on Mercator's projection, in which meridians are parallel to each other, and parallels are straight lines whose distance from each other increases from the equator. At all places the degree of latitude and longitude have to each other the same ratio as on the sphere. The projection has great value in navigation, since a bearing line is always a straight line, i.e., it makes equal oblique angles with all meridians.

MERIDIAN, LOCAL. Great circle of the celestial sphere, passing through its poles and the zenith of the observer; the zenith is the point in the celestial sphere directly above the observer.

MUCK-SWAMP LAND. Low-lying wet ground, which is a mixture of muck-soil and swamp. Muck is wet, decayed, dark vegetable matter mixed with minerals. Swamp is wet, spongy ground with live vegetation, in which water collects. Some muck-swamp areas can be drained for cultivation.

NEW CALEDONIA. Area named by Simon Fraser in 1806. A trading district of the North West Company, and then of the Hudson's Bay Company. Fort St. James on Stuart Lake was the district centre. Its indefinite boundaries were the Rocky Mountains and Coast Range from east to west, and 50 degrees to 57 degrees north.

NORTHWEST PASSAGE. Water route through the Arctic Ocean, north of the Canadian mainland, desired as the route from the Atlantic to the Pacific Ocean. It was first traversed in 1903–6 by Amundsen of Norway in the ship *Gjoa*.

OREGON TERRITORY. Formerly applied to the valley of the Columbia River and adjacent territory on the Pacific coast. Between 1818 to 1846 there was a dispute over the international boundary, which was finally fixed at the 49th parallel. Oregon Territory was created in 1848, extending from California to the 49th parallel.

OUTFIT. Hudson's Bay Company term for the annual trade goods and other supplies brought to the factories by boat each year and distributed to the fur trade district centres, from which the goods were dispersed to the fur trade posts for the trade.

PARALLAX. Difference in angle between the altitude of the sun or moon, viewed from an observer's position on the surface of the earth, and the angle as if viewed from the centre of the earth. Parallax is at a minimum when the moon or sun is directly overhead.

PARALLEL GLASSES. Two rectangular pieces of glass set into a hinged metal frame, shaped like a ridge roof. If attached to the artificial horizon, the glasses could be folded flat. If not attached, they could be folded and kept in a case.

PARCHMENT. The skin of a sheep or goat, or other animal, flensed, and prepared to receive ink.

PLAN. Drawing, sketch, or diagram of an object, or of a small area on a large scale, with considerable detail. Plans usually are of architectural details.

PLANIMETRIC. Representation of cultural or natural features on a map without any indication of terrain features or relief.

PLAT. Map at large scale showing the accurately scaled boundaries and subdivisions of a piece of property, usually indicating measurements.

PORTLEDGE. Account of the names and claims for wages and allowances of a ship's crew.

POST. Term applied to the complex of building and grounds of a fur trading station. At its simplest, it may have been only one building.

PRECAMBRIAN SHIELD. A major geographical feature of the North American continent, composed of ancient, hard, basement Precambrian rocks. The original mountain belts therein have been almost completely eroded. A slightly convex area extending from the Arctic islands around Hudson and James bays, strewn with innumerable lakes, rivers, ponds, and marshes.

PRICKING OFF A COURSE. In laying out the course of a boat, the navigator would take bearings from point to point and measure distances. On the chart, using compasses, the course points would be pricked through the paper, leaving holes as markers.

PRISMATIC COMPASS. Hand compass with sights and triangular glass prisms used to aim the instrument at a point, and at the same time to read the compass direction of the point.

PROJECTION GRID. *See* GRATICULE.

PUGET'S SOUND AGRICULTURAL COMPANY. Founded by the HBC in 1839 to develop settlement and farms in area north of Columbia River to strengthen British claims to area and to supply produce to the HBC. It was operated by its own directors, officers, and employees, but its shareholders were also HBC shareholders.

QUADRANT. Instrument in the shape of a quarter circle, with a graduated arc of 90 degrees and a movable radius for measuring angles, used to measure the altitude of the sun.

RATING (OF A WATCH OR CHRONOMETER). The daily gain or loss of time of a watch or chronometer as compared with the true time.

REFLECTING TELESCOPE. Telescope in which the image is produced by a mirror and then magnified by a lens or lenses.

RELIEF. Difference in elevation between the lowest and highest point in an area. Also, the actual configuration of the earth's surface, in the sense of differences in altitude, of slope, of shapes and forms of the surface.

RHUMB LINE. *See* MAGNETIC BEARING.

ROADSTEAD. Open anchorage area where ships may lie safely or conveniently near the shore.

ROCKY MOUNTAIN TRENCH. Narrow elongated valley between the Rocky Mountains on the east and the Cariboo, Selkirk, and Purcell Ranges on the west. The site of the Columbia, Kootenay, and upper Fraser Rivers.

SCALE. Ratio between the distance between two points on a map and the actual distance of the same two points on the ground.

SCHOONER. Boat having two masts, rigged fore and aft, with a smaller sail on the foremast, the mainmast nearly amidships, and sometimes with square topsails on one or both masts.

SEGMENTAL SKETCH. As used in this study, a sketch map at large scale, part of a successive series of sections of a river and lake network drawn in the journals of a journey. Usually included distances, compass directions, as well as particulars of the shoreline, rapids, falls, and portages. When joined together, the sketches could form a composite map at a smaller scale.

SEMI-DIAMETER (OF THE SUN). Apparent radius of a generally spherical heavenly body, in this case, the sun.

SEXTANT. Instrument used to measure the angular distance between two objects, such as the horizon and a heavenly body (usually the sun), by a double reflection from two mirrors. The position of the observer in degrees of latitude and longitude may then be calculated.

SKETCH MAP. Normally quickly drawn, rough, and simple map, prepared without using precise distance and direction measurements. Sometimes used in preparation of a more finished map.

SLOOP. Small, one-masted, fore-and-aft rigged vessel, with a mainsail and a headsail.

STRAIT OF ANIAN. Mythical strait dividing North America from Asia, which became, in reality, the Bering Strait.

SURVEY. Process for measuring a tract of land or a coastline, whereby a description and a plan may be obtained; applied also to a body of persons or a department engaged in such work.

TAIGA (BOREAL FOREST). Forest region encircling the northern hemisphere south of the *tundra*. It is dominated by conifers (spruce, fir, pine, cedar, larch), although frequent fires favour the invasion of aspen, poplar, birch. It includes extensive peatlands.

THEODOLITE. Surveying instrument that measures both horizontal and vertical angles. Using a vertical circle or arc, measurement of angles of altitude and depression are made. With horizontal circle or arc, graduated plate bearings are read. The instrument is equipped with a telescope, level, *vernier*, and micrometer.

TOPOGRAPHIC. Depiction of both natural and cultural features on maps at larger scale. Relief and landform features are shown by some form of map symbol, most commonly contours.

TRACING PAPER. Semi-transparent paper used for copying drawings, maps, etc., by placing it over the original, and "tracing."

TRACKING. To tow, drag, pull, or tug a canoe or boat along a river course, usually from the bank but, sometimes, necessarily in the river course.

TRIANGULATION. Surveying technique whereby a network of triangles is laid out on precisely determined points and lines of known lengths. From this structure, lesser distances and various directions can be established.

TUNDRA. Large circumpolar region north of the treeline, with discontinuous vegetation such as shrubs, mosses, lichens, and dwarf herbs; usually flat to rolling terrain with abundant rock outcrops; often includes permafrost.

VERNIER. Device by which minute measurements may be obtained, consisting of a short, moveable scale which is passed over the divisions of the graduated scales on surveying, astronomical, and other instruments.

WASHINGTON TERRITORY. Created in 1853 out of Oregon Territory, extending from the Columbia River north to the 49th parallel. Its capital was Olympia.

WESTERN OCEAN (SEA). The Pacific Ocean, especially the part west of Alaska, Canada, and the northwest of the United States.

WEST OF THE MOUNTAINS. General term applied in the HBC to the area extending west of the Rocky Mountain divide to the Pacific Ocean.

WHATMAN'S PAPER. Kind of paper derived from the name of its English maker, available in several qualities and used for drawings, engravings, and map making.

WOODRUNNER (COUREUR-DE-BOIS). Itinerant, unlicensed fur traders during the French regime. Eventually, some were licensed. Many associated with the interior trading posts.

WRITER. Employment category in HBC. Essentially an office worker who acted as secretary to the senior trader, writing and copying letters, reports, journals and account books. The writer was sometimes used in other capacities at the will of his superior.

XY COMPANY. In 1795 some former North West Company partners joined together as the XY Company in opposition to their former partners. In 1804 the two companies joined together as the North West Company.

Notes

CHAPTER ONE

1 Jeanette D. Black, "Mapping in the English Colonies in North America: the Beginnings," in *The Compleat Plattmaker*, ed. Norman Thrower (Berkeley: University of California Press 1978), 104.

2 See D. Wayne Moodie and John C. Lehr, "Macro-Historical Geography and the Great Chartered Companies: the Case of the Hudson's Bay Company," *Canadian Geographer* 25, no. 3 (1981): 267–71. See also appendix 1 for a discussion of HBC organizational structure.

3 See location maps for the extent of this territory.

4 The records of the company do not show consistent use of terms applied to cartographic materials. The words, map, chart, plan, sketch, draft, survey, or some other differently spelled variant (plain, skitch, draught) were often used indiscriminantly.

5 Philip Turnor (c. 1751–c. 1800), appointed as surveyor-cartographer in 1778, served with the company as a surveyor, map-maker, trader, and consultant until his retirement in 1795. See chapter 8. See also J.B. Tyrell, ed., *Journals of Samuel Hearne and Philip Turnor Between the Years 1774 and 1792* (Toronto: The Champlain Society 1934), especially 60–94, and the article on Turnor in the *DCB*.

6 "Orders and Instructions to John Martin for his Upland Expedition ..." from John Favell Jr, Henley House, 14 June 1776. HBCA, B86/b/1,fo.13. See foreword for an explanation of HBCA references.

7 See chapter 6 for details of Coats's expedition and maps.

8 On the relationship of this company to the Hudson's Bay Company, see chapters 11 and 12. See also HBRS, 32 (Fort Victoria Letters, 1846–1851), lxxxii, lxxxii,411.

9 The beginning of the company's coal mining operations on Vancouver Island is dealt with in HBRS 32 (Fort Victoria Letters), *passim*.

10 D.W. Moodie, "Early British Images of Rupert's Land," in *Man and Nature on the Prairies*, ed. Richard Allen (Regina: Canadian Plains Research Center 1976), 3. The competition included French traders during the French regime. After the defeat of the French in 1759, both English and French independent traders brought trade goods to the Indians in the west and north and built posts. Hudson's Bay Company men called them indiscriminately "Canadians" and "Pedlars." The Canadians or Pedlars traded individually or merged into small groups. After 1784, when the North West Company was formed, the term Pedlar became defunct, and either the "Northwester" or "Canadian" was used. In fact, by then "Canadian" had become a generic term applied to free traders, that is, to those men who were not attached to either the HBC or the NWC. Canadians largely operated out of Montreal.

11 For a discussion of the cooperation between the company and the Royal Society of London For Improving Natural Knowledge, see: Richard I. Ruggles, "Governor Samuel Wegg, 'Winds of Change'," the *Beaver*, Outfit 307, no. 2, (1976): 10–20; Richard I. Ruggles, "Governor Samuel Wegg, Intelligent Layman of the Royal Society, 1753–1802," *Notes and Records of the Royal Society of London* 32, no. 2 (1978): 181–9; Raymond P. Stearns, "The Royal Society and the Company," *The Beaver*, Outfit 276 (June 1945): 8–13. Of the eighteen original "adventurers" named in the charter of 1670, Stearns identified six who were fellows of the society. I provide a list of names in "Governor Samuel Wegg, Intelligent Layman," 181.

12 The paper on the magnetic needle was presented by Captain Christopher Middleton in 1731. This same year he gave the society the journal of his voyage to Hudson Bay that was later published in the society's *Philosophical Transactions*, vol. 34. In 1737, he was elected a fellow, and in 1742 was the recipient of the society's Copley Medal. See Stearns, "The Royal Society and the Company," 9–11.

13 E.E. Rich, *Hudson's Bay Company 1670–1870*, 2 vols. (London: HBRS 1958), 1:299.

14 See Glyndwr Williams, "The Hudson's Bay Company and the Critics in the Eighteenth Century," *Transactions of the Royal Historical Society*, 5th ser., 20 (1970): 149–71; Glyndwr Williams, *The British Search for the Northwest Passage in the Eighteenth Century* (London: Longmans 1962), 31–57. For Arthur Dobbs, an Irish landowner and writer, member of the Irish House of Commons from 1727 to 1730, see Desmond Clark, *Arthur Dobbs, Esquire, 1689–1765; Surveyor-general of Ireland, Prospector and Governor of North Carolina* (London: Bodley Head 1958).

15 This policy was initiated by Henry Kelsey. See chapter 5.

16 The two men sailed on the company's supply ship, the *Prince Rupert*, to Fort Churchill, arriving in August 1768, where "Cabbins" for their accommodation, and an "Observatory" were built (B42/a/70,fos.61,62). They returned to England in October 1769 and presented papers to the Royal Society soon after. Wales's journal was printed in the Society's *Philosophical Transactions* (vol. 60, 1771), and a copy forwarded to the company (A1/43,fo.180d). Concerning Wales's later contact with the company, consult chapter 4.

17 Alexander Dalrymple (1737–1808) had entered the service of the East India Company at age

fifteen. He was proposed as the leader of an expedition to the south Pacific to observe the Transit of Venus in 1769; the appointment, however, went to James Cook. Dalrymple was the author of a number of books and pamphlets on voyages of discovery in the south Pacific as well as on India and proposals for imperial trade. He was appointed the first hydrographer when the Admiralty established a hydrographic office in 1795. He held the post until 1808 and was involved in "collecting, collating and publishing a large number of charts." See article in *DNB*.

18 Ruggles, "Governor Samuel Wegg, Intelligent Layman," 193.

19 Ibid., 195–7; James Cook and James A. King, *A Voyage to the Pacific Ocean*, ed. Rev. J. Douglas (Edinburgh: R. Morison and Son, 1793); Alexander Dalrymple, *Memoir of a Map of the lands around the North Pole*. 1789; Samuel Hearne, *A Journey from Prince of Wales's Fort, in Hudson Bay, To the Northern Ocean* (London: A. Strahan & T. Cadell 1795).

20 Aaron Arrowsmith (1750–1823), came to London at age twenty and published his first map in 1790. John Arrowsmith (1790–1873) was one of the founders of the Royal Geographical Society. In 1860, Sir George Simpson informed the Smithsonian Institution that John Arrowsmith was "regularly supplied with every chart or journal" received by the company, (D4/79,fo.326). See, also, Coolie Verner, "The Arrowsmith Firm and the Cartography of Canada," *Proceedings of the Association of Canadian Map Libraries* 4, (1970): 16–21.

21 See appendix 2.
22 See appendix 4.
23 See appendix 3.
24 Ibid.
25 See chapter 5.
26 The holdings of the business were auctioned on 28 July 1874 after John Arrowsmith's death.

27 For example, the oldest extant chart in the HBCA, that of Hudson and James bays, dated 1709, was drawn by Samuel Thornton. There are two copies in the HBCA, G2/1 and G2/2 (1^A).

28 For example, an 1847 map of the north shore of Lake Huron in the vicinity of La Cloche came into the company's records with correspondence from the Crown Lands Department of the Province of Canada. D5/20,fo.414.

29 See appendices 7–10.

30 Under half of the men produced only one map; a little over a third of them accounted for two to five maps. Fourteen men prepared seven or more maps (Philip Turnor, eight; John Hodgson and John Rae, eleven each; and George Taylor Jr, twelve, in addition to his series of segmental sketches.)

31 In 1974, the company transferred its archival holdings from London to the Provincial Archives of Manitoba in Winnipeg.

32 See Richard I. Ruggles, "Canada's First 'National' Map Agency: The Hudson's Bay Company," *Bulletin*, Association of Canadian Map Libraries, no. 55, (June 1985): 1–18.

CHAPTER TWO

1 See chapter 7 for a discussion of Henday's inland journey and map making.

2 See chapter 8. See also Shirlee A. Smith, "James Sutherland, Inland Trader 1751–97," *The Beaver*, Outfit 306 (Winter 1975): 18–23.

3 See Richard I. Ruggles, "Hospital Boys of the Bay," *The Beaver*, Outfit 308 no. 2, (1977) 4–11. Christ's Hospital or the Blue Coat School was granted Letters Patent in 1553 by Edward VI. Located near St. Paul's Cathedral, its mathematical school, one of five subdivisions, was endowed in 1673 by Charles II and had as its main aim that of preparing boys to become qualified naval and maritime personnel. The school still operates on the outskirts of London. The Royal Foundation of Queen Anne or the Grey Coat Hospital or School, was established for boys and girls in Westminster in 1698, but received its Royal Charter in 1706. Some boys entered the mathematical class and were apprenticed to the navy, the maritime service, to trading companies, and to trades where their training had a practical advantage. The school was operated until recently for girls in the same building where it had been located for a large part of its existence.

4 For example, Anthony Henday received a gratuity of £20 for his first inland journey, commencing in 1754, and for his sketch of the Saskatchewan River area. He would, the company promised, receive additional reward for like service (A6/9,fo.33d). In 1742, George Howy was granted five guineas for his "care and diligence" in exploring the Moose River estuary and lower river, and preparing maps for the company. (A6/7,fo.4d). Future benefits were promised Moses Norton if he succeeded in the examination of the coast north of Churchill, and especially the investigation of the opening that was later named Chesterfield Inlet (A1/41,fo.152). John Thomas was praised for his readiness to go inland on useful discoveries, and promised a gratuity of £10 per annum, above his normal salary, for the remaining two years of his contract, provided he made further discoveries (A6/12,fo.75). David Thompson, the very eager and useful Grey Coat apprentice, who asked for surveying instruments to be granted him at the end of his apprenticeship, rather than the usual gift of a suit of clothes, was given not only a sextant, magnifying glass, and a pair of parallel glasses by the grateful company, but also the apparel; moreover, it promised him other encouragement to convince him of its "honest intentions" toward him (A11/117,fo.54d, A5/3,fo.56d).

5 Some of these were Thomas Beads, Thomas Corcoran, Alexander G. Dallas, George Gladman, Philip Good, Anthony Henday, John Mannal, Alexander H. Murray, John Spencer, and James Yale.
6 He found that Taylor had died. The company arranged for Rae to go to Toronto to be trained by John H. Lefroy, director of the magnetic observatory there.
7 Ruggles, "Hospital Boys of the Bay." See also note 3 above.
8 Joseph Robson, who had worked as a builder and stone mason earlier during the construction of Fort Prince of Wales at Churchill, could handle simple land surveying. Adolphus Lee Lewes, F.W. Green, William Newton, and H.O. Tiedemann were all practiced draftsmen, with some surveying experience. George Taylor Jr was educated to handle navigation instruments, learned basic mathematics, and possibly some elements of surveying during his schooling in Britain. He gained further experience as a sloop master and in inland surveying tasks, particularly when he became land surveyor for the Red River colony. George Barnston, another employee, was educated as a surveyor and engineer, but although he drew several sketch maps, he did not engage in these professions for the Hudson's Bay Company.
9 In 1791, Jarvis complained that the failure to establish a post at Lake St. Anns could be attributed to "the Indians always deceiving them from their disinclination to meet the St. Ann Indians with whom they were in a state of hostility." A11/4,fo.179d.
10 See appendices 9 and 10 for lists of native draftsmen.
11 Thomas Simpson, *Narrative of the Discoveries on the North Coast of America. Effected by the Officers of the Hudson's Bay Company During the Years 1836–1839* (London: 1843), 360.

CHAPTER THREE

1 Various company cartographers developed their own graticules from that of colleagues and from printed maps available to them. The Arrowsmiths based their maps partly on company information, but used other sources, such as Admiralty hydrographic charts. The projection graticules were of considerable reciprocal value to company map makers.
2 Pemberton and his crew at Fort Victoria obtained elevations when they were laying out the trigonometric system using triangulation as the basic surveying technique, in which the southeast part of Vancouver Island was divided into a network of triangles, based on precisely determined points and lines of known length, and from which lesser distances and various directions could be established.
3 See A14/1,fos.99d,100; A14/3,fo.99d; A15/1,fos.11d,18d,19,22d,24. Seller, who had a shop in Exchange Alley near the Royal Exchange, also compiled, published and sold maps, charts and almanacs. In 1675, he published "A Breviate" of Captain Zacariah Gillam's journal of the 1668 journey in the *Nonsuch* to James Bay, in his series of publications known as the *English Pilot*.
4 The lens greatly increased the value of the telescope since it eliminated spectral colours from the refracted light.
5 See D.H. Sadler, *Man is Not Lost, a Record of Two Hundred Years of Astronomical Navigation with the Nautical Almanac. 1767–1967* (Greenwich: HMSD 1978).
6 Some of Peter Fidler's set of almanacs for the period 1789–1809 are in the HBCA Library.
7 The drafting room in Fort Victoria was the most meticulous as to qualities and types of paper. For example, in one order, they specified paper should be double elephant size (a sheet 40 × 26.5 inches).

8 John Robertson (1712–1776), who had been appointed master of the mathematical school in Christ's Hospital in 1748, became in 1755 the first master of the Royal Naval Academy, Portsmouth. He was clerk and librarian to the Royal Society from 1768–1776. *The Elements of Navigation* was first published in 1754 and went through seven editions in fifty years.
9 The normal procedure used to obtain latitude on land is given in appendix 5.
10 The general procedure used for obtaining longitude by the method of lunar distances is given in appendix 6.
11 John Work, with a party, exploring the Winisk River in 1819.
12 Robert Campbell, reporting at Fort Halkett in June, 1839.
13 John Work, with a party, exploring the Winisk River in 1819.
14 The problems travellers faced were compounded if the native guides proved unreliable or, for whatever reason, decided mid-journey to return to their homes. The chief factor at Albany on May 10, 1775 explained to the committee that Edward Jarvis's first attempt to travel inland between the Albany and Moose Rivers via the Chepy Sepy was foiled when the two Indian guides refused to complete the journey. He wrote that "Nothing could exceed ... [his] surprize to see Mr. Ed. Jarvis returning ... without accomplishing his Expedition ... he is both grieved and mortified at his disappointment ... but none of the Indians can be prevailed to show the way" (A11/4,fo.2d).

CHAPTER FOUR

1 The two men were Joseph Lindley and George Beck. The latter's name is recorded as "Back" in *The Journals of Samuel Hearne and Philip Turnor*, 94, but it is clearly "Beck" in the primary sources cited. In correspondence concerning Beck, the

committee gave instructions that he was to be instructed in astronomy and should be able to measure distances. He was also to practice making charts until the ship left for Hudson Bay, A64/1,fo.18d.

2 Laleham, Middlesex is situated on the north bank of the Thames river just above Staines.

3 No copy of the contract has been located in the HBCA.

4 Humphrey Marten (ca.1729–ca.1790) entered the company's service as a writer. He became chief at York in 1775. See Tyrrell, *Journals of Samuel Hearne and Philip Turnor*, 592–7, and DCB.

5 G1/109, "A Ground Plan of York Fort in Hayes River America taken August 1778 by Philip Turnor." Large scale plans like this one of Turnor's, which were drawn specifically to portray building structures, have not been included in the map catalogues. (However, some sixteen plans are listed there because their essential purpose was to show characteristics of the site or of company property within a short distance of a post.) For a reproduction of Turnor's plan, see appendix D of *The Journals of Samuel Hearne and Philip Turnor*.

6 It was not until the nineteenth century that it became the usual custom to make a short reference in the minutes to the receipt of a map.

7 The two copies of Turnor's first map, that is, the one of the interior from York Fort up the Saskatchewan, are recorded in the 1796 catalogue as item 42 in a drawer of Map Chest 1 (A64/45,p.129). In the later catalogue of ca. 1847, they were designated as item 11 in Chest 1 (A64/65,p.177). See appendix 4.

CHAPTER FIVE

1 "A Polar Card, wherein are all the Maines, Seas and Ilands, herein mentioned," 1635, in Luke Foxe, *North-West Fox; or, Fox from the North-West Passage* (London: 1635); "The Platt of Sayling For the discoverye of a Passage into the South Sea. 1631 1632," in Thomas James, *The Strange And Dangerous Voyage of Captaine Thomas Iames …* (London: John Leggatt, for John Partridge 1633).

2 The journey claimed by these two men is still a matter for scholarly contention, as there has been no satisfactory evidence of it having been undertaken.

3 John Nixon (ca.1623–92), overseas governor of the company's territories from 1679 to 1683.

4 Between 1680 and 1717 the company hired several Christ's Hospital apprentices, who, since they had navigation and charting training, might have made nautical charts. But no evidence of their work appears in the HBCA, and none of them became involved in exploration or mapping on land.

5 See George E. Thorman, "An Early Map of James Bay," *The Beaver*, Outfit 291, (Spring 1961): 18–22.

6 Moore was promoted from sailor to trader, which indicates that he knew the Cree language to some degree. Radisson and Groseilliers were at James Bay with the same group for several years. These men knew the country in the region of Lakes Superior and Nipigon, as well as the west shore of James Bay. If one accepts their claim that they had travelled earlier between these two regions previous to joining the adventurers, the Albany route would have been familiar to them. For details of Moore's relationship with the company, see Thorman, "An Early Map."

7 References to these two map makers are found in Norman Thrower, ed., *The Compleat Plattmaker, Essays on Chart, Map, and Globe Making in England in the Seventeenth and Eighteenth Centuries* (Berkeley: University of California Press 1978). See also Introduction to HBRS, 11 (Hudson's Bay Company Letters Outwards, 1679–94).

8 Most of the chart makers of the city were members of this London Company. See Thrower, *The Compleat Plattmaker*, 45–100.

9 Portions of the map are reproduced in HBRS, 11 (Hudson's Bay Company Letters Outwards, 1679–94), facing xi; see also ibid., xxvii, xxviii.

10 A photographic copy is in the map library of the Royal Geographical Society. See HBRS, 11 (Hudson's Bay Company Letters Outwards), xxvii, xxviii.

11 See A1/22,fo.17, A1/23,fo.9d, and A2/1,fo.21 for accounts of these meetings.

12 Vernon was secretary to the Lords Commissioners of Trade and Plantations.

13 Two "Mapps of hudsons Bay," 1701 (5^c); Copy of "Mapp of Hudson Bay," 1701 (6^c); and "Mapp of Hudsons Bay," 1702 (7^c). See A1/23,fo.35d and A1/24,fo.9.

14 John Churchill, first Duke of Marlborough (1650–1722), Governor of the Company from 1685 to 1692.

15 Also designated as the Signe of England Scotland and Ireland.

16 This is one designation given to a group of chart-makers who had their shops in the several hamlets adjacent to the Tower of London, and especially down river to the east. See Thrower, *The Compleat Plattmaker*, 45–100.

17 Bridgar was in company service from 1679 to 1686. Earlier stationed at Albany, he became governor at Port Nelson in 1682, and governor of the James Bay posts (the Bottom of the Bay) in 1686.

18 Sergeant succeeded Nixon as governor of the Bottom of the Bay in 1683 and retained the position until 1686 when Bridgar succeeded him.

19 In 1674, Thomas Gorst had started up the Nottaway River to contact Indians of the area but ascended only a few miles to the first main waterfalls. See Rich, *History of Hudson's Bay*, vol. 1, 136.

20 Until 1713, when hostilities were settled by treaty, the company had to defend itself from

French incursions into the bay. It lost all posts in James Bay for some years and York Fort for a period of time. Other posts were burned. There were also diplomatic attacks upon the British right to the region. In order to strengthen its position in situ, and politically to strengthen its charter claims, the company opened posts at Churchill and Severn, and attempted to contact both Inuit and inland Indians north of Churchill in order to bring them into the trading system.

21 Rich, *History of Hudson's Bay*, vol. 1, 297. For an extended biographical note on Kelsey, see *DCB*. See also HBRS, vol. 20 (Hudson's Bay Company Letters Outwards, 1688–96).

22 Kelsey's versified account of his inland journey has given students of exploration many difficulties in interpretation. See Arthur G. Doughty and Chester Martin, eds., *The Kelsey Papers* (Ottawa: Public Archives 1919). See also *DCB*.

23 William Stewart (Stuart) (ca.1678–1719) was only about thirteen years old when he came into the company's service in 1691. See *DCB*.

24 Richard Norton (1701–41) was apprenticed at age thirteen. He became a chief trader in 1723 and the following year was placed in charge at Churchill. He remained there until 1741. See *DCB*.

25 James Knight (ca.1640–ca.1720) joined the company as a carpenter in 1676. In 1682, he was appointed chief factor at Albany and deputy governor under Nixon; in 1692, he became governor of the company's territories. A redrawn version of 2^A, prepared by Richard I. Ruggles, is found in John Warkentin and Ruggles, *Manitoba Historical Atlas 1612–1969* (Winnipeg: Historical and Scientific Society of Manitoba 1970), 84–9 (map on p. 87).

26 In 1769 Hearne visited the island and was told by Inuit that the men died over a two-year period, the last five dying in the summer of 1721. The impression given by the Inuits was that the men gradually became debilitated and died of starvation, but it is possible that Inuit attacks were a contributing factor as well.

CHAPTER SIX

1 See HBRS, 21 (E.E. Rich, *History of Hudson's Bay Company*, vol. 1, 1670–1763), 552, and Charles A. Bishop, "The Henley House Massacres," *The Beaver*, Outfit 307, no. 2 (1976): 36–41.

2 Thomas White appears to have joined the company in 1719, and was referred to as a "good" and "very serviceable man." He was second in command and "book-keeper" at York after 1723 until 1734, when he was placed in charge.

3 The Indians referred to would be the Cree and Assiniboine of the Manitoba lakes area.

4 Christopher Middleton (d.1770), made sixteen annual voyages to company posts around the bay. His scientific observations, published by the Royal Society, aroused the interest of Dobbs in the discovery of a Northwest Passage. He resigned from the company in 1741 to enter the British navy. See *DCB*. William Moor joined the company as a boy, served under Captains Middleton and Coates, and left its service in 1741 to join an Admiralty expedition.

5 See Glyndwr Williams, *The British Search for the Northwest Passage in the Eighteenth Century* (London: Longmans 1962).

6 The map has been reproduced in Williams, *The British Search*.

7 Although Seale was an experienced cartographer as well as an engraver, he was probably not the cartographer of this map. This is the conclusion reached by Glyndwr Williams, who in "A Remarkable Map," *The Beaver*, Outfit 293 (Winter 1962): 30–36, argues that Seale usually signed his maps and that the geographical content was out of line with the earlier maps he produced.

8 Walker joined the company in 1749 as master of the Churchill sloop, a position he held for five years.

9 See Glyndwr Williams, "Captain Coats and Exploration along the East Main," *The Beaver*, Outfit 294 (Winter 1963): 5.

10 This opening was shown on several maps of the coast as far as 55° North, e.g., John Senex and John Maxwell's "North America," 1710.

11 A section of Middleton's chart is printed in Williams, "Captain Coats," 4.

12 Also humorously named "Sr. Atwls Lake," after Sir Atwell Lake, deputy governor of the company. This map is the first one depicting Richmond Gulf.

13 With the expedition were three miners, brought over to investigate the metal deposits (they hoped to find lead) located by Mitchell in 1744 toward the Little Whale River. The miners dug some four tons of lead ore, which were sent back to Britain on the *Mary*.

14 The site in Richmond Gulf was named Richmond Fort by Coats and Mitchell in 1749.

15 It is not clear whether a revision or a new map was prepared. There is a reference in the minutes to a "Draft ... which he had revised." A1/38,p.331. It is more likely that a new chart was drafted, for the two maps handed in were drawn on parchment, whereas the first map (9^A) was on paper.

16 There are vestiges of the formerly inked shape of Metesene Island visible on the chart.

17 See Williams, "Captain Coats," 6. On 13 March 1751, Coats was called into a committee meeting and informed that the members were pleased with his overall contribution, and therefore that a gratuity of £80 would be granted to him in addition to the £100 which he had previously received.

18 The statement appears in a cartouche titled "Explanation" on the map.

19 At the same time the committee, "as proof" of its "good opinion" of Howy agreed to "Continue" him three years longer, and to advance his wages to "Thirty Pounds pr Annum for that time." He had been awarded the not inconsiderable sum of five guineas for his earlier work.

20 Robson entered the company's service in 1733 to assist in the building of Prince of Wales Fort at Churchill. In 1752, he published *An Account of Six Years Residence in Hudson's Bay from 1733 to 1736, and 1744–1747* (London: J. Payne and J. Bousquet 1752), in which he was critical of the company and noted its failure to explore inland. It had, he said, "slept at the edge of a frozen sea." According to Glyndwr Williams in "Arthur Dobbs and Joseph Robson: New Light on the Relationship between Two Early Critics of the Hudson's Bay Company," *Canadian Historical Review* 40, no. 2 (1959): 132–6, parts of Robson's account were written or revised by Arthur Dobbs.

21 The two sections of it fit together in one corner, and have, in the past, been stuck together with sealing wax.

22 No further information is given as to the map's identity.

23 The detailed instructions for field observations that Henday was to follow are given in chapter 3, p. 16.

CHAPTER SEVEN

1 Only twelve are located in the HBCA (about 25%); one is in the University of Minnesota Library; thirty-four have not been located.

2 Two sons of La Verendrye, Louis-Joseph and Francois, are believed to be the first to see the Rocky Mountains. In the summer of 1742, they passed west of the Black Hills in what became South Dakota.

3 Only four maps were drawn by the wintering servants themselves, in spite of the opportunities for observing and sketching these new landscapes. Samuel Hearne prepared three of them.

4 Andrew Graham, *Observations on Hudson's Bay* 27 (London: Hudson's Bay Record Society, 1969).

5 Matthew Cocking (1743–1799) spent much of his career at York and Severn and inland in the Saskatchewan region.

6 The governor and committee to the chief and council at Churchill, 24 May 1753.

7 See John Warkentin and Richard I. Ruggles, *Manitoba Historical Atlas, A Selection of Facsimile Maps, Plans, and Sketches from 1612 to 1969* (Winnipeg: The Historical and Scientific Society of Manitoba 1970), 89, for a transcription of this map. See also plate 8.

8 A large amount, equal to his annual salary.

9 The stone Prince of Wales Fort was being built on Eskimo Point on the north shore at the mouth of the Churchill River during the years 1731 to 1740, when it was occupied by Governor Richard Norton. Work continued sporadically until 1771, and it was the company post until 1782 when it was captured and blown up by the French. Churchill had been built in the usual style after 1717, about eleven kilometers from the river mouth on the south shore. It was reestablished as the company post in 1783 by Hearne.

10 A transcription of this map is in Warkentin and Ruggles, *Manitoba Historical Atlas*, 91. Also see Richard I. Ruggles, "The West of Canada in 1763: Imagination and Reality," *The Canadian Geographer* 15, no. 4 (1971), 235–61, for a redrawn version and discussion of this map.

11 This was the same as the Kisk-stack-ewen, earlier sought north up the bay coast.

12 Samuel Hearne, *A Journey from Prince of Wales's Fort, in Hudson's Bay, to the Northern Ocean ...* (London: A. Strahan and T. Cadell 1795); Reprinted, Richard Glover, ed., *A Journey to the Northern Ocean* (Toronto: Macmillan 1958).

13 The name Assinibouels was used to refer to Lake Winnipeg or to the Manitoba lakes as a whole. In this instance, the likely reference is to Lake Manitoba.

14 See B86/a/29, fos. 16d, 17, 19d, 20d, 21d, 22d.

15 The quadrant was returned in the autumn of 1778.

16 For a detailed account of Turnor's appointment, see chapter 4.

CHAPTER EIGHT

1 John Marley, for thirty years at least a sailor on company ships, prepared three charts between 1781 and 1783. One showed the north Atlantic route to Hudson Bay (49^C); one was a chart centred on Hudson Strait (29^A); and a third was of a small section of the bay shore (48^C). Captain Charles Duncan of the Royal Navy was supported by the company from 1790 to 1793 in its last attempt to find and follow a passage west out of northern Hudson Bay. Two maps resulted, but nothing else of great significance for exploration (60^C, 61^C).

2 Two copies of the map, similar in detail, but with minor differences in lettering, are extant.

3 William Wales also sent him the *Requisite Tables*, which would have been of great use to him.

4 There is a second copy of this map located in the British Museum (5^B).

5 Delays occurred after the company's decision because an epidemic of small-pox had spread into the Saskatchewan region in 1781 and because French forces, having captured York and Churchill forts in 1782, held them for a year. Thus the 1783 supplies did not reach the interior.

6 As Gaddy did not record much information

about his trip, nothing more can be said about his route.

7 Robert Longmoor (Longmore) had been with the company since 1771. He spent most of his career out of York Fort in the Saskatchewan-Assiniboine area and took part in establishing the Cumberland, Lower Hudson, and Manchester houses.

8 Malcolm Ross (c. 1754–99), an Orkneyman, had been employed by the company since 1774. Excelling as a canoe maker, trader, and hunter, he served mainly out of York and Churchill until he was drowned in 1799.

9 Over half were made during the day when they took double meridian altitudes of the sun; otherwise, they shot the star Aquilae at night. The journals which indicate the character of the records kept by Turnor, Fidler and Ross are as follows: Turnor: B9/a/3 and B239/b/52; Fidler: E3/1 and E3/2; and Ross: B9/a/1.

10 Tyrrell, *Journals of Samuel Hearne and Philip Turnor*, 353.

11 See chapter 9 for a discussion of Fidler's journals and map books.

12 The height of land between Trade Lake, which is part of the Churchill River, and Wood Lake, which drains through Sturgeon-Weir River into the Saskatchewan River.

13 It was so entered in the two historical catalogues, but no further references were made to it.

14 From his earlier maps G1/21, G1/22 (27A), and G2/11 (34A).

15 He must have been referring to a lack of drafting instruments. He had received his sextant, a watch, and a compass earlier, and would have had all he needed in that regard (B145/a/6,fos.28d,30).

16 At the time that Sutherland was at Piskocoggan Lake in the spring, he met a young French Canadian trader who gave him a rough sketch of the route to Lake St. Anns on which he had marked the various portages which would be encountered on the way (51C). Apparently, it did not enter the collection in London, nor is it listed in the catalogues.

17 Although this map is not in the HBCA, the details appear on two later maps by Hodgson and Fidler, who used the original journals for their information, G1/2 (36A), and G1/37 (164A). See below for a discussion of these maps.

18 As was the case with the 1784 journeys, this detail was collated on the two later maps by Hodgson and Fidler. It may be that the map was drafted a little later after Osnaburgh House was constructed. The construction crew was on its way when Sutherland passed them. Also, the lake on which they were to build was called Piskocoggan by Sutherland in his journal, but on the map he called it Lake St. Joseph, the name given it when Osnaburgh was completed. It could be that the details he placed on his map were known to Sutherland late in the summer.

19 No further map by Hudson is known.

20 The Little North is the country lying to the east of Lake Winnipeg and between Lake Superior and Hudson and James bays. See Victor Litwyn, *The Fur Trade of the Little North: Indians, Pedlars, and Englishmen East of Lake Winnipeg, 1760–1821* (Winnipeg: Rupert's Land Research Centre 1986).

21 This map could be the one indexed as #13, "J. Hodgson, a Map of Hudsons Bay & Interior Westerly," since the map also has the notation, "1/13", on it. See also Warkentin and Ruggles, *Historical Atlas of Manitoba*, 105.

22 It is interesting that Aaron Arrowsmith's 1795 composite map copied the red marked area almost exactly.

23 These included a sextant, the characteristics of which he carefully defined. He also hoped to obtain a good magnifying glass, a pair of parallel glasses, and nautical almanacs for the years 1791–94. He invited the committee to debit his account for that part of the cost of the equipment which went above the usual allowance for clothing.

24 Thompson asked the committee to bill him and send the next year a brass boat compass and a Fahrenheit thermometer, so that, if the members approved, he could more satisfactorily undertake further surveying in the interior. In its May 1792 response, the committee expressed its continued approbation of his diligence and attention and notified him that it was further rewarding him with a gift of both these instruments and a case of drawing tools.

25 Confirmation of Thompson's authorship may be had from the inclusion in the 1796 catalogue of a note regarding a map of the rivers and lakes above York Fort, with the communication of the Port Nelson River with Churchill River, including part of Churchill River "by David Thompson 1794 & 1795." No other map of 1795 or later by Thompson has appeared in the HBC record, so that one can only conclude that the appearance of "1795" in this entry was merely a mistake by the clerk.

CHAPTER NINE

1 See chapter 8 for an account of Fidler's activities as Turnor's assistant. See also J.G. MacGregor, *Peter Fidler: Canada's Forgotten Surveyor, 1769–1822* (Toronto: McClelland and Stewart 1966).

2 Swan River, Somerset, and Marlborough Houses were the three York Factory houses in this area. The ridiculous situation in which two company houses were in competition a few miles from each other was eventually eliminated when the company reorganized its regional structure.

3 His other surveying instruments were a twelve-inch brass sextant made by Cary, an artificial horizon, and a good Jolly watch with a second hand (E3/2,fo.42).

4 The map has some aspects which might indicate that Fidler was not the author. One is that the shapes of the main lakes do not coincide with Fidler's segmental sketches. There are some stylistic differences in drafting from his later maps, but his methods were, in 1795–6, just in the process of gestation.

5 The maps produced were: a sketch of part of Cumberland House Lake, 1798, (62A); a map (in five sections) of part of the Burntwood River area, based on a sketch by Mr. Flew, a colleague, 1798, (63A); and a map of a route from York Fort to Edmonton House, prepared for his superior Ballenden, the York Factor, 1796–7 (68C).

6 For example, he made eight latitude and thirty-four longitude calculations at Greenwich House. He also recorded observations usually three times per day for temperature, wind direction, and general weather conditions when at the house and once per day at least while on the move.

7 Arrowsmith's 1795 map had shown the Red Deer River incorrectly as an upper portion of the Battle river. It also indicated that Fidler had reached an upper branch of the Missouri River in 1792–3, whereas he had been on a South Saskatchewan tributary.

8 He had in fact determined twelve latitude and two longitude locations from Chesterfield House to the main Saskatchewan River.

9 See D.W. Moodie and B. Kaye, "The Ac Ko mok ki Map," *The Beaver*, Outfit 307:4 (Spring 1977):4–15.

10 The Hudson's Bay Company was thereby deprived of an explorer, geographer, surveyor, and cartographer of great talent, who could have been of immense value in their northwest and western expansion. Unfortunately for the company, Thompson, at this vigorous and effective stage of his career, was about a decade out of synchronization with company needs. It was not ready to extend exploration far northwest in the Mackenzie River basin or across mountain passes into the fastness of the Cordillera.

11 The sketch (E3/4,fos.16d,17) in Fidler's map book had the notation on it, "Drawn by Jean Findley 1806." Jacques-Raphael (Jaco) Finlay, born of a Scots father and Salteaux Indian mother, was a guide, interpreter, clerk, and fur trader with the North West Company. Most of his career with the company was spent in the Rocky Mountain and Columbia and Snake river areas from 1774 to 1821. After 1821 he worked as a trapper and free trader, but in 1824 he accompanied Peter Skene Ogden on his Snake Country expedition. He was associated with David Thompson on several occasions while with the NWC. See *DCB*. vol. 6, 253–4.

12 The Saskatchewan River is misnamed the "N[orth] Branch Saskatchewan" on the map.

13 In addition to those mentioned, Fidler transcribed twelve sketches made by other Indians and two by one Inuit informant between 1807 and 1810.

14 At least eight additional sketches were transcribed from colleagues' sketches by Fidler.

15 For a copy of the map and a discussion of the Red River settlement see Barry Kaye, "Birsay Village on the Assiniboine," *The Beaver*, Outfit 312:3 (1981): 18–21.

16 In the spring letter, 1807, to York Factory, the committee expressed the hope that the South Branch would be explored again (A6/17,fo.105.). Whether this meant that Howse had been up the river in 1804–5 or in later years is difficult to determine since the Carlton House journals of 1805–10 are missing.

17 The river and pass were named after Howse, after his 1810–11 expedition to the Columbia River, although he was not the first European to use this route. Several employees of the North West Company, including David Thompson, had crossed Howse Pass previously. The "Jean Findley" map of 1806, drawn by Fidler, indicates the route through the pass. Fidler may even have suggested this route to Howse when they were together.

18 In fact, Howse's Pass lay astride the conflict zone of the warring parties.

19 Joseph Howse was born in Cirencester, England, where he also died. He gained an international reputation as an expert in linguistics in his later life. He wrote the first grammar of the Cree language, a book still recognized as an outstanding accomplishment. See *DCB*, vol 8, 411–14, and the *Canadian Encyclopedia*, 2nd edit. vol. 2, 1018.

20 The map was improperly titled "Hudson's Bay Chart of …", in the old HBC catalogue (A64/52,Portfolio B,p.185,#75), for rather than being a chart of that body of water, it is a river-lake map. Nor was the unsigned map drafted in 1811, but rather in 1812. I believe it was produced then by Howse and Auld.

21 The cartography is much more precise than that in evidence on the Nelson-Hayes estuary chart produced by Auld (B42/b/57,1812).

22 Although company officers had been asked to make proposals regarding the instruction of children in the rudiments of education, and the company had been sending schoolbooks to the factories, Clouston was the first schoolmaster to be appointed.

23 He operated out of Neoskweskau House for about nine seasons, Mistassini House for almost three seasons, and Rupert's House for parts of two seasons. For two years he was at Fort George at Big River (Fort George River), leaving there when the fort was closed.

24 Berens River, Carlton, Cumberland, Eastmain, Gloucester, Osnaburgh, Lesser Slave Lake, Red River, Severn, Whale River, and York.

25 In the later years of the eighteenth century, the company had received charts of a variety of

types from John Marley, Captain Charles Duncan, and George Donald, ranging from a chart of part of the Eastmain coast (32^A) and of the west shore from Churchill to Cape Tatnam (48^C), to larger area charts of Hudson Strait (29^A) and the northern section of Hudson Bay (60^C).

CHAPTER TEN

1 George Simpson, born in Scotland circa 1787, was employed by the HBC in 1820 in the Athabasca district. In 1821 he was appointed governor of the Northern Department, and in 1826 he became governor in charge of trading in North America. From 1833 to 1860, he resided in Lachine, the company's overseas headquarters.

2 Although Hendry was hired as a surgeon, he was also to "make himself useful in the Accounts department" (A6/21,fo.90d). He was stationed at Moose Fort throughout most of his career.

3 The maps indicate that originally the track was to follow the Leaf River. But among the few Indians Hendry met on the coast, only one could be found who could remember any route leading across to the Koksoak, and this led from Richmond Gulf to Lac à l'Eau Claire, to Lac des Loups Marins and the Larch River. The party was in such a hurry on their return for an expected meeting at the coast that they had no time to search to the north for the Leaf River. See the section "James Clouston, Inland from the Eastmain," in chapter 9.

4 One map (228^A) was sent out from Chimo on the ship *Ganymede* (A4/15,fo.30). The other (229^A) reached London in the packet assigned to the Esquimeaux brig at Fort Chimo. See *N. Quebec and Labrador Journals and Correspondence*, 24 (London: Hudson's Bay Record Society), 247n.

5 For example, the Moose Factory sloop would face heavy ice and strong tidal currents in Hudson Strait, or if one of the regular supply ships from Britain called in, it could be wrecked in the shallow waters.

6 The map was not signed by Davies, but it would appear to have been drawn in 1840. The drafting paper has an 1839 watermark. It may be that the map was a separate, perhaps larger scale version of the Kaipokok section of his 1838 map.

7 Alexander Roderick McLeod was born circa 1782. After employment with the NWC, he became a chief trader with the HBC in the Athabasca and Mackenzie districts from 1821 to 1824. He moved to the Columbia District in 1825. John McLeod also joined the HBC after working for the NWC. He was first in service at Ile à la Crosse; from 1823 to 1835 he was a clerk and then chief trader in the Mackenzie region, being posted to Fort Simpson, Fort Liard, and Fort Halkett. He was moved to the Columbia District in 1835.

8 Murdoch McPherson was described by Governor Simpson as "A fine, active fellow and tolerable Clerk & good trader" (D4/89,fos.85d–86). After he had helped build the new post of Fort Liard, he remained there as post master. He became chief trader at Fort Simpson in 1834.

9 Samuel Black, born circa 1785 in Scotland, was apprenticed to the XY Company in 1802, and to the NWC in 1804, with whom he remained until 1821. By 1823 he was a clerk in the HBC and by 1824, a chief trader for the company. From 1825 to 1830 he was in charge of the Walla Walla post in the Columbia District, and after 1830 he was chief at Kamloops. He was killed there in 1841.

10 The name Frances was also attached to the Stikine River by McLeod, but it was dropped from use for this waterway later. In this area, interior and coastal Indians met in an annual trade concourse.

11 Robert Campbell, born circa 1808, joined the company in 1834, after being hired in 1830 as sub-manager of an experimental farm to be developed at Red River. He became clerk at Fort Simpson and in 1837 volunteered to go to Dease Lake to establish a post and to explore the west side of the mountains beyond the lake. There had been a previous attempt at this in 1836, but the HBC group had retreated from the area in fear of a possible "Russian Indian" attack.

12 The Lewes is now known as the upper Yukon River.

13 In 1843 Adam McBeath had tried unsuccessfully to force his way through the mountains using the Gravel (Keele) River. It seems from the journal of the journey that no reasonable supply track could be made via this approach.

14 John Bell, born circa 1799 in Scotland, first worked for the NWC as a clerk from 1818. Transferring to the HBC, he spent most of his career in the Mackenzie and Athabasca districts, at Fort Good Hope, Peel River (Fort McPherson), Fort Liard, Fort Simpson, and Fort Chipewyan. He ended his career in the Montreal District. Alexander Kennedy Isbester was born in 1822 at Cumberland House. At the age of 16 he was employed by the HBC and sent to Fort Simpson from where in 1840 he helped to establish the Peel River Post. In 1842 he left the company, attended the universities of Aberdeen and Edinburgh, and became a highly respected schoolmaster. In 1866 he became a barrister of the Middle Temple in London.

15 The map was redrawn by John Arrowsmith and illustrated Isbester's paper on the region published in the *Journal of the Royal Geographical Society* 15 (1845), facing p. 332.

16 In 1823, company traders reached the lower course of the Mackenzie River.

17 Barnston was a clerk at York Fort. He was born in Scotland circa 1800, where he was trained in surveying. He entered the service of the NWC in 1820 and that of the HBC in 1821. He was described as a "young Gentleman of considerable promise who is enthusiastic in the cause and qualified to take the necessary observations for

the purpose of ascertaining the Latitude and Longitude of the country" (A12/1,fo.49d). He spent over forty years with the company in the Red River region, the Columbia District, at Albany Factory, in charge of the King's Posts centered at Tadoussac, at Norway House, and at Michipicoten.

18 Peter Warren Dease was born at Mackinac in 1788, the son of Dr. John B. Dease, captain and deputy superintendent of the Indian Department. Dease first joined the XY Company at age 13, was transferred to the NWC, and finally in 1821, moved to the HBC, who immediately sent him to the Athabasca District. He was seconded to the Franklin expedition from 1824 to 1827. Later he was at Fort Good Hope on the Mackenzie, and in charge of the New Caledonia District. He retired in the Montreal area. Thomas Simpson, cousin of Governor Simpson, was born in Scotland in 1808. After graduating with an M.A. from King's College Aberdeen, he joined the HBC in 1829 as secretary to the governor. He spent most of his time at Red River until he was appointed to the Arctic exploration expedition.

19 In fact, the map might not have been completed until the winter or spring of 1838, although the date 1837 appears on it. For a discussion of Arctic maps, see Kathryn M. Harding, *Discovery Maps of the Canadian Northwest: The Hudson's Bay Company, The Royal Navy and the British Map Trade, 1819–1857*, M.A. thesis, Queen's University, 1986.

20 "An Account of the Recent Arctic Discoveries by Messrs. Dease and T. Simpson. Communicated by J.H. Pelly, Esq., Governor of the Hudson's Bay Company," *The Journal of the Royal Geographical Society of London* 8 (1838): 213–225; map facing p. 224.

21 For some time it was believed that he was murdered by his Metis companions or by someone unknown during a melee in which two Metis were also killed. Later, it was suggested that he had killed his companions and then committed suicide while in a depressed state of mind.

22 *The Life and Travels of Thomas Simpson, the Arctic Discoverer* (London: Richard Bentley, 1845).

23 They were not located in the library of the Royal Geographical Society during a search in 1981.

24 "Narrative of the Progress of Arctic Discovery on the Northern Shore of America, in the Summer of 1839. By Messrs. Peter W. Dease and Thomas Simpson. Communicated by Sir J. H. Pelly, Bart., Governor of the Hudson's Bay Company," *The Journal of the Royal Geographical Society of London* 10 (1841): 268–74; map facing p. 274.

25 See J. M. Wordie, "Introduction," *John Rae's Correspondence with the Hudson's Bay Company on Arctic Exploration 1844–1855* 16 (London: Hudson's Bay Record Society 1953), for a full discussion of this part of Rae's career.

26 *Narrative of an Expedition to the Shores of the Arctic Sea in 1846 and 1847* (London: T. & W. Boone 1850).

27 HBCA has these charts (G3/166 and B3/23).

28 Reproduced in Wordie, *John Rae's Correspondence*, facing p. 98.

29 "Journey from Great Bear Lake to Wollaston Land. By Dr. John Rae. Communicated by the Hudson's Bay Company," *The Journal of the Royal Geographical Society of London* 22 (1852): 73–81; map facing p. 73. Also in Wordie, *John Rae's Correspondence*, 180–192. A further paper was read, "Recent Explorations along the South and East Coast of Victoria Land. By Dr. John Rae. Communicated by the Hudson's Bay Company," *The Journal of the Royal Geographical Society of London* 22 (1852): 82–96. Also in Wordie, *John Rae's Correspondence*, 194–214. Rae was awarded the Society's Founder's Medal in 1852 for his combined Arctic geographical explorations to that date.

30 The map was turned in with his report to the committee in late October 1854.

31 "Arctic Exploration, with Information respecting Sir John Franklin's missing Party. By Dr. John Rae, F.R.G.S. (Gold Medallist)," *The Journal of the Royal Geographical Society of London* 25 (1855): 246–56; map facing p. 256. Also in Wordie, *John Rae's Correspondence*, 265–86.

32 Many of Rae's mementoes were given to the University of Edinburgh, but personal enquiry has elicited the information that the three missing maps from the Victoria Land expedition were not included.

33 Eight maps and plans were drafted by William Kempt, two being copies of originals. Only four are in the HBCA. He made a plan of York Fort for the company which concerns proposals on surface and subsurface drainage of the factory and factory area. It was likely drawn as his first assignment upon arrival there in 1822 before he went inland to Red River.

34 PAM, Bulger Correspondence, III,M.151,pp.156–161. The chief company official at the colony, John Clarke, was given the map, but decided against this site, choosing the Image Plain as a better location. The map Kempt drew may then have been discarded.

35 An "index to the Plan of Red River Settlement, describing the Inhabitants, their Country, Number and Religion; the Boundaries of their Lots, with the extent of the Wood, Pasture and Cultivated Land by William Kempt, Surveyor 1824" (E6/11) indicates that a completed map existed by this date.

36 At least two copies were prepared.

37 A pencil notation, "A copy of this Sketch was left at York Factory with Mr McTavish," indicates that this large-scale map was a working plan to be used by officials to carry out certain of the recommendations. See 97[c].

38 These were corduroy causeways, built of logs laid side by side on top of fill.

39 See J.A. Alwin, "Colony and Company, Sharing

the York Mainline," *The Beaver*, Outfit 310:1 (1979): 4–11.

CHAPTER ELEVEN

1 A tracing from photostats of the original was made by Michael R. Crosier. It is available in facsimile from Friends of the Ellensburg Public Library, Ellensburg, Washington.
2 On this map, according to the catalogue (A64/52,#98,Portfolio B), the routes taken by early trapping parties, both of the NWC and the HBC, into Snake Country were plotted.
3 It is difficult to judge which Thompson map was referred to since it was apparently available to McDonald in the Columbia area (D4/88,fo.41). In 1813–14, Thompson had prepared a large map for the NWC and delivered it to William McGillivray. It hung in the hall of the NWC headquarters at Fort William. It may have become HBC property after the union with the NWC. It was not entered into the collection of maps in London. Simpson likely knew of this map and could have had the base details of the western area available to McDonald in 1824–5 at Fort Vancouver. Other Thompson maps were sent by him to the Foreign Office in 1843. The NWC map is now in the Ontario provincial archives.
4 1824–5, 1825–6, 1826–7, 1827–8, 1829–9, 1829–30. Peter Skene Ogden was born circa 1790 in Quebec, and died in Oregon City in 1854. He was with the NWC until he was taken into the HBC at the time of the union. He led the Snake Country expeditions from Fort Vancouver from 1824 to 1830, and then was chief trader and chief factor from 1831 to 1844 at Fort Simpson on the northwest coast. For ten years he was in charge of the New Caledonia District, but he ended his career back at Fort Vancouver.
5 Alexander Ross described Kittson as "a smart fellow … full of confidence and life," A. Ross, *Fur Hunters of the Far West* (London: Smith & Elder 1855). Simpson called him a "very spirited active little fellow" (D4/88,fo.143).
6 A64/52,Portfolio B,p.186,#113. See, however, a later tracing of his route on the map prepared for *Peter Skene Ogden's Snake country Journal 1826–7* (London: Hudson's Bay Record Society 1961).
7 This map is reproduced in *Peter Skene Ogden's Snake Country Journals 1827–8 and 1828–9* 28 (London: Hudson's Bay Record Society 1971).
8 Ogden wrote a brief account of the journey on 12 March 1831 for the governor, chief factors, and chief traders (D4/125,fos.85d–86). The account is given in *Peter Skene Ogden's Snake Country Journals 1827–8 and 1828–9*, Appendix C, 177–181.
9 It would have shown the route via the lower Willamette River, across the mountains through the Nestucca valley, and the Pacific coast south to the Siuslaw River, some twenty miles north of the Umpqua, with the many parallel stream crossings (B223/a/2).
10 John Work had been engaged at Severn and Island Lake from 1815 to 1823 and had just arrived in the Columbia district after having examined the Winisk River area in early 1824 (see chapter 10). McKay, who was Dr. McLoughlin's stepson and a former NWC employee, had, between 1819 and 1821, been across the Coast Range from the Willamette valley to the Pacific coast. Annance, an Abenaki mixed-blood employee from near Quebec City, was described by Governor Simpson as being well-educated, firm with Indians, a good shot, and qualified to lead the life of an Indian, but not worthy of belief even upon oath – altogether a bad character although a useful person (A34/2,fo.20). He fulfilled Simpson's more negative assessment when sexual misconduct resulted in his being dismissed from the service in 1834. James McMillan was born around 1783 in Scotland. He entered HBC service in 1821 after being with the NWC since 1803 or 1804. He accompanied Simpson from York Factory to the Columbia District, led the Fraser River expedition, and was stationed at Fort Assiniboine. He established Fort Langley and ended his career at Red River and then in the Montreal District in 1839.
11 A64/52,Portfolio B,p.187,#119–130.
12 The map was prepared in two copies (F25/1,fo12; F25/1,fo.13).
13 This map was then (in 1857) in Tolmie's possession. Douglas himself also claimed to have had the map drafted when he was shown it in 1865 and was giving evidence as to its origin for the commission established to decide upon the compensation to be paid to the two companies by the government of the United States for the loss of lands and facilities in the Columbia region (British-American Joint Commission, vol. 3, p. 80.)
14 F25/1,fo.10; and F26/1,fo.67. Two copies were provided, one being an exact replica.
15 Henry Newsham Peers, born in England in 1821, attended the Royal Military Academy, Woolwich, and joined the HBC in 1841 in the Montreal area as apprentice clerk. He was a clerk at Fort Vancouver from 1843 to 1848. His later career was spent at Fort Kamloops, Fort Yale and Cowlitz; he was in charge of establishing Fort Hope in 1848. Peers retired in 1859. A second map was prepared by an unknown draftsman on the same scale, although it was not an exact copy (265^A).
16 Vavasour's chart is based on one drawn by James Scarborough in 1837. See section "Mapping the Fort Victoria Area" below.
17 F25/1,fos.31,33–36.
18 In what is now Canada, the place name is Okanagan, but in the American part of this region, it is spelled Okanogan.
19 This map was given by Anderson's son to the

Surveyor General of British Columbia in 1910.
20 Anderson in 1846 had met a local Indian chief, Black eye, who described a shorter and better Indian trail bypassing a large section of the Tulameen River. A map was sketched showing this route (12B).
21 In 1848 a copy of this map was made for some purpose (301A), but several alterations were made by the draftsman. Among these were the eradication of the small-scale inset map. The name "Pt. Ogden" has been added at the harbour entrance and several other stylistic changes included, largely related to the simplification of compass symbol, bar scale, and border. An annotation states that the tracing was made by R.A., about whom nothing is known.

CHAPTER TWELVE

1 About ninety percent of the approximately 215 maps prepared in this decade concerned regions or locales of the present Canadian and American cordilleran and coastal regions. The other ten percent were various types of maps of areas east of the mountains and in the Arctic.
2 The island had remained British as a result of the Oregon Treaty.
3 W. Colquhoun Grant, "Description of Vancouver Island. By its first Colonist, W. Colquhoun Grant," *Journal of the Royal Geographical Society* 27 (1851): 268–320.
4 That is, Metchosin and Sooke.
5 Calendar of the Royal Agricultural College, Cirencester, England, incorp. 1845, p. 4.
6 Pemberton, born in Dublin, had been at Trinity College for just over one session. Since there was no school of engineering there at that time, he was supervised by George E. Hemans during his apprenticeship.
7 Calendar of the Royal Agricultural College, 1850, 11.
8 Except where specific knowledge of authorship by another department employee is known, the maps prepared in this map office are attributed to Pemberton. I have used this procedure also for maps drawn in 1855, even though Pemberton was on leave and Pearse was in charge, since Pemberton had final authority over production from the map office.
9 John Chapman was on the staff in 1851 and 1852; James Newbird worked through most of the 1851 through 1854 seasons; William Norquay was on the staff from 1854 through 1857.
10 Two books Pemberton mentioned are *The New British Province of South Australia*, written by Edwart Gibbon Wakefield, and published in 1835, and *Survey of New Zealand*, by a "Captain Dawson," published in 1840.
11 Pemberton acknowledged that it would be necessary to re-measure certain areas later after the instruments had been received.
12 Victoria, Surveys and Land Records, 5 Locker L is likely the original manuscript topographic map.
13 Pemberton suggested that if the map was to be redrawn in London the draftsman should "consult the execution of an Irish ordnance map, of the same scale" (A11/73,fo.187) for stylistic guidance, because the vegetation symbols on his map were too large. His grouping of vegetation should be preserved, he insisted.
14 Douglas reported that Pemberton had adopted the Indian name "Nanaimo" for the harbour, using it instead of Wentuhuysen Inlet.
15 In the autumn of 1852; the ship was the *Mary Dare*.
16 John Muir, a Scottish coal miner, became oversman (foreman) of coal miners at Fort Rupert from 1849 to 1851, and later was oversman for several years at Nanaimo. In 1854 he retired to a land holding at Sooke.
17 Hamilton Moffatt and Captain William Brotchie. In July 1859 Moffatt sent a copy of his journal of the trip and an unfinished chart of the Koskimo Inlet, the portage, and coast of the island to the Nimpkish River to Pemberton.
18 Pemberton named the mine and settlement on the Nanaimo peninsula, Colvile Town, after the Deputy Governor Andrew Colvile.
19 Several were duplicates. There likely were other copies made, but these have not remained in the collection.
20 As noted above, maps prepared in the surveying department are attributed to Pemberton even during the period he was on leave.
21 There seems to have been more than one copy made (A11/80,fo.154). Hermann Tiedemann, employed as a surveyor and draughtsman later, remarked in 1865 that the town maps of 1855 differed somewhat by the time he saw them, only one having the Indian reserve marked on it.
22 Two map copies were prepared, (A11/76a, fo.669,672). The financial evaluation was £4220 for the whole establishment.
23 Alexander G. Dallas, born in the West Indies, educated in Scotland, became a successful businessman in England and the Far East. Elected as a director of the HBC in 1856, he was almost immediately sent to Fort Victoria to reorganize PSAC operations and to assist Douglas with the business affairs of the company and colony.
24 There is a second copy, 28T2 Victoria Town (31B), and a photostat copy in B.C. Archives, Map Division, CMD48 1858. This was called the official map of 1858, although it was not completed that year.

CHAPTER THIRTEEN

1 This river was later named the Anderson River in honour of the man who had directed the company's attention to it, and who, before he was transferred from the district, had begun to organize Roderick MacFarlane's exploratory expedition to that region in 1858–9.

2 John S. Galbraith, "Conflict on Puget Sound," *The Beaver*, Outfit 281 (March 1951): 21.
3 The governor of Washington Territory, on a visit to Nisqually in January 1858, was shown a map of the Nisqually land claim, which was likely the original Huggins's map of 1852.
4 A simplified copy was traced in 1855 by an employee in the surveyor general's office (26^B). The copy is now in the National Archives (Record Group 76, Series 71, Map 3).
5 The original is not available in the HBCA, but a true copy of the map handed in was made in 1861 in the surveyor general's office, and is now in the National Archives (Record Group 76, Series 71, Map 5).
6 Washington Territory was established in 1853 from the former and larger Oregon Territory.
7 Earlier, in 1853, John Ballenden, the chief factor at Fort Vancouver indicated that he had refused the request of Surveyor General Preston for a survey of company claims in the vicinity of the fort (B223/b/42,fo.57d). Mactavish's refusal was sent in a letter of 9 May 1855 to Surveyor General Tilton.
8 There is a further tracing (369^A) of the Fremont-Preuss map in the HBCA, possibly prepared by John Arrowsmith for the committee, as it has his name attached to it. The published Fremont map is the "Map of Oregon and Upper California From the Surveys of John Charles Fremont And other Authorities Drawn by Charles Preuss Under the Order of the Senate of the United States Washington City 1848."
9 During this time, outside of its chartered territory of Rupert's Land and the Colony of Vancouver's Island, the company had to deal with colonial governments or American territorial officials.
10 These were of Mississanque, Nipigon, La Cloche, Michipicoten, Pic, Fort William and the Auguewang River (194^C). Upon completion of his contract, McDonald was hired as clerk and surveyor by the company and assigned to the Eastmain.

CHAPTER FOURTEEN

1 In September 1862, another set of tracings was transferred to London, but they have not remained in the collection.
2 This sketch, No. 5, is missing from the HBCA.
3 A.G. Dallas provided a small sketch of this area in February 1863 (503^A), which indicated his understanding of the Beckley Farm area, the unsold portion in 1861, and the general area from which the fifty acres was to be chosen.
4 One went to the colonial secretary's office in Victoria (likely (35^B)) and one to A.G. Dallas, who had been involved in the transfer.
5 This map and a second copy (215^C) were based on an indenture map of lot 24, Section XVIII, of 1853 (362^A).
6 This map and a copy (541^A) were based on an indenture map of lot 1, Section VI, of 1851 (313^A).
7 Two copies were made.
8 Two copies were made.
9 Two copies were made. Alterations had been made in the Indian Reserve boundaries; there had been an encroachment of the government land office in 1858; there had been changes in the location of the government reserve boundaries as shown on the 1859 town map and in Mactavish's boundaries shown on the 1863 map.
10 Two copies were made.
11 A second copy was sent to the colonial secretary's office.
12 These developments included possible British naval use of land and the Hudson's Bay Company's interest in moving the headquarters of the board of management to the Constance Cove area.
13 Through a series of misadventures, this map and two possible copies did not remain in the collection. The original map, completed in 1866, was sent to London by mistake, a new original to be made by Tiedemann in Victoria was apparently not finished before he decamped in 1868. It is not known whether, as was suggested, the tracing of the original in London, was ever undertaken.
14 Although it is not dated, it would appear to have been traced in 1860. The name of the draftsman is not indicated.
15 It is not possible to identify the other maps purportedly taken to Britain by Dallas. However, on (485^A) there is an endorsement "Plans of Nanaimo brought home by Mr. A.G. Dallas. July 1861."
16 Two copies were made.
17 This plan was drawn in either 1861 or 1862.
18 One of these buyers was Captain William H. Franklyn.
19 The government asked also that all claims at each post be fenced or boundary posts erected. Dallas replied that this would involve great expenditures and that such fences and boundary stakes had not been sufficient to the present to protect company land (A11/77,fo.586d). However, he agreed to have corner posts erected.
20 The map was enclosed in Rae's letter of 29 July 1864 to the committee, received by them on 30 November (E15/13,fo.20d).
21 This map was enclosed in Rae's letter of 23 August 1864 to the committee, received by them on 28 November E (15/13,fo.27).
22 Rae may have inserted a "sketch," with associated bearings, of passes over the mountains to the east of the Fraser River. It may not have been a map sketch, but a drawing of the mountain front as seen from the Fraser River area (E15/13,fo.34).
23 His report (B226/c/2,fos.369–372) was not written until 29 July 1865, in Victoria.

Bibliography

Alwin, John A. "Colony and Company, Sharing the York Mainline." *The Beaver*, Outfit 310:1 (1979): 4–11.

Back, George. *Narrative of the Arctic land expedition to the mouth of the Great Fish River and along the Shores of the Arctic Ocean in the Years 1833, 1834 and 1835.* Rutland, Vermont: Charles E. Tuttle Co. Inc. Edmonton: M.G. Hurtig Ltd., 1970.

Bishop, Charles A. "The Henley House Massacres." *The Beaver*, Outfit 307:2 (1976):36–41.

Black, Jeanette D. "Mapping in the English Colonies in North America: the Beginnings." In *The Compleat Plattmaker*, edited by Norman Thrower, 101–25. Berkeley: University of California Press, 1978.

British and American Joint Commission. *British and American Joint Commission for the Final Settlement of the Claims of the Hudson's Bay and Puget's Sound Agricultural Company.* 11 volumes. Washington, D.C.: McGill & Withrow, 1865–7.

Bulger Correspondence, III. M151, 156–61. Provincial Archives of Manitoba.

Campbell, Robert. *Two Journals of Robert Campbell (Chief Factor, Hudson's Bay Company) 1808 to 1853.* Typescript. Seattle, Washington, 1958.

Campbell, Tony. "The Drapers' Company and Its School of Seventeenth Century Chart-Makers." In *My Head Is a Map*, edited by Helen Wallis and Sarah Tyacke, 81–106. London: Francis Edwards and Carta Press, 1973.

Clarke, Desmond. *Arthur Dobbs, Esquire, 1689–1765; Surveyor-general of Ireland, Prospector and Governor of North Carolina.* London: Bodley Head, 1958.

Cole, Jean Murray. *Exile in the Wilderness: The Biography of Chief Factor Archibald McDonald, 1790–1853.* Don Mills, Ontario: Burns & MacEachern, 1979.

Cook, James, and James A. King. *A Voyage to the Pacific Ocean.* Edited by Rev. J. Douglas. Edinburgh: R. Morison and Son, 1793.

Davies, K.G., ed. *Northern Quebec and Labrador Journal, and correspondence, 1918–35.* Hudson's Bay Record Society Publication no. 24. London, 1963.

— ed. *Ogden's Snake Country Journal, 1826–27.* Hudson's Bay Record Society Publication no. 23. London, 1961.

— *Letters from Hudson Bay, 1703–40.* Hudson's Bay Record Society Publication no. 25. London, 1965.

Dawson, Captain. *Survey of New Zealand.* 1840.

Dease, Peter W., and Thomas Simpson. "An Account of the Recent Arctic Discoveries by Messrs. Dease and T. Simpson. Communicated by J.H. Pelly, Esq., Governor of the Hudson's Bay Company." *Journal of the Royal Geographical Society* 8 (1838):213–25.

Dease, Peter W., and Thomas Simpson. "Narrative of the Progress of Arctic Discovery on the Northern shore of America, in the Summer of 1839. By Messrs. Peter W. Dease and Thomas Simpson. Communicated by Sir J.H. Pelly, Bart., Governor of the Hudson's Bay Company." *Journal of the Royal Geographical Society* 10 (1841):268–74.

Doughty, Arthur G., and Chester Martin, eds. *The Kelsey Papers.* Ottawa: Public Archives of Canada, 1919.

Fox, Luke. *North-West Fox; or, Fox from the North-West Passage.* London: 1635.

Galbraith, John S. "Conflict on Puget Sound." *The Beaver*, Outfit 281 (March 1951):18–22.

Glover, Richard, ed. *A Journey to the Northern Ocean By Samuel Hearne.* Toronto: Macmillan, 1958.

Grant, W. Colquhoun. "Description of Vancouver Island. By its first Colonist, W. Colquhoun Grant." *Journal of the Royal Geographical Society* 27 (1851):268–320.

Harding, Kathryn M. "Discovery Maps of the Canadian Northwest: The Hudson's Bay Company, the Royal Navy and the British Map Trade 1819–1857." Masters thesis, Queen's University, 1986.

Hearne, Samuel. *A Journey from Prince of Wales's Fort, in Hudson's Bay, to the Northern Ocean… 1796. 1770. 1771 & 1772.* London: A. Strahan and T. Cadell, 1795.

Hudson's Bay Company. *Hudson's Bay Company A Brief History.* London: 1934.

Ireland, Willard E. "Captain Walter Colquhoun Grant Vancouver Island's First Independent Settler." *British Columbia Historical Quarterly* 17 (1953):87–125.

James, Thomas. *The Strange And Dangerous Voyage of Captain Thomas Iames…* London: John Leggatt, for Partridge, 1633.

Kaye, Barry. "Birsay Village on the Assiniboine." *The Beaver*, Outfit 312:3 (1981):18–21.

Lytwyn, Victor. *The Fur Trade of the Little North: Indians, Pedlars, and Englishmen East of Lake Winnipeg, 1760–1821.* Winnipeg: Rupert's Land Research Centre, 1986.

MacGregor, J.G. *Peter Fidler: Canada's Forgotten Surveyor, 1769–1822.* Toronto: McClelland and Stewart, 1966.

Mancke, Elizabeth. *A Company of Businessmen, The Hudson's Bay Company and Long Distance Trade, 1670–1730.* Winnipeg: Rupert's Land Research Centre, 1988.

McLean, John. *John McLean's Notes of a Twenty-five Year's Service in the Hudson's Bay Territory.* Edited by W.S. Wallace. The Champlain Society Publication no. 19. Toronto, 1932.

Mickle, S, (Introduction). *The Hudson's Bay Company Expedition in Search of Sir John Franklin. Journal of Chief Factor Anderson, Commander of the H.B. Expedition in Search of Sir John Franklin.* Toronto:

Canadiana House, 1969.

Moodie, D. Wayne, and Barry Kaye. "The Ac ko mok ki Map." *The Beaver*, Outfit 307:4 (1977):4–15.

Moodie, D. Wayne. "Early British Images of Rupert's Land." In *Man and Nature on the Prairies*, edited by Richard Allen. Regina: Canadian Plains Research Center, 1976.

Moodie, D. Wayne, and John C. Lehr. "Macro-Historical Geography and the Great Chartered Companies: the Case of the Hudson's Bay Company." *Canadian Geographer* 25, no. 3 (1981):267–71.

Morton, Arthur S. *A History of the Canadian West to 1870–71*. London: 1939. 2nd. ed. Edited by Lewis G. Thomas. Toronto: University of Toronto Press, 1973.

Norcross, E. Blanche, ed. *The Company on the Coast*. Nanaimo: Nanaimo Historical Society, 1983.

Pearse, Benjamin W. "Early Settlement of Vancouver Island; reminiscences 1900." Provincial Archives of British Columbia, EB.

Pemberton, Joseph Despard. *Vancouver's Island, Survey of the Districts of Nanaimo and Cowichan Valley*. London: Groombridge & Sons, 1859.

Rae, John. "Arctic Exploration, with Information respecting Sir John Franklin's missing Party. By Dr. John Rae, F.R.G.S. (Gold Medallist)." *Journal of the Royal Geographical Society* 25 (1855):246–56.

— "Journey from Great Bear Lake to Wollaston Land. By Dr. John Rae. Communicated by the Hudson's Bay Company." *Journal of the Royal Geographical Society* 22 (1852):73–81.

— "Recent Explorations along the South and East Coast of Victoria Land. By Dr. John Rae. Communicated by the Hudson's Bay Company." *Journal of the Royal Geographical Society* 22 (1852):82–96.

— *Narrative of an Expedition to the shores of the Arctic Sea in 1846 and 1847*. London: T. & W. Boone, 1850.

Rich, E.E., *The History of the Hudson's Bay Company 1670–1870*. Hudson's Bay Record Society Publications 21, 22. London, 1958–9.

— ed. *Copy-Book of Letters Outward. Begins 29th May 1680 and ends 5 July, 1687*. Hudson's Bay Company Series 11. Toronto: Champlain Society, 1948.

— ed. *Hudson's Bay copy booke of letters, commissions, instructions outward, 1688–1696*. Hudson's Bay Record Society Publication 20. London, 1957.

— ed. *James Isham's Observations on Hudsons Bay 1743*. Hudson's Bay Company Series 12. Toronto: Champlain Society, 1949.

— ed. *John Rae's Correspondence with the Hudson's Bay Company on Arctic Exploration 1844–1855*. Hudson's Bay Record Society Publication 16. London, 1953.

— ed. *Journal of occurrences in the Athabasca department by George Simpson, in 1820 and 1821 and report*. Hudson's Bay Company Series 1. Toronto: Champlain Society, 1938.

— ed. *A Journal of A Voyage From Rocky Mountain Portage in Peace River To the Sources of Finlays Branch and North West Ward in Summer 1824*. Hudson's Bay Record Society Publication no. 18. London, 1955.

— ed. *The Letters of John McLoughlin From Fort Vancouver to the Governor and Committee*. Hudson's Bay Company Series 4, 6, 7. Toronto: Champlain Society, 1944.

— ed. *Minutes of the Hudson's Bay Company 1671–1674*. Hudson's Bay Company Series 5. Toronto: Champlain Society, 1942.

— ed. *Minutes of the Hudson's Bay Company 1679–1682*. Hudson's Bay Company Series 8. Toronto: Champlain Society, 1945.

— ed. *Minutes of the Hudson's Bay Company 1682–1684*. Hudson's Bay Company Series 9. Toronto: Champlain Society, 1946.

— ed. *Peter Skene Ogden's Snake Country Journals 1824–25 and 1825–26*. Hudson's Bay Record Society Publication no. 13. London, 1950.

Robertson, John. *The Elements of Navigation*. 1754.

Robson, Joseph. *An Account of Six Years Residence in Hudson's Bay from 1733 to 1736, and 1744–1747*. London: J. Payne and J. Bousquet, 1752.

Ross, Alexander. *Fur Hunters of the Far West*. London: Smith & Elder, 1855.

Royal Agricultural College. *Calendar of the Royal Agricultural College, Cirencester, England*. 1850.

Ruggles, Richard I. "Beyond the 'Furious Over Fall': Map Images of Rupert's Land and the Northwest." In *Rupert's Land A Cultural Tapestry*, edited by Richard C. Davis, 13–50. Waterloo: Wilfred Laurier University Press for The Calgary Institute for the Humanities, 1988.

— "Canada's First National Mapping Agency: The Hudson's Bay Company." *Bulletin of the Association of Canadian Map Libraries* 55 (June 1985):1–18.

— "Governor Samuel Wegg, Intelligent Layman of the Royal Society, 1753–1802." *Notes and Records of the Royal Society of London* 32, no 2 (1978): 181–89.

— "Governor Samuel Wegg, 'Winds of Change'." *The Beaver*, Outfit 307:2 (1976):10–20.

— "Hospital Boys of the Bay." *The Beaver*, Outfit 308:2 (1977):4–11.

— "Hudson's Bay Company Mapping." In *Old Trails and New Directions, Papers of the Third North American Fur Trade Conference*, edited by Carol M. Judd and Arthur J. Ray, 24–36. Toronto: University of Toronto Press, 1980.

— "Mapping the Interior Plains of Rupert's Land by the Hudson's Bay Company to 1870." In *Mapping the North American Plains, Essays in the History of Cartography*, edited by Frederick C. Luebke, Frances W. Kaye, and Gary E. Moulton, 145–60. Norman: University of Oklahoma Press, 1987.

— "The West of Canada in 1763: Imagination and Reality." *The Canadian Geographer* 15, no. 4 (1971):235–261.

Sadler, D.H. *Man is Not Lost, a Record of Two*

Hundred Years of Astronomical Navigation with the Nautical Almanac, 1767–1967*. Greenwich: HMSO, 1978.

Schooling, William. *The Governor and Company of Adventurers of England Trading into Hudson's Bay during Two Hundred and Fifty Years 1670–1920*. London: The Hudson's Bay Company, 1920.

Sellers, John. "A Breviate of Captain Zacariah Gillam's journal of the 1668 journey in the *Nonsuch* to James Bay." *The English Pilot*, London, 1671.

Simpson, Alexander. *The Life and Travels of Thomas Simpson, the Arctic Discoverer*. London: Richard Bentley, 1845.

Simpson, Thomas. *Narrative of the Discoveries on the North Coast of America. Effected by the Officers of the Hudson's Bay Company During the Years 1836–1839*. London: Richard Bentley, 1843.

Smith, Shirlee A. "James Sutherland, Inland Trader 1751–97." *The Beaver*, Outfit 306:3 (1975):18–23.

Stearns, Raymond P. "The Royal Society and the Company." *The Beaver*, Outfit 276 (June 1945):8–13.

Thorman, George E. "An Early Map of James Bay." *The Beaver*, Outfit 291 (Spring 1961):18–22.

Thrower, Norman J.W., ed. *The Compleat Plattmaker, Essays on Chart, Map, and Globe Making in England in the Seventeenth and Eighteenth Centuries*. Berkeley: University of California Press, 1978.

Tyrrell, J.B., ed. *Journals of Samuel Hearne and Philip Turnor Between the Years 1774 and 1792*. Toronto: The Champlain Society, 1934.

Verner, Coolie. "The Arrowsmith Firm and the Cartography of Canada." *Proceedings of the Association of Canadian Map Libraries* 4 (1970):16–21.

Wakefield, Edward Gibbon. *The New British Province of South Australia*. London: Charles Knight, 1835.

— ed. *A View of the Art of Colonization With Present Reference to the British Empire; In Letters Between a Statesman and a Colonist*. London: John W. Parker, 1849.

Warkentin, John and Richard I. Ruggles. *Manitoba Historical Atlas, A Selection of Facsimile Maps, Plans, and Sketches from 1612 to 1969*. Winnipeg: The Historical and Scientific Society of Manitoba, 1970.

Williams, Glyndwr. "Arthur Dobbs and Joseph Robson: New Light on the Relationship between Two Early Critics of the Hudson's Bay Company." *Canadian Historical Review* 2 (1959):132–6.

— *The British Search for the Northwest Passage in the Eighteenth Century*. The Royal Commonwealth Society Imperial Studies, no. 24. London: Longmans, 1962.

— "Captain Coats and Exploration along the East Main." *The Beaver*, Outfit 294 (Winter 1963):4–13.

— "East London Names in Hudson Bay." *East London Papers* 7, no. 1 (1964):23–30.

— "The Hudson's Bay Company and the Critics in the Eighteenth Century." *Transactions of the Royal Historical Society* 5th Ser., 20 (1970):149–71.

— "A Remarkable Map." *The Beaver*, Outfit 293 (Winter 1962):30–36.

— ed. *Andrew Graham's Observations on Hudson's Bay, 1767–91*. The Hudson's Bay Record Society Publication no. 27. London, 1969.

— ed. *Hudson's Bay Miscellany, 1670–1870*. The Hudson's Bay Record Society Publication no. 30. Winnipeg, 1975.

— ed. *London Correspondence Inward From Sir George Simpson 1841–42*. The Hudson's Bay Record Society Publication, no. 29. London, 1973.

— ed. *Peter Skene Ogden's Snake Country Journals, 1827–28 and 1828–29*. The Hudson's Bay Record Society Publication no. 28. London, 1971.

Index

Plate numbers (in italics) and map catalogue numbers are given at the end of entries. Maps are not indexed under their titles. The location of textual discussions of maps is given in the catalogues.

Abenaki Indians, 283n10
Abitibi, Lake, 47, 51, 55, 69, 76, 86; 28[C]
Abitibi River, 51
Absaroka Range, 63
Ac ko mok ki (The Feathers), 63, 64, 280n9; *19*
Acton House, 66, 203; 123[A], 81[C]
Ageena (Ageenah), 66
Akimiski (Akkawmiskee) Island, 86; 148[C]
Ak ko wee ak, 63
Alaska, 91, 106; 419[A]
Albany Factory (Fort), 11, 16, 21, 28, 29, 35, 37, 45, 46, 47, 48, 50, 51, 55, 56, 57, 58, 59, 66, 72, 86; 28[A], 36[A], 150[A], 157[A], 29[C], 31[C], 35[C], 36[C], 46[C], 55[C], 56[C], 79[C]
Albany River, 7, 26, 32, 35, 36, 37, 45, 46, 47, 48, 50, 51, 54, 55, 56, 57, 58, 275n14, 276n6; *11*, *13*, 28[A], 36[A], 30[C]–32[C], 37[C], 39[C], 45[C]
Alberni Canal, 104
Alder, Thomas, 70, 71
Alexander, Cape, 79, 80
Alexandria, Fort, 94, 115
Allard, Ovide, 114; 196[C], 197[C]
Anderson, Alexander C., 94–5, 108, 283n19; *49*, 273[A], 351[A], 12[B], 13[B], 146[C], 151[C]
Anderson, James (A), 82, 106, 107, 115, 284n1; *58*, *63*; 367[A], 412[A], 479[A]
Anderson Lake, 94, 108
Anderson River (Northwest Territories), 107, 284; *58*, 412[A]; geographical concept of (James (A) Anderson), 107

Anderson River. *See* Bridge River
Anian, Strait of, 30
Annance, Francis (François), 90, 283n10; 102[C]
Anthropological Institute, 4
Antiquaries, Society of, 5
Apprentices, mathematical, 5, 11, 12, 38, 45, 47, 48, 49, 52, 59, 66, 120, 274n3
Arctic: Circle, 79; Franklin overland expedition to, 72, 79; search for Franklin in 79, 81–2; HBC expedition to, 12, 79–80; Hearne expedition to, 42–4, 278n12; islands, 80; magnetic pole in, 17; maps of coasts of, 7, 79–82, 282n18, 282n22, 282n24; *34*, *35*, 143[A], 237[A], 238[A], 149[C]; natives along coast of 30–1, 42–3
Arrowsmith, Aaron, 5, 58, 60, 61, 62, 64, 65, 66, 67, 68, 71, 274n20, 275n1, 279n22, 280n7
Arrowsmith, John, 5, 80, 81, 82, 99, 108, 274n20, 274n26, 281n15, 285n8
Arrowsmith map firm, 60, 66, 68, 71, 75, 79, 90, 118, 121, 261;
Arrowsmith, Samuel, 5, 148[A], 85[C]
Artiwinipeck (Artuirnipeg). *See* Richmond Gulf
As se an Lake, 13[A]
Assiniboine, District of (Assiniboia), 82, 115; 148[A]
Assiniboine, Fort, 283n10
Assiniboine Indians, 29, 49, 277n3
Assiniboine River 38, 57, 58, 62, 65, 67, 82, 83, 84, 115, 279n7, 280n15; 175[A], 284[A], 18[C], 95[C]
Assinibouels Lake, 45
Assinipoet Indians. *See* Assiniboine Indians
Athabasca expedition, 12, 52–4, 59, 61
Athabasca Lake, 29, 30, 49, 51, 53, 54, 58, 64, 65, 66, 77, 79; 40[A], 43[A], 88[A], 100[A], 106[A], 115[A], 140[A], 191[A], 73[C]
Athabasca River, 30, 42, 43, 62, 64, 77, 79; 72[A], 95[A], 121[A], 67[C], 82[C]
Athapascow (Athapescow) River. *See* Athabasca River
Athapaskan (Athapeeska) Indians, 29, 42
Atkinson, Christopher, 36
Atkinson, George II, 47, 48, 69, 70, 74; 169[A], 179[A]

Atkinson, George III, 69; 87[C]
Atlantic Ocean, 3, 14, 31, 69, 70, 75, 278n1; 57[A]
Atlantic-Pacific ship tracks; 472[A], 474[A], 480[A]
Attawapiskat River, 86
Auguewang River and trading post, 285n10; 194[C]
Auld, William, 64, 68, 72, 280n20, 280n21; 149[A], 153[A]

Babine Lake, 93–4; 129[C]
Babine Range, 93–4
Back, Cape, 81; 160[C]
Back River, 43, 80, 82; 190[C]
Baffin Bay, 43
Baker Lake, 41
Ballenden, John, 108
Banks, Sir Joseph, 64
Barkerville, 114; 572[C], 218[C]
Barkley (Barclay) Sound, 104; 413[A], 192[C]
Barnston, George, 79, 83, 86; 232[A], 303[A]
Barren Ground (Meadow) Lake, 62; 64[A], 65[A]
Barren Grounds, 17, 29, 30, 37, 43, 80; 9[B], 22[C], 25[C]
Barrow, Point, 80; 237[A], 238[A]
Basquia (Basquiau, Basquiaw). *See* Pas, The
Batchewana Bay, 188[C]
Batt, Isaac, 51
Battle River, 51, 63, 280n7; 123[A]
Beads, Thomas, 76–7, 275n5; *31*, 247[A], 133[C], 134[C]
Beale, Cape, 104
Bear River (British Columbia), 116
Bear River (Idaho, Utah), 89
Beaufort, Admiral Francis, 81
Beaufort Sea, 79
Beaver, 77, 85, 86, 88
Beaver conservation preserve, 4, 74, 86; 42, 240[A], 251[A], 256[A], 268[A], 275[A], 276[A], 135[C], 140[C]
Beaver Harbour, 103; 152[C]
Beaver (Amisk) Lake, 163[A]
Beaver River (British Columbia-Yukon), 77
Beaver River (Saskatchewan), 42, 54, 62; 66[A], 67[A], 75[A], 120[A], 122[A], 145[A], 67[C]

Beck, George, 275
Beckley Farm, 111, 285n3
Beghulatesse River. *See* Anderson River (Northwest Territories)
Bell, Benjamin, 220[A]
Bell, John, 79, 81; 147[C]
Bell River, 79, 106; 175[C]
Bellingham Bay, 102, 103; 365[A], 178[C], 182[C]
Bentinck, Captain John Albert, 5
Berens, Fort, 94, 108; 199[C]
Berens, Henry, 107, 114
Berens River, 38, 71, 280n24
Berens River district, 71; *27*, 178[A]
Bering Sea, 577[A]
Bersimis River, 558[A]
Best, John, 57; 51[A]
Bienville, Lac, 69
Big River (Quebec), 69, 280n23
Bighorn Mountains, 63
Bighorn River (Montana-Wyoming), 63; 196[A]–198[A]
Big Lake House, 89[C]
Bigstone River, 44
Bird, James, 62; 171[A]
Birsay (Assiniboine River area), 280n15
Bissett, James, 117
Black Eye, 284n20; 12[B]
Blackfoot Indians, 63, 64
Black Hills (South Dakota), 278n2
Black, Samuel, 77–8; 8[B]
Bligh, Captain William, 5
Bloodvein River, 47
Bloody Falls, 43
Blue Coat School (Christ's Hospital), 12, 18, 274n3, 275n8, 276n4
Blue Mountains, 89
Boats: horse scows, 95; instrument damage, 18; Lachine Canal, 85; loads on rivers, 54; lock proposed on river, 84; navigation on rivers, 68; steamship use, 116–17; using dead reckoning on, 14; *Beaver*, 106; *Brig Wear of London*, 72; *Cadboro*, 90, 91, 96; *Churchill* sloop, 32, 33, 38, 41, 42, 277n8; *Eastmain* sloop, 34, 46; *Enterprise*, 114, 116; *Ganymede*, 281n4; *King George*, 19, 20; *Labouchere*, 109; *Marten*, 116, 117; *Mary*, 34; *Moose* sloop, 34, 35, 281n5; *Phoenix*, 34; *Prince Rupert*, 45; *Princess*

Royal, 109; *Sea Horse*, 36; *Severn* sloop, 20, 40, 50; *Success* sloop, 34, 35; *Whalebone*, 31; *William and Ann*, 90; *York*, 36
Boise, 88
Bolsover House, 62; 64^A, 65^A
Books, 16; astronomical, 19, 50, 56; colonial surveys, 99, 284n10; geographical, 67; library, 3, 15, 67; mathematical, 3, 15, 19
Boothia, Gulf of, 80
Boothia (Boothia Felix) Peninsula, 31, 81; 246^A, 184^C
Bottom of the Bay, 25, 26, 49, 50, 51, 276n17, 276n18; 2^C
Bougainville, Louis Antoine de, 5
Boundary Bay, 90
Boundary Treaty 1846, 87, 92–3
Bouthillier, Thomas, 299^A
Bow River, 53, 59, 63, 64; 138^C
Bowden's Inlet. *See* Chesterfield Inlet
Brandon House, 62, 65, 67, 83
Bridgar, John, 28, 276n17, 276n18
Bridge River, 95
Brigade trails, 74, 93–5; 49, 273^A, 304^A, 12^B, 13^B, 111^C, 151^C, 153^C, 154^C
British-American Indemnity (Land Claims) Commission, 93, 108, 110, 114, 283n13
British Columbia, Colony of, 4, 7, 14, 87, 110, 113, 116, 121
British Columbia, Commissioner of Lands and Works, 116
British Columbia, Surveys and Land Records Office, 121; 484^A, 487^A, 514^A, 515^A–533^A, 14^B, 15^B, 18^B, 19^B, 25^B, 28^B–32^B, 34^B, 36^B
British Museum, 25, 33, 278n4
Broad River, 137^A
Broadback River, 69, 70
Brome, John, 262
Brotchie, Captain William, 284n17
Broughton, Captain William Robert, 5
Brunswick House, 51, 54, 55
Buckingham House, 54, 59, 62, 64; 52^A, 75^A, 122^A, 67^C, 69^C, 70^C
Buckingham (Quebec), 109
Buffalo, 41, 49
Buffalo Country, 49
Bulger, Governor Andrew, 82, 282n34
Bulkley (Simpson's) River, 94; 227^A

Bunn, Thomas, 66
Burntwood Lake, 63^A
Burntwood River, 63^A
Bute Inlet, 117
Button, Captain Thomas, 25
Bylot, Captain Robert, 25

Cadastral maps, 14, 60, 61, 74, 82, 83, 86, 87, 91–2, 93, 96, 97, 98, 99, 100, 104, 105, 110, 111, 113, 114, 285n5, 285n6
Cadboro Bay, 95, 100; 318^A
Cairn Island. *See* Metesene Island
California, 87; 368^A
California, Gulf of, 88, 89
Cameron, Angus, 83; 253^A
Cameron Island, 112
Camosun (Cammusan, Camosack, Comoosan) Inlet, 95–6. *See also* Victoria Harbour
Campbell, Robert, 78, 79, 106; 261^A
Canadian traders: competition with, behind Eastmain, 70; in Abitibi area, 47, 55; in Athabasca, 52–3, 64; in Manitoba lakes area, 38, 57, 58, 62, 65; north of Lake Superior, 45, 46, 47, 55, 56, 279n16, 279n20; north of St. Lawrence River, 69, 76; in area of North Saskatchewan River to foothills 49, 51, 59, 68; in upper Churchill River area, 62, 64, 66; general trade of, 58; information to Turnor, 54; trading alcohol, 49
Caniapiscau House, 76; 247^A
Caniapiscau Lake, 70
Caniapiscau (Caniapuscaw) River, 70, 74, 75, 76; 188^A, 214^A, 215^A
Canoe Building Lake, 132^A
Canoe River, 116
Canoes: accidents, 18, 50, 51, 54; building of, 69; expertise with, 10, 51; surveying from, 14, 102; travelling in, 101, 103, 115
Capenocoggamy, 157^A
Capush Cushee Lake, 184^A
Cariboo area, 117
Carlton district, 280n24; 171^A
Carlton House, 68, 92, 115, 116, 280n16, 280n24; 156^A, 285^A
Carrot River, 65

Carson Sink, 89
Cassini family, 5
Castor and Pollux River, 80, 81
Cat Lake, 57; 113^A, 165^A, 76^C
Cat Lake House, 57
Cauc-chi-chenis, 86; 135^C
Cedar Lake, 38, 50, 62, 65; 58^A, 109^A
Cha chay pay way ti, 66
Champoeg, 114; 565^A, 566^A
Chantrey (Bay) Inlet, 80, 81, 82
Chapman, John, 285n9
Charity Hospital. *See* Blue Coat School (Christ's Hospital); Grey Coat Hospital (School)
Charles, George, 52, 65
Charles, John, 66, 120; 71^C
Charlton House, 62
Charlton Island, 72, 74, 86; 91^A, 92^A, 150^A, 240^A, 251^A, 268^A, 275^A, 276^A, 135^C, 140^C
Charlton Sound, 72
Charts (hydrographic, marine, nautical, navigation), 3, 4, 6, 7, 9, 11, 12, 14, 18, 25, 87, 276n4; Arctic, 80, 81, 82; commercial chart makers, 14, 25, 26, 28, 275n3, 276n7, 276n8; of Eastmain, 34, 35, 36, 277n15, 277n16, 280n25; of Hudson and James bays, harbours, inlets, 33, 34, 35, 36, 38, 41, 42, 71–2; of Hudson and James bays, Hudson Strait, 38, 40, 72, 278n1, 280n25; of Labrador coast, 75, 76; of Magellan Strait, 109; of Northwest coast, 89–91, 92, 95, 99, 106, 283n16
Charter territory. *See* Hudson's Bay Company
Chase River, 177^C
Chatham House, 54; 132^A
Chepy Sepy, 45, 275n14; 29^C
Chesterfield House, 63, 64, 65, 68, 280n8
Chesterfield Inlet, 33, 41, 42, 43, 44, 53; 7, 17^A, 18^A, 135^A, 136^A, 19^C, 60^C, 66^C
Chidley, Cape, 28
Chimo, Fort, 74, 75, 76, 281n4; 229^A–231^A, 242^A
Chipewyan (Athapaskan, Northern) Indians, 29, 30, 31, 41, 42, 43, 66
Christopher, William, 41; 17^A, 19^C
Christ's Hospital. *See* Blue Coat School (Christ's Hospital)

Chuck,e,ta,naw River, 40^C
Churchill Cape, 133^A, 134^A, 48^C
Churchill Factory: capture, 278n5; Dobbs, 32–3; Fidler, 64; founding, 28, 29, 30, 276n20; Graham, 41; Hearne, 42, 43, 44; Kelsey, 29; letter book, 72; local maps, 61; Northern Indians, 29, 30, 42, 43; Norton, Moses, 41, 274n4; Norton, Richard, 277n24; Prince of Wales Fort, 278n9; Robson, 275n8, 278n20; Ross, Malcolm, 279n8, 279n9; Thompson, David, 52; Transit of Venus, 5, 273n16; 34^A, 61^A, 79^A, 100^A, 101^A, 115^A, 134^A, 155^A, 185^A, 259^A, 25^C, 73^C
Churchill Lake, 66; 142^A
Churchill River: Canadian traders in region of, 49, 51, 64; Crees move into region of, 29; factory at mouth of, 28, 29; HBC traders in region of, 51, 52, 53; mapping of 7, 42, 51–2, 53, 54, 64–5, 66, 279n25; route inland from, 51, 279n12; seventeen rivers north of, 30; Thompson in upper area of, 59; 9, 16, 79^A, 106^A, 127^A, 135^A–137^A, 140^A, 21^C, 22^C, 73^C;
Cirencester, England, 68, 98, 280n19
Clapham House, 66
Clarke, John, 282n34
Clarke, Lawrence, 420^A
Clearwater Lake (British Columbia), 117
Clearwater Lake (Quebec), 34
Clearwater River (British Columbia), 117; 568^A
Clearwater River (Saskatchewan), 54, 64, 66; 39^A, 45^A, 142^A
Clouston, James, 60, 68–70, 71, 74, 75, 77, 120, 280n22, 280n23, 281n3; 24, 25, 168^A, 170^A, 173^A, 180^A, 188^A, 195^A, 83^C
Clouston, John, 140^C
Clover Point, 100
Coast Range, 89, 110, 118, 283n10
Coats, Captain William, 3, 34–5, 273n7, 277n4, 277n9, 277n11, 277n14, 277n17; 5, 9^A, 10^A, 11^A, 13^A, 14^A, 3^B, 10^C
Cochrane River, 66; 141^A
Cocking, Matthew, 40, 41, 44, 278n5
Coeur d'Alene, 138^C
Cold Lake, 54; 122^A, 145^A

Colen Joseph, 33[A]
Colorado River, 89
Columbia River and district, 7, 14, 60, 61, 65, 68, 74, 87–9, 90, 91, 92, 93, 95, 96, 98, 106, 107, 108, 114, 116, 117, 280n17, 283n3, 283n10, 283n13; 93[A]–95[A], 218[A], 239[A], 248[A], 270[A], 272[A], 281[A]–283[A], 305[A], 28[B], 84[C], 100[C], 102[C], 105[C], 112[C], 144[C]
Colvile, Andrew, 101, 103, 284n18
Colvile district, 128[C]
Colvile, Eden, 95
Colvile, Fort, 92, 108; 269[A], 138[C]
Colvile River (Alaska), 78, 79, 107
Colvile River (British Columbia), 36[B]
Colvile Town. *See* Nanaimo
Committee Bay, 80, 81; 293[A], 149[C]
Committee, HBC: operation and procedures, 18–19, 259–60
Compass, 11, 12, 14–18, 20, 44, 47, 101, 273n12, 279n15; aberration, 17, 44, 89; prismatic, 101; rose, 35, 40, 42, 85, 284n21; variation, 35, 44, 84, 106
Compton, P.N., 113; 61, 510[A]
Concord, Fort, 86
Confidence, Fort, 80, 81
Connolly, Henry, 86; 240[A]
Coochenau, 28[C]
Cook, Captain James, 5, 273n17, 273n19
Cook, William, 66; 161[A], 86[C]
Copper (natural), 30, 31
Copper Indians, 30
Copper River. *See* Coppermine River
Coppermine Hills, 43
Coppermine River, 30, 42, 43, 44, 66, 79, 80, 81, 107; 10, 2[A], 143[A], 306[A], 4[B], 9[B]
Coquihalla (Quaqualla) River, 94, 95; 304[A], 146[C], 153[C]
Coquille River, 89
Corbets Inlet. *See* Chesterfield Inlet
Corcoran, John, 184[A]
Corcoran, Thomas, 275n5; 148[C]
Cordillera, 14, 63, 68, 77, 79, 87, 110, 118, 280n10, 284n1
Cordova Bay, 319[A], 360[A], 18[B]
Coronation Gulf, 81
Corrigal, Jacob, 57; 55[A]
Cossack Harbour, 223[A], 121[C]
Cot aw ney yaz zah, 66
Couteau Country (Fraser River), 108

Covington, Richard, 92, 99, 108; 277[A], 471[A]
Cowichan (Cowitchin, Cowetchin, Cowichin), 102, 104, 105; 321[A], 387[A], 426[A], 179[C]
Cowichan Head, 100; 172[C]
Cowichan Indians, 104
Cowichan Lake, 105
Cowichan River and valley, 105; 55 427[A]
Cowlitz (Cowelitz) Farm, 91, 92; 44, 250[A], 257[A], 258[A], 262[A], 274[A], 141[C]
Cowlitz River, 4, 91, 108; 249[A], 17[B], 174[C]
Cranberry Portage, 44, 49
Cranberry River, 116
Crawford, P.W., 302[A]
Cree Indians, 26, 29, 30, 277n3, 280n19
Cree Lake and River, 66; 140[A]
Cree language, 26, 29, 30, 276n6
Cross Lake, 40, 50; 104[A], 126[A], 72[C]
Crown Lands Department, 83, 109
Crown Timber office, 83
Cumberland district, 280n24; 162[A]
Cumberland House, 12, 40, 42, 44, 45, 47, 49, 50, 52, 53, 54, 59, 62, 63, 64, 65, 66, 279n7; 38, 46[A] 50[A], 62[A], 106[A], 124[A], 186[A], 58[C], 59[C]
Cumberland House Lake, 44, 50, 280n5; 163[A]
Current River, 421[A]

Daer, Fort, 67
Dallas, Alexander Grant, 105, 107, 108, 110, 111, 113, 114, 115, 275n5, 285n3, 285n4, 285n15, 285n19; 59, 462[A], 464[A], 465[A], 481[A], 503[A], 199[C]
Dalrymple, Alexander, 5, 64, 273n17, 274n18
Dauphin, Fort, 67
Dauphin Lake, 40, 65; 109[A]
Dauphin River, 40
Davies, W.H.A., 76, 281n6; 243[A], 244[A], 130[C], 136[C]
Dean Channel, 117
Dease Lake, 78, 117; 233[A], 220[C]
Dease, Peter Warren, 79–80, 282n18, 282n20; 34, 237[A]
Dease River (British Columbia), 78; 233[A]
Dease River (Northwest Territories), 80, 81

Dease Strait, 246[A]
Deer's River. *See* Reindeer River
Derbyshire, England, 61, 66
Dering (Deer) River, 29
Desert River, 109; 60, 411[A], 437[A]
Diggs, Cape, 34, 35, 74; 13[A], 10[C]
Disappointment, Cape, 90, 92; 218[A], 283[A], 114[C]
Dismal Lakes, 81
Districts, HBC 60, 61, 62, 67, 69, 71, 76, 86, 87, 93; topographical account or plan of, 71. *See also* individual districts, e.g., Cumberland district
Dobbs, Arthur, 5, 32–3, 273n14, 277n4, 278n20
Dog's Head (Lake Winnipeg), 65; 114[A]
Dolland, John, 15
Donald, George, 47, 48, 50, 51, 54, 55, 120, 281n25; 32[A], 42[C]
Douglas, Fort, 67
Douglas, James: chief factor, 96, 99–100; and Gilmour, 103, 104, as governor, 99; and Grant, 97–8; and Lewes, A.L., 95–6, 97; and McKay, J.W., 103; and maps of: Camosun, 95–6; Columbia region, 283n13, Coquihalla, 95, Georgia Strait, 108; Nanaimo, 101, 284n14; Newcastle Island, 102; Nisqually-Cowlitz, 91–2, 93; Queen Charlotte Islands, 106, Victoria, 110; and Pemberton, 99, 100, 101, 103, 104; obtaining rights to Indian land, 99; property of, 100, 111; regarding site of Victoria, 95, 51, 252[A]; search for surveyors, 98; survey of Fort Langley, 114; survey of Yale, 114
Drake, Montague W.T., 565[A], 566[A]
Drapers' Company, 26, 28, 276n8
Dubawnt Lake and River, 43
Duberger, George, 76; 131[C]
Duncan, Captain Charles, 278n1, 280n25; 60[C], 61[C]
Dymond, Joseph, 5, 273n16

Eagle Creek, 117; 567[A]
Eagle Hills, 40
Eagle Pass, 117
East Branch (Liard River). *See* Fort Nelson River
East India Company, 7

Eastmain, 33, 120, 277n10; Beads and Spencer in, 76–7; Coats in, 16, 34–5, 277n9; Clouston, James in, 68–70; exploration of, 33–4, 35, 50, 60, 69–70; Hendry in, 281n2; mapping of, 7, 21, 33–5, 60, 68–70, 75, 109, 281n25; McDonald, A., in, 109, 285n10; Mitchell in, 16, 34; overtrapping of, 86; Turnor in, 71; Yarrow in, 35; 25, 169[A], 176[A], 179[A], 195[A], 215[A], 467[A], 576[A], 11[C]
Eastmain House, 21, 50, 51, 69, 72; 87[A], 150[A]
Eastmain River, 26, 69, 70, 74, 76; 176[A]
Eastmain and Rupert district, 71, 280n24
Edmonton, Fort, 62, 92, 115, 116, 280n5; 45, 123[A], 286[A], 67[C], 68[C], 210[C], 211[C]
Eetow wemammis Lake, 62
Egan township (Quebec), 109; 411[A], 438[A]
Egg Lake (Methy Portage), 117[A]–119[A]
Ekwan (Equan) River, 11, 71, 86
Ellice, Fort, 92, 115; 284[A]
Englefield Bay (Queen Charlotte Islands), 106; 163[C]
English River, 57
Equan River. *See* Ekwan River
Erlandson, Erland, 75, 77; 30, 228[A]–231[A]
Eskimo. *See* Inuit
Eskimo Point, 29, 32, 278n9
Esnagamie River, 46
Esquimalt: mapping of, 4, 101, 102, 103, 110, 112; surveying of, 100, 101, 102, 103; Village Bay 101, 103, 112; water sources in, 101; 56 321[A], 335[A], 336[A], 339[A]–342[A], 349[A], 354[A], 355[A], 374[A], 375[A], 378[A]–381[A], 388[A], 401[A]–403[A], 414[A], 415[A], 416[A], 434[A], 443[A]–446[A], 457[A], 459[A]–461[A], 523[A], 524[A], 526[A], 527[A], 530[A], 533[A], 537[A], 14[B], 20[B]–23[B], 167[C], 169[C], 213[C], 214[C], 216[C], 217[C]. *See also* Puget's Sound Agricultural Company
Esquimalt Harbour, 100, 101, 103, 112; 322[A], 570[A], 571[A], 578[A], 176[C]
Esquimalt Bay. *See* Melville Lake-Hamilton Inlet
Etawney Lake, 52

Falconer, William, 38, 40, 48; 23^C, 24^C, 27^C
Favell, John, 46, 273n6
Fidler, Peter, 7, 12, 60-7, 68, 70, 120, 279n1, 280n4; Assiniboine River-Lake Winnipeg, 62, 65, 67; astronomical observations, 264, 279n9, 280n6, 280n8; Athabasca expedition, 52 4; Beaver River region, 62; books, 16, 67, 275n6; Churchill River, 64-5; on the high plains, 59, 61, 280n7; individual maps, 57, 60, 61, 71, 279n17, 279n18, 280n4; lot surveys of Red River, 67, 82; map transcriptions, 65-6, 280n5, 280n13, 280n14; route to Athabasca, 64-5; segmental sketches, 9, 49, 53-4, 60, 61, 62, 64, 65; South Saskatchewan River, 63-4; Sturgeon River, 59; Swan River, 62, 65; 16, 19, 20, 21, 22, 23, 39^A-41^A, 43^A-46^A, 48^A-50^A, 52^A-54^A, 58^A, 59^A, 62^A-86^A, 90^A, 95^A-135^A, 137^A-145^A, 164^A, 165^A, 174^A, 175^A, 181^A, 182^A, 187^A, 65^C, 67^C-70^C, 73^C, 79^C-81^C, 90^C
Findley, Jean, 65, 280n11, 280n17
Finlay, Jaco. See Findley, Jean
Finlay River, 77
Finlayson Arm, 102; 359^A, 179^C
Fnlayson, Nicol, 75
Finlayson, Roderick, 95
Fires (forest and grassland), 17, 100
Flamboro House, 65; 125^A
Flathead Lake, 65, 68, 88; 138^C
Flathead post, 88
Flathead River, 16, 68
Flett, Andrew, 113^A, 76^C
Flew, Mr., 280n5
Ford, Captain John, 28
Ford, Richard, 36; 15^C
Fort George River, 34, 4, 280n23
Fort Nelson River, 77
Fowler, Captain, 36
Fox (Foxe, Foxes) River, 44, 85; 39, 226^A
Foxe Basin, 25
Foxe (Foxes) Lake, 66; 130^A, 202^A
Fox, Captain Luke, 25, 276n1
France: territorial claims, 26, 28, 29, 277n20; war with England, 28, 29, 277n20, 278n5

Frances Lake, 78, 107
Frances River (Liard River tributary), 78; 33
Frances River. See Stikine River
François Lake, 117; 569^A
Franklin, Sir John: award for relics of, 81, 82; and Dease and Simpson, 80, 81; early expeditions, 72, 79; map of Nelson-Saskatchewan rivers, 72; search for, 79, 80, 81-2; and Thompson, Thomas, 72
Franklyn, William H., 285n18
Fraser canyon, 95, 108; 195^C
Fraser Lake, 117
Fraser Lake post, 94
Fraser River, 74, 90, 93, 94, 95, 105, 108, 114, 115, 116, 117, 285n22; 8^B, 28^B, 29^B, 100^C, 102^C, 111^C, 124^C, 195^C
Fr. Augt., 104; 387^A
Frederick House, 51, 54, 55
Frederick House Lake, 55
Fremont, John Charles, 108, 285n8
French River, 50, 51
Frenchman's River. See Ogoki River
Frog Plain, 67
Frog Portage, 64; 79^A, 100^A, 106^A, 73^C
Fullerton Cape, 33
Fur trade competition (French), 29, 32, 36, 37, 38, 49
Fury and Hecla Strait, 80

Gaddy, James, 51, 53, 278n6
Gardner Canal, 117
Garry, Fort, 115, 262-3; 300^A, 536^A, 210^C
Gastineau, John, 114
Gatineau River, 109; 411^A
Geneau, Lac, 53
Geographical field observation, 3, 10, 11, 16-18, 20-1, 29, 38, 42, 45, 49, 50, 51, 53, 59, 61, 64, 67, 75, 76, 78, 80, 85, 88, 97, 98, 101, 102
Geographical information: to Royal Society, 5; from inland winterers, 38, 40, 42, 43-4, 120; from native people, 12, 13, 120; concerning fur trade districts, 71
Geographical knowledge, increase of 4-5, 12, 14, 28, 33, 35, 37, 38, 40, 41, 43, 49, 53, 59, 60, 64, 65, 69, 70, 71, 75, 76, 78, 81, 82, 84, 85, 87, 89, 97, 101, 104, 106, 107
Geographical knowledge, lack of, 25, 28, 30, 34, 57, 69, 76, 79, 106
Geological study, 101, 103
George, Fort (Columbia), 21, 90, 91, 94
George, Fort (Eastmain), 280n23
George, Fort (Fraser River), 115
George River, 75; 245^A
Georgetown (Minnesota Territory), 115; 475^A, 559^A
Georgia Strait (Gulf of), 96, 102, 103; 107^C, 185^C
Gillam, Zachariah, 4, 275n3
Gilmour, Boyd, 103, 104; 53, 324^A, 325^A, 326^A, 385^A, 181^C
Gladman, George, 83, 275n5; 176^A, 177^A, 307^A, 315^A, 316^A
Gladman, Joseph, 86; 268^A, 275^A, 276^A
Gloucester House, 46, 47, 49, 50, 54, 55, 56, 57, 58; 28^A, 45^C, 50^C
Gloucester House district, 280n24; 166^A
Gods (Good Spirit) Lake, 62
Gold: gold fields map, 108; on prairies, 115, 116; on Queen Charlotte Islands, 106; rush (Fraser River), 99, 105, 108; 195^C
Gonzales Hill, 98
Good, Philip, 55, 275n5; 64^C
Good Hope, Fort, 79, 107
Good Spirit Lake. See Gods Lake
Gorst, Thomas, 276n19
Graham, Andrew, 38, 40, 41, 42, 44, 48, 64, 120, 278n4; 6, 21^A-24^A
Grand Falls (Saskatchewan River), 44, 65; 50^A, 114^A, 75^C
Grand Fish River. See Back River
Grand Lake Miscosinke. See Mistassini, Lake
Grand Portage, 48, 58
Grand Rapid. See Grand Falls
Grant, Walter Colquhoun, 97-8, 118, 284n3; 308^A, 155^C, 156^C
Grass River, 44, 49, 54, 66
Graticule (grid), 14, 16, 35, 44, 48, 50, 54, 61, 67, 68, 70, 71, 75, 79, 81, 105, 118
Gravel River. See Keele River
Great Bear Lake, 66, 77, 80. 81, 82, 107, 282n29; 143^A, 306^A, 9^B

Great Britain, 28, 33, 34, 51, 52, 90, 92, 97, 98, 99, 115, 277n13, 283n15, 284n5
Great Britain, Admiralty: charts, 81, 82, 275n1; and Coats, 35; and Dobbs, 33, 277n4; and Franklin, 61, 72, 81; and Rae's maps, 81, 82; Royal navy, 12, 42, 72, 90, 104
Great Central Lake, 104
Great Lakes, 25, 33, 46, 58, 85
Great plains, 14, 29, 31, 38, 41, 51, 59, 61, 63, 64, 68, 110, 115; 18^C
Great (Old) River, 4^A
Great River. See Fort George River
Great Salt Lake, 87, 88, 89
Great Slave Lake, 29, 30, 42, 43, 44, 53, 54, 58, 264; 41^A, 44^A, 191^A
Great Slave River, 43; 41^A, 42^A, 90^A
Great Valley of California, 89
Great Whale (Great White Whale) River, 34, 35, 69, 70; 169^A, 467^A, 3^B, 87^C
Green, F.W., 111, 275n8; 483^A, 504^A, 206^C-208^C
Green Lake, 145^A
Green River, 198^A
Greenwich House, 62, 280n6; 69^A
Grenville Canal, 117; 569^A
Grey Coat Hospital (School), (Royal Foundation of Queen Anne), 12, 15, 38, 45, 47, 48, 49, 52, 66, 274n3
Grey Deer's Lake, 106^A, 73^C
Griffin, C.J., 202^C
Groseilliers, Medard Chouart, Sieur de, 25, 26, 276n2, 276n6
Gulf Islands (Strait of Georgia), 101
Gull Lake, 99^A

Haggart, Duncan, 126^C
Hairy Hill, 206^A
Hairy Lake, 99^A, 203^A
Halkett, Fort, 78, 275n12; 233^A
Hamilton Inlet. See Melville Lake-Hamilton Inlet
Hannah Bay, 85
Hannah Bay House, 219^A
Hansom, Joseph, 49, 50
Hanwell, Captain Henry Jr, 72, 90, 91; 150^A, 151^A, 105^C-107^C
Hanwell, Captain Henry Sr, 45, 72, 90; 28, 89^A, 91^A, 33^C, 34^C
Harbours, 16, 25, 26, 28, 32, 33, 34, 35,

42, 71, 72, 74, 87, 89, 90, 91, 95, 101, 106, 116
Harding, Robert, 259[A]
Hardisty, William L., 106–7; 57, 366[A]
Haro (Arro) Archipelago, 368[A]
Haro Strait (Canal de Harro), 101; 319[A], 180[C]
Harpur's House, 63[A]
Harricanaw River, 47, 54
Harrison Lake, 94, 108; 273[A], 146[C]
Harrison's River, 94; 273[A]
Hawaii (Sandwich Islands), 98
Hawkridge, Captain William, 25
Hayes River, 10, 29, 36, 37, 38, 40, 42, 50, 54, 59, 67, 71, 72, 84, 276n5, 280n21; 22[A], 54[A], 111[A], 129[A], 146[A], 149[A], 203[A], 209[A], 220[A], 9[C], 12[C], 13[C], 74[C]
Hazard Gulf (Gulph). *See* Richmond Gulf
Hearne, Samuel, 40, 42, 43, 44, 45, 52, 53, 79, 120, 277n26, 278n3, 278n12; 7, 9, 10, 18[A], 20[A], 25[A], 4[B], 25[C]
Hemans, George E., 284n6
Henday, Anthony, 10, 16, 31, 36–7, 38, 120, 274n1, 274n4, 275n5, 278n23; 16[C]
Hendry, William, 74, 77, 281n2; 29, 214[A], 215[A]
Henley House, 32, 45, 46, 47, 50, 55, 56, 58, 273n6, 277n1; 26[A], 35[A], 157[A], 31[C], 32[C], 37[C], 53[C]
Henrietta Maria, Cape, 25, 26
High plains. *See* Great plains
High River, 59, 64
Hillier, William, 72; 152[A], 154[A]
Hill River, 66, 85; 48[A], 130[A]
Hobson Lake, 117
Hodgson, John, 45, 46, 47, 50, 56, 57, 58, 59, 120, 274n30, 279n17, 279n18, 279n21; 13, 36[A], 38[A], 30[C]–32[C], 36[C]–38[C], 45[C], 53[C], 55[C], 56[C]
Holy Lake, 84; 203[A]
Home (Homeguard) Indians, 29
Homphrey (Homfray), Robert, 99; 32[B]
Hong Kong, 98
Honolulu, 98
Hope, Fort, 94, 95, 107, 108, 113–14, 283n15; 464[A], 29[B], 33[B], 153[C], 154[C], 195[C], 198[C], 205[C]
Hopkins, Edward M., 300[A]

Horne, Adam, 104
Horne Lake, 104
Howse, Joseph, 60–1, 68, 72, 280n16, 280n17, 280n18, 280n19, 280n20, 280n21; 153[A], 78[C], 82[C], 84[C]
Howse Pass, 68, 280n17, 280n18
Howse's House, 66, 68; 81[C]
Howy, George, 35–6, 274n4, 278n19; 3[A], 8[C]
Hudson Bay: charts of, 7, 25, 26, 38, 40, 61, 66, 72, 278n1, 280n20, 280n25; commercial charts of, 26, 28, 276n13; early voyages to, 25; inaugural HBC voyage to, 4; northwest coast voyages, 28, 31, 32, 33, 41, 79; possible Northwest Passage from, 5, 32–3, 41, 42, 277n4; 2, 3, 6, 7, 8, 14, 15, 1[A], 8[A], 12[A], 16[A], 21[A], 23[A], 24[A], 27[A], 30[A], 32[A], 34[A], 36[A], 42[A], 57[A], 89[A], 92[A], 103[A], 133[A], 135[A]–138[A], 153[A], 185[A], 259[A], 2[B], 5[B], 1[C], 4[C]–7[C], 21[C], 26[C], 27[C], 47[C]–49[C], 52[C], 60[C], 77[C], 79[C], 80[C], 134[C], 149[C]
Hudson, George, 49, 52
Hudson, Captain Henry, 25
Hudson House, 49, 51, 54, 279n7
Hudson, James, 56, 59, 279n19; 35[A]
Hudson-James bays watershed, 26, 28, 46, 47, 50, 55, 56, 57, 85
Hudson Bay Company, 3, 25, 31, 36, 37, 38, 41, 42, 44, 48, 56, 57, 60, 62, 64, 66, 69, 85, 259–60, 261, 277n4, 278n5, 280n10, 280n22; annual ships, 19–20; catalogues of maps, 59, 68, 91, 92, 261–2, 279n13, 279n16, 279n25, 280n20, 283n2, 283n6; charter, 25, 26, 28, 33, 118; charter territory, 3, 5, 14, 26, 28, 32, 33, 115, 276n20; fisheries, 83, 109; fur trade departments, 71, 84, 259, 260, 281n1; General Court, 26, 259; gifts to museums, 4, 5, 34; Lachine headquarters, 83, 85; North West Company, union with, 6, 74, 87, 283n3, 283n4; property matters, 83, 93, 98, 106, 107, 108, 109, 110, 111, 112, 113, 114, 115, 118; scientific involvement, 4, 5, 14; secrecy, 4, 5; secretary, 19, 20, 26, 38, 48, 60, 71, 259, 261, 262; trading policies, 28, 29, 36, 37, 38, 42, 43, 44, 45, 46, 49, 51, 52,

55, 57, 64, 68, 69, 71, 72, 74, 75, 76, 77, 79, 83, 88, 90, 95, 107, 279n2, 280n10, 281n11; transport planning, 83–5, 116–18. *See also* Committee, HBC
Hudson Bay Company Archives, 6–7, 33, 35, 42, 49, 62, 87, 91, 261–2, 274n31, 278n1
Hudson's Bay House, 3, 4, 18, 37, 60, 64, 65, 72, 74, 85, 95, 105, 110, 115, 118, 261
Hudson Strait, 7, 14, 16, 25, 34, 40, 74, 278n1, 281n5, 281n25; 2, 3, 1[A], 8[A], 12[A], 24[A], 29[A], 60[C]
Huggins, Edward, 108, 114, 285n3; 317[A], 16[B], 165[C], 173[C]
Humboldt River, 89
Huron, Lake, 85, 274n28; 210[A], 299[A], 194[C]
Huron, Lake district, 85
Hutchins, Thomas, 45, 46, 47, 48; 40[C]

Idaho, 88; 511[A]
Idotlyazee, 42; 22[C]
Ile à la Crosse, 53, 54, 66, 68; 45[A], 46[A], 59[C]
Inconnue River. *See* Anderson River (Northwest Territories)
Indemnity negotiations (Columbia), 93
Indians: beavers conservation program, 86; captains, 29, 36; cartographic aid to HBC, 7, 30–1, 37, 42, 54, 61, 63–4, 65, 66, 67, 70, 71, 76, 77, 85, 94; exploration, aid, 10, 12, 13, 16, 29, 30, 36, 42, 43, 45, 47, 48, 51, 69, 75, 85, 94, 95, 106, 275n14; fear of, 104; gifts to, 77; and gold mining, 106; inland from Eastmain, 33, 69; land rights, 99; linguistics, 60–1, 280n19; maintaining peace between, 29, 31; as middlemen, 29; relations with, 12–13; reports by, 36, 38, 41, 95; reserves, 99, 104, 109, 111, 284n21, 285n9; trade journeys, 29, 36, 37, 49; trail, 70, 284n20; translators, 30, 41; tribal regions, 9, 40, 41, 63–4, 85, 93, 6, 9, 19, 57; warfare, 68, 280n18
Inland wintering, 5, 10, 29, 30, 36, 37–8, 40–1, 42–4, 47, 51, 53, 120, 278n3
Instruments: drawing, 14, 15–16, 44, 45,

55, 56, 69, 86, 98, 279n15, 279n24; lack of, 18, 50, 56, 97, 100, 101, 102; magnifying glass, 16, 279n23; nautical, 14, 41, 82; repair of, 14; surveying, 10, 14–18, 19, 40, 43, 44, 47, 50, 51, 52, 53, 54, 59, 97, 98, 99, 101, 102, 104, 105, 279n23, 279n24
Interior plains. *See* Great plains
Interior (upland) plateau, 94, 95
International border, 88, 92–3, 107, 110, 114
Inuit, 5, 29, 30, 31, 32, 33, 43, 65, 69, 74, 82, 277n19, 277n26, 280n13
Ireland, 12, 98
Isbester, Alexander Kennedy, 79, 281n14; 11[B]
Isbister, Joseph, 32
Isham, Charles, 62
Isham, James, 10, 36, 37, 38, 54; 13[C], 14[C], 17[C], 18[C]
Island Lake (Methy portage), 117[A], 119[A]
Island Lake, 104[A], 72[C]
Island Lake post, 86, 283n10

Jack Lake and River, 105[A]
James Bay, 4, 7, 11, 14, 19, 25, 26, 38, 45, 46, 47, 54, 66, 67, 72, 85, 275n3, 276n5, 276n6, 276n17, 276n18, 279n20; 1, 2, 14, 15, 28, 92[A], 150[A], 219[A], 1[B], 34[C], 39[C], 47[C], 79[C]
James Bay (Victoria), 100, 110, 111
James, Captain Thomas, 25, 276n1
Jarvis, Edward, 45, 46, 47, 56, 57, 58, 59, 120, 275n9, 275n14; 14, 26[A], 37[A], 6[B], 29[C], 35[C]
Jasper House, 84, 115; 211[A]
Johnston, Magnus, 41–2; 20[C], 21[C]
Jones, Cape, 34
Juan de Fuca, Strait of, 91, 97, 100, 105, 111, 112, 426[A]
Judith River, 63

Kabinakagami Lake, 55
Kabinakagami River, 46
Kaipokok Inlet, 76, 281n6; 243[A], 130[C], 136[C]
Kamchatka, 577[A]
Kamloops, Fort, 88, 90, 94, 108, 116, 283n15; 146[C], 153[C]
Kamloops Lake, 94, 116; 552[A]

Keele River, 281n13
Kellett, Captain, 102
Kelsey, Henry, 29, 30, 31, 32, 38, 120, 273n15, 277n21, 277n22
Kemano River, 117
Kempt, William, 82, 83–4; 37, 192[A], 193[A], 200[A], 95[C], 96[C], 97[C], 108[C]
Kendall River, 80, 81
Kennedy, Alexander, 162[A], 163[A]
Kennedy, Robert, 71; 183[A], 93[C]
Kennedy, William, 245[A]
Kennewap, 86; 135[C]
Kenogami River, 46, 55, 56, 57
Kenogamissi (Kenogamisi, Kinogamesee) Lake, 54, 55, 85; 44[C], 63[C]
Kenogamissi River, 75
Kequeloose, 95
Kesagami Lake and River, 50, 51
Kettle Falls, 130[A]
Kibokok Inlet. See Kaipokok Inlet
Kidala Arm and River, 117
King Charles II, 14, 25
King William Island (Land), 80, 82
Ki oo cus (The Little Bear), 63; 20
Kipling, John, 46, 47
Kisk-stack-ewen (Kiscachewan, Kiscatch-ewen River) River, 41, 43, 278n11; 19[C]
Kitchin, Eusebius, 47, 48
Kitimat Channel, 117
Kittson, William, 88; 43, 199[A]
Klamath Lake, 89
Knee Lake, 40, 50, 59, 85; 48[A]
Knight, James, 25, 29, 30–1, 33, 277n25; geographical concept of western Canada, 30–1; 2[A]
Knowles, John, 56–7; 56[C]
Koksoak River, 70, 74, 75, 281n3
Kootenay Lake, 266[A], 36[B]
Kootenay post, 94
Kootenay River, 68; 266[A]
Koskimo Inlet, 284n17; 171[C]
Krusenstern, Cape, 81

Labrador, 7, 26, 28, 34, 69, 70, 74, 75, 76, 87, 120; 30, 12[A], 242[A]–244[A], 130[C], 131[C], 136[C]
Lac à l'Eau Claire, 70, 281n3
Lac des Français. See François Lake
Lac des Loups Marins, 70, 281n3

Lac d'Iberville, 34
Lachine, 21, 83, 85, 281n1; 41, 221[A]
Lac la Biche, 54, 62; 66[A]–70[A]
La Cloche, 85, 274n28, 285n10; 40, 210[A], 299[A], 194[C]
Lake, James Winter, 92[A]–94[A]
Lake of the Woods, 32, 45, 48, 57, 58; 39[C], 57[C]
Laleham, Middlesex, 19, 52, 276n2
Lands Office, Province of Canada, 109
Langley Farm, 114; 563[A], 564[A]
Langley Fort, 21, 90, 94, 95, 108, 114, 283n10; 273[A], 13[B], 146[C], 209[C]
LaPuew, Lake. See Ogoki Lake
Larch River, 281n3
La Reine, Fort, 67
Latitude, 11, 14, 16, 17, 35, 45, 52, 53, 55, 66, 70, 71, 72, 79, 81, 84, 85, 93, 102, 104, 262–3, 275n9, 279n9, 280n6, 280n8
Launders, J.B., 116; 66, 560[A]
La Vérendrye, Pierre, Louis-Joseph, François, 32, 278n2
Leaf Portage, 128[A]
Leaf River, 281n3
LeFroy, John H., 275n6
Legace, Pierre, 106
Lesser Slave Lake, 62, 71; 72[A]–74[A], 121[A], 93[C]
Lesser Slave Lake district, 71, 280n24; 183[A]
Lesser Slave River, 62
Lewes, Adolphus Lee, 91, 92, 95–6, 97, 108, 275n8; 50, 249[A], 252[A], 263[A], 267[A], 289[A], 352[A], 137[C]
Lewes, John, 91
Lewes River, 78, 106, 281n12
Liard, Fort, 77; 191[A]
Liard River, 17, 77, 78, 117; 33, 190[A], 217[A], 233[A], 261[A]
Lillooet. See Berens, Fort
Lillooet-Harrison Lake trail, 94; 273[A], 13[B], 146[C]
Lillooet Lake and River, 94
Limestone Falls, 75
Limestone River, 127[A]
Lindley, Joseph, 275n1
Little North, 58, 279n20
Little Sea. See Winnipeg, Lake
Little Slave River. See Lesser Slave River

Little Whale River, 35, 69, 70, 74, 109, 277n13; 10[A], 13[A], 177[A], 188[A], 468[A], 10[C]
Logan Lake, 84
Logan, William, 115
Longitude, 11, 14, 15, 16, 17, 35, 44, 45, 52, 53, 55, 56, 66, 72, 79, 81, 84, 93, 102, 104, 263–4, 275n10, 280n6, 280n8
Long Lake, 57
Long Lake district, 172[A]
Longland, John, 34
Longmoor (Longmore), Robert, 51–2, 279n7
Lord Mayor Bay, 80
Lords Commissioners of Trade and Plantations, 28, 276n12
Lower Seal Lake, 34

McBean, John, 85; 40, 210[A]
McBeath, Adam, 281n13
McCulloch, Robert, 91[C]
McDonald, Alexander, 109, 285n10; 467[A], 468[A], 186[C], 187[C], 188[C], 194[C]
McDonald, Angus, 36[B]
McDonald, Archibald, 87–8, 93, 283n3; 48, 201[A], 266[A], 99[C], 104[C], 111[C], 138[C]
McDonald, Finan, 89
McDonald, John, 65
McFarlane, Roderick, 107, 284n1
McGillivray, Simon, 93–4; 227[A]
McGillivray, William, 283n3
McKay, Donald, 56, 57–8, 62, 120; 14, 37[A], 57[C], 62[C]
McKay, John, 56–7; 55[C]
McKay, Joseph, W., 103, 116, 118; 65, 364[A], 550[A], 551[A], 157[A], 182[C]
McKay, Thomas, 89, 90, 283n10; 103[C]
Mackenzie district, 72, 77–9, 106–7, 115, 117; 32, 191[A]
Mackenzie, John, 421[A]
Mackenzie Range, 74, 77–9
Mackenzie River, 7, 43, 51, 54, 66, 74, 77–80, 81, 106–7, 281n16; 143[A], 190[A], 367[A]
McKenzie's River district. See Mackenzie district
McLean, John, 75, 77; 242[A], 254[A]
McLeod, Alexander R., 77, 89; 112[C]
McLeod, John, 77, 78, 281n7; 33, 217[A], 233[A]

McLoughlin, Fort, 94
McLoughlin, Dr. John, 88, 89, 91, 92, 95, 96, 114, 283n10
McMillan, James, 90, 283n10
McNab, Thomas, 72[C]
McNaughton, Archibald, 109; 60, 411[A], 437[A]
McNeill, Captain William, 91, 95, 103, 106; 305[A]
McNeill's Harbour, 326[A], 152[C]
McPherson, Fort, 106
McPherson, Murdoch, 77; 32, 190[A], 191[A]
MacTavish, Dugald, 111, 113; 35[B], 205[C]
Mactavish, W., 559[A]
McTavish, William, 115; 535[A]
Magellan, Strait of, 109
Malade River, 189[A]
Malheur Lake, 89
Manchester House, 51, 52, 53, 54, 279n7; 54[C]
Manicouagan Lake, 76
Manitoba district, 67; 23, 187[A]
Manitoba, Lake, 38, 40, 65, 67; 109[A], 182[A]
Manitoba lakes, 29, 32, 38, 40, 57, 62, 277n3
Mannal, John, 55, 275n5; 63[C]
Man, ne, tow, oo, pow, Lake. See Manitoba, Lake
Manson, William, 117–18; 220[C]
Mantouapau Hills. See Riding Mountain
Manuscript maps, 6–7, 118, 261
Maps: attrition, 6, 261; as business documents, 3, 37–8, 261; cartouche on, 9, 40, 42, 45, 61; colour on, 9, 16, 26, 35, 40, 42, 50, 59, 61, 75, 81, 82–3, 84, 85, 93, 95, 100, 101, 102, 111, 112: compilation, 10, 42, 44, 49, 50, 51, 54, 56, 59–60, 61, 64, 65, 66, 70, 79, 81; design techniques, 7, 9, 26, 28, 35, 40, 41, 44, 50, 59, 61, 64, 66, 68, 69, 76, 80, 81, 82, 84, 85, 97, 280n4, 284n21; examination by HBC committee, 3, 20; lettering, 40, 42, 61, 82, 85; military, 92–3; storage, 3, 21; watermarks, 262, 281

Marble Island, 31, 33, 41, 277n26
Marlborough House, 62, 279n2
Marlborough, John Churchill, Duke of, 28, 276n14

Marley, John, 278n1, 280n25; 29^A, 48^C, 49^C
Marten, Humphrey, 19–20, 276n4
Martin, John, 45, 46, 273n6
Martin Fall House, 59, 67; 90^C
Matchousin. See Metchosin
Matonabbee (Meatonabee), 42, 43; 22^C
Mattagami Lake, 47, 54, 85
Mattagami River, 54, 55, 85
Maurepas, Fort, 32
Maxwell, John, 277n10
Meadow Lake. See Barren Ground Lake
measurement of area, 14
measurement of depth (sounding), 14, 34, 35, 37, 41, 42, 72, 92, 101, 116
measurement of direction, 10, 11, 14, 16, 17, 18, 34, 35, 36, 38, 44, 50, 52, 53, 61, 64, 66, 67, 69, 70, 71, 84, 85, 88, 89, 104, 105, 106, 285n22
measurement of distance, 10, 11, 14, 16, 17, 18, 34, 35, 36, 38, 44, 45, 46, 47, 50, 52, 53, 56, 61, 62, 63, 64, 66, 67, 68, 69, 72, 84, 85, 87, 89, 102, 104, 106, 116, 276n1
measurement of height, 14, 70, 101, 275n2
measurement of location, 14, 49, 50, 53, 66, 67, 84, 87, 89, 101, 102, 106
measurement of time, 63, 264
Melville Lake-Hamilton Inlet, 75, 76; 30, 229^A, 230^A, 242^A, 244^A, 254^A, 255^A, 130^C, 136^C
Melville Peninsula, 31, 43; 149^C
Mepiskawaukau Lake, 45, 47, 56, 57; 56^C
Mercury. See Quicksilver
Mesackame Lake, 48
Meshickemau Lake, 231^A
Meshippicoot Lake. See Michipicoten Bay
Metals, 30, 31, 42, 43
Metchosin district, 97, 100, 102, 284n4;; 356^A, 361^A, 458^A
Metesene (Cairn) Island, 34, 277n16
Methy Portage, 51, 53, 54, 64, 66; 39^A, 117^A–119^A, 142^A
Metis, 83, 99, 263
Micabanish House. See New Brunswick House
Micabanish Lake, 55

Michipicoten Bay, 46, 47, 50, 51; 26^A, 35^C, 36^C
Michipicoten post, 285n10; 194^C
Middleton, Christopher, 31, 32–3, 273n12, 277n4, 277n11
Miles, Robert, 76, 86; 42, 251^A, 256^A, 135^C
Mille Vaches Seigneury, 83
Millstone River, 113
Minago River, 66
Mingan, 75, 76, 115; 63, 479^A
Ministickwattam Island, 86; 256^A
Minnesota Territory, 115
Miscosinke (Mistassiny) Lake. See Mistassini Lake
Missinaibi House, 55
Missinaibi Lake, 47, 50, 55; 42^C, 43^C
Missinaibi River, 20, 45, 47, 50, 55; 41^C–43^C
Missinnipe River. See Churchill River
Mississanque, 285n10; 194^C
Mississippi River, 108
Missouri (Mis sis sury) River, 7, 54, 63, 64, 65, 68, 280n7; 19, 80^A, 196^A, 197^A, 70^C, 82^C
Mistassini House, 280n23
Mistassini Lake, 26, 28, 60, 69, 70, 76; 168^A, 91^C, 132^C
Mitchell Harbour, 106; 163^C
Mitchell, Thomas, 34, 35, 277n13; 4, 4^A–6^A
Moffatt, Hamilton, 284n12; 171^C
Mjave Desert, 89
Molson (Winepegucies) Lake, 84
Monotogga (Monotoggy, Monotai), 55
Montana, 63, 88; 511^A
Montreal, 58, 283n10, 283n15
Moor, William, 33, 277n4
Moore, Thomas (Eastmain sloop captain), 46, 47
Moore, Thomas, 25–6, 276n6; 1, 1^B
Moose Factory, 20, 34, 35–6, 37, 45, 47, 48, 50, 51, 55, 69, 85, 281n2; 3^A, 150^A, 157^A, 219^A, 8^C, 28^C, 29^C, 41^C, 63^C
Moose Lake, 98^A, 126^A
Moose River, 7, 20, 26, 35, 37, 42, 45, 47, 48, 50, 51, 54, 55, 58, 85, 274n4, 275n14; 3^A, 8^C, 39^C, 44^C
Moravian missions, 74
Morin, A., 386^A

Moshewanaugan. See Muschowiaugan
Mosquito Bay, 467^A
Mossy Point (Lake Winnipeg), 108^A
Mouat, Captain William A., 116
Mountain Portage (Back River), 190^C
Mudge, Cape, 103
Muir, John, 103, 284n16
Munck, Jens, 25
Murray, Alexander Hunter, 79, 106, 275n5; 175^C
Muschowiaugan River, 132^C, 133^C
Musquaro (Masquaro), 76, 131^C
Mynd, James, 12^A

Nabowisho, 86; 42, 256^A
Nahanni Indians, 77
Nanaimo: coal mining at, 4, 103, 104, 113, 54; company property at, 112; Episcopalian property at, 113; exploration of, 101–2; government reserve at, 112–13; harbour, 102, 103; Indian reserve at, 113; Methodist property at, 113; naming of, 284n14, 284n18; maps of, 101–2, 103, 327^A–329^A, 358^A, 364^A, 370^A–372^A, 385^A, 406^A, 407^A, 409^A, 410^A, 413^A, 417^A, 435^A, 436^A, 463^A, 478^A, 485^A, 486^A, 488^A, 489^A, 495^A, 501^A, 502^A, 506^A–509^A, 561^A, 562^A, 170^C, 172^C, 177^C, 180^C, 181^C, 192^C, 193^C
Naosquiscaw House. See Neoskweskau House
Naskaupi River, 75
Nass Inlet, 90
Nass River (Simpson's River), 90, 91, 94, 117, 125^C
Nautical Almanac and Astronomical Ephemeris, 15, 16, 18, 20, 21, 51, 53, 262–3, 264, 275n3, 275n5, 279n23
Navigation methods, 12, 15, 41, 72, 82
Nayhektil lok, 136^A
Nechako River, 117
Nelson House, 63^A
Nelson River, 7, 10, 26, 28, 29, 36, 37, 42, 44, 49, 52, 54, 59, 65, 66, 68, 72, 85, 279n25, 280n21; 17, 7^A, 56^A, 124^A, 125^A, 127^A, 128^A, 137^A, 144^A, 146^A, 149^A, 152^A–154^A, 202^A, 9^C, 12^C–14^C
Neoskweskau House, 69, 280n23; 24, 170^A, 173^A

Nestucca River, 283n9
Nevada, 88, 89
Newbird, James, 284n9
New Britain, 188^A
New Brunswick House, 54, 55, 86; 157^A, 184^A, 64^C
New Brunswick Lake, 184^A
New Caledonia, 88, 90, 93, 115, 116, 117, 118, 283n4; 273^A, 92^C
Newcastle Island, 102, 103, 112; 385^A, 408^A, 181^C
New Dungeness, 91; 213^A
New England, 12^A
Newfoundland, 12^A
Newton, John, 17, 36
Newton, William H., 102, 103, 114, 116, 117, 275n8; 552^A, 563^A, 564^A, 567^A–569^A, 572^A
New Westminster, 114
New York City, 21
Nez Perces House, 88, 89, 92, 108; 270^A
Nichicun Lake, 69
Nicol, Charles, 113
Nicola Lake, 93, 95; 153^C
Nimescaw Lake, 180^A
Nimpkish River, 103, 284n17; 171^C
Nipigon Lake, 25, 55, 58, 59, 276n6; 50^C, 51^C, 53^C, 55^C, 57^C
Nipigon post, 109, 285n10; 194^C
Nipissing, Lake, 85
Nisqually (Nasqually), 4, 91, 92, 93, 107–8, 114, 285n3; 46, 47, 249^A, 272^A, 279^A, 287^A, 294^A–298^A, 317^A, 26^B, 137^C, 150^C, 162^C, 165^C, 189^C
Nitchequon House, 69, 70, 76; 173^A, 133^C
Nitinat Inlet, 104–5
Nixon, John, 25, 276n3, 277n25
Nodaway River. See Nottaway River
Nootka Sound, 103
Norbury, Brian, 25
Norquay, William, 284n9
Northern Indian Lake (Churchill system), 66; 42^A, 97^A, 71^C
Northern Indians. See Chipewyan (Athapaskan, Northern) Indians
North Thompson River, 116, 117, 118; 568^A
Northwest of Canada, 77, 106, 107; 62^C
Northwest coast, 7, 90, 91, 117–18; 212^A, 100^C, 103^C

North West Company, 67, 68, 70, 72, 74, 77, 87, 280n17, 283n2, 283n3, 283n4, 283n10
Northwest Passage, possible entrance from Hudson Bay, 5, 12, 25, 31, 32–3, 41, 42, 43, 44, 276n1, 277n4, 277n5, 278n1
North West River (Meshickemac) 228[A], 131[C]
North West River House. *See* Smith, Fort
Northwest Territories, 7
Norton, Moses, 41, 42, 43, 44, 274n4; 8, 16[A], 17[A], 19[A]
Norton, Richard, 30, 36, 120, 277n24, 278n9
Norton Sound, 107
Norway House, 67, 83, 84; 211[A]
Norwood, Mr., 25, 26; 1[C]
Nottaway River, 46, 47, 69, 76, 86, 276n19; 38[C]
Nottingham House, 66, 264; 88[A]
Nutt, George M., 106; 163[C]

Observations on Hudson's Bay (Graham), 40, 41, 120, 278n4
Observatory Inlet, 106[C]
Ogden, Peter Skene, 88–9, 283n4, 283n6, 283n7, 283n8; 216[A], 109[C], 110[C], 113[C]
Ogilvie, John, 198[C]
Ogle Point, 82
Ogoki Lake, 56
Ogoki River, 32, 56
Okanagan, 87, 93, 283n18; 351[A]
Okanagan Indians, 93
Okanagan Lake, 93; 111[C]
Okanogan, Fort, 108
Okanogan (Okinagan). *See* Okanagan
Olds (Alberta), 38
Olympia (Washington Territory), 114
Olympic Peninsula, 91
Ordnance Survey of Ireland, 98, 284n13
Oregon City, 92, 283n4; 263[A], 143[C]
Oregon Territory, 88, 89, 92, 107, 110, 114, 285n6; 368[A], 369[A], 511[A], 565[A], 566[A], 112[C]
Oregon Treaty, 92, 94, 100, 284n2
Orkneymen, 69, 72
Osnaburgh district, 280n24; 167[A]

Osnaburgh House, 56, 57, 58, 59, 279n18, 280n24; 51[A], 55[A], 113[A], 57[C], 76[C]
Ottawa River, 58, 109
Otter Lake, 116
Owl River, 137[A], 138[A]
Oxford House, 64, 84; 39, 226[A], 127[C]
Oxford Lake, 40, 50, 84; 39

Pacific-Atlantic ship tracks, 109; 473[A], 480[A]
Pacific coast, 4, 14, 21, 54, 63, 74, 87, 88, 89, 90, 91, 107, 110, 117, 283n9; Pacific drainage divide, 78; Pacific-flowing rivers, 78, 89, 93–4, 117
Pacific Ocean, 25, 30, 31, 32, 78, 94, 274n19, 276n1; 57[A], 93[A], 94[A], 281[A]
Pacific Fur Company, 89
Pacific Northwest, 65, 77, 88, 117
Packets, 20, 41, 46, 47, 48, 51, 52, 58, 64, 67, 71, 281n4
Painted Stone Lake, 99[A]
Panache, Lake, 85
Park belt, 63, 64, 115
Parsnip River, 77
Pashkokogan (Pascocoggan) Lake, 56, 279n16, 279n18; 50[C], 51[C]
Pas, The (Basquiaw, Pasquia) 40, 44; 25[A], 33[C], 40[C]
Passes through the mountains, 65, 68, 110, 115, 116, 117, 118, 285n22
Patagonia (Messier Channel), 466[A]
Pauqua-thacow a scow a River. *See* South Knife River
Peace River, 42, 54, 77; 93[A], 191[A]
Pearl Harbour, 225[A]
Pearse, Benjamin W., 99, 103, 104, 110
Pedder Bay, 100
Pedlars. *See* Canadian traders
Peel River, 79, 106, 281n14; 11[B], 147[C], 175[C]
Peers, A.R., 106
Peers, Henry Newsham, 92, 95, 283n15; 264[A], 265[A], 153[C]
Pelly, Fort, 115
Pelly, Sir John Henry, 80, 95, 97
Pelly Point, 81; 160[C]
Pelly River, 78, 79; 261[A]
Pemberton, Joseph Despard, 7, 9, 12, 14, 15, 96, 98–105, 110, 117, 118, 121, 275n2; 51, 52, 53, 56, 310[A]–314[A], 318[A]–323[A], 327[A]–350[A], 353[A]–363[A], 370[A]–384[A], 388[A]–405[A], 413[A]–416[A], 422[A]–434[A], 439[A]–461[A], 469[A], 470[A], 14[B], 15[B], 18[B]–25[B], 27[B], 30[B], 31[B], 161[C], 167[C], 168[C], 170[C], 172[C], 176[C]–180[C], 185[C], 192[C], 198[C]
Pembina, 83; 209[A]
Pembina Hills, 67
Pemmican, 49, 63
Pend Oreille Lake, 138[C]
Pend Oreille River, 114
Pennant, Thomas, 5
Pennycutaway River, 36
Pequatisahaw, 48; 43[C]
Perkin, Richard, 56, 57, 67
Petaibish (Peataibish), 132[C]
Peter Pond Lake, 66; 142[A]
Petitsikapau Lake, 75
Pic (Pique) River, 55, 285n10; 64[C], 194[C]
Piegan Indians 53, 63
Pierce County (Washington Territory) 16[B], 173[C]
Pigeon River, 38, 47
Pike Lake, 66; 119[C]
Pine Island Lake. *See* Cumberland House Lake
Pine River, 126[A]
Pipmouagan Lake, 558[A]
Piskocoggan Lake. *See* Pashkokogan Lake
Pitt, Fort, 92; 145[C]
Place names (toponyms), 35, 44, 45, 58, 63, 66, 75, 77
Plains Indians, 115
Platte River, 196[A], 197[A], 198[A]
Playgreen Lake, 54, 65, 84; 48[A], 107[A]
Pond, Peter, 51, 58
Poplar River, 202; 108[A], 112[A], 75[C]
Porcupine River, 79, 106; 147[C]
Porpoise fishery, 109
Portage de l'Isle, 57, 58; 57[C]
Portage du Traite, 54
Portage Inlet, 95, 100, 103; 484[A]
Portages, 11, 13, 16, 18, 20, 50, 51, 54, 55, 57, 61, 62, 64, 66, 68, 83; engineering work at, 67, 84
Port Bull, 120[C]
Porter, Cape, 82; 184[C]
Port George, 223[A], 121[C]

Portin, Captain, 26
Portland Canal (Inlet), 90, 91, 118
Portland House, 66
Portland Point, 34
Port Nelson, 26, 32; 3[C]. *See also* York Factory.
Portneuf, 83
Port San Juan, 104, 413[A], 192[C]
Potts, John, 15[A]
Preuss, Charles, 108, 285n8
Prince of Wales, Fort, 31, 42, 275n8, 278n9, 278n20; 9, 2[A], 20[A], 106[A]
Prince of Wales Lake, 58
Projection. *See* graticule (grid)
Protection Island, 102, 112
Prudhoe Bay, 79
Puget Sound, 7, 90, 91, 95, 112, 285n2; 272[A], 279[A]
Puget's Sound Agricultural Company: Columbia area farms, 4, 91–2, 93, 107–8, 114; Esquimalt farms, maps of, 99, 101, 110, 112, sale of land to HBC, 112, selection of land for farms, 97, subdivision into lots, 112, surveying of farms, 100, 101, 102–3, 112; Constance Cove Farm, 103, 112, 375[A], 570[A], 214[C]; Craigflower Farm, 103, 112, 375[A], 376[A], 570[A], 214[C]; Langford Farm 102, 103, 105, 112, 56, 375[A], 476[A]; Viewfield Farm 103, 112, 375[A], 570[A], 571[A], 213[C], 214[C]

Qualicum River, 104; 413[A], 191[C], 192[C]
Qu'appelle River, 62, 65
Quatsey, 326[A]
Quatsino Sound, 103
Quebec, 7, 28, 74, 75, 76, 77; 94[C], 132[C]
Quebec City, 283n4, 283n10
Queen Charlotte Islands, 106; 224[A], 163[C], 164[C], 183[C]
Quesnel, 114, 117; 573[A]
Quesnel (Quesnelle) Lake, 117; 568[A]
Quesnel River, 117
Quicksilver, 15, 20, 50, 51, 53, 262

R.A., 284n21; 301[A]
Radisson, Pierre-Esprit, 25, 26, 276n2
Rae, Dr. John, 12, 80–2, 115–16, 274n30, 275n6, 285n20, 285n21, 285n22,

285n23; 35, 292[A], 306[A], 159[C], 160[C], 166[C], 184[C], 211[C], 212[C]
Rae Isthmus, 80, 81, 82; 293[A]
Rainy Lake, 45, 48, 55, 56, 57, 58; 39[C], 57[C]
Rankin Inlet, 33, 41,, 42; 18[A], 20[C]
Rat Country, 59
Rat River, 106
Red Deer River (Alberta), 38, 63, 280n7; 95[A], 123[A]
Red Deer River (Manitoba), 38, 40, 62
Red Deer River. *See* Yellowstone River
Red Deer's Lake. *See* Lac la Biche
Red Lake, 47, 57, 58; 51[A]
Red Lake House, 57, 58
Red River, 12, 57, 58, 62, 65, 82, 83, 84, 115; 107[A], 110[A], 175[A], 192[A], 10[B]
Red River district, 71, 280n24; 181[A]
Red River settlement, 12, 14, 60, 61, 67, 72, 74, 82, 83, 84, 115, 262–3, 275n8, 280n15, 281n11; 36, 203[A], 204[A], 234[A]–236[A], 241[A], 260[A], 300[A], 535[A], 90[C], 96[C], 212[C]
Red Sucker Lake, 104[A], 72[C]
Reference (composite) maps, 9, 49, 54, 58–9, 60, 61, 62, 67, 69, 70, 71, 72, 76–7, 84, 85, 95, 96, 107, 111–12
Reindeer Lake, 64–5, 66; 115[A], 141[A]
Reindeer River, 59, 64; 106[A], 73[C]
Repulse Bay, 33, 80, 81, 82; 259[A], 293[A], 149[C], 184[C]
Requisite Tables, 15, 16, 20, 21, 263, 278n3
Return Reef, 79, 80; 237[A], 238[A]
Rewards (gratuities, premiums), 3, 10, 11, 28, 35, 38, 41, 45, 52, 59, 60, 68, 69, 72, 274n4, 277n17, 278n19, 279n24
Richardson Range, 74, 78, 79, 106
Richardson, Sir John, 81, 107
Richfield, 551[A]
Richmond Fort, 34–5, 277n14; 3[B]
Richmond Gulf, 16, 17, 34, 35, 70, 74, 76, 277n12, 277n14, 281n3; 5, 6[A], 9[A], 10[A], 11[A], 14[A], 176[A], 11[C]
Riding Mountain, 38, 40
Rivers: lower courses, 32, 35, 36, 37, 41, 43, 44, 45, 46, 68, 70, 72, 77, 90, 92; mouths, 25, 26, 28, 29, 30, 32, 34, 36, 38, 40, 42, 43, 44, 46, 66, 70, 72, 74, 77, 79, 80, 88, 90, 91, 92, 95; networks, 37, 38, 40, 76, 84, 85, 88

Roberts, George, 72; 77[C]
Roberts, Point, 90
Robertson, James, 168[A], 119[C]
Robertson, John, 16, 275n8
Robinson, George, 104; 54, 406[A]–410[A], 417[A], 418[A], 435[A], 436[A], 463[A], 193[C]
Robinson Lake, 84
Robson, Joseph, 36, 275n8, 278n20; 7[A], 9[C]
Rock Bay, 101
Rocky Mountain House, 52[A]
Rocky Mountains, 7, 9, 21, 38, 51, 59, 61, 62, 63, 64, 65, 66, 68, 74, 77, 78, 87, 115, 118, 278n2; 77[A], 80[A], 93[A], 95[A], 121[A], 191[A], 211[A], 536[A], 65[C], 69[C], 70[C], 81[C], 82[C], 98[C], 99[C], 138[C]
Rocky Mountain Trench, 68, 116
Roe's Welcome Sound, 31
Rogue River, 89
Rose Point, 106
Ross, Alexander, 87, 283n5; 189[A], 7[B], 101[C]
Ross, Bernard, 107; 419[A]
Ross, Malcolm, 52, 53, 65, 279n8, 279n9
Royal Agricultural College, Cirencester, 98, 284n5
Royal Geographical Society, 4, 80, 121, 274n20, 276n10, 282n23
Royal Geographical Society, Journal of the, 81, 82, 97, 281n15, 282n20, 282n24, 282n29, 282n31
Royal Military Academy, Woolwich, 283n15
Royal Society, 64, 275n8, 277n4; dining club, 5; relationship with HBC, 4, 5, 121, 273n11; Transit of Venus, 5, 19
Rupert and Eastmain district, 71, 76, 86
Rupert, Fort, 99, 103, 104, 113; 61, 325[A], 422[A], 423[A], 510[A]
Rupert House, 70, 86, 280n23; 188[A], 219[A]
Rupert River, 26, 69, 70, 71, 75, 76; 180[A], 83[C]
Rupert's Land: boundaries of, 84; charter territory, 3; company property claims in, 106; exploration, 86; Fidler arrival in, 61; Fidler at Red River, 67; governor of, 115; Inuit of, 5; senior officers in, 37; surveying and mapping, 7, 46, 60, 71; Turnor arrival in, 19; Turnor leaves, 52, 54

Russell, James, 87[A]
Russians, 77, 78

Saanich, 100, 102, 104; Inlet, 104; 321[A], 179[C]; Peninsula, 100, 101, 102, 104; 359[A], 179[C]
Saaquash, 324[A]
Sabeston, Hugh, 74[C]
Sacramento River, 89
Saguenay district, 303[A]
Saguenay River, 558[A]
St. Anns (Nipigon) Lake, 45, 46, 55, 56, 57, 58, 275n9, 279n16; 50[C], 51[C], 53[C], 55[C]
St. Boniface, 83
Saint Cloud (Minnesota Territory), 475[A]
St. John, Fort (Peace River), 77
St. Joseph Lake, 47, 56; 31[A]
St. Lawrence Gulf and River, 70, 74, 75, 76, 115; 254[A], 255[A], 558[A], 131[C], 133[C], 134[C]
St. Martin Lake, 65; 109[A]
St. Maurice River, 76
St. Paul (Minnesota Territory), 80
St. Peters (Minnesota Territory), 10[B]
Salmon Arm, 117
Salmon River, 117; 189[A]
Salt River, 90[A]
Sand Hill Bay, 190[C]
Sandwich Islands. *See* Hawaii
Sandy Lake, 113[A], 76[C]
San Francisco, 88, 89, 105, 112
Sangster, James, 152[C]
San Juan Islands, 90, 102, 103, 108; 178[C], 202[C]
Saskatchewan River, 20, 29, 32, 37, 38, 40, 44, 45, 49, 51, 52, 54, 58, 59, 63, 64, 65, 66, 68, 72, 84, 115, 116, 274n4, 276n7, 278n5, 279n7, 279n12; 49[A], 50[A], 53[A], 56[A], 75[A], 76[A], 120[A], 122[A], 123[A], 285[A], 286[A], 16[C], 17[C], 54[C], 65[C], 78[C], 82[C]; South Saskatchewan branch, 29, 40, 51, 61, 63, 64, 280n7, 280n8, 280n12, 280n16
Saul, Lake. *See* Seul, Lac
Sault Ste Marie, 83, 85, 109; 142[C], 186[C], 187[C]
Savana's Ferry. *See* Savona
Savona, 116; 560[A]

Scarborough, Captain James, 96, 283n16; 248[A], 139[C]
Schwieger, A.W., 12, 115–16; 64, 536[A]
Scotland, 82, 99, 283n10
Scroggs, Captain John, 31
Seal River, 32, 38, 43, 65; 101[A], 61[C]; track, 66; 141[A]
Seale, R.W., 33, 261, 277n7; 3, 8[A]
Seine River, 67
Selkirk, Fort, 79, 106, 107
Selkirk, Lord, 67, 87
Selkirk settlement. *See* Red River settlement
Seller, John, 14–15, 26, 275n3
Senex, John, 277n10
Sept Iles, 76
Sergeant, Henry, 28, 276n18
Seton Lake, 94, 108
Setting River, 120[A]
Seul (Saul), Lac, 47, 56, 57, 58; 31[A]
Severn district, 71, 280n24; 26, 158[A], 159[A], 160[A], 194[A]
Severn Fort, 36, 37, 38, 40, 41, 47, 48, 50, 57, 66, 82, 85, 86, 276n20, 278n5, 283n10; 23[A], 103[A], 129[A], 139[A], 165[A], 23[C], 24[C], 79[C], 80[C]
Severn River, 26, 28, 29, 36, 38, 40, 61, 66, 71, 85; 15[C]
Seymour, 560[A]
Shamattawa River, 104[A], 72[C]
Shepherd, Fort, 114; 62, 534[A]
Shew ditha da, 53
Shuswap Lake, 93, 116, 117, 118; 552[A]
Shuswap River. *See* Thompson River
Similkameen River, 94; 111[C], 146[C]
Simpson, Aemilius, 90–1; 213[A], 114[C], 115[C], 120[C]–125[C]
Simpson, Alexander, 80
Simpson, Fort (Mackenzie River), 77, 78, 79, 80, 106, 107; 367[A]
Simpson, Fort (Nass), 90, 91, 117, 283n4
Simpson, Sir George: Arctic exploration, 79, 80, 81; beaver conservation, 86; Columbia affairs, 87; comment re Annance, 283n10; company-colonial surveyor, 97, 98; Great Lakes affairs, 85, 109; Kempt engineering work, 84; Lachine, 85; military mapping, 92; Mingan, 115; New Caledonia brigade trails, 90, 94, 95; personal data, 281n1;

PSAC farms, Columbia, 91, 92; report on Northwest, 106; Snake Country expeditions, 88; to Sitka, 91; trade and exploration behind Eastmain, 74–77; Vancouver Island colony, 95; Yukon map, 107
Simpson (Simpson's) Lake, 78
Simpson's River. *See* Bulkley River
Simpson's River. *See* Nass River
Simpson, Thomas, 79–80, 282n18; *34, 237*[A], *9*[B], *10*[B]
Simpson's Strait, *246*[A]
Sinclair, Thomas, 91; *222*[A]*–225*[A]
Sinclair, William, *60*[A], *61*[A]
Sitka, 90, 91
Siuslaw River, 283n9
Skeategats (Skiddigates, Skidgetts). *See* Skidegate
Skeena River, 94
Skidegate, 106; *224*[A], *123*[C], *164*[C]
Slave Indian Lake. *See* Lesser Slave Lake
Slave Indian River. *See* Lesser Slave River
Slave River, 29, 30, 43, 53
Sleds, 11, 18, 50, 62, 81, 82
Slude River and post, 33, 42, 69; *83*[C]
Smallpox, 56, 278n5
Smith, Edward, 74
Smith, Fort, 75, 76; *231*[A], *242*[A], *244*[A], *136*[C]
Smith, Joseph, 37, 38
Snake Country expeditions, 88–9, 283n2, 283n4; *101*[C], *109*[C], *110*[C], *113*[C]
Snake River, 7, 87, 88, 89; *43; 199*[A], *104*[C]
Snow blindness, 17, 47, 50, 117
Soke. *See* Sooke
Somerset House, 62, 279n2
Sooke, 97, 100, 102, 284n4; *387*[A], *18*[B], *179*[C]
Sooke Bay, *360*[A]
Sooke district, 102, 103; *373*[A], *521*[A]
Souris River, 62
Southampton Island, 31, 33
South Branch House, 51, 52, 63, 64; *54*[C]
Southern Indian Lake, 52, 65; *101*[A]
South Knife River, 52
South Nahanni River, 77, 78
South River. *See* Caniapiscau River
South River House, 75
South Sea. *See* Pacific Ocean

Soweawaminica (Sowowominicaw), 55; *63*[C]
Souweska sepy, *41*[C]
Spanish River, 85; *210*[A]
Special purpose maps, 4, 6, 61, 74, 85–6, 99, 109
Spence, George, *75*[C]
Spencer, John, 76–7, 275n5; *31, 247*[A], *134*[C]
Spilamacheen River, 117
Split Lake, 44, 66, 68; *130*[A], *131*[A], *153*[A]
Spurrell, Captain George, 36, 45
Sr. Atwls Lake. *See* Richmond Gulf
Starvation, 11, 45, 47
Staunton, Richard, 35
Stayner, Thomas, 52, 65; *54*[C]
Steel River, 36, 66, 85; *128*[A], *202*[A]
Stewart, James, 82
Stewart (Stewart's) Lake, 93
Stewart, William, 29, 30, 36, 120, 277n23
Stikine River, 78, 117, 281n10; *233*[A], *220*[C]
Stony Mountains. *See* Rocky Mountains
Strutton Sound (Charlton Island), *151*[A]
Stuart, Captain Charles E., 102, 104, 106; *164*[C], *183*[C], *191*[C]
Sturgeon Bay, 65
Sturgeon Lake, 56, 57; *55*[A], *129*[A]
Sturgeon River, 59, 85; *118*[C]
Sturgeon River Indians, 85
Sturgeon-Weir River, 279n12
Summit Lake, 116
Superior, Lake, 21, 25, 46, 47, 48, 51, 55, 56, 57, 276n6, 279n20; *26*[A], *39*[C], *43*[C], *126*[C], *194*[C]
Surveying: astronomical observations, 11, 12, 14, 15, 16, 17, 20, 40, 44, 45, 47, 48, 49, 50, 52, 53, 55, 56, 61, 62, 66, 70, 72, 81, 84, 85, 101, 102, 105, 262–4, 280n6, 280n8; cadastral, 14, 19, 67, 82–3, 87, 92, 93, 97, 98, 100, 101, 102, 104, 109, 110, 111, 112, 113, 114; instruments, 3, 12, 14–15, 16–17, 19, 20, 36, 41, 43, 44, 47, 51, 53, 59, 69, 97, 102, 104, 105, 262, 264, 275n4, 279n3, 279n15, 279n23, 279n24; surveys, American, British, Spanish, 89; survey methods, 14, 34, 44, 61, 98, 100, 101, 102, 105, 275n2; topographic, 98, 100, 102, 275n2
Surveyor: colonial, 96, 97, 98, 99, 110, 261; inland, 5, 6, 19, 48, 49, 66–7, 118; Red River, 67, 82–3; Surveyor-General, British Columbia, 116, 261, 283n19, Oregon Territory, 107–8, 285n7, Washington Territory, 108, 114, 285n4, 285n5
Sutherland, Donald, 71; *27, 178*[A]
Sutherland, George, 46, 47; *46*[C]
Sutherland, James, 11, 55–6, 58, 67, 274n2, 279n16, 279n18; *31*[A], *50*[C]
Sutherland, John, 57, 58
Swain, James Sr, 66, 71; *26, 158*[A], *159*[A]
Swain, Thomas, 66; *88*[A]
Swamp River, 116
Swampy Bay River, 75
Swampy Portage, 62
Swan Lake, 38, 57, 62; *58*[A], *59*[A]
Swan River, 38, 62, 65; *59*[A]
Swan River House, 65, 279n2
Swanson, William, 85; *219*[A]
Syusum. *See* Sooke

Taché, E.P., 109; *438*[A]
Tacoutche River. *See* Columbia River
Tadoussac, 76, 83; *307*[A], *315*[A], *316*[A], *386*[A], *133*[C]
Tar sands, 30
Tate, James, *172*[A]
Tatnam, Cape, 281n25
Taylor, George Jr, 12, 61, 82–3, 84–6, 120, 262–3, 274n30, 275n6, 275n8; *36, 38, 39, 202*[A]*–209*[A], *211*[A], *226*[A], *234*[A]*–236*[A], *116*[C], *127*[C]
Taylor, George Sr, 66, 82, 120; *66*[C]
Telegraph Creek, 116; *550*[A]
Telegraph line, 4, 12, 110, 115–16, 118; *64, 536*[A], *210*[C]*–212*[C]
Temagami Lake, 85
Terrewill, *129*[C]
Tête Jaune, *92*[C]
Tête Jaune Cache. *See* Yellowhead Pass
Thames-side cartographers. *See* Drapers' Company
Thayendenaga, Chief, 5
Thickwood Hills, 49
Thlewycho desse, *420*[A]
Thomas, John, 47, 55, 274n4
Thomas, Thomas, 66
Thompson, David, 52–3, 59–60, 64, 65, 66, 68, 70, 87, 88, 120, 274n4, 279n24, 279n25, 280n10, 280n17, 283n3; *17, 56*[A], *58*[C]
Thompson River, 74, 90, 93, 94, 95, 108, 116, 118; *550*[A], *552*[A], *8*[B], *111*[C], *195*[C]
Thompson's River district, 7, 93; *48, 201*[A], *273*[A]
Thompson, Thomas, 72; *185*[A], *186*[A]
Thornton, John, 26, 274n26; *2*[B], *2*[C]*–7*[C]
Thornton, Samuel, 26, 28, 274n27; *2, 1*[A]
Three Point lake, *63*[A]
Three Rivers, 76
Thunder Bay, *421*[A]
Thunder Hill, 62
Tiedemann, Hermann Otto, 99, 105, 111, 112, 275n8, 284n21, 285n13; *505*[A], *506*[A], *537*[A]*–545*[A], *553*[A], *570*[A], *571*[A], *30*[B], *31*[B], *213*[C]*–216*[C]
Timiskaming (Temiscamingue) district, 83, 85; *210*[A]
Timiskaming, Lake, 51, 55, 83, 85; *253*[A]
Timiskaming post, 76
Tolmie, William Fraser, 92, 93, 97, 107, 108, 111, 112, 114, 117, 283n13; *46, 47, 62, 294*[A]*–298*[A], *534*[A], *162*[C]
Tomison, William, 38, 40, 49, 66
Tongue Point, 92; *282*[A]
Topographic maps, 3, 9, 38, 68, 69, 71, 75, 85, 96, 98, 99, 100, 101, 102, 105
Toponyms. *See* Place names
Toronto, 115, 275n6
Touchwood Hills, 115
Tower of London, 28, 276n16
Town plans, 9, 99, 100, 101, 103, 104, 111, 112, 113, 114, 284n21, 284n24, 285n9
Trade Lake, 279n12
Trade routes, 53, 54, 55, 56, 60, 61, 64, 65, 69, 74, 75, 83–5, 87, 94–5, 106, 108, 116, 117, 280n5
Traders (Canadian, independent, Pedlar). *See* Canadian traders
Trading posts: astronomical locations of, 16, 50, 71, 101–2; building of, 44, 46, 47, 51, 55, 57, 58, 62, 63, 68, 69, 74, 75, 77, 78, 86, 87, 90, 91, 95; geographical locations of, 37–8, 46, 75, 83, 85, 106; land use at, 3–4, 114; potential sites for, 42, 57, 58, 62, 78, 90; sites of, 3, 34, 35, 36, 58, 74, 95–6, 114
Transit of Venus, 5, 19, 273n16
Transport (animals, arrangements, difficulties), 20, 37, 43, 68, 69, 83–4, 92,

94–5, 114, 115, 116, 117
Trivett, Captain J. F., 109; 466[A], 472[A]–474[A], 480[A]
Trout Fall, 84; 193[A], 97[C]
Trout Lake, 57; 102[A]
Trowbridge, W.P., 365[A]
Tuck, go, my, 44[C]
Tulameen River, 284n20; 153[C], 154[C]
Tumgass (Tumgaise) Harbour, 222[A], 122[C]
Tundra. See Barren Grounds
Turnagain River, 77
Turnor, Philip, 12, 19–21, 48, 49, 61, 70, 118, 273n5, 279n9, 279n10; hiring, arrival at York, 18–19; in Albany, Moose, Abitibi, Eastmain area, 50–1; in Nelson–Saskatchewan area, 49–50, 52–3, 276N7; to Churchill River-Athabasca area, 52–4; maps, 54, 59–60, 120, 11, 12, 18, 27[A], 28[A], 30[A], 34[A], 42[A], 47[A], 57[A], 5[B], 52[C], 59[C]
Turtle Mountain, 67, 83, 84, 263; 205[A]–208[A]

Umpquah River, 88, 89; 112[C]
Ungava Bay, 7, 34, 70, 74, 75, 76, 87; 229[A], 230[A], 254[A], 255[A], 134[C]
Unijah River. See Peace River
United States: army, 92, 93, 114; Corps of Engineers map, 103; indemnity for land, 283n13; land claims, 92, 106, 107–8, 114; military pressure from, 92–3; Oregon Treaty, 92–3; settlers north of Columbia River, 92, 93, 107–8; traders in Snake Country, 88, at Turtle Mountain, 83
Unknown (Inconnue) River. See Anderson River (Northwest Territories)
Upashewa Lake, 46, 47; 45[C]
Upishingunga Lake. See Seul, Lac
Upper Canada, 5
Usquemay. See Inuit
Utah, 87
Utrecht, Treaty of, 28, 29

Valdez Inlet, 102; 353[A]
Vancouver Coal Mining Limited, 113
Vancouver, Fort, 21, 88, 89, 90, 91, 92, 93, 98, 108, 114, 283n3, 283n4, 283n15; 264[A], 265[A], 267[A], 271[A], 277[A], 278[A], 281[A], 288[A], 289[A], 302[A], 352[A], 471[A], 557[A], 114[C], 115[C]
Vancouver, Captain George, 102
Vancouver Island, 7, 9, 92, 93, 95, 96, 97, 98, 102, 103, 104, 110, 118, 121, 284n2, 284n3; coal mining on, 4, 12, 99, 102, 103, 104, 112, 113, 273n9; settlers, 96, 97, 98, 99, 100, 101, 102, 104, 105; 252[A], 301[A], 308[A], 309[A], 329[A], 353[A], 360[A], 369[A], 387[A], 413[A], 426[A], 462[A], 120[C], 152[C], 156[C]–158[C], 168[C], 170[C]–172[C], 179[C], 185[C], 191[C], 192[C], 212[C]
Vancouver's Island, Colony of, 7, 9, 12, 96, 97, 98, 99, 104, 105, 110, 111, 113, 118, 121, 285n11
Vavasour, Mervin 92–3, 283n16; 45, 269[A], 270[A]–272[A], 279[A]–286[A], 143[C]–145[C]
Vermilion Lake, 132[A]
Vernon, Mr. Secretary, 26, 276n12
Victoria: Beckley Farm, 111; church reserve, parsonage, cemetery, 111; districts, 102; Fort Victoria, 21, 92, 94, 95, 100, 104, 110, 115, 50; government properties, 110, 111, 112, 285n9; harbour, 87, 98, 100, 110; Indian reserve, 99, 111; Lowenburg lot Z, 110–11; Oak Bay, 100; Ogden Point, 284n21; surveying from hills, 98, 100; water supplies, 99, 101; 52, 280[A], 290[A], 320[A], 323[A], 353[A], 424[A], 425[A], 469[A], 476[A], 481[A]–483[A], 490[A]–494[A], 496[A]–500[A], 503[A]–506[A], 512[A], 542[A]–549[A], 574[A], 575[A], 15[B], 19[B], 25[B], 27[B], 32[B], 35[B], 139[C], 155[C], 170[C], 180[C], 201[C], 203[C], 204[C], 206[C]–208[C], 219[C]
Victoria District, 100, 101, 102, 103, 105, 111; 51, 308[A], 310[A]–314[A], 330[A]–334[A], 337[A], 338[A], 343[A]–348[A], 350[A], 353[A], 362[A], 363[A], 374[A], 375[A], 377[A], 382[A]–384[A], 389[A]–400[A], 404[A], 405[A], 428[A], 429[A], 439[A]–442[A], 447[A]–450[A], 455[A], 456[A], 487[A], 513[A]–520[A], 522[A], 525[A], 528[A], 529[A], 538[A]–541[A], 553[A]–556[A], 580[A], 581[A], 14[B], 24[B], 30[B], 31[B], 34[B], 161[C], 167[C], 215[C], 219[C]
Victoria HBC operations: field crew, 99; Fur Trade Reserve, 97, 98, 99, 100, 102; land registry, 96, 99, 100, 101, 103, 104, 105; land sales, 98, 99, 100, 101, 103, 104, 110, 111, 112; surveying department, drafting office, 7, 9, 14, 15, 99, 101, 102, 104, 105, 110
Victoria Island (Land), 80, 81, 282n29; 246[A], 159[C], 160[C], 166[C]
Victoria Strait, 81
Village Bay (Esquimalt), 101, 103, 112; 176[C]
Vincent, Thomas 69; 157[A]

Wager Bay, 32, 42, 43, 44, 80
Waggoner, Joseph, 37, 38
Wales, William, 5, 18, 19, 20, 273n16, 278n3
Walker, James, 33, 277n8
Walla Walla. See Nez Perces House
Wanapitei Lake, 85; 210[A]
Wappiscogamy House, 47, 49, 50, 51; 41[C], 42[C]
Wappiscogamy Lake, 47
Warre, Henry, 92; 272[A]
Washington Territory, 4, 108, 110, 114, 285n3, 285n6, 285n19
Waswanipi (Waswanapy) Lake, 47, 70; 180[A]
Waswanipi River, 70
Waterhen River, 145[A]
Watkins, Edward, 115
Waupissattiga Lake, 184[A]
Weenisk. See Winisk River
Wegg, Samuel, 5, 12, 19, 121, 261, 273n11
Wegg's House, 60[A]
Wentuhuysen Inlet. See Nanaimo
Wepiscauacaw Lake. See Mepiskawaukau Lake
West Branch. See Liard River
Western District board of management, 94, 96, 97, 112, 113, 114, 116, 117, 118, 205n12
Western Sea (Ocean). See Pacific Ocean
Weymouth, George, 25
Whale Cove, 41
Whale River, 228[A]
Whale River district, 280n24
Whale River post, 71
White Falls portage, 84; 37, 200[A], 108[C]
White Horse Plain, 263
White, Thomas, 32, 277n2
White Whale River, 5[A]
Willamette River, 89, 92, 114, 283n9, 283n10; 281[A], 565[A], 566[A]

William, Fort, 283n3, 285n10
Williams Creek, 115, 116
Wind River, 198[A]
Winipegucies Lake. See Molson Lake
Winisk River, 71, 86, 275n11, 283n10; 232[A]
Winnipeg, 261
Winnipeg (Winipic, Winipeg) Lake, 32, 38, 40, 44, 45, 47, 48, 50, 54, 58, 65, 67, 71, 84, 279n20; 21, 48[A], 49[A], 56[A], 94[A], 107[A]–109[A], 114[A], 144[A], 147[A], 203[A], 209[A], 24[C], 39[C], 40[C], 46[C], 75[C]
Winnipeg River, 45, 56, 57, 58, 59; 112[A]
Winnipegosis, Lake, 38, 40, 62, 67; 58[A], 109[A]
Wintering Lake, 44
Wollaston Lake, 66; 106[A], 115[A], 141[A], 73[C]
Wollaston Peninsula, 81, 282n29; 159[C]
Wood Lake, 279n12
Wop poominnakek River, 99[A]
Work, John, 86, 90, 93, 103, 106, 110, 275n13, 283n10; 194[A], 100[C], 128[C]
Wyoming, 88

Yale, Fort, 107, 108, 114, 283n15; 59, 465[A], 29[B], 196[C], 197[C], 200[C]
Yale, James, 95, 108, 275n5; 29[B]
Yarrow, John, 35; 11[C]
Yellowhead (Tête Jaune Cache) Pass, 12, 115, 116; 550[A], 551[A], 211[C]
Yellowstone (Red Deer) River, 63; 196[A], 197[A]
York district, 280n24; 161[A], 86[C]
York Factory: building of, 28, 36; capture of, 29, 276n20, 278n5; inland travel from, 16, 17, 20, 29, 30, 38, 40, 44, 45, 49, 51, 52, 53, 278n5; plans of: Kempt, 282n33; Robson, 36; Turnor, 20, 276n5; river-estuary measurements, charts from, 10, 36, 68, 72; Turnor at, 19–20, 49, 50, 52, 53, 54; 23[A], 25[A], 33[A], 34[A], 48[A], 54[A], 103[A], 124[A], 139[A], 155[A], 185[A], 186[A], 192[A], 202[A]–204[A], 209[A], 220[A], 13[C], 14[C], 16[C]–18[C], 23[C], 33[C], 58[C], 68[C], 80[C], 116[C], 117[C]
Yukon, 7, 74, 87, 106; 57, 366[A], 419[A]
Yukon, Fort, 79, 106, 107
Yukon River, 74, 78, 79, 86, 106, 107, 281n12; 147[C]

70463

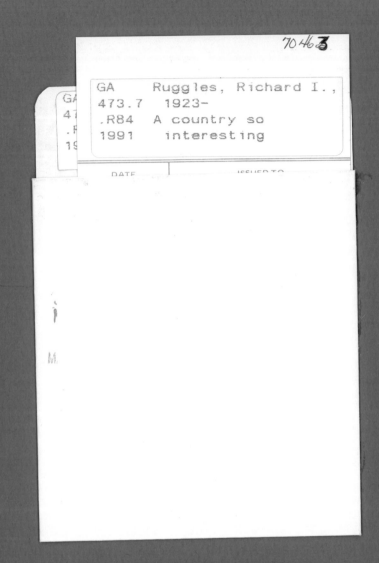

```
GA      Ruggles, Richard I.,
473.7   1923-
.R84    A country so
1991    interesting
```